DISEASES OF THE NERVOUS SYSTEM

ELSEVIER

science &
technology books

Companion Web Site:

http://booksite.elsevier.com/9780128002445/

Diseases of the Nervous System
Harald Sontheimer

Resources for Readers:

Please access the Companion website, hosting the figures from the volume.

ELSEVIER

ACADEMIC
PRESS

DISEASES OF THE NERVOUS SYSTEM

HARALD SONTHEIMER
University of Alabama Birmingham, AL, USA

ELSEVIER

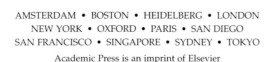
AMSTERDAM • BOSTON • HEIDELBERG • LONDON
NEW YORK • OXFORD • PARIS • SAN DIEGO
SAN FRANCISCO • SINGAPORE • SYDNEY • TOKYO
Academic Press is an imprint of Elsevier

Academic Press is an imprint of Elsevier
125 London Wall, London EC2Y 5AS, UK
525 B Street, Suite 1800, San Diego, CA 92101-4495, USA
225 Wyman Street, Waltham, MA 02451, USA
The Boulevard, Langford Lane, Kidlington, Oxford OX5 1GB, UK

Notices
Knowledge and best practice in this field are constantly changing. As new research and experience broaden our understanding,
changes in research methods, professional practices, or medical treatment may become necessary.

Practitioners and researchers must always rely on their own experience and knowledge in evaluating and using any
information, methods, compounds, or experiments described herein. In using such information or methods they should
be mindful of their own safety and the safety of others, including parties for whom they have a professional responsibility.

To the fullest extent of the law, neither the Publisher nor the authors, contributors, or editors, assume any liability for any injury
and/or damage to persons or property as a matter of products liability, negligence or otherwise, or from any use or operation of
any methods, products, instructions, or ideas contained in the material herein.

ISBN: 978-0-12-800244-5

British Library Cataloguing-in-Publication Data
A catalogue record for this book is available from the British Library

Library of Congress Cataloging-in-Publication Data
A catalog record for this book is available from the Library of Congress

For information on all Academic Press publications
visit our website at http://store.elsevier.com/

Working together
to grow libraries in
developing countries

www.elsevier.com • www.bookaid.org

Acquisition Editor: Melanie Tucker
Editorial Project Manager: Kristi Anderson
Production Project Managers: Karen East and Kirsty Halterman
Designer: Matthew Limbert

Typeset by TNQ Books and Journals
www.tnq.co.in

Printed and bound in the United States of America

Dedication

To the most important people in my life:
My wife Marion and my daughters Melanie and Sylvie.
Their encouragement is my motivation;
their love and their smiles are the greatest reward.

Contents

III

SECONDARY PROGRESSIVE NEURODEGENERATIVE DISEASES

VI

COMMON CONCEPTS IN NEUROLOGICAL AND NEUROPSYCHIATRIC ILLNESSES

14. Shared Mechanisms of Disease

HARALD SONTHEIMER

VII

BENCH-TO-BEDSIDE TRANSLATION

15. Drug Discovery and Personalized Medicine

HARALD SONTHEIMER

VIII

NEUROSCIENCE JARGON

16. "Neuro"-dictionary

HARALD SONTHEIMER

Acknowledgments

English is a second language for me. To make up for my shortcomings, I am indebted to my Assistant, Anne Wailes, who tirelessly edited and polished every sentence in this book. She also tracked down the copyrights for hundreds of figures that were reproduced in this book. Anne did all this while attending to the many daily tasks of administrating a large research center and looking after my trainees in my absence. This was a monumental undertaking and words cannot describe how fortunate I feel to have had her support throughout this journey.

Each chapter went through two stages of scientific review. The first stage of review was conducted by a tremendously gifted young scientist, Dr Alisha Epps, who, for an entire year, spent almost every weekend reading and correcting book chapters as I completed them. Alisha had a talent to simplify and clarify many difficult concepts, and, if needed, she found suitable figures or even drew them from scratch. Her contributions to this book were tremendous and I am indebted to her generous support.

The second stage of review involved experts in the respective disease. I am privileged to have a number of friends who are clinicians or clinician–scientists and who were willing to selflessly spend countless hours correcting the mistakes I had made. While I am acknowledging each person with the very chapter they reviewed, I like to acknowledge all of them in this introduction by name.

Alan Percy, MD, PhD, University of Alabama Birmingham
Amie Brown McLain, MD, University of Alabama Birmingham
Anthony Nicholas, MD, PhD, University of Alabama Birmingham
Christopher B. Ransom, MD, PhD, University of Washington
Erik Roberson, MD, PhD, University of Alabama Birmingham
James H. Meador-Woodruff, MD, University of Alabama Birmingham
Jeffrey Rothstein, MD, PhD, Johns Hopkins University
Leon Dure, MD, University of Alabama Birmingham
Louis Burton Nabors, MD, University of Alabama Birmingham
Richard Sheldon, MD, University of Alabama Birmingham
Stephen Waxman, MD, PhD, Yale University
Steven Finkbeiner, MD, PhD, The Gladstone Institute for Neurological Disease
Thomas Novack, PhD, University of Alabama Birmingham
William Britt, MD, University of Alabama Birmingham

To be able to spend a year writing a book is a luxury and privilege that, even in academia, only a few people enjoy. I am grateful for the support of my employer, the University of Alabama at Birmingham, for allowing me to devote much of my professional time to writing this book. I thank the Dean, President, and my Chairman David Sweatt for enthusiastically supporting this endeavor.

During the spring semester of 2014, I became a visiting Professor, embedded among the wonderful faculty of Rhodes college in Memphis TN, a picturesque small liberal arts college. I am thankful for the hospitality and support of all the Rhodes administrators and faculty, many of whom I engaged in inspirational discussion during lunch or coffee breaks. I am particularly grateful to the Neuroscience program, Doctors Kim Gerecke, David Kabelik, Rebecca Klatzkin, and Robert Strandburg, for letting me participate in their curriculum and take residence in Clough Hall.

Many students provided invaluable feedback toward this book, some formal, using a prescribed feedback form, other informal during office hours. I am thankful to all the students who attended the Rhodes Spring 2014 NEU365 course, as I have received feedback from all of you. The following students took a particular interest and regularly provided recommendation for improvements:

Jessica Baker, Morgan Cantor, Shelley Choudhury, Jason Crutcher, Sarah Evans, Nancy Gallus, Kyle Jenkins, Megan LaBarreare, Mallory Morris, Swati Pandita, Hayden Schill, Nathan Sharfman, and Sara Anne Springfellow.

I trust that all of them are either in Graduate or Medical school by now, and I wish them well.

My final acknowledgment goes to my publisher, Elsevier Academic Press for their tremendous work editing, publishing, and marketing this book. Particularly to the editorial project manager Kristi Anderson, the senior acquisitions editor Melanie Tucker and the production team.

Introduction

The study of nervous tissue and its role in learning and behavior, which we often call neuroscience, is a very young discipline. Johannes Purkinje first described nerve cells in the early 1800s, and by 1900, the pathologist Ramón y Cajal generated beautifully detailed histological drawings illustrating all major cell types in the brain and spinal cord and their interactions. Cajal also described many neuron specific structures including synaptic contacts between nerve cells, yet how these structures informed the brain to function like a biological computer remained obscure until recently. Although Luigi Galvani's pioneering experiments in the late-1700s had already introduced the world to concept of biological electricity, ion channels and synaptic neurotransmitter receptors were only recognized as "molecular batteries" in the late-1970s and early 1980s. The first structural image of an ion channel was generated even more recently in 1998, and for many ion channels and transmitter receptors such information still eludes us.

Surprisingly, however, long before neuroscience became a freestanding life science discipline, doctors and scientists had been fascinated with diseases of the nervous system. Absent any understanding of cellular mechanisms of signaling, many neurological disorders were quite accurately described and diagnosed in the early to mid-1800s, including Epilepsy, Parkinson disease, Schizophrenia, Multiple Sclerosis, and Duchenne's muscular dystrophy. During this period and still today, the discovery process has been largely driven by a curiosity about disease processes. What happens when things go wrong? Indeed, much of the early mapping of brain function was only possible because things went very wrong. Had it not been for brain tumors and intractable epilepsy, surgeons such as Harvey Cushing and Wilder Penfield would have had no justification to open the human skull of awake persons to establish functional maps of the cortex. Absent unexpected consequences of surgery, such as the bilateral removal of the hippocampi in H.M. that left him unable to form new memories, or unfortunate accidents exemplified by the railroad worker, Phineas Gage, who destroyed his frontal lobe in a blast accident, we would not have had the opportunity to learn about the role of these brain structures in forming new memories or executive function, respectively. Such fascination with nervous system disease and injury continues to date, and it is probably fair to say that neuroscience is as much a study of health as that of disease.

For the past 15 years, I have been teaching a graduate course entitled "Diseases of the Nervous System" and more recently I added an undergraduate course on the same topic as well. Every year, almost without fail, students would ask me whether I could recommend a book that they could use to accompany the course. I would usually point them to my bookshelf, filled with countless neuroscience and neurology textbooks ranging from *Principles in Neuroscience* to *Merritt's Neurology*. I concluded that there was no such book and there really should be one. For the next year I kept my eyes peeled for this textbook to appear. Surely, sooner or later some brave neuroscientist would venture to write a book about neurological illnesses. Surprisingly, as of this writing, this has not happened so two years ago I decided to fill this void. My initial inclination was to produce a multiauthor edited book. By calling on many friends and colleagues to each write a chapter on their favorite disease this should be a quick affair. However from own experience I know that book chapters are always the lowest priority on my "to do" list, and I really was eager to pester my colleagues monthly to deliver their goods. Ultimately they would surely ask a senior postdoc to take the lead and in the end, the chapters would be heterogeneous and not necessarily at a level appropriate for a college audience. For my target audience this book needed to be a monograph. While I did not know at the time what I was getting into, roughly 18 months later, having read over 2500 scientific papers and reviews and after writing for about 7–10 h daily, I feel exhausted but also quite a bit more educated than before.

The target audience for this book is any student interested in neurological and neuropsychiatric illnesses. This includes undergraduates, early graduate students, and medical students taking a medical neuroscience course. I also expect the material to be of benefit to many health professionals who are not experts in the field. The book may even appeal to science writers or simply a science minded layperson, possibly including persons affected by one of the illnesses. Purposefully, the book lacks a basic introduction to neuroscience and I would expect the reader to have a basic understanding of neurobiology. Many excellent textbooks have been written, each of which would prepare one well to comprehend this text. I feel that I could not have done justice to this rapidly expanding field had I attempted to write a short introduction. However, to at least partially make up for this, I include an extensive final chapter that is called "Neuroscience Jargon." I consider this more than just a dictionary. It has a succinct summary of approximately 500 of the most important terms and is written as nontechnically as possible. I hope that this will assist the reader to get his/her bearings as needed.

The book makes every effort to cover all the major neurological illnesses that affect the central nervous system though it is far from complete. My intention was to go fairly deep into disease mechanisms and this precluded a broader coverage of small and less well-known conditions. I found it useful to group the diseases into five broad categories that provided some logical flow and progression. Specifically, I begin with static illnesses, where an acute onset causes immediate disability that typically does not worsen over time. This group is best exemplified by stroke and CNS trauma but also includes genetic or acquired epilepsy (Chapters 1–3). I next covered the classical primary progressive neurodegenerative diseases including Alzheimer, Parkinson, Huntington, and ALS (Chapters 4–7). For each of these chapters I added some important related disorders. For example, the chapter on Alzheimer includes frontal temporal dementia; for Parkinson I included essential tremors and dystonia, and for Huntington I touch on related "repeat disorders" such as spinocerebellar ataxia. The chapter that covers ALS includes a variety of disease along the motor pathway essentially moving from diseases affecting the motor neurons themselves (ALS), their axons (Guillain–Barre Syndrome), to the presynaptic (Lambert Eaton Myotonia), and postsynaptic (myasthenia gravis) neuromuscular junction.

Next, I progressed to neurodegenerative diseases that are secondary to an insult yet still cause progressive neuronal death. I call these secondary progressive neurodegenerative diseases and the examples I am covering include Multiple sclerosis, Brain tumors, and infections (Chapters 8–10). It may be unconventional to call these secondary neurodegenerative diseases yet in multiple sclerosis the loss of myelin causes progressive axonal degeneration, brain tumors cause neurological symptoms by gradually killing neurons, and infection causes progressive illnesses again by progressively killing neurons. Nervous system infection could have quickly become an unmanageable topic since far too many pathogens exist that could affect the nervous system. I therefore elected to discuss important examples for each class of pathogen (prion proteins, bacteria, fungi, viruses, single- and multicellular parasites). While none of these pathogens are brain specific, I chose examples in which the nervous system is primarily affected including meningitis, botulism, tetanus, polio- myelitis, neurosyphilis, brain-eating ameba, neurosistercosis, neuroaids, and prion diseases. I also used this chapter as an opportunity to highlight the tropism displayed by some viruses for the nervous system and how this can be har- nessed to deliver genes to the nervous system for therapeutic purposes.

For the section on neurodevelopmental disor- ders I similarly chose four important examples including Down syndrome, Fragile X, Autism, and Rett syndrome. These disorders have so many commonalities that it made sense to cover them in a single chapter (Chapter 11).

No contemporary book of nervous system disease would be complete without coverage of neuropsychiatric illnesses and I elected to devote one chapter each to depression (Chapter 12) and Schizophrenia (Chapter 13).

Taken together, I believe the material covers the "big" brain disorders that any neuroscien- tist or medical student should know. However, anyone looking for more detailed information on rare disorders or disorders primarily affect- ing the peripheral nervous system or sensory organs is referred to some of the excellent neu- rology textbooks that I cite as my major sources throughout the book.

To assure that the material is presented in an accessible, yet comprehensive format, the book was developed in a uniquely student-centered way, using my target audience as a focus group. To do so, I wrote the book as accompanying text to an undergraduate course, writing each chapter as I was teaching it to a class of neuro- science majors. Rather than embarking on this project on my home turf, I elected to enter a self-imposed exile, free from the distraction of family and friends, which would allow for sub- mersive reading, writing, and teaching for lit- erally every awake hour of every day. In short,

I took a sabbatical leave. This strategy assured that I would stay motivated and on task.

Rhodes College in Memphis TN, a small and highly selective Liberal Arts college, became my temporary academic home. Rhodes has been offering a neuroscience major for the past 5 years, and it has grown to be among the more popular majors at the college. I was elated to learn that 25 brave Rhodes students elected to take my NEUR365, Diseases of the Nervous sys- tem class, in spite of not knowing a shred about their professor who, being a medical school edu- cator, was not listed on the "rate my college pro- fessor" Web site.

With 5 chapters completed prior to my arrival, 14 of the 16 chapters came together while teach- ing the class. Each week, I handed out a new disease chapter, and after giving a 75-minute lecture, small groups of students had to prepare independent lectures that they delivered to the class based on recent influential clinical and basic science papers that I assigned (and list in this book with each chapter). Each week, using a questionnaire, the students provided detailed feedback on how accessible, interesting, and complete my chapters were, and how well the book prepared them for the assigned papers that they had to present in class. I took their com- ments very seriously, frequently spending days incorporating their suggestions. I am thankful to all of them and acknowledge a number of exceptionally helpful students in the acknowl- edgment section by name.

A challenge that became immediately evi- dent was the sheer magnitude of the available literature. Moreover, writing about a disease that is outside ones' personal research specialty leaves one without a compass to decide which facts are important and which are not. Narrow- ing literature searches to just "diseases" and "review articles" did not help much and only marginally reduced the number of hits from the tens of thousands into the thousands. While it was gratifying to see the enormous amount of information that has been published, it was daunting to filter and condense this material into a manageable number of sources. In the end, I developed a strategy to first identify the "opinion leaders" in each field, and then, using their high impact reviews, widen my search to include reviews that appeared to cover the most salient points on which the entire field appears to largely agree upon, while staying largely out of more tentative emerging and controversial topics. This was important since the objective of this textbook was to introduce current accepted concepts rather than speculations.

Another challenge I faced was to keep the material interesting. As teacher of medical neuroscience I have long recognized the value of clinical cases. I decided to start each disease chapter with case story, which is either an actual case or one close to cases that I have actually witnessed in some form or other. The students liked this format, particularly since many of the cases I describe involve young people. To offer perspective on each disease, I also elected to provide a brief historic review for each disease. How long has society been dealing with stroke, epilepsy, Huntington, or autism? What were early interpretations on the disease cause, how was disease treated, and what were the most informative milestones. This was possibly the hardest section for me to write, since good sources were difficult to find. Yet it was also the most fascinating. The students initially had little appreciation for these sections and really did not see much value in them. However, this changed after we discussed the value of what I call "science forensics" and the historic insight that could be gleaned. We discussed how the history of disease, when viewed in the context of the history of mankind, allows us to dismiss or consider human endeavors and exposure to man-made chemicals as disease causes. After we discussed how Mexican vases made over 600 years ago already depicted children with Down syndrome, or how Polio crippled children were portrayed on Egyptian stilts that were over 2000 years old, it became clear that neither of these conditions were modern at all. Historic accounts similarly suggest that environmental exposures are unlikely contributors to stroke or epilepsy. Yet, by contrast, the earliest accounts of Parkinson disease align perfectly with the early industrial revolution of the mid-1800 making industrial pollutants potential disease contributors. Even more extreme, no historic account for autism exists prior to the 1930s. Clearly, for some of those diseases, human influences must be considered as contributory factors.

The historic adventures also allowed me to examine diseases in the context of society at a given time in history, clearly important lessons when teaching neuroscience at a Liberal Arts college. Our classes included how patients with epilepsy were labeled witches and burned in medieval Europe; how the heritability of diseases such as Huntington corrupted even doctors to subscribe to the reprehensible teachings of the eugenics movement; or how the infamous Tuskegee syphilis studies served as the foundation for the protection of human subjects participating in human clinical trials, measures that we take for granted today. Another lesson learned from the ancient accounts of Down syndrome is that child birth late in a mother's life occurred throughout history, but more importantly that those children were cared for in many societies with the same love and compassion we have for them today.

The majority of pages in this book are devoted to the biology of each disease. It is remarkable how much we know and how far we have come in just the past few decades, from the historic disease pathology focused approach to contemporary considerations of genes and environmental interactions causing disease in susceptible individuals. It is fascinating to note how cumbersome the initial positional cloning efforts were that identified the first candidate genes for disease compared to today's large genome-wide association studies that identify large networks of gene and their interactions. Clearly, we experience a transformational opportunity to study and understand disease through the study of rare genetic forms of familial diseases that can inform us about general disease mechanisms and allow us to reproduce disease in genetic animal models. At the same time, it is sobering to see how often findings in the laboratory fail to subsequently translate into better clinical practice. I devote a considerable amount of discussion to such challenges and end each chapter with a personal assessment of challenges and opportunities. After completing the disease chapters, it was clear that there were many cross-cutting shared mechanisms and features of neurological disease that I elected to devote an entire chapter solely to shared mechanisms of neurological illnesses (Chapter 14).

Not surprisingly, almost all the class discussions sooner or later gravitated toward ways to translate research findings from the bench to the bedside. Yet few of the students had any idea what this really entails or the challenges that clinical trials face. Having been fortunate enough to develop an experimental treatment for brain tumors in my laboratory that I was able to advance from the bench into the clinic through a venture capital supported biotech start-up, I felt well equipped to discuss many of the challenges in proper perspective. So I devoted an entire chapter (Chapter 15) to this important, albeit not neuroscience specific, topic. The class included important discussions on the placebo effect and frank conversations as to why many scientific findings cannot be reproduced, and why most clinical trials ultimately fail.

I also added several provocative topics to class discussions such as the questionable uses of neuroscience in marketing and advertising and the controversial use of neuroscience in the courtroom. Since neither relates to specific neurological diseases, I elected to leave this out of the book but encourage neuroscience teachers to bring such topics into the classroom as well.

One thing that troubled me throughout my writing was the way in which sources are credited in textbooks. As a scientist, I reflexively place a source citation behind every statement I make. In the context of this book, however, I could only cite a few articles restricting myself to ones that I felt were particularly pertinent to a given statement. A list of general sources that most informed me in my reading is included at the end of each chapter. I am concerned, however, that I may have gotten a few facts wrong, and that some of my colleagues will contact me, offended that I ignored one of their findings that they consider ground breaking; or if I mentioned them, that I failed to explicitly credit them for their contribution. It was a danger that I had to accept, albeit with trepidation and I hope that any such scientists will accept my preemptive apologies. To mitigate against factual errors, I reached out to many colleagues around the country, clinical scientists whom I consider experts in the respected disease, and asked them to review each chapter. I am indebted to these colleagues, whom I credit with each chapter, who selfishly devoted many hours to make this a better book. Their effort has put me at greater ease and hopefully will assure the reader that this book represents the current state of knowledge.

Given that the book was developed as an accompaniment to a college course, I expect that it may encourage colleagues to offer a similar course at their institution. I certainly hope that this is the case. To facilitate this, I am happy to share PowerPoint slides of any drawings or figures contained in this book, as well as any of the 1000 + slides that I made to accompany this course. I can be contacted by email at Sontheimer@uab.edu. Also, for each chapter I am listing a selection of influential clinical and basic science articles that I used in class. These are just my personal recommendations and not endorsements of particular themes or topics. These papers have generated valuable discussion and augmented the learning provided through the book.

Finally, as I finish editing the book, I keep finding more and more articles reporting exciting new scientific discoveries that I would have liked to include. However, if I did, this book would have never reached the press. It is refreshing to see that neuroscience has become one of the hottest subjects in colleges and graduate schools and even the popular press. Neuroscience research is moving at a lightning pace. It is therefore unavoidable that the covered material will only be current for a brief moment in time, and, as you read this book, that time will have already passed.

Harald Sontheimer
September 15, 2014

STATIC NERVOUS SYSTEM DISEASES

Cerebrovascular Infarct: Stroke

Harald Sontheimer

1. CASE STORY

Although I was barely 6 years old at the time, I remember as if it were yesterday. I had anxiously waited for the day, Easter Sunday. It was an annual ritual that the entire family gathered at grandma's house for the largest Easter egg hunt in the neighborhood. Counting well over 30 cousins, this was no small event, and there had always been intense competition for finding

3

the most treats. But per protocol, the hunt would not get on its way until after Sunday Mass. Excited, after breakfast I ran to grandma's house next door to meet her for the drive to church. Surprisingly, she did not answer her doorbell, and her window curtains were still drawn. This was unusual, for she was always up by the crack of dawn. Puzzled, I ran back to tell mom, who retrieved grandma's spare key to check on her. As we entered the dark hallway, I heard strange labored breathing coming from upstairs. Mom rushed up the stairs and I followed at a close distance shouting "Grandma! Grandma!," but we did not hear a response. The bedroom was empty, her bed untouched. Mom ran to the bathroom. There she was, stretched out on the floor, laying on her side, barely conscious. Grandma was trying to speak, but was unable to vocalize anything intelligible. Wearing only her nightgown, she was shivering. Mom ordered me to run back to the house and ask dad to call an ambulance. When I returned a short while later, mom had carried grandma to her bed. She was clearly not well. Her left face was drooping and she just gazed into space, her shallow breaths interrupted by occasional attempts to vocalize. As the paramedics arrived I was ordered out of the room. They carried grandma down the stairs, and when I caught her eyes, she seemed very afraid. Mom traveled with her to the hospital by ambulance, which sped away with blaring sirens. Hours later, as the rest of the family arrived, there was nothing festive about this Easter. Without grandma, we picked up treats without much interest and without any laughter. Mom soon returned, reporting that grandma had suffered a stroke but was in stable condition. She would have to stay in the hospital for a few days but would probably recover. Grandma did come home the following weekend, but she clearly was not herself. She could barely stand and had to use a walker to make even a few steps around the house. Her face was still drooping and her speech was unintelligible. However, she clearly understood everything I said, and while

trying to answer, eventually gave up in frustration after several attempts. Throughout the following week her speech gradually returned, and by a month after her stroke, she was sitting at the dinner table eating by herself, although mostly using only her right hand. As she slowly pieced together words, we were able to have at least a rudimentary conversation. Mom had to help grandma with just about any task, from dressing to bathing. Every evening, mom and I would help her to bed, and before she retired, mom would check in on her once more for good measure. This evening, Mom did not return for a long time. I woke to sirens screaming and I feared for the worst. Mom was still not home for breakfast and called in the afternoon to report that grandma had passed away. She had suffered another stroke almost exactly a month after the first one disrupted our annual Easter gathering.

2. HISTORY

Without recognizing its underlying cause, Hippocrates (460 BC), the "father of medicine," provided the first clinical report of a person being struck by sudden paralysis, a condition he called apoplexy. This Greek word, meaning "striking away," refers to a sudden loss of the ability to feel and move parts of the body and was widely adopted as a medical term until it was replaced by cerebrovascular disease at the beginning of the twentieth century. Most patients and the general public prefer the term *stroke*, which first appeared in the English language in 1599. It conveys the sudden onset of a seemingly random event.

Hippocrates explained apoplexy using his humoral theory, according to which the composition and workings of the body are based on four distinct bodily fluids (black bile, yellow bile, phlegm, and blood), which determine a person's temperament and health. Accordingly, diseases result from an imbalance in these four humors, with apoplexy specifically affecting the flow of humors to the brain. Humors were rebalanced through purging and bloodletting, which became the treatment of choice for stroke throughout the middle ages. The first scientific evidence that a disruption of blood flow to the brain causes stroke came through a series of autopsies conducted by Jacob Wepfer in the mid-1600s.

The humoral theory of Hippocratic medicine ruled until the German physician Rudolf Virchow discredited it in his "Theories on Cellular Pathology," published in 1858. Virchow made countless impactful contributions to medicine. With regard to stroke, he first explained that blood clots forming in the pulmonary artery cause vascular thrombosis and that fragments arising from these thrombi can enter the circulation as emboli. These emboli then are carried along with blood into remote blood vessels, where they can occlude blood flow or rupture vessels. His theory was initially based only on patient autopsies. However, together with his student, Julius Cohnheim, Virchow went on to test this idea by injecting small wax particles into the arteries of a frog's tongue to show that the wax acted as an embolus, thereby causing embolization that shut off blood flow to the parts of the tongue supplied by this vessel. In subsequent studies Cohnheim showed that an embolus can cause either blockade (ischemic stroke) or rupture (hemorrhagic stroke), contradicting competing views at the time that suggested that only blood vessel malformations or aneurisms could hemorrhage. It is worth noting that in the early twentieth century, the recognition that emboli cause the selective abolition of blood flow in cortical blood vessels gave neurologists the first insight into functional neuroanatomy, showing selective and predictive deficits in sensory and motor function depending on where an embolus occluded a vessel.

Throughout history and well into the nineteenth century it was common to view stroke as a divine intervention, a summons to duty. Stroke was regarded as God's punishment for unacceptable behavior. Shockingly, in spite of Virchow's discoveries on thromboembolism,

even major medical textbooks continued to blame the patient for the disease. For example, Osler's medical textbook (1892) suggested that "the excited action of the heart in emotion may cause a rupture." Others even suggested that a patient's physical attributes, namely, a short, thick neck and a large head, were predisposing factors.[1] Yet a diet "high in seasoned meat, poignant sauces and plenty of rich wine" was already accurately predicted by Robinson as a risk factor in 1732. (For further reading on the history of stroke please consult refs 1 and 2.)

We have obviously come a long way in the past 100 years. The routine medical use of X-rays, introduced in 1895, ultimately led to the development of the now widely available computed tomography (CT), with which it is possible to quickly and accurately localize blood clots or bleeds to guide further intervention. Surgical, mechanical, or chemical recanalization are now standard procedures, and various forms of image-guided stenting procedures, adopted from cardiac surgery, are now able to open cerebral vessels. The discovery of tissue plasminogen activator (tPA), approved as a chemical "clot buster" in 1996, was a major advance in the clinical management of acute stroke. Together with widely adopted rehabilitation, the outlook for many stroke patients has improved considerably.

3. CLINICAL PRESENTATION/ DIAGNOSIS/EPIDEMIOLOGY

Cerebrovascular infarct is defined by the sudden onset of neurological symptoms as a result of inadequate blood flow. This is commonly called a stroke because the disease comes on as quickly as a "stroke of lightning" and without warning; we use the terms *stroke* and *cerebrovascular infarct* interchangeably throughout. We typically distinguish three major stroke types that differ by their underlying cause and presentation. Focal ischemic strokes make up the vast majority of cases (~80%) and result from vessel occlusion by atherosclerosis or blockage by an embolus or thrombus that causes a focal neurological deficit. Global ischemic strokes, often called hypoxic-ischemic injury, are more rare (10%) and result from a global reduction in blood flow, for example, through cardiac insufficiency. The neurological deficit affects the entire brain and is typically associated with a loss of consciousness. Finally, hemorrhagic strokes result from rupture of fragile blood vessels or aneurysm. These account for 10% of all strokes and can present with focal deficits if a small vessel is affected or global deficits if massive intracranial bleeding occurs. For the majority of patients who suffer an ischemic stroke, maximal disability occurs immediately after the blockage forms, without further worsening unless secondary intracranial bleeding occurs. Once the obstruction clears, the patient's symptoms improve. However, a stroke patient has a greatly increased likelihood of recurrence: 20–30% of patients experience a second stroke within a year after the first insult. Hemorrhagic stroke is a severe medical emergency, with mortality approaching 40%. Symptoms are often progressive as bleeding continues, and a loss of consciousness is common.

Stroke is the most common neurological disorder in the United States and affects close to 800,000 people each year. Behind only heart disease and cancer, it is the third leading cause of death, with 200,000 stroke-related deaths annually. Many patients survive but remain permanently disabled, making stroke the leading cause of permanent disability in the United States. In 2013, there were 4 million stroke survivors in the United States.

Although neurological symptoms vary depending on the brain region affected, most strokes are focal and affect only one side of the body with muscle weakness and sensory loss. Telltale signs (Table 1) include a drooping face, change in vision, inability to speak, weakness and sensory loss (preferentially on one side of the body), and severe sudden-onset headaches. Many of these symptoms clear once blood flow to the affected brain region is restored. Therefore rapid diagnosis and immediate medical

intervention are of the essence, and anyone suspected of suffering a stroke should immediately call for emergency medical services: "Time lost is brain lost." In the case of hemorrhaging stroke, symptoms may worsen rapidly because intracranial bleeding affects vital brain functions, and patients may lose consciousness. To encourage rapid admission of potential stroke victims to a hospital, the National Stroke Association devised the Act FAST campaign, which aids the public in quickly identifying the major warning signs of the disease (Table 2).

TABLE 1   Common Symptoms of a Focal Stroke

Alteration in consciousness; stupor or coma, confusion or agitation/memory loss seizures, delirium

Headache, intense or unusually severe often associated with decreased level of consciousness/neurological deficit, unusual/severe neck or facial pain

Aphasia (incoherent speech or difficulty understanding speech)

Facial weakness or asymmetry, paralysis of facial muscles (e.g., when patients speak or smile)

Incoordination, weakness, paralysis, or sensory loss of one or more limbs (usually one half of the body and in particular the hand)

Ataxia (poor balance, clumsiness, or difficulty walking)

Visual loss, vertigo, double vision, unilateral hearing loss, nausea, vomiting, photophobia, or phonophobia

TABLE 2   Act FAST Emergency Response Issued by the National Stroke Association

Use the FAST test to remember warning signs of stroke.	
F = FACE	Ask the person to smile. Does one side of the face droop?
A = ARMS	Ask the person to raise both arms. Does one arm drift downward?
S = SPEECH	Ask the person to repeat a simple sentence. Does the speech sound slurred or strange?
T = TIME	If you observe any of these signs (independently or together), call 9-1-1 immediately.

Once the patient is receiving medical care, a diagnostic decision tree is typically followed to guide treatment, as illustrated in Figure 1. Immediately upon admission to a hospital, the distinction between ischemic (occluding) stroke and hemorrhagic stroke must be established because treatment for the two differs completely. CT, essentially a three-dimensional X-ray, is the preferred test. It is quick, relatively inexpensive, and widely available, even in small hospitals or community clinics. Moreover, it is very sensitive for detecting intracranial bleeding because iron in the blood's hemoglobin readily absorbs X-rays. Examples of CT scans from two patients, one with an ischemic stroke and one with a hemorrhagic stroke, are illustrated in Figure 2. The ischemic lesion contains more water because of edema and therefore presents with reduced density on CT, whereas the absorption of X-rays by blood creates a hyperdense image on CT in a hemorrhagic stroke.[3]

If bleeding is detected, any attempt to stop blood entering the brain (including surgical, if possible) must be considered. In the absence of bleeding, restoring blood flow to the affected brain region as quickly and effectively as possible is imperative. Major advances to restore blood flow using chemical or mechanical recanalization of occluded blood vessels have been made and are extensively discussed in "Treatment/ Standard of Care/Clinical Management," below.

If the symptoms resolve spontaneously and quickly, within less than 24 h, we typically consider the insult a transient ischemic attack (TIA) as opposed to a stroke. However, the distinction between TIA and stroke is less important regarding treatment decisions because we do not have the luxury to wait 24 h before providing treatment. If, as is often the case, symptoms resolve within minutes to an hour, the diagnosis of TIA is an important risk factor for the patient, who has an elevated risk of developing a stroke in the future (5% within 1 year).

It is possible to misdiagnose a stroke in an emergency room setting, where time is of the essence and diagnoses must be made quickly. A number of conditions can mimic stroke symptoms, including migraine headaches, hypoglycemia (particularly in diabetic patients), seizures, and toxic-metabolic disturbances caused by drug use. Some of these can be ruled out by simple laboratory tests; hypoglycemia is a good example. Others can be excluded through

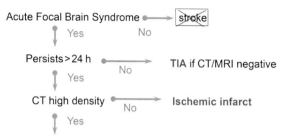

FIGURE 1 Stroke diagnosis. Upon admission to a hospital, a physician uses this decision tree to establish the most likely diagnosis and aid in subsequent treatment. First, a focal neurological deficit must be established. If it resolves spontaneously, it suggests a transient ischemic attack (TIA). If the deficit persists for more than 24 h, a stroke is suspected. The imaging results determine whether the infarct is hemorrhagic, with evidence of blood that results in high-density areas on the computed tomographic (CT) scan. If this is not the case, an ischemic infarct is the most likely diagnosis.

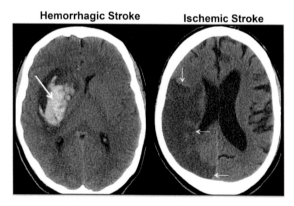

FIGURE 2 Representative examples of computed tomography (CT) scans from two patients. The left image illustrates a hemorrhagic stroke in the basal ganglia; the increased signal (hyperdensity, arrow) signifies bleeding. The right is a characteristic ischemic stroke with reduced density in the infarct region (hypodensity, arrows) suggestive of stroke-associated edema. *Images were kindly provided by Dr Surjith Vattoth, Radiology, University of Alabama Birmingham.*

a detailed patient history and, in particular, the ability of the physician to establish a definitive history of focal neurological symptoms, ideally corroborated by an eye witness.

Rapid diagnosis is facilitated by the use of a simple, 15-item stroke assessment scale established by the National Institutes of Health. This assessment, called the National Institutes of Health Stroke Scale, assesses level of consciousness, ocular motility, facial and limb strength, sensory function, coordination, speech, and attention.[4]

The pathophysiology of stroke is well understood, and the treatments available to date are effective for many patients. Unfortunately, the underlying disease causes, including atherosclerosis and hypertension, can rarely be completely removed, although a combination of lifestyle changes and chronic medical management of risk factors can reduce the likelihood of recurrence.

Numerous risk factors have been identified, many of which are modifiable through changes in lifestyle or medication. By far the largest risk factor is a person's age, which increases incidence almost exponentially, doubling with every 5 years of life. Put in perspective, only 10 in 100,000 persons are at risk of suffering a stroke at age 45; that number climbs to 1 in 100 by age 75. The second leading risk factor is hypertension, which increases stroke risk about 5-fold, followed by heart disease (3-fold), diabetes (2- to 3-fold), smoking (1.5- to 2-fold), and drug use (1- to 4-fold). Note that these risk factors are additive, and thus a 75-year-old diabetic smoker who drinks and has heart disease has a greatly compounded risk. African Americans are twice as likely to suffer a stroke than Caucasians. Although men are slightly more likely to suffer a stroke, women are twice as likely to die from a stroke.

Epidemiological data established what is often called the "stroke belt," namely, a geographic region within the United States where annual stroke deaths are highest (Figure 3). This is readily explained by the confluence of risk factors of race, diabetes, and obesity among the population in the southern and southeastern United States.

4. DISEASE MECHANISM/CAUSE/ BASIC SCIENCE

Stroke is conceptually a relatively simple disease wherein the brain's "plumbing" is defective. We have a fairly good understanding of causes and remedies. In its most elementary form, a stroke is the direct result of inadequate blood flow to a region of the brain, with ensuing death of neurons as a consequence of energy loss. To fully appreciate the vulnerability of the brain to transient or permanent loss of blood flow, it is important to discuss the unique energy requirements of the brain and the cerebral vasculature that delivers this energy.

The brain is the organ that uses the largest amount of energy in our body. At only 2% of body mass, an adult brain uses 20% of total energy, whereas a child's brain uses as much as 40%. This equates to about 150 g glucose and 72 L of oxygen per day. The cellular energy unit is adenosine triphosphate (ATP), the majority of which is produced by the oxidative metabolism of glucose to carbon dioxide (CO_2) and water. The adult brain has very little synthetic activity because few cells and membranes are replaced. Thus, the vast majority of energy is used to shuttle ions across the cell membrane to establish and maintain ionic gradients necessary for electrical signaling (Figure 4). Of greatest importance is the extrusion of Na^+ and the import of K^+ through Na^+/K^+ ATPase. This pump not only establishes the inward gradient for Na^+ needed to generate an electrical impulse or action potential but also maintains a negative resting membrane potential that neurons assume between action potentials. Moreover, the electrochemical gradient for Na^+ is harnessed to transport glucose and amino acids across the membrane and to

Annual Stroke Deaths for Adults over 35 by Region

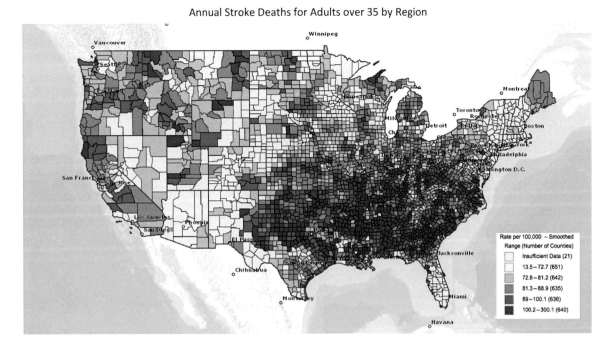

FIGURE 3   Color-coded annual stroke deaths by region shows an elevated incidence in the Southern United States, a region often dubbed the "stroke belt." Stroke death rate for adults over 35 and including all races all genders for the time period 2008–2010. *Produced with data from Centers for Disease Control and Prevention, 1600 Clifton Rd, Atlanta, GA 30333, USA.*

regulate intracellular pH. Therefore, these transport systems are indirectly coupled to the ATP used by the Na^+/K^+ ATPase. Additional important consumers of cellular ATP are Ca^{2+}-ATPases that transport Ca^{2+} against a steep concentration gradient either out of the cell or into organelles. Intracellular Ca^{2+} is maintained around 100 nM, which is 10,000-fold lower than the 1 mM concentration of Ca^{2+} in the extracellular space. Ca^{2+} functions as a second messenger in only a very narrow concentration range of 100–1000 nM and therefore must be carefully regulated by the Ca^{2+}-ATPases. Any increase above this range activates enzymes and signaling cascades that are largely destructive (discussed in more detail later in this chapter). Finally, ATP serves as an important source for high-energy phosphates that can attach to proteins and enzymes through phosphatases that act as on/off switches to regulate the activity of these proteins and enzymes.

ATP stores in neurons are exhausted after only 120 s. Therefore, neurons must continuously produce ATP from glucose via oxidative metabolism of glucose in the mitochondria. Glucose is the most readily available energy source throughout the body, and most cells can store some readily available glucose in the form of glycogen granules. These glycogen granules are a polysaccharide of glucose that can be quickly converted back to glucose when needed for energy. Unfortunately, neurons do not contain glycogen stores and therefore rely on a constant, uninterrupted supply of glucose from the blood. From an evolutionary point of view, the cellular space saved by giving up energy stores allows an important benefit of an increased number of nerve cells packed into a finite cranial space.

To meet its high energy demands, it is also essential that the brain metabolize all glucose in the most effective way possible. To do

FIGURE 4 Cellular energy use of neurons in the brain. Adenosine triphosphate (ATP) produced in the mitochondria directly fuels ATP-driven pumps such as the Na⁺K⁺ ATPase and the Ca²⁺-ATPases and indirectly provides the energy for Na⁺-coupled transporters.

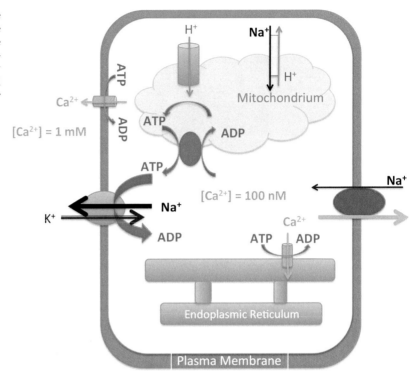

this, it metabolizes it aerobically, which yields 36 mol ATP/mol glucose. This far exceeds the anaerobic glycolytic production of ATP, which only yields 2 mol ATP/mol glucose. Unlike most other cells in the body, neurons are not able to switch to glycolysis in the absence of oxygen, necessitating a constant delivery of sufficient oxygen. The convergence of high energy demand, the absence of glycogen stores, and exclusively aerobic metabolism makes the brain uniquely vulnerable to injury in situations where glucose or oxygen supply is disrupted. Rare conditions limit only one of these substrates. For example, hypoglycemia may occur in a diabetic patient who receives an excess amount of insulin, and anoxia can occur in a patient in a near-drowning situation who stops breathing. In general, however, cerebrovascular infarction is the result of reduced blood flow that limits both glucose and oxygen delivery; this condition is called ischemia.

It is important to note that while energy consumption throughout the body varies with activity, most notably in skeletal muscles and the heart, the brain's metabolic activity is fairly constant and not measurably affected by changes in mental state. Therefore, the overall regulation of blood flow to the brain simply ensures a constant flow of oxygenated blood to the brain. However, regional differences in energy consumption occur and give rise to the blood oxygen concentrations measured by functional magnetic resonance imaging. For every region with enhanced regional blood flow, there is another that has reduced blood flow, effectively canceling each other for a constant metabolic activity.

The cerebral vasculature receives its main supply of oxygenated blood via the two common carotid arteries on each side of the neck, which branch into the internal carotid artery (ICA) and external carotid artery (ECA), respectively (Figure 5). The ICA is the predominant supply

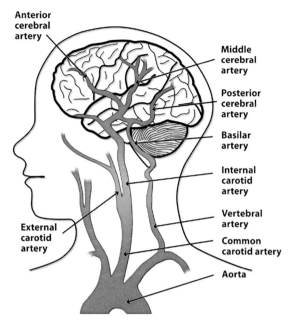

FIGURE 5 The cerebral vasculature in a schematic view. The main supply of oxygenated blood to the brain is through the two common carotid arteries on each side of the neck, which branch into the internal (ICA) and external carotid arteries (ECA), respectively. The ICA is the predominant supply line, carrying ~75% of the total blood volume to the brain, while the ECA primarily feeds the neck and face. The ICA ends by dividing into the middle (MCA) and anterior cerebral (ACA) arteries. Two vertebral arteries at the back of the neck provide an additional minor supply pathway for the brain; this pathway becomes important in situations where the carotids are narrowed or blocked. *Image by Ian Kimbrough, Department of Neurobiology, University of Alabama Birmingham.*

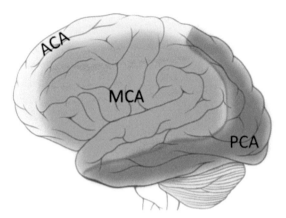

FIGURE 6 Perfusion fields of the major cerebral arteries. ACA, anterior cerebral artery; MCA, middle cerebral artery; PCA, posterior cerebral artery.

line, carrying ~75% of the total blood volume to the brain, whereas the ECA primarily feeds the neck and face. Two vertebral arteries at the back of the neck provide an additional minor supply pathway for the brain; this pathway becomes important in situations where the carotids are narrowed or blocked. The ICA ends by dividing into the middle cerebral artery (MCA) and anterior cerebral artery (ACA). The MCA is the largest branch and divides into 12 smaller branches; together, these 12 branches supply almost the entire cortical surface, including the frontal, parietal, temporal, and occipital lobes (Figure 6).

At its stem, the MCA gives rise to additional vessels that supply the midbrain, including the globus pallidus and caudate nucleus. The ACA supplies primarily the frontal lobes. The vertebral arteries supply the cerebellum and medulla. The basilar artery branches off the vertebral artery and supplies the pons and lower portions of the midbrain, hypothalamus, and thalamus. The posterior cerebral artery branches off the basilar artery and feeds the occipital lobe.

Many of the major vascular branches are interconnected and form a network that allows blood to circumvent obstructions if present. One particular structure that deserves mentioning is the circle of Willis. This ring-like connection of the cerebral vasculature is established by the anterior commissure connecting the left and right branches of the ACAs and the posterior commissures connecting the posterior cerebral arteries. By contrast, smaller arteries less than 100 μm in diameter are end arteries that are not interconnected, and any blockage results in loss of perfusion to the innervated brain region.

Each heartbeat delivers about 70 mL of oxygenated blood to the aorta, 10–15 mL of which are allocated to the brain. Every minute about 500 mL of blood circulate through the brain. To ensure constant perfusion, pressure and blood

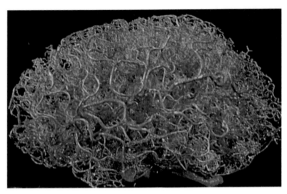

FIGURE 7 Vascular cast of a human brain shows the extensive branching of vessels into finer and finer structures. The cast was prepared by injection of a plastic emulsion into the brain vessels, and, upon hardening, the brain parenchymal tissue was enzymatically dissolved. *Reproduced with permission from Ref. 5.*

diffusion of gas in tissue is very limited, capillaries reach within about 50 μm of any neuron throughout the brain.

4.1 Causes of Vessel Occlusions: The Thrombolytic Cascade

The loss of blood flow in stroke patients is the result of insufficient arterial flow due to vessel narrowing, complete blockage, or rupture. The underlying causes are directly related to biochemical events involved in the formation of blood clots. The ability of blood to coagulate and stop bleeding from wounds (hemostasis) is essential for survival. Coagulation involves two steps: the initial formation of a cellular plug by circulating platelets and a secondary reinforcement of this plug by an aggregate of fibrin fiber strands. The end product is a blood clot or thrombus that seals the vessel wall. The clotting process is well understood and involves numerous clotting factors. Clotting proceeds along two pathways: the intrinsic (contact activation) pathway and the extrinsic (tissue factor) pathway. The latter is the pathway we are most concerned with in the context of stroke. Here, following damage to the vessel wall, the serine protease thrombin causes the production of fibrin from fibrinogen. Fibrin fibers then form the hemostatic plug. Eventually, blood clots need to be dissolved through a process called fibrinolysis. The main protein mediating the dissolution of the plug is plasmin. Plasmin is derived from a precursor molecule, plasminogen (PLG), which is a component of blood serum produced in the liver. The cleavage of the precursor PLG to the active plasmin is catalyzed by tPA, a serine protease produced by endothelial cells in blood vessels (Figure 8). Plasmin in turn helps degrade the clot. Therefore tPA is an important regulator of clotting, and recombinant tPA emerges as an effective thrombolytic agent, often called a "clot-buster." We discuss its clinical use later in this chapter.

flow are highly regulated. The first line of regulation is via the arterial walls of the major arteries, which constrict in response to increases in blood pressure. Arterioles are exquisitely sensitive to changes in the partial pressure of CO_2 such that when the CO_2 content increases, indicating high metabolic activity, arterioles dilate. This dilation causes increased blood flow and enhances delivery of oxygenated blood. When CO_2 decreases, vessels constrict to reduce blood flow. As noted above, functional activity within subregions of the brain adjusts regional blood flow without affecting the overall delivery of blood to the brain, which remains about 500 mL/min.

To illustrate just how sensitive the brain is to a loss in blood pressure and flow, consider the following values. The normal perfusion is 20–30 mL/100 g tissue. A decrease to 16–18 mL/100 g tissue causes infarction within 1 h, and any further reduction kills brain tissue in just minutes.

As illustrated in a vascular cast of a human brain (Figure 7), arterioles branch extensively, giving rise to capillaries so small that erythrocytes have to bend to fit through their lumen. This site is where the major exchange of glucose and blood gases with brain tissue occurs. Since

In many stroke patients the obstruction of blood flow is the result of atherosclerosis, a

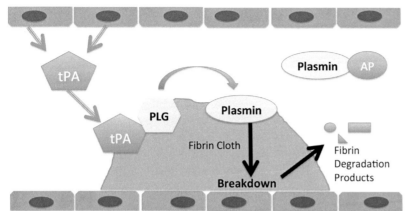

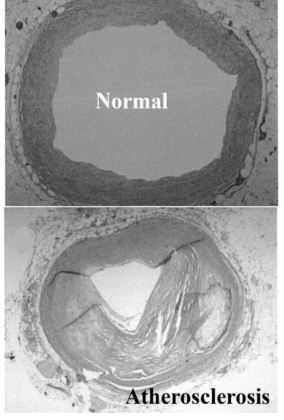

FIGURE 8 Breakdown of thrombotic deposits or blood clots by tissue plasminogen activator (tPA). Schematically shown is the lumen of an arterial blood vessel formed by capillary endothelial cells. tPA activates the conversion of plasminogen (PLG) to plasmin, which catalyzes the breakdown of fibrin clots into fibrin degradation products. Plasmin can be inhibited by antiplasmin (AP).

chronic vascular disease that causes a thickening of the arterial wall with plaque deposits that narrow the vessel lumen (Figure 9). Plaques contain Ca^{2+} and fatty acids, including triglycerides and cholesterol, as well as macrophages and white blood cells. Atherosclerotic plaque formation is promoted by low-density lipoproteins, often called "bad cholesterol," which enters into the vessel wall and causes an immune response. This in turn sends macrophages and T-lymphocytes to the affected vessel, causing them to aggregate locally. In addition to a narrowing of the arterial lumen, this causes a hardening of the vessel wall and inflammation of the smooth muscle. Atherosclerosis can affect essentially all arterial vessels in the body and is considered a chronic condition that typically proceeds asymptomatically for many years or decades. Even a small narrowing of the vessel lumen has a profound effect on blood flow and hence energy delivery to the brain. The weakened vessel wall becomes prone to rupture and may cause a hemorrhagic stroke, particularly in patients with hypertension and chronically increased blood pressure. Plaques are covered by a fibrous cap of collagen, which is highly clot promoting. This cap is typically weak and prone to rupture. If it enters the bloodstream, it forms a thrombus that attracts additional platelets and white blood cells. It eventually attaches at a vessel branch point,

FIGURE 9 Examples of atherosclerosis critically narrowing vessel lumen and thereby significantly reducing blood flow. *Images kindly provided by Edward C. Klatt, MD, Mercer University, School of Medicine, Savannah, GA, USA.*

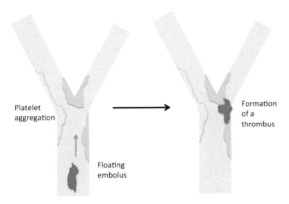

FIGURE 10   Schematic view of thromboembolic ischemia showing a reduction of blood flow by the buildup of thrombolytic plaque material on the inside of the vessel wall. This thrombus can break off and fragment; so-called emboli travel with the blood and can plug finer vessels. Occlusions therefore often have been called thromboembolic because no distinction is made between stationary buildup and floating debris.

where it occludes blood flow. A small particle breaking off from a thrombus or the plaque is called an embolus. It floats along with the blood into smaller and smaller penetrating arteries where it eventually becomes lodged. This process is illustrated in Figure 10. We refer to ischemic strokes caused by thrombi or emboli as thromboembolic strokes.

4.2 The Ischemic Cascade

The loss of blood flow resulting from either leakage or blockage of a blood vessel starts what is known as the ischemic cascade. From decades of experimentation in animals and humans, we have an excellent understanding of the cellular and molecular changes that occur in the affected brain. The time course and the major cellular changes are illustrated graphically in Figures 11 and 12. Following the fairly rapid exhaustion of ATP, the neuronal membrane slowly depolarizes as the Na^+/K^+ ATPase fails to maintain the resting membrane potential. After just a few minutes, most affected neurons reach the activation threshold for Na^+ channel-mediated action potentials, causing a transient period of

hyperexcitability that further depolarizes the membrane. The depolarization opens voltage-gated Ca^{2+} channels and causes the removal of the Mg^{2+} block that normally safeguards the N-methyl-D-aspartate (NMDA) receptor. This in turn causes further uncontrolled influx of Ca^{2+}. Note that while NMDA receptors typically cause a brief excitatory postsynaptic current that is largely due to Na^+ influx, the channel per se has fivefold higher permeability to Ca^{2+} than Na^+. At this stage all floodgates for Ca^{2+} are open. The Ca^{2+}-ATPase that normally sequesters Ca^{2+} into organelles is inactivated by the loss of ATP. To aggravate matters further, surrounding astrocytes also exhaust their small glycogen stores and their failure to produce ATP results in astrocytic depolarization. This dissipates the transmembrane gradient for Na^+ that is required for uptake of glutamate (Glu) into astrocytes. The depolarized neurons release more Glu, which is no longer cleared by astrocytes. This causes a "perfect storm" scenario whereby sustained activation of neuronal Glu receptors causes Ca^{2+} influx, which feeds forward on further vesicle fusion and Glu release. Ultimately, within 5–10 min, irreversible neuronal death begins in the core of the ischemic lesion. Neuronal and glial cell death occurs primarily through a necrotic pathway, where the influx of ions, including Na^+ and Cl^-, causes cytotoxic edema or cell swelling as water follows these ions into the cells. This in turn ruptures cell membranes, causing spillage of cytoplasm and creating a toxic extracellular space. The Ca^{2+} increase also activates a secondary programmed cell death cascade, which is explained further below.

4.3 The Ischemic Penumbra

The ischemic lesion is surrounded by brain tissue that still receives some blood flow via collaterals; however, this blood flow is not of sufficient quantity to maintain normal brain function. This hypoperfused brain tissue, schematically illustrated in Figure 13, is called the ischemic penumbra and is the focus of essentially all

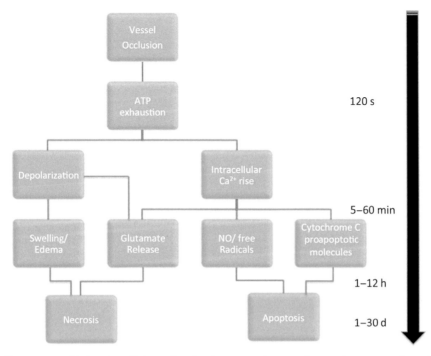

FIGURE 11   Time course of ischemia; cellular and molecular events that contribute to injury are indicated. Different pathways lead to either apoptotic or necrotic cell death.

intervention strategies. Here, neurons and glial cells are in a latent state, paralyzed from participating in proper neuronal function but clinging to life. If perfusion occurs within a reasonable amount of time, the ischemic penumbra can recover completely. If not, the penumbra is gradually assumed by the expanding ischemic core. The viability of the ischemic penumbra has been extensively studied in rodents and monkeys, and our best estimates suggest that neurons in the penumbra remain viable for at least 3–6 h, although some studies suggest that this time period may be as long as 24–48 h. However, based on consensus data from many studies, 3 h has been adopted as the critical time period during which reperfusion provides maximal benefit, and this is often called the "window of opportunity." In agreement with this suggestion, chemical recanalization using tPA, discussed in the following paragraph, provides maximal benefit to patients who are only less than 3 h removed

from the insult, with declining benefits between 3 and 6 h and essentially none thereafter. Realizing this window of opportunity, clinicians are urging patients to seek rapid attention because "time lost is brain lost."

Our understanding of the ischemic penumbra is not quite as clear as that of the ischemic core. It seems, however, that neuronal death in the penumbra primarily occurs via programmed cell death, called apoptosis. Increases in Ca^{2+} to supraphysiological, but not catastrophic, levels activate several of the apoptosis-promoting pathways. Activation of these pathways then culminates in the release and activation of caspases, a family of cysteine proteases that cleave cellular proteins, resulting in slow cellular disassembly. This goes hand in hand with changes to the mitochondrial membrane, which releases proapoptotic molecules such as cytochrome C. The opening of a large ion channel, called the mitochondrial transition pore, breaks down the

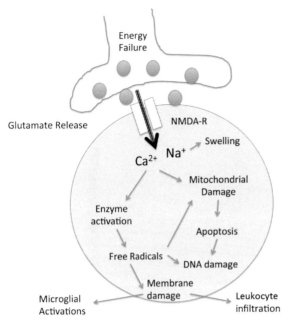

FIGURE 12 The ischemic cascade and pathways activated as a result. Initiated by energy failure, presynaptic release of glutamate aberrantly activates *N*-methyl-ᴅ-aspartate receptors (NMDA-Rs), permitting the uncontrolled influx of Ca^{2+} and Na^+ and causing mitochondrial damage, cell swelling, and activation of destructive enzymes. These processes culminate in apoptosis and membrane damage. This may cause microglial activation and the recruitment of leukocytes from peripheral blood.

voltage gradient across the mitochondrial membrane required for the production of ATP. This in turn further compromises the production of energy by brain tissue, further contributing to a vicious cycle of neuronal cell death.

The ischemic penumbra is also a site of extensive inflammation. Both astrocytes and microglial cells are activated during a stroke, and blood-borne immune cells, including leukocytes, T-lymphocytes, and natural killer cells, infiltrate the region as well. These inflammatory cells release cytokines, which in turn exacerbate the neural injury by stimulating nitric oxide production and enhancing NMDA-mediated excitotoxicity.

In both the ischemic core and the penumbra the integrity of the blood–brain barrier is compromised. This is partly due to the activation of matrix metalloproteinases and the retraction of astrocytic endfeet. The breakdown of the blood–brain barrier enhances penetration of blood-borne toxins and immune cells, each of which contribute to the further decline of tissue health.

It is commonly assumed that the ischemic core cannot be rescued, but neurons and glial cells in the penumbra can survive for some time and therefore are the therapeutic targets of essentially all acute treatment/neuroprotection strategies. The objectives of treatment are to maximize survival of cells in the ischemic penumbra by limiting excitotoxic or inflammatory injury and to restore normal blood flow. Therefore, noninvasively defining whether a penumbra exists in a given patient and determining the size of this area is of great importance. A combination of imaging techniques now allows some insight into this question in patients. Specifically, magnetic resonance imaging studies can readily distinguish between perfusion and diffusion within tissue. Diffusion-weighted images show structural, and hence permanent, destruction, thereby identifying the ischemic core. Perfusion images, on the other hand, show changes in relative perfusion and presumably the brain region with abnormally low, but not absent, perfusion. By overlaying these two images, one can establish whether there is an area in which there is a mismatch between diffusion and perfusion—this area represents the penumbra. Example images in Figure 14 illustrate this approach. While a potentially powerful predictor of vulnerable brain tissue, the clinical value of this approach may be debated. Unfortunately, given the current limitations on reperfusion of the infarcted tissue, there simply is not enough time to establish these images in all patients admitted with a stroke, for whom treatment has to begin immediately.

Figure 12 shows some of the various cellular changes that contribute to either necrotic or apoptotic cell death, illustrating many pathways, receptors, channels, and proteins involved. For the past 40 years, scientists have been extensively

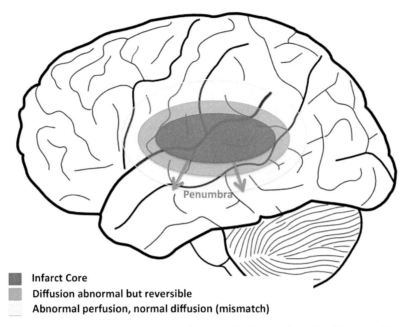

Infarct Core
Diffusion abnormal but reversible
Abnormal perfusion, normal diffusion (mismatch)

FIGURE 13   The ischemic penumbra is the area surrounding a stroke lesion where blood flow is still reversible and where tissue can potentially be rescued if blood flow is restored in a timely manner. It can be divided into two regions: one in which diffusion is abnormal, suggesting structural damage, and one in which diffusion is normal but perfusion of tissue with blood is abnormal. This area in particular can be rescued through rapid restoration of blood flow.

studying the contribution of each of these to cellular and animal models of disease. Many of these approaches predicted cures, yet, as further discussed below, none translated into effective treatments in humans. The one pathway that received the most attention is the NMDA receptor, believed to be the pivotal pathway for the entry of toxic Ca^{2+} in ischemia (Figure 15).

4.4 The NMDA Receptor and Glutamate Excitotoxicity

The NMDA receptor (NMDA-R) is a Glu-gated cation channel that is a member of the ionotropic Glu receptor family. Like the other family members, it is a heteromeric channel composed of four subunits, namely, two GluN1 and two GluN2 subunits. The GluN2 subunit comes in four isoforms named GluRN2A–D. The channel is gated by the binding of Glu, but it requires a modulatory site to be occupied by glycine. The channel distinguishes itself from all other Glu receptors by its multiple modulatory sites. In addition to the glycine site, which can bind D-serine instead of glycine, there are sites for modulation via nitric oxide, H^+, and phencyclidine. The channel is the target of numerous anesthetics, such as ketamine, which are often used as recreational hallucinogenic drugs. Other common agonists of differing potency include ethanol, memantine, phencyclidine ("Angel Dust"), dextrorphan, and methadone.

The feature that is of greatest relevance for stroke is the unique gating of the channel by voltage and intracellular Mg^{2+}. Normal, fast, excitatory synaptic transmission occurs via the α-amino-3-hydroxy-5-methyl-4-isoxazolepropionic acid (AMPA) and kainate-type Glu receptors. These too are cation-permeable, ionotrophic Glu receptors that open in response to Glu. NMDA-Rs are normally not available for activation by Glu because, at the resting

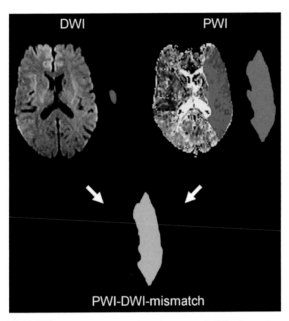

FIGURE 14   Mismatch of perfusion and diffusion using magnetic resonance imaging can identify the presence and size of the penumbra. Red demarcates the infarct zone, with irreversible injury, blue demarcates the zone with reduced blood flow containing at-risk tissue, and the green mismatch zone shows the area in which at-risk tissue can be saved. *Reproduced with permission from Ref. 6.*

potential of the postsynaptic membrane, an Mg^{2+} ion is lodged in the pore of the channel, preventing it from allowing ions to permeate. This Mg^{2+} binding is dependent on voltage, and with depolarization of the membrane, Mg^{2+} is released from the pore. Under normal physiological conditions Mg^{2+} keeps the NMDA-R closed at all times. Should the postsynaptic terminal be depolarized, however, the Mg^{2+} ion pops out, causing cations to flux through NMDA-Rs. While the vast majority of the current is carried by Na^+, which is the most abundant extracellular cation, the NMDA pore is fivefold more permeable to Ca^{2+} than to Na^+. As a consequence, a considerable amount of Ca^{2+} fluxes through the NMDA-R. This Ca^{2+} influx plays an important role during learning and memory, allowing the NMDA-R to function as a coincidence detector. As more extensively discussed in Box 1, Chapter 4, only if multiple signals arrive on the same

terminal within 50 ms of each other is the Mg^{2+} removed. Removal of the Mg^{2+} ion causes Ca^{2+} influx, thereby signaling the cell of the coincident occurrence of two events. This is believed to be an essential component of learning and memory. In the situation of a stroke, the coincident activation is bypassed by the chronic, long-lasting depolarization of the postsynaptic membrane due to energy failure. This energy failure permanently removes the Mg^{2+} block, opening the flood gates for Ca^{2+} to enter unimpeded through this receptor. Hence the very protein that is so critical in allowing for human learning can also make us vulnerable to cell death by excitotoxicity following a stroke.

The concept of Glu excitotoxicity was originally documented by Olney in the 1960s and has been extensively studied since. It seems to be the final common death pathway in numerous disorders, including many neurodegenerative diseases. The precipitating cause of the chronic depolarization of the neuronal membrane may differ in each case, yet the principal involvement of the NMDA-R as an influx pathway for Ca^{2+} is shared.

While NMDA-R plays the most important role in ischemic neuronal death, other family members participate or can substitute. The depolarization that occurs as a result of activation of AMPA or kainate-type Glu receptor depolarizes the postsynaptic cell, thereby allowing Ca^{2+} to enter via voltage-gated Ca^{2+} channels. Furthermore, at least one variety of the AMPA receptor is also permeable to Ca^{2+}.

4.5 Role of Glutamate

While Olney suspected Glu as a major toxin in disease, showing its contribution to stroke in animals and humans suffering a stroke was necessary. Many studies have measured Glu directly by microdialysis, a technique through which the extracellular milieu in the brain can be sampled continuously and with high accuracy. These studies demonstrated that occlusion of the MCA causes a 60- to 100-fold increase in Glu within a

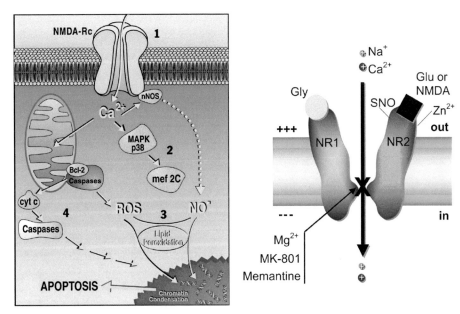

FIGURE 15 *N*-Methyl-D-aspartate receptor (NMDA-R) as mediator of excitotoxicity and target for therapeutic drugs. Permeation of Ca^{2+} via NMDA-Rs (left) causes activation of the mitogen-activated protein kinase (MAPK) cascade as well as release of cytochrome c from the mitochondria, each culminating in apoptotic cell death. The NMDA-R harbors many regulatory sites that can be exploited for therapy (right), including sites for the drugs memantine and MK-801 as well as the coagonist sites for Zn^{2+} and glycine (Gly). *Reproduced with permission from Ref. 7.*

relatively short time. The increase extends to the ischemic penumbra and even affects the composition of the cerebrospinal fluid (CSF) bathing the brain. CSF can be examined by lumbar puncture in patients in whom microdialysis would rarely be feasible unless the patient underwent surgery. Glu in the CSF rises eightfold in stroke patients and has been suggested to have predictive value for disease severity.[8] It is important to note that persistent increases in Glu, such as that measured after a stroke, also require the astrocytic reuptake of Glu to fail as well and suggest that energy failure in astrocytes in the penumbra significantly contributes to the progression of disease.

4.6 NMDA Inhibitors to Treat Stroke

Having discussed extensively the role of the NMDA-R in stroke, examining the many modulatory sites of this receptor and how they may be exploited therapeutically is worthwhile. In principle, blocking NMDA-R ameliorates neuronal death. However, it also impairs most higher cognitive function. This is exemplified by ketamine, used as a powerful anesthetic, or by phencyclidine, a potent, mind-altering substance, both of which have been associated with cognitive impairment. The ideal drug would therefore block only those NMDA-Rs that experience a pathophysiological depolarization by catastrophic increases in Glu, while sparing those receptors exposed to physiological concentrations. Such drugs actually exist and are, generally speaking, poor NMDA antagonists. One of these, memantine, is an open-channel blocker that inhibits NMDA channels poorly under physiological conditions but becomes more effective as Glu concentrations increase. Memantine has been used in experimental clinical trials for amyotrophic lateral sclerosis and dementia. It is very effective in animal models of stroke, where it almost completely protects the brain after MCA occlusion; hopefully it or related open-channel NMDA blockers may be examined for future use in the clinical treatment of stroke.

4.7 Effect of Temperature

Most biological processes are dependent on temperature. In the brain, ion transport via the various ATPases changes two- to threefold with every 10 °C change in temperature. Because the ischemic cascade is driven by a loss of function of these ATPases, one would expect that increases in temperature would accelerate damage, whereas lowering temperature would slow the injury. This is indeed observed in rats after MCA occlusion, where the development of an infarct can be significantly reduced by hypothermia[9] (Figure 16). Interestingly, some of the early neuroprotective trials using rats and the NMDA antagonist MK-801 that led to the development of many other NMDA antagonists for use in stroke are now being reinterpreted after it was found that MK-801 lowers the brain temperature by about 2°. Hence the neuroprotective effect seen with MK-801 may have been partially attributable to the hypothermia rather than just the blockage of NMDA-Rs. A correlation with fever and poor stroke outcome was reported as early as the eighteenth century,[9] and clinical studies now conclusively show that even slightly elevated temperature after an acute stroke leads to a much worse outcome and is associated with significantly higher risk of death.[11] From a mechanistic point of view, it is interesting that clinical studies have also shown a twofold increase in Glu in the CSF of ischemic stroke patients who had elevated body temperature. Moreover,

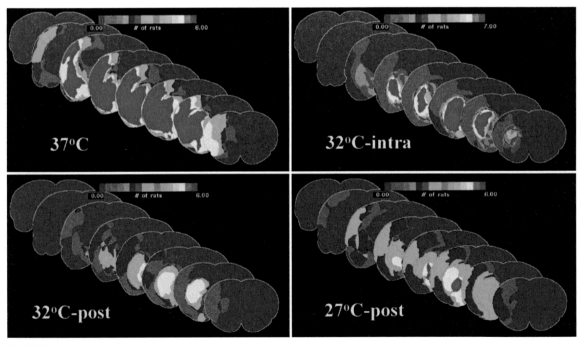

FIGURE 16   Effect of hypothermia on stroke volume after middle cerebral arterial occlusion in a rat. Reduction in temperature to 32 °C greatly reduces the size of the ischemic lesion with no further benefit when reduced to 27 °C. Tissue damage is indicated by more intense colors. Normothermic (37 °C) animals show a large consistent cortical and subcortical infarct at multiple coronal levels. Both intraischemic and postischemic hypothermia to 32 °C are associated with marked reductions in the frequency of cortical infarction. intra = intraischemic; post = postischemic. *Reproduced with permission from Ref. 10.*

clinical studies of stroke patients showed that a stroke per se can cause a local increase in brain temperature, suggesting active brain inflammation in the ischemic penumbra. These findings provide ample reasons to consider lowering brain temperature in patients with a stroke to improve outcome. At a minimum, one must control fever when present in stroke patients using drugs such as acetaminophen or ibuprofen.

5. TREATMENT/STANDARD OF CARE/CLINICAL MANAGEMENT

From previous discussion of mechanisms underlying stroke injury, it is clear that the greatest therapeutic benefit can be achieved by rapid reperfusion of the infarcted brain, particularly to rescue the ischemic penumbra that is at high risk of tissue destruction. Time is of the essence, and immediate admission to the hospital is required to achieve a maximum therapeutic benefit. Once admitted to the emergency room the door-to-needle time, that is, the time from admission to receiving an infusion of tPA (see below), must be as short as possible. During the initial evaluation the decision tree in Figure 1 is followed to categorize the underlying event as either ischemic or hemorrhagic. The latter excludes both mechanical and chemical recanalization.

5.1 Treatment of Stenosis Using Intravenous tPA

The only US Food and Drug Administration-approved drug for chemically opening vessels is the recombinant human plasminogen activator (alteplase/Activase). Its use is now considered standard of care provided certain inclusion criteria are met.[12] Most important of these is the time that has elapsed since the stroke. Treatment with tPA must begin within 3 h from the beginning of stroke symptoms. If the time of the event cannot be conclusively determined, treatment cannot be given. Unfortunately, this is often the case. For example, particularly in elderly patients,

the stroke victim may not be found until several hours after the stroke has occurred. If no one witnessed the attack, the last time the patient was seen in a normal condition will be used as a substitute.

Once in the hospital, a number of additional tests must be completed for a patient to be eligible within this 3-h time window. These include a noncontrast CT scan of the head to rule out hemorrhage and a blood glucose test through fingerstick to rule out hypoglycemia. Other factors, such as recent surgery, a history of head trauma or stroke, or high blood pressure (>185 mmHg systolic), would rule out tPA use as well. As a result of these stringent criteria, only a small percentage of stroke patients are eligible to receive tPA. Thus in 2009 only 5.2% of all stroke patients were able to receive tPA based on these criteria.

Following the outcome of two large phase III clinical studies conducted by the National Institute of Neurological Disorders and Stroke,[13] tPA is administered at 0.9 mg/kg, with 10% injected as a single bolus and the remaining dose administered through continuous infusion over the next 60 min. During this infusion period, the patient is monitored for any signs of intracerebral hemorrhage, determined by neurological signs such as headache, nausea, or vomiting. These signs would be cause for immediate termination of tPA infusion. The largest risk associated with tPA is intracranial hemorrhage, which occurs in 6–7% of cases and carries a very high mortality of 45%.[13] Data reported in the original study suggest that patients treated with tPA were at least 30% more likely to have minimal or no disability 3 months after treatment. Other studies have since explored the use of tPA up to 4.5 h after the incident and report smaller but still significant improvements of outcome with an acceptable increase in risk.[14]

A fascinating vignette from the original tPA study is the "Lazarus effect" (named after the miraculous resurrection of Lazarus 4 days after his death). Of patients treated with tPA, 20% showed a dramatic, indeed almost miraculous, improvement 24 h after treatment. Such a phenomenon has been previously reported following cardiac arrest, where spontaneous, unexplained reperfusion of the heart occurred in some patients after resuscitation efforts were suspended.

Unlike acute thrombolysis with tPA, the prevention of clotting using systemic treatment with antiplatelet agents has not been shown to be effective and indeed carries an increased risk. This is not to be confused with the proven benefit of such drugs to prevent stroke recurrence, as further discussed below.

5.2 Rehabilitation

Most stroke patients are left with permanent disability, and up to 50% of patients never regain functional independence. While Hippocrates would have recommended rest and immobility, today stroke recovery calls for active rehabilitation. The benefit of rehabilitation was first recognized in 1942 by Howard Rusk, an internist in St. Louis who joined the Army Air Corps as the chief of medical services. He pioneered rehabilitation after acute wartime injuries based on exercise and continuous movement. After the war ended, he applied these approaches to the treatment of stroke victims for whom little help was available at the time. Since then the benefit of rehabilitation has been well established. A typical rehabilitation regimen includes physical, occupational, and sensory therapy but is tailored to the specific disability of each patient. A promising experimental strategy constrains the functioning arm or leg in a hemiplegic stroke victim over periods of many weeks to months, thereby forcing them to use their impaired limbs. This constraint-induced therapy has shown remarkable benefit in some patients even when introduced years after the insult,[15] although early intervention produces the best results.[16] Since the ability to articulate language is greatly impaired in many stroke victims, an important part of rehabilitation involves treatment by a specialized stroke language pathologist.

5.3 Stroke Prevention

From the above discussion of primary causes of stroke and risk factors for the disease, it is clear that for many people stroke is a preventable disorder. The National Stroke Association has embarked on a campaign to educate physicians and the public to make every effort possible to reduce an individual's risk for a first stroke. This begins with a wellness visit with a primary care physician, who will be able to establish whether a person is at an elevated risk for stroke. The physician will consider in particular a person's age, diabetes, cholesteremia, high blood pressure, cardiovascular health, smoking, alcohol or drug consumption, and physical activity. Based on this information, a physician will educate the patient to address all modifiable risk factors through changes in lifestyle and diet. Moreover, cholesterol-lowering statin drugs and blood pressure-lowering beta-blockers can address chemically modifiable risk factors. Prophylactic aspirin use (80 mg/day) has been a common recommendation by many physicians in people with elevated risk. Aspirin irreversibly inhibits the formation of thromboxane A2, which is important for platelet aggregation and vasoconstriction. Once-daily, low-dose aspirin reduces the risk of stroke by about 25–30% and has minimal side effects. An alternative blood thinner, warfarin (2 mg), is typically used to prevent recurrence in patients who have already had a stroke.

6. EXPERIMENTAL APPROACHES/ CLINICAL TRIALS

Following a period of intense study of neuronal cell death throughout the 1960–1990s, which identified several core pathways involved in ischemic injury, tremendous enthusiasm ensued with not only hope but also conviction that effective neuroprotective drugs were just around the corner. Indeed, during the period of 1960–1980, basic researchers published 1026 different experimental treatments for stroke.[17] Drug companies invested billions of dollars into translating these findings into clinically useful neuroprotective drugs, and some representative drugs that were tested clinically are listed in Table 3. In total, 114 neuroprotective drugs were tested in humans and failed. Below we review some of these trials and provide a critical discussion of why they might have failed. In addition, we look at alternative strategies that are showing clinical promise.

6.1 Neuroprotection

Given the prominent role of NMDA-R-mediated Ca^{2+} entry, numerous modulators of the NMDA-R have been studied in humans. For example, dextrorphan is a noncompetitive NMDA antagonist that unfortunately causes unacceptable side effects, including

TABLE 3 Experimental Neuroprotective Strategies That Are the Subject of Clinical Trials

Proposed Neuroprotection	Drug	Mechanism of Action	Results
NMDA antagonists	Selfotel (CGS1975S)	Competitive antagonist	Complete/no benefit
	Eliprodi	Modulator of polyamide site	Terminated
	Aptiganel (Cerestat, CNS1102)	Noncompetitive blocker	Complete/no benefit
	$MgSO_4$	Blocks Ca^{2+} permeability of NMDA-R	Complete/no benefit
AMPA antagonist	YM872	AMPA antagonist	Complete/no benefit
Calcium channel blockers	Nimodipine	L-type Ca^{2+} channel blocker	Complete/no benefit
	Flunarizine	Nonselective, blocks Ca^{2+} entry, has calmodulin binding properties	Complete/no benefit
Sodium channel blockers	Fosphenytoin	Phenytoin prodrug	Complete/no benefit
Potassium channel activator	Maxipost (BMS-204352)	Activates KCNQ "M-channels"	Complete/no benefit
Anti-inflammatory agents	Erilimomab	Intercellular adhesion molecule-1 (ICAM)	Complete/worsening
	rNIF	Recombinant neutrophil inhibitory factor	Terminated
Free radical scavengers	Recombinant neutrophil lr	Blocks interaction between neutrophils (white blood cells) and fibrinogen	Complete/no benefit
	Citicoline (cytidyl diphosphocholine)	Counteracts inflammatory cytokines such as TNFa	Complete/no benefit
	Ebselen	Peroxidase with antioxidant activity	Ongoing/result pending
	NXY-059	Free radical trapping agent	Terminated

hallucinations and serious hypotension, which limited its use. The competitive NMDA-R antagonist, Selfotel, increased rather than decreased mortality, and consequently trials were halted.

To limit the hallucinogenic side effects experienced with NMDA antagonists, a new drug called GV150526 was designed. It acts on the modulatory glycine site on the NMDA-R. This drug was safe and well tolerated in 1367 patients but failed to show clinical improvements.

Intracellular Mg^{2+} blocks Ca^{2+} influx via NMDA-R and in theory would be a great modulator. This provided the rationale for a large multicenter study examining intravenous Mg^{2+} given as magnesium sulfate to 2589 patients 12 h after an acute stroke. Unfortunately, those receiving treatment did not show improvement over the placebo group. Currently, a follow-up trial is evaluating a much earlier treatment given within less than 2 h.

Targeting the AMPA receptor may be beneficial by limiting the depolarization of the postsynaptic membrane following a stroke. The AMPA receptor antagonist YM-872 was safe in patients, but no benefits were detected in a placebo-controlled phase II study.

Given the role of Ca^{2+} channels in synaptic Glu release as well as postsynaptic entry of Ca^{2+} during the ischemic cascade, Ca^{2+} channel blockers have been evaluated in numerous clinical studies, but, disappointingly, none of these have shown efficacy.

Yet another attractive strategy that emerged from our understanding of the ischemic cascade was reducing oxidative damage. The free radical trapping reagent NXY-059 was studied in a phase III study that enrolled 1722 patients with acute stroke. While it initially provided promising results, a follow-up clinical trial failed to reproduce these findings.

particularly the importance of avoiding elevated temperature, and clinical studies clearly establish that increased body and/or brain temperatures cause significantly worse outcome.[9,11] Together with promising studies in animal models (Figure 16), a strong rationale to explore the therapeutic benefit of hypothermia in stroke exists. The key idea behind hypothermia treatment is to slow the rate of tissue decline and extend the therapeutic time window for treatment. Ideally, a patient suffering from a stroke would be cooled very quickly until they reach a hospital. To do so, the whole body can be cooled using cold blankets, ice packs, and the like. However, a number of more sophisticated cooling devices have been developed, including catheters inserted into the inferior vena cava to cool the blood in circulation. At present, feasibility studies have been conducted only in patients beyond the time limit to receive tPA treatment; these studies showed that hypothermia was a safe procedure with only minimal side effects.[18] However, efficacy data will have to wait for an upcoming phase II study comparing cooled and uncooled patients in the same setting. The logistical challenges of widely administering cooling in an ambulance, and even in the emergency room, are not trivial. For this reason, researchers have been exploring other strategies. Some drugs, for example, the Parkinson's drug Talipexole, cause a reduction of body temperature as a side effect. These presumably act on dopamine receptors that are involved in temperature regulation in the hypothalamus. Researchers are now exploring ways to harness this side effect to therapeutically reduce body and brain temperature in an effort to confer similar neuroprotection to that observed by physically cooling the body or blood flow through the carotid arteries.[19]

6.2 Neuroprotection by Hypothermia

We have already discussed the effect of temperature on tissue damage after stroke,

6.3 Improved Clot Busters

Recombinant tPA remains the only approved chemical agent to recanalize vessels, and having second-generation drugs with enhanced ability to do so is desirable. Not surprisingly, there are a number of candidates at various stages of clinical testing. Some have an improved half-life; others promise to offer faster and more complete clot-busting activity.

An interesting experimental approach aimed at improving outcome is the combination of tPA with transcranial ultrasound. The ultrasound mechanically assists in the disintegration of the clot, which is chemically weakened by tPA. This approach, called "clotbust," showed dramatic results, with 49% of patients achieving complete recanalization.[20] A further refinement of this approach involves nano-bubbles that are infused into the bloodstream. These bubbles adhere to the clot surface and are activated to burst through ultrasound energy. Because they burst like small microexplosions, they help to mechanically degrade the clot.

6.4 Mechanical Revascularization

Microcatheters have long been used to place stents into cardiac blood vessels. These devices are now small enough to be inserted into cerebral arteries as well and have been further developed into sophisticated clot dissolution and retrieval devices. In their simplest form, repeated passage of a wire through a thrombus helps to disintegrate it mechanically. In a more advanced form, these devices contain a soft silicon balloon that can be expanded to open a collapsed artery to allow placement of a stent to keep it open. This approach is particularly promising for opening larger arteries such as the MCA or ICA; successful opening of an artery and placement of a stent is shown in Figure 17.

6.5 Why Have So Many Promising Drugs Failed in Human Clinical Trials?

Between 1955 and 2000, 178 stroke trials enrolled a total of 73,949 patients. Of these, 88 trials pursued neuroprotective strategies, 59 pursued antithrombotic approaches, and 26 used a combination of the two. These trials were based on solid preclinical evidence typically obtained using established and validated rodent

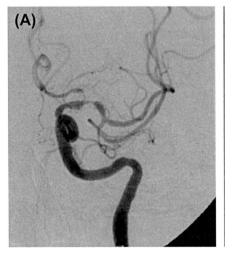
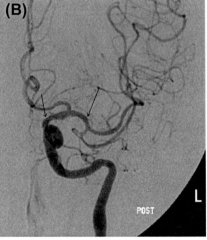

FIGURE 17 Successful recanalization of the anterior cerebral artery (ACA) and middle carotid artery (MCA) by balloon angioplasty: (left) before and (right) after angioplasty and placement of a stent. A, Left internal carotid artery cerebral angiography demonstrates severe stenoses of the ACA and MCA, which are confluent at the carotid terminus. B, After angioplasty alone of the ACA and stent-assisted angioplasty of the MCA, there is normal caliber of the previously stenosed ACA and MCA. Arrows denote the proximal and distal ends of the stent. *Figure and legend reproduced with permission from Ref. 21.*

models of stroke. A partial list of the drugs and their presumed targets is listed in Table 3. Not one of the 88 neuroprotective trials had a positive outcome, and only two drugs emerged from the revascularization trials, namely, tPA and aspirin.

How can this be explained? Is there any reason to pursue this any further? A careful analysis of both clinical trials conducted during the past 40 years[22] and the preclinical data on which they were based[23] provides some important lessons.

First, the vast majority of preclinical studies are done on healthy, young adult mice or rats, which do not suffer from diabetes, heart disease, or hypertension. They neither smoke nor drink and have little genetic predisposition to disease. Each receives an identical vascular insult to induce a stroke, most commonly ligation of the MCA. Treatment is administered quickly, within 30 min to a few hours at most, and each subject is treated at the same time after the insult. The effectiveness of treatment is then measured by the relative decrease in infarct volume, not by neurological or cognitive status, with little long-term follow-up. By comparison, patients enrolled in stroke studies tend to be old and have a host of health conditions ranging from heart disease to atherosclerosis. The arterial obstructions vary greatly in size and region, and the delay until treatment ranges from a few hours to many days after the insult. A patient's response to treatment is not measured by infarct size but by behavioral outcomes that are assessed 3–6 months later. Clearly, the preclinical studies of rodents do not adequately model the human condition.

Second, most clinical studies conducted to date were statistically underpowered to detect the effect size they attempted to measure. In other words, they never enrolled a sufficient number of patients to show a significant improvement.[23] As an example, showing a 5% reduction in patient deaths or disability 6 months after a stroke with typical 80% power requires enrollment of 3148 patients. Only 2% of all clinical trials conducted were appropriately powered to show this 5% clinical benefit.

Third, a review of the 1026 drugs that showed therapeutic benefit in preclinical studies[13] concluded that the 114 drugs that were selected for use in patients were not any more effective than the 912 drugs that were not used. The rationale for selecting the drugs that were advanced to the clinic is therefore unclear.

Based on this sober assessment, several changes are required. Foremost, making studies using animal models more relevant to human disease and measuring similar behavioral outcomes at different times after the insult are imperative. Also, consideration should be given to conducting trials using rodents that are not in good health but mimic the fragility of the human stroke population. We must ensure that clinical trials have the statistical power to be able to detect the subtle improvements they set out to find. If we do trials, we must do them right! Moreover, the notion that a single reagent is sufficient to treat a complex disorder must be abandoned and replaced with polytherapy that combines drugs with different mechanisms of action, possibly administered at different stages of disease. The combination of early tPA, followed by an NMDA inhibitor such as dextrorphan, with a caspase inhibitor such as ZVAD-fmk to limit neuronal cell death may be an example.

7. CHALLENGES AND OPPORTUNITIES

During human evolution, some vulnerabilities were accepted to meet the demand of packaging an increasing number of neurons into the cranial space. Wasteful glycolytic generation of ATP and energy storage in the form of glycogen were abandoned, making neurons reliant on the uninterrupted delivery of oxygen and glucose. In parallel, the NMDA-R evolved as a coincidence detector, utilizing a voltage-dependent mode of Ca^{2+} entry that fails under conditions of energy loss. Hence, as the cellular protein fundamental to learning and memory, the NMDA-R became the executioner of the cell in a process called excitotoxicity. Our ability to learn came at the cost of suffering strokes.

In spite of heroic efforts by basic and clinical researchers and billions of dollars in expenditures by many drug companies, most efforts to protect neurons from certain death after a stroke have failed. All remaining effective treatments correct the brain plumbing by busting clots chemically or mechanically to quickly restore blood flow. Those approaches are much more effective and safer than they were two decades ago, and they provide life-altering benefit to many patients. However, for over 94% of stroke victims who do not qualify for antithrombotic treatment, current remedies remain inadequate.

Given the staggering numbers of stroke-related deaths (5.5 million worldwide/200,000 in the United States annually) and over 4 million persons living with stroke-related disability, an urgent call to action is in order. Unless scientists discover multipronged approaches to prevent neuronal death after periods of ischemia, our greatest challenge and promise will be in stroke prevention. Since the disease is intrinsically linked to heart disease, diabetes, and poor lifestyle choices, we are fighting the same uphill battle as those dealing with rampant diabetes and obesity. Throughout this book I have elected to remain optimistic that clinical and basic research will ultimately prevail, and against all odds, this should certainly be the case for stroke as well. Decades of stroke research has not been wasted but has afforded us incredibly detailed insight into the cellular pathways that make life-and-death decisions. The protective mechanisms that neurons have developed, and their interdependence with supporting glial and vascular cells, are now much better understood. Yet many of the preclinical experiments were done prematurely, long before we fully understood the ischemic

process in earnest. Many prior studies of animal models of stroke must be revisited under conditions that more adequately mimic human disease. Multimodal treatment designs, where a combination of targets are hit simultaneously, have to be explored, and preclinical data must be scrutinized much more fully before introducing new drugs into human trials. Given the magnitude of the problem, complacency is not an option.

Acknowledgments

This chapter was kindly reviewed by Christopher B. Ransom, MD, PhD, Director, Epilepsy Center of Excellence and Neurology Service, VA Puget Sound & Department of Neurology, University of Washington.

References

1. Pound P, Bury M, Ebrahim S. From apoplexy to stroke. *Age Ageing.* September 1997;26(5):331–337.
2. Schiller F. Concepts of stroke before and after Virchow. *Med Hist.* April 1970;14(2):115–131.
3. Xavier AR, Qureshi AI, Kirmani JF, Yahia AM, Bakshi R. Neuroimaging of stroke: a review. *South Med J.* April 2003;96(4):367–379.
4. Barrett KM, Levine JM, Johnston KC. Diagnosis of stroke and stroke mimics in the emergency setting. *CONTINUUM: Lifelong Learn Neurol.* 2008;14(6, Acute Ischemic Stroke):13–27. http://dx.doi.org/10.1212/1201. CON.0000275638.0000207451.ae.
5. Zlokovic BV, Apuzzo ML. Strategies to circumvent vascular barriers of the central nervous system. *Neurosurgery.* 1998;43(4):877–878.
6. Rimmele DL, Thomalla G. Wake-up stroke: clinical characteristics, imaging findings, and treatment option - an update. *Frontiers in neurology.* 2014;5:35.
7. Lipton SA. Failures and successes of NMDA receptor antagonists: molecular basis for the use of open-channel blockers like memantine in the treatment of acute and chronic neurologic insults. *NeuroRx.* 2004;1(1):101–110.
8. Davalos A, Shuaib A, Wahlgren NG. Neurotransmitters and pathophysiology of stroke: evidence for the release of glutamate and other transmitters/mediators in animals and humans. *J Stroke Cerebrovasc Dis: Off J Natl Stroke Assoc.* November 2000;9(6 Pt 2):2–8.
9. Campos F, Blanco M, Barral D, Agulla J, Ramos-Cabrer P, Castillo J. Influence of temperature on ischemic brain: basic and clinical principles. *Neurochem Int.* April 2012;60(5): 495–505.

10. Huh PW, Belayev L, Zhao W, Koch S, Busto R, Ginsberg MD. Comparative neuroprotective efficacy of prolonged moderate intraischemic and postischemic hypothermia in focal cerebral ischemia. *J Neurosurg.* 2000;92(1):91–99.

11. Azzimondi G, Bassein L, Nonino F, et al. Fever in acute stroke worsens prognosis. A prospective study. *Stroke.* November 1995;26(11):2040–2043.

12. Khatri P, Levine J, Jovin T. Intravenous thrombolytic therapy for acute ischemic stroke. *CONTINUUM: Lifelong Learn Neurol.* 2008;14(6, Acute Ischemic Stroke):46–60. http://dx.doi.org/10.1212/1201.CON.0000275640.0000284580.0000275610.

13. The national institute of neurological disorders and stroke rt-PA stroke study group. Tissue plasminogen activator for acute ischemic stroke. *N Engl J Med.* December 14, 1995;333(24):1581–1587.

14. Hacke W, Kaste M, Bluhmki E, et al. Thrombolysis with alteplase 3 to 4.5 hours after acute ischemic stroke. *N Engl J Med.* September 25, 2008;359(13):1317–1329.

15. Taub E, Morris DM. Constraint-induced movement therapy to enhance recovery after stroke. *Curr Atheroscler Rep.* July 2001;3(4):279–286.

16. Wolf SL, Thompson PA, Winstein CJ, et al. The EXCITE stroke trial: comparing early and delayed constraint-induced movement therapy. *Stroke.* October 2010;41(10):2309–2315.

17. O'Collins VE, Macleod MR, Donnan GA, Horky LL, van der Worp BH, Howells DW. 1026 experimental treatments in acute stroke. *Ann Neurol.* March 2006;59(3):467–477.

18. Horn CM, Sun CH, Nogueira RG, et al. Endovascular Reperfusion and Cooling in Cerebral Acute Ischemia (ReCCLAIM I). *J Neurointerventional Surg.* 2014;6(2):91–95.

19. Johansen FF, Hasseldam H, Rasmussen RS, et al. Drug-induced hypothermia as beneficial treatment before and after cerebral ischemia. *Pathobiology: J immunopathol, Mol Cell Biol.* 2014;81(1):42–52.

20. Alexandrov AV, Molina CA, Grotta JC, et al. Ultrasound-enhanced systemic thrombolysis for acute ischemic stroke. *N Engl J Med.* November 18, 2004;351(21):2170–2178.

21. Layton KF, Hise JH, Thacker IC. Recurrent intracranial stenosis induced by the Wingspan stent: comparison with balloon angioplasty alone in a single patient. *AJNR. American journal of neuroradiology.* 2008;29(6):1050–1052.

22. Kidwell CS, Liebeskind DS, Starkman S, Saver JL. Trends in acute ischemic stroke trials through the 20th century. *Stroke.* June 2001;32(6):1349–1359.

23. Gladstone DJ, Black SE, Hakim AM. Toward wisdom from failure: lessons from neuroprotective stroke trials and new therapeutic directions. *Stroke.* 2002;33(8):2123–2136.

General Readings Used as Source

1. Sacco RL. Pathogenesis, classification, and epidemiology of cerebrovascular disease. Chapter 35. In: Lewis P, Rowland (ed.), Merritt's textbook of Neurology, 10th edition, Baltimore, MD: Lippincott Williams & Wilkins; 2000.

2. Smith WS, English JD, Johnston S. Cerebrovascular Diseases. Chapter 370. In: Longo DL, Fauci AS, Kasper DL, Hauser SL, Jameson J, Loscalzo J (Eds.) Harrison's Principles of Internal Medicine, 18th edition. New York, NY: McGraw-Hill; 2012.

3. Bhardwai A, Alkayed NJ, Kirsch Richard JR, Traystman J, eds. *Acute Stroke: Bench to Bedside.* CRC press; 2013.

Suggested Papers or Journal Club Assignments

Clinical Papers

1. Hacke W, et al. Thrombolysis with alteplase 3 to 4.5 hours after acute ischemic stroke. *N Engl J Med.* 2008;359(13):1317–1329.

2. The national institute of neurological disorders and stroke rt-PA stroke study group. Tissue plasminogen activator for acute ischemic stroke. *N Engl J Med.* 1995;333(24):1581–1587.

3. Alexandrov AV, et al. Ultrasound-enhanced systemic thrombolysis for acute ischemic stroke. *N Engl J Med.* 2004;351(21):2170–2178.

4. Xavier AR, et al. Neuroimaging of stroke: a review. *South Med J.* 2003;96(4):367–379.

5. Wolf SL, et al. The EXCITE stroke trial: comparing early and delayed constraint-induced movement therapy. *Stroke.* 2010;41(10):2309–2315.

Basic Papers

1. Johansen FF, et al. Drug-induced hypothermia as beneficial treatment before and after cerebral ischemia. *Pathobiol.* 2014;81(1):42–52.

2. Campos F, et al. Influence of temperature on ischemic brain: basic and clinical principles. *Neurochem Int.* 2012;60(5):495–505.

3. Huttner HB, Bergmann O, Salehpour M, et al. The age and genomic integrity of neurons after cortical stroke in humans. *Nat Neurosci.* 2014;17(6):801–803.

4. Shih AY, Blinder P, Tsai PS, et al. The smallest stroke: occlusion of one penetrating vessel leads to infarction and a cognitive deficit. *Nat Neurosci.* 2013;16(1):55–63.

5. Wu F, Catano M, Echeverry R, et al. Urokinase-type plasminogen activator promotes dendritic spine recovery and improves neurological outcome following ischemic stroke. *J Neurosci.* 2014;34(43):14219–14232.

Central Nervous System Trauma

Harald Sontheimer

1. CASE STORY

How much Jamie had missed the snow. Although Tennessee receives a few inches of snow on occasion, nothing compares to the dry, powdery coat that covers the high mountains in winter. Since moving to the southern United States some 20 years earlier, she had not been on skis once. On occasion, when asked to join friends on skiing trips to the Appalachians, she dismissed the low hills of Virginia as inadequate for her skills. Growing up in Innsbruck, near the Austrian Alps, she fondly remembers being on the slopes on every possible occasion. When the opportunity presented to join the management of her investment firm for a retreat in Snowbird, Utah, there was no hesitation. Waking up to blue skies, she took the day's first gondola to Hidden Peak, quietly moving high above the hills covered in virgin powder. From the top of the mountain she started out on Cirque Travers, a gentle glide toward the valley. Shaking off her initial hesitation and rustiness, she quickly got into a smooth, rhythmic motion as the skis carved into the snow below. It seemed like she had never stopped her passion and felt right at home in the powder. Her colleagues would soon lose touch with her as, clearly, no one came close to her skiing skills. Even after all those years, she still knew how to go hard and fast. Not knowing the mountain, she turned toward Wilbere Bowl, a double diamond slope winding through the trees. She still had her jumps down and was going faster and faster, soon hitting the moguls hidden under a foot of fresh powder. As the hill got steeper, her concentration started to fade and as her legs tired the skiing became labored. She lost her footing jumping over the next mogul, accelerated out of control, and panicked before she could react, with a tree coming toward her fast. She doesn't remember what happened next.

"Can you move your finger? Can you wiggle your toe? Stay right where you are, don't move." She was feeling anxious and her head hurt. She panicked, since she clearly couldn't move her feet or hands. She didn't even feel them. Where were her colleagues? Would they be searching for her? Who called the ski patrol? "I am Jim, and this is Josh. You wiped out good. Just stay still and don't try to move. We will take you down the mountain." Jim and Josh carefully lifted her into a toboggan, while she faded in and out of consciousness. Jamie barely remembers the ride to the valley, where an ambulance was already waiting for her transfer to Intermountain Medical Center in Murray. When she awoke, she was disoriented, in a hospital emergency room, hooked up to all sorts of monitors and wearing a neck brace. Her whole body was motionless, trapped in a plastic mold, with straps over her torso and legs. That day she learned that skiing would most likely be in her past. Her collision had fractured her spinal column and injured her spinal cord between the fourth and fifth cervical spine. Had the lesion been any higher, she could have died on the spot, unable to breathe on her own. Lucky enough to survive, she was left quadriplegic, unable to use either her arms or legs and with little hope for even a modest recovery.

2. HISTORY

As long as man has roamed the earth, injuries to the head or spinal cord have been a common and often deadly occurrence. As hunters, humans faced many injuries that were inflicted chasing wild animals or resulted from brawls with unwelcome neighbors. The transition to walking upright on two legs made *Homo erectus* able to travel long distances and quickly expand throughout the globe, but it also came with an increased incidence of falls, which, when they occurred, were more volatile.

Judging from archaeological evidence, head injuries were common in ancient times, and typically the injured individuals did not survive. An old Egyptian medical papyrus from 1700 BC, purchased by Edwin Smith in 1862, revealed after its translation surgical procedures and recommended treatment utilized in 48 cases of traumatic injury (Figure 1). Of these, six involved the spinal cord. These are believed to be the earliest known medical documents describing head and spinal cord trauma, along with proposed treatments. Sadly, for those suffering from spinal cord injury (SCI), the author recommended that these should not be treated at all: "an ailment not to be treated." This lesson would guide physicians well into the twentieth century, sealing the fate of James A. Garfield (1831–1881), 20th president of the United States. Shortly after taking office he was shot; a bullet lodged just below the first lumbar vertebra and left his legs paralyzed. As complications developed from what should have been a very treatable injury, his physician simply repeated the dreaded words that his was "an ailment not to be treated."[1]

However, forensic evidence also tells us that humans have long attempted to treat head or spinal cord injuries. For example, in 1518 the surgeon Berengario da Carpi published his work on head injury, including a report on the care of Lorenzo de Medici, one of the members of the influential Medici family in Florence, Italy. He had suffered a fracture to the occipital lobe from a gunshot, on which da Carpi successfully operated. Hans von Gerssdorff (c. 1455–1529) published the "Feldbuch der Wundartzney" ("Field Book of Wound Therapy"), depicting surgery on a patient with a depressed skull fracture, who, judging from the drawing, is suffering from right third cranial nerve palsy and left facial paralysis (Figure 2). Whether the patient survived the procedure is anyone's guess.

Throughout history, surgeons have experimented on just about any kind of injury and many have carefully described their results. By and large, survival from head or spinal surgery had a 50% chance at best, with the most common complication being infection. Arguably

FIGURE 1 The Edwin Smith Medical Papyrus (c. 1700 BC) is the oldest known document describing trauma surgery (public domain).

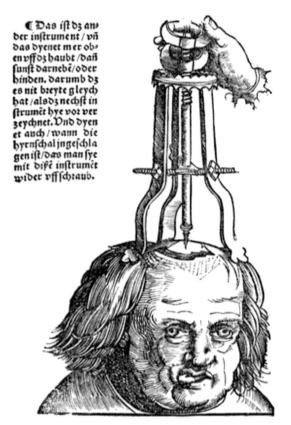

❡ Das ist dz an-
der instrument / vñ
das dyenet mer ob-
en vff oz haubt / daſ
sunst darnebē / oder
hinden. darumb dz
es nit breyte gleych
hat / als oz nechst in
strumēt hye vor ver
zeychnet. Vnd dyen
et auch / wann die
hyrnschal jngeschla
gen ist / das man sye
mit diſē instrumēt
wider vff schraub.

FIGURE 2   Operation on a depressed skull fracture that resulted in a palsy of the third cranial nerve and left facial paralysis (public domain).

such as knives and bayonets were added. These were ultimately replaced with penetrating high-velocity projectiles such as bullets and shrapnel, and now by nonpenetrating yet equally devastating blast waves. Historically, the majority of head and spinal injuries occurred on the war field, and much of our knowledge was gained from soldiers. Unfortunately, although wartime injuries are no longer the leading cause of CNS trauma, these still present the largest challenge.

Possibly the greatest achievement in CNS trauma to date is not our ability to heal and restore function but instead the approach taken toward an individual after he or she suffers an injury. Donald Munro (1898–1978), often called "the father of paraplegia," realized that for a patient suffering from CNS injury, he must be more than just a neurosurgeon. He had to accept responsibility to provide for the whole person, tending to all the involved organ systems from bladder function to psychological well-being.[1] He had to coordinate patient rehabilitation, instruction on self-care, and mobility and ultimately facilitate the patient's reintegration into family and society. To do so, he established the first spinal cord unit at the Boston City Hospital in 1936, which became the model for specialized integrated physical medicine and rehabilitation clinics around the world.

Equally influential was the landmark legislation known as the Americans with Disabilities Act. Passed in 1990, it provided equal rights, access, and opportunities to individuals with disabilities. It is responsible for the countless augmentative services and accommodations we take for granted today, not least important of which are wheelchair ramps and wide elevators and bathroom doors.

Sadly, however, from the Smith papyri of 1700 BC to the late twentieth century, a "nihilistic" acceptance of the inevitable outcome from injury remained common among physicians and scientists. As a consequence, comparatively little research had been conducted on mechanisms of injury and possible strategies to recover and restore function. I would argue that the tragic

the greatest success in surgical interventions of any kind came on the heels of Louis Pasteur's (1832–1895) "germ theory," which ultimately resulted in the development of Joseph Lister's (1827–1912) aseptic techniques and Alexander Fleming's (1881–1955) discovery of the antibiotic penicillin. Open head and spinal cord injuries in particular had a significantly improved outcome after systemic administration of antibiotics.

Over the course of history, the precipitating causes of central nervous system (CNS) injuries evolved significantly,[2] and so did the nature of the injury. Originally, blunt instruments such as stones or clubs crushing the cranial or vertebral bones largely caused head and spinal trauma. Then sharp, penetrating low-velocity objects accident of "Superman," Christopher Reeve, on May 30, 1995 was a sad but much-needed catalyst for change. An expert equestrian, Christopher participated in the Commonwealth Dressage finals in Culpeper, Virginia. As he approached the third of 18 jumps, a 3½-foot-high triple bar, his horse stopped abruptly, unable to find the right footing for the jump. Christopher rolled along the horse's neck and fell head first to the ground. Although he was wearing a helmet and protective vest, he fell unconscious immediately. Emergency personnel initiated mouth-to-mouth resuscitation and Christopher first survived the medical transport to Culpeper Medical Center and then the transfer to the University of Virginia Medical Center. Here, he was diagnosed with a life-threatening complex fracture to the first and second vertebrae, leaving him with no movement and without the ability to breathe on his own. The location of Christopher's fracture was essentially incompatible with life and, in hindsight, it is astonishing that he ever made it to the hospital alive. Against all odds, Christopher was able to gradually acquire the ability to breathe on his own for periods of time, allowing him to get off his respirator to speak. And speak he did. He became the most powerful voice for research on SCI. He would not accept "no" as an answer, and his main crusade was against complacency. His public efforts have been transformative and have injected much-needed energy, enthusiasm, and hope into the medical and scientific community.

Sadly, traumatic brain injury (TBI) has never had a comparable spokesperson and still carries a significant stigma. However, coincident with Reeve's tragic SCI, reports of children suffering from the consequences of sports-related concussions injected palpable interest in understanding how repeated, seemingly small injuries may aggregate to produce severe physiological changes that impair cognitive function. The public interest is fueled by recent prominent cases, including the Pro Football Hall of Famers Tony Dorsett and Joe DeLamielleure and the former NFL All-Pro Leonard Marshall, who suffer from chronic traumatic brain inflammation as a result of repeated sports injuries. For the first time we recognize injury prevention as a major health challenge.

3. CLINICAL PRESENTATION/ DIAGNOSIS/EPIDEMIOLOGY

We will discuss TBI and SCI together, because both present with common underlying pathology and neurophysiological challenges. Both are part of the CNS and are enclosed by the same meningeal coverings that separate the CNS from the rest of the body (Box 1). They share the same vascular supply and are equally protected by the blood–brain barrier (Box 1). Finally, many of the sensory or motor pathways affected in each form of injury have axonal cell processes in both structures. However, since TBI and SCI are clinically well-delineated health conditions, we will discuss epidemiology and treatment approaches separately and later converge on common themes under disease mechanisms.

3.1 Traumatic Brain Injury

TBI constitutes a major health problem, with an estimated 57 million people experiencing TBI worldwide. Approximately 1.7 million new cases are reported in the US annually. Of these, 500,000 patients report to the emergency room, 275,000 are admitted to the hospital, and 52,000 die each year. The actual numbers may indeed be much larger, as many mild injuries go unreported. Although TBI spans a large range of severity from mild to severe, it can be a lethal condition that is responsible for roughly one-third of all injury-related deaths in the US. Although head trauma can occur in any individual at any time, it is much more common in males than in females (3:1) and is most prevalent in very young children (<4 years), followed by adolescents and older adults (>65 years). The increased TBI prevalence in adolescents is largely due to increased risk-taking behavior, motor vehicle accidents, and sports-related injuries. In the

BOX 1

CEREBROSPINAL FLUIDS AND THE BLOOD–BRAIN BARRIER

All parts of our body are perfused by circulating blood delivering oxygen, nutrients, and immune cells. Venous blood removes many waste products, including CO_2. In contrast to the rest of the body, the brain and spinal cord tissue are not in direct contact with blood but instead are bathed in cerebrospinal fluid (CSF), a clear fluid that is essentially a cell-free ultra-filtrate of blood. The reason for separating brain tissue from direct contact with blood is that many molecules found in blood are potentially harmful to neurons. Such molecules include, for example, amino acids such as glutamate, GABA, and glycine, which are used in micromolar quantities as neurotransmitters in the brain but reach millimolar concentrations after digestion of protein in the blood.

The separation of blood and CSF occurs at the level of individual cerebral blood vessels. Endothelial cells fusing with each other into a hollow tube form blood vessels. The fusion sites throughout our body are fenestrated junctions, which leave enough space for macromolecules, including immune cells, to leave the vessel and enter the interstitial space of an organ (Box 1-Figure 1). By contrast, in the central nervous system (CNS), endothelial cells are fused to each other through tight junctions, which do not permit any leakage to occur. As a consequence, all substrates, including glucose and even gases, must go through the endothelial cells. Lipophilic molecules such as ethanol or many hormones readily cross the endothelial lipid membrane. Gases such as O_2 and

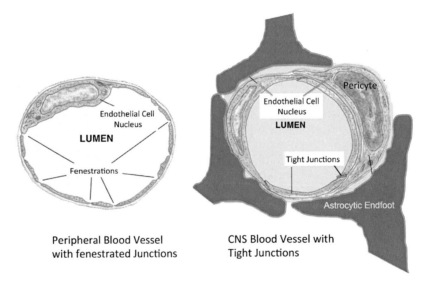

Peripheral Blood Vessel
with fenestrated Junctions

CNS Blood Vessel with
Tight Junctions

BOX 1-FIGURE 1　Comparison of peripheral and central blood vessels. Peripheral vessels lack tight junctions that seal the endothelial cells, thereby preventing leakage of molecules across the vessel wall. Central vessels contain pericytes and astrocytes that induce and maintain tight junctions and formation of the blood–brain barrier. (Redrawn with permission from a cartoon by Dr Thomas Caseci, Virginia Maryland Regional College of Veterinary Medicine, and Dr Samir El-Shafey, of the Faculty of Veterinary Medicine, Cairo University, Giza, Egypt.)

BOX 1   *(cont'd)*

CO_2 also diffuse relatively unimpeded. However, many amino acids, vitamins, and even glucose must be actively transported across the endothelial membrane. Note that since the blood vessel walls are living cells, they contain a cytoplasmic lumen. Therefore, molecules must cross two lipid membranes and the cells, cytoplasm where enzymes can break down unwanted molecules, providing a second, enzymatic barrier.

During development, the tight junctions that form the blood–brain barrier (BBB) are induced by pericytes. Subsequently, continued attachment of astrocytic endfeet is required for the maintenance of an impervious BBB. Many diseases, and trauma in particular, disrupt either the vessel or the astrocytes and pericytes required to maintain the BBB intact. As a result, blood-borne molecules including toxic amino acids and immune cells enter the brain and contribute to inflammation.

The choroid plexus, a secretory epithelium at the roof of the third and fourth ventricle, produces CSF (Box 1-Figure 2). The brain contains approximately 150 ml of CSF in the interstitial and ventricular space (Box 1-Figure 2). CSF is replaced ~3 times daily. It drains in the subarachnoid space, where it is absorbed into veins.

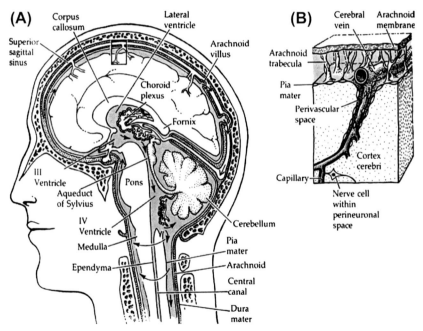

BOX 1-FIGURE 2   A. Flow of CSF from the choroid plexus where it is produced, through the ventricles and into the subarachnoid space, where CSF is absorbed into the venous system thereby clearing waste products from the brain. The CSF essentially surrounds the entire brain, and through the buoyancy effect reduces the brain's effective weight and provides a cushion effect. B. Meningeal covers of the brain include the pia mater that is directly in contact with the brain surface, the dura mater on the skull side, and the arachnoid mater in between. It contains the arachnoid trabecula, a space that is filled with CSF. This fluid-filled space serves as a cushion for the brain as the head moves. *Reproduced with permission from Ref. 3.*

Continued

BOX 1 *(cont'd)*

This space is between two connective tissue layers on the outside of the brain called the dura mater, against the skull, and the pia mater, attached to the brain. The small fluid-filled arachnoid space provides a liquid cushion for the nervous system. Because it is floating in fluid, the brain's effective mass is reduced 30-fold, thereby reducing the impact experienced by accelerations and decelerations. CSF plays an important role in clearing debris after injury and removing cellular waste products, and has been compared to the lymphatic system found elsewhere in the body.[4] Any obstruction to its flow can generate pressure within the brain, causing headaches and possibly hydrocephalus. Tapping CSF through the lumbar spinal cord has become invaluable to assess the CNS for evidence of infection or disease. For example, the presence of leukocytes would indicate breakdown of the BBB as a result of inflammation.

---

young and elderly, the increased incidence is due to falls and age-related instability. Overall, falls are responsible for the majority of head injuries (35%), followed by motor vehicle accidents (17.3%), being hit by an object (16.5%), and assaults (10%). However, of the moderate to severe TBIs, motor vehicle accidents remain the largest cause. The total cost to society is immense and not fully quantifiable. Acute healthcare costs alone for an individual with a mild form of TBI are estimated at $80,000–100,000; that of a patient with severe TBI at ~$3 million. Even mild forms of TBI cause significant disability and loss of mental capacity, leaving many patients who are in otherwise good health unable to reach their full economic potential. Those with severe TBI may never participate in work and require life-long care.

Injuries are often classified as either closed or open head injuries based on whether the integrity of the skull was or was not compromised. Open head injuries naturally have an increased potential to present with bleeding and may be associated with extensive inflammation, as pathogens can readily enter the brain through an open wound.

TBI can be divided into three grades of severity, which reflect the relative impairment of a patient irrespective of the underlying cause. These impairments can be quickly assessed by the Glasgow Coma Scale (GCS, Table 1), which measures the level of consciousness on a 15-point scale. Values in three categories are added, yielding a final score range of 3 for a comatose individual to a 15 for a normal functioning person.

Mild TBI (mTBI) is often used synonymously with concussion and represents a score range of 13–15 on the GCS. Consciousness is only lost for a very short period (<30 min), if at all, and memory loss is typically mild and transient, lasting for less than 24 h. Patients often have headaches, are confused, and may have difficulty sleeping. Dizziness, vertigo, irritability, impulsiveness, and difficulty concentrating are other common symptoms, which typically resolve within a few hours or days. Concussions are particularly common among athletes, with football, soccer, hockey, and boxing among the leading sports causing concussions. We now recognize that even if there are no visible changes on brain scans, the injuries can have profound effects on brain function. Importantly, experiencing a first concussion increases the risk of repeat concussions, which incurs additive damage to the brain and can cause long-lasting deficits in memory and cognition.

Moderate TBI has a GCS of 9–12 and is typically associated with prolonged loss of consciousness and neurological deficits that do not readily disappear on their own. Minor bleeding is not uncommon, and patients require immediate medical attention. Postinjury rehabilitation is typically required after the acute phase of the injury.

Severe TBI, with a GCS of 8 or less, presents with visible damage to the brain, including

TABLE 1   The Glasgow Coma Scale Rates a Person's Consciousness from 3 (comatose) to 15 (normal)

Response	Score
Eye opening	
Opens eyes spontaneously	4
Opens eyes in response to speech	3
Opens eyes in response to painful stimulation	2
Does not open eyes in response to any stimulation	1
Motor response	
Follows commands	6
Makes localized movement in response to painful stimulation	5
Makes nonpurposeful movement in response to painful stimulation	4
Flexes upper extremities/extends lower extremities in response to pain	3
Extends all extremities in response to pain	2
Does not show any motor response to any stimulus	1
Verbal response	
Is oriented to person, place, and time	5
Converses but may be confused	4
Replies with inappropriate words	3
Makes incomprehensible sounds	2
Gives no response at all	1

structural lesions, intracranial hemorrhage, and skull fractures. It causes profound, long-lasting personality changes and often leaves a patient comatose for extended periods of time. Many patients with severe TBI require resuscitation and other stabilizing measures by experienced emergency medical technicians to allow transfer to a specialized trauma center.

Repeated injury, including repeated concussions, has additive damaging effects.[5] This was first recognized in boxers in the 1930s who developed "punch-drunk syndrome," also called dementia pugilistica. This has lately been termed a "chronic traumatic encephalopathy" (brain inflammation) caused by repeated hits to the head. It is a chronic neurodegenerative condition with some of the pathologic hallmarks of dementia. Autopsies in boxers and football players show accumulation of tau protein in the form of fibrillary tangles like those seen in Alzheimer's disease or frontotemporal dementia.

Patients who survive moderate or severe TBI may present with dementia later in life. One-third of trauma survivors show diffuse amyloid beta deposits in the tissue immediately surrounding a brain lesion.[6] Furthermore, 20–30% of Alzheimer patients have experienced a traumatic incident preceding the onset of their disease. Therefore, the hallmarks of dementia—fibrillary tangles and amyloid beta deposits—are each linked to different forms of TBI, each presenting with dementia.

Time course of TBI: The injury itself can be divided into two temporal phases. The primary injury event is simply the direct impact on the head, such as that from a collision in football or the head hitting a dashboard in a car accident. The force that is experienced by the brain causes immediate structural damage that may be invisible. There is currently no treatment for the primary injury, only prevention. In fact, essentially all protective strategies, such as helmets and seat belts, are aimed at reducing the impact that the brain experiences during the primary insult.

The secondary phase of the injury begins immediately after the impact and is associated with the gradual functional loss of neurons, glial cells, and the vasculature. This delayed death continues for months after the impact and is the primary target of current research aimed at reducing the extent of cell loss and retaining as much function as possible. We will discuss this more extensively below.

3.2 Spinal Cord Injury

SCI currently affects 250,000 persons in the US, with 11,000 new cases reported each year. Males are four times more likely to suffer SCI than females, with a mean age of 37 at the time of injury.[7] Approximately 40% of patients die within 24 h after the injury. By far the most common causes are automobile accidents (46%), followed by falls (20%), violence (18%), and sporting activities (13%). Unlike TBI, SCI typically leaves a person's cognitive health unchanged.

Like the brain, the spinal cord is surrounded and protected by the bones that form the vertebral column. Unlike the cranium protecting the brain, the vertebral column has 24 individual ring-like segments chained together as a somewhat flexible strand interlaced with pliable disks. This arrangement is necessary to allow us to flex our back to reach to the ground or extend our back to reach upward. However, these flexible joints make the spinal cord vulnerable to injury by dislocation or fracture, each able to compress the spinal cord it contains and/or the nerves that exit between the vertebrae. Typical injury involves sudden flexion, hyperextension, or rotation of the vertebral column, causing stretching, shearing, or laceration of the spinal cord in its center. The spinal cord is an extension of the brain stem and contains both ascending sensory and descending motor fibers connecting the brain with the body (Figure 3). These tracts run on the outside of the spinal cord and are named

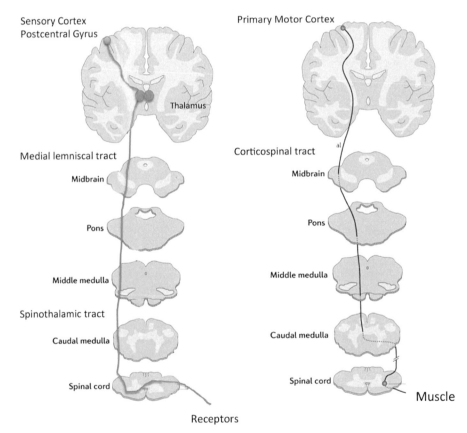

FIGURE 3   Sensory and motor pathways through the spinal tracts. Sensory information enters the spinal cord via the dorsal horn, and after crossing to the opposite site, travels via the spinothalamic and medial lemniscal tract toward the thalamus and/or sensory cortex. Motor commands travel from the primary motor cortex along the corticospinal tract to ventral horn of the spinal cord. Motor axons cross in the lower half of the brain stem at the level of the caudal medulla. *Drawing adapted from Ref. 8.*

according to the site of their origin and where they terminate. Examples include the descending lateral corticospinal tracts, which originate in the motor cortex and innervate motor neurons in the spinal cord, and the spinothalamic tract, which conducts ascending sensory information from the periphery through the spinal cord to the thalamus. Neuronal cell bodies are contained in the H-shaped center of the spinal cord, which is also the site of synaptic contacts for reflex pathways. All motor nerves exit on the ventral side while sensory nerves enter on the dorsal side. The segmentation divides the spinal cord into distinct functional segments. These predict the degree of impairment that might occur through a lesion, with decreasing involvement along

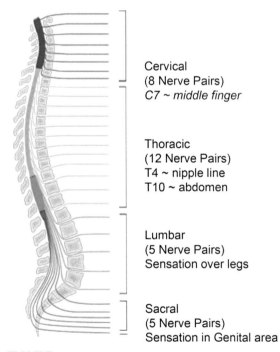

Cervical
(8 Nerve Pairs)
C7 ~ middle finger

Thoracic
(12 Nerve Pairs)
T4 ~ nipple line
T10 ~ abdomen

Lumbar
(5 Nerve Pairs)
Sensation over legs

Sacral
(5 Nerve Pairs)
Sensation in Genital area

FIGURE 4   Segmentation of the spinal cord. There are 8 cervical spinal nerves but 7 cervical vertebrae (C1–C7), 12 thoracic spinal nerves and vertebrae (T1–T12), 5 lumbar and 5 sacral nerves that are contained in a fused (S1–S5) sacrum. *Adapted from an image at Children's Hospital of Philadelphia.* http://www.chop.edu/healthinfo/acute-spinal-cord-injury.html.

the anterior–posterior axes (Figure 4). There are eight cervical spinal cord segments but only seven vertebrae (C1–C7). Cervical spinal nerves 1 through 7 exit above the corresponding vertebrae, whereas cervical spinal nerve 8 exits below the C7 vertebrae. The most common level of injury is at the C5–C6 vertebral level and leaves an individual quadriplegic (tetraplegic) and unable to move his or her arms and legs. Lesions in the chest region, which has 12 thoracic and 5 lumbar vertebrae, result in paraplegia, where patients are unable to move the legs. Finally, the sacral part of the spinal column has five vertebrae. Lesions in this region cause some loss of movement and sensation in the back of the legs and sacrum.

Nature of the injury: As with TBI, the primary injury in SCI is either the penetration of a foreign object that causes partial or complete dissection or, much more commonly, blunt force trauma that contuses or compresses the cord. An example of the former would be a knife or bullet entering the spinal cord. More typical are blunt force injuries as a result of vehicle accidents or sports injuries. It is important to point out that regardless of the insult, many patients initially present with incomplete lesions that leave a part of the spinal cord functional. An example is shown in Figure 5. Unfortunately, the cord may initially show significant tissue preservation, as is the case in the example in Figure 5, yet over time the residual axons thin out, leaving little functional connectivity. This period of time presents a therapeutic time window during which neuroprotective treatments may reduce functional loss. Any remaining connectivity can make a tremendous difference for an affected individual's quality of life. It may provide some limited sensory or motor control and, importantly, may preserve bowel, bladder, or sexual function in spite of overall immobility. Interestingly, incomplete lesions are more common (60%) than complete lesions (40%) in quadriplegic individuals, while the reverse is true for paraplegic individuals. Since 2010, the most frequent neurological category at discharge is incomplete tetraplegia (40.6%), followed by incomplete paraplegia (18.7%), complete paraplegia (18%), and complete tetraplegia (11.6%). Fewer than 1% of persons experience complete neurological recovery by hospital discharge.

3.3 Shaken Baby Syndrome

Unfortunately, countless children are suffering seemingly harmless, yet often deadly head or SCI at the hands of their caregivers. This is nothing new. In 1860, the French forensics expert Auguste Ambroise Tardieu described babies and small children that frequently presented with nonaccidental head and spinal cord injuries consistent with abuse, or shaken baby syndrome.[10] Surprisingly, such a diagnosis with its legal ramifications was not routinely reached until the 1960s, prior to which time it had been assumed that the underlying cause was inflammation or infection and thus went without any ramifications for the perpetrator.

3.4 Value of Imaging: Diffusion-Tensor Magnetic Resonance Imaging

The value of obtaining magnetic resonance imaging (MRI) image of a patient who has suffered brain or SCI is obvious. Computed tomography (CT) or MRI scans can readily resolve tissue lesions and can assess the changes in lesions with time, as illustrated in the example in Figure 5. However, mild forms of injury, particularly concussions, as well as blast injuries can cause significant diffuse axonal injury that falls below the detection threshold for common neuroimaging techniques. A new methodology called diffusion-tensor MRI (DT-MRI) allows visualization of axonal fiber tracts with great fidelity. Importantly, it allows a noninvasive functional assessment of fibers.[11] DT-MRI takes advantage of the fact that the movement of water molecules within an axon will preferentially occur along its axis. This directed or coherent diffusion can be visualized for entire nerve fiber bundles; an example where this has been applied to a child with an SCI is illustrated in Figure 6. A recent meta-analysis of over

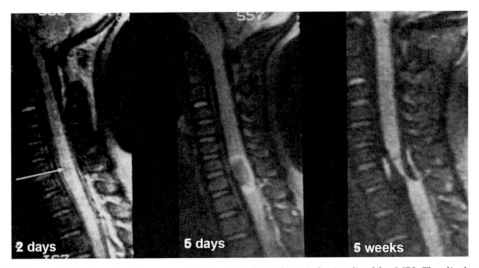

2 days   5 days   5 weeks

FIGURE 5   Incomplete lesion after injury, worsening over a 5-week period, visualized by MRI. The displayed images are T2-weighted images of a patient who was injured in a motor vehicle accident. (left) Acute spinal cord hemorrhage has produced a zone of decreased signal intensity (arrow) 2 days after the injury. (middle) Five days later, the central hypointensity is surrounded by peripheral hyperintensity. (right) Five weeks after the accident the hemorrhage has resolved, revealing nearly complete transection of the cord. *Reproduced with permission from Ref. 9.*

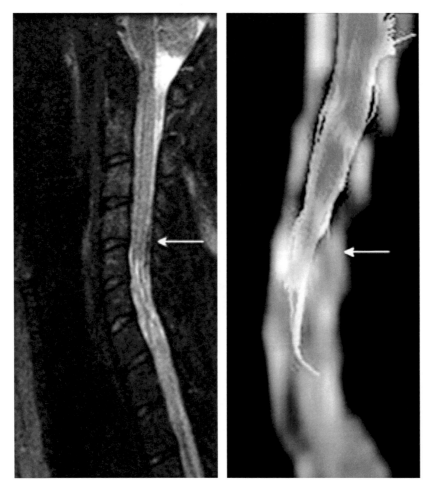

FIGURE 6 Left, T2-weighted MRI of a child with a spinal cord lesion with questionable residual connectivity. Right, DT-MRI shows lack of functional axons below the level of the lesion. *Reproduced with permission from Ref. 12.*

100 studies using this approach concludes that DT-MRI is able to detect clinically meaningful changes in axonal function[11] that were invisible or inconclusive by structural imaging alone.

4. DISEASE MECHANISM/ CAUSE/BASIC SCIENCE

4.1 Anatomy of the CNS

The brain and spinal cord each show macroscopically distinguishable gray and white matter tissue. In the brain the outermost layer, the cortex, contains primarily cell bodies and dendritic processes, along with astrocytes. This structure appears darkish gray in coronal sections and is called the gray matter. Underneath is the white matter, which contains the myelinated axonal processes that project to and from the cortex. These are the fibers detected by DT-MRI above. Its pale appearance is due to the lipid-rich myelin that defines the white matter. Here, the axonal processes are surrounded by oligodendrocytes that form the insulating myelin sheath (Box 2). The white matter of both brain hemispheres is connected via a central fiber bundle, the corpus callosum.

BOX 2

MYELINATION AND SALTATORY CONDUCTION

Among the important achievements in human evolution is the ability to quickly communicate between a centralized brain and a distant body. Sensory and motor commands travel at a speed of up to 200 m/s through the spinal cord from peripheral sensory organs and into skeletal muscles. A second evolutionary pressure was packing as many of these nerve fibers as possible into the small central canal of the vertebral column. Both problems were solved in an elegant way through a symbiotic relationship between glial support cells and neuronal fibers.

The axons found in invertebrate species, for example, mollusks, have bare axons where the membrane contains clusters of Na^+ and K^+ channels spaced equidistantly (Box 2-Figure 1, top). The membrane depolarization that follows the opening of Na^+ channels at the site of the action potential (AP) initiation reaches just far enough to activate the next cluster of Na^+ channels. As a result, the AP propagates continuously along the axon at relatively slow speed, typically 0.5–10 m/s. For humans, that would be much too slow. Consider how long it would take for the pain signal from a hand placed on a hot stove to reach the brain, and again travel down to the skeletal muscles to remove the hand. Making fast movements or driving a car would be entirely impossible. A solution to the problem was to insulate the axon with sheaths of myelin, produced by oligodendrocytes, leaving only small hot spots, so-called nodes of Ranvier, where the Na^+ channels cluster at high density ($10,000/\mu m^2$). These nodes are spaced about 200 μm apart (Box 2-Figure 1, bottom). Now, instead of spreading gradually, the AP jumps in a "saltatory fashion" from node to node, allowing for a faster conduction of signals over large distances.

The depolarization at one node of Ranvier will generate an electromagnetic field (arrows in Box 2-Figure 1) strong enough to depolarize the next node, and this process repeats itself. As a result, APs can travel at up to 200 m/s, and the speed is largely limited by the time it takes the Na^+ channels to become activated. This arrangement has several additional advantages. First, far fewer Na^+ channels are needed, since they are only found at the nodes of Ranvier. Second, fewer channels result in less Na^+ entering the axon and, as a result, much less energy is consumed by the Na^+/K^+ ATPase to reequilibrate these ions. Finally, since the fibers are relatively small, tens of thousands readily fit into the narrow spinal canal. Interspaced with these nerve bundles are oligodendrocyte cell bodies with processes extending toward the axons (Box 2-Figure 1, right). Their flattened processes wrap up to 50 axons with a myelin segment. This 50:1 relationship again maximizes the space occupied by neurons rather than by support cells. Note that in the peripheral nervous system, where Schwann cells form the myelin sheath, there is a 1:1 relationship between Schwann cells and axons. This is possible because there is sufficient space available.

The myelinated CNS axon is an incredibly efficient system when it works properly. However, in any situation where the oligodendrocytes become injured or die, such as in cases of traumatic injury or in multiple sclerosis, up to 50 axons will become nonfunctional from just a single oligodendrocyte being lost. Consequently, evolutionary optimization in the CNS came at a significant risk.

BOX 2 *(cont'd)*

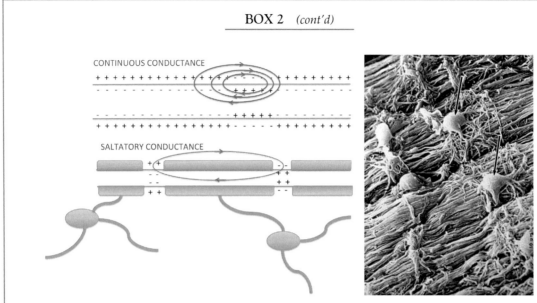

BOX 2-FIGURE 1 Comparison of unmyelinated (top) and myelinated axons (bottom), propagating in a continuous or saltatory fashion. Up to 50 axons are myelinated by a single oligodendrocyte whose cell bodies are interspaced between the axons (right).

The spinal cord has the reverse layering (Figure 7, right). Here, the cell bodies are found in the center, which appears as a figure "H" in cross section. The gray matter primarily houses the lower motor neurons that innervate the skeletal muscles and interneurons that participate in local signal processing. All the myelinated fibers surround the central gray matter. This arrangement may provide some added protection to the central neuronal cell bodies. In light of this anatomy it is clear that any force applied to the brain first damages the gray matter, whereas any damage to the spinal cord will first and primarily affect the white matter.

4.2 How Does Trauma Cause Injury?

With regard to the acute phase of tissue damage, both TBI and SCI present quite similarly. The primary injury consists of either partial or complete dissection of nerve bundles, damage to neuronal cell bodies, or blunt force that contuses or compresses the brain or spinal cord. An example of the former would be a knife or bullet entering the brain or spinal cord, which fortunately are both relatively uncommon occurrences. The majority of injuries are caused by blunt force resulting from a vehicle accident, or from being hit with a blunt object. Whether the person is in motion and collides with a stationary object (e.g., a wall) or whether a stationary person is hit by a moving object (e.g., a pedestrian being hit by a car), both receive the same negative force on impact. The primary injury may cause bone of the skull or vertebra to fracture and enter the nervous tissue. For TBI, the rapid displacement of the bone will cause the brain to collide with the bone, causing an impact at the side that the force is applied to (coup) with a delayed "contre-coup" on the opposite side as the force abruptly stops and the head snaps

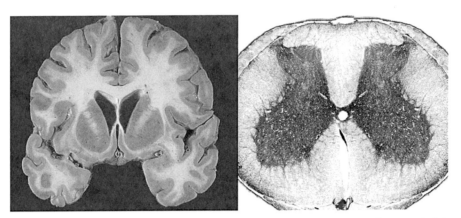

FIGURE 7   Gross anatomy of the brain and spinal cord shows clear gray and white matter layering. *Left image kindly provided by Edward C. Klatt, MD, Savannah, Georgia, USA; right image kindly provided by Dr Thomas Caceci.*

back (Figure 8). A classic example for such a coup–contre-coup scenario is the whiplash injury sustained by the driver of a car during a head-on collision. First the head snaps forward into the steering wheel, and then it snaps back into the headrest. A hidden force we need to consider is the shear force. In this instance, the shear force is applied on the spinal cord in the neck region, where the quick extension and retraction can pull on axonal nerve bundles, causing stretching and axonal damage. While less visible, shearing of nerve tracts occurs in the cortical white matter and along all white matter projections through the midbrain and brain stem. Whether by means of local impact or by shearing, these forces affect not only neuronal processes but the vasculature as well. Rupture of blood vessels is common and exposes the brain to harmful blood-borne molecules, and will also give immune cells, macrophages, and leukocytes access to the brain. Tissue downstream of the affected vessels will be cut off from blood supply, and quickly becomes ischemic. This is especially the case with SCI, where the force at the time of trauma damages the vasculature in the center of the spinal cord, with the secondary ischemic damage gradually affecting the surrounding tissue. The secondary injury cascade is

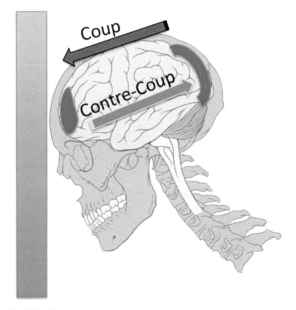

FIGURE 8   Typical head injury with coup and contre-coup force applied to the brain, schematized. The resulting bruises (red) are on opposite ends of the head. *Modified after Patrick J. Lynch, medical illustrator.*

similar to that discussed for stroke in Chapter 1 (Figure 9). It includes a loss of ATP production, which causes a run-down of ionic gradients across neuronal and glial membranes. This in turn causes an uncontrolled release

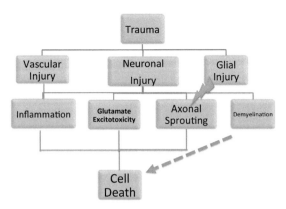

FIGURE 9 Schematized sequence of events following nervous system trauma: macroscopic injury, cellular injuries, and resulting neuronal cell death. Injury affects neurons, glia, and the vasculature indiscriminately. Inflammation of tissue occurs rapidly, as does energy failure giving rise to glutamate excitotoxicity. Demyelination is a delayed response as is axonal sprouting which tends to be unsuccessful in reestablishing lost connections.

of amino acid neurotransmitters, including glutamate, which leads to excitotoxicity. The uncontrolled rise in Ca^{2+} entering neurons causes the formation of free radicals, which damage phospholipids in the cell membrane, including the myelin sheaths. Finally, release of proapoptotic molecules, including cytochrome C from the mitochondria, induces programmed cell death or apoptosis. In addition to this injury cascade that develops on the first day after injury, trauma has an additional delayed chronic phase of so-called secondary injury. This continues for many months and is characterized by progressive demyelination, the dissolution of gray matter, formation of cysts, and extensive glial scarring (reactive gliosis, Box 3).

4.3 Amyloid Deposits and Tau Aggregates Following Trauma

Trauma appears to be a risk factor for Alzheimer disease (AD), and severe trauma can cause dementia. The molecular link is not entirely understood, but both neurofibrillary tangles and amyloid beta deposits, the molecular hallmarks of AD, are variably associated with TBI (Figure 10). Although it is clear that these deposits are intrinsically associated with dementia, we still do not fully understand how tangles or plaques cause dementia. It is possible that they are a consequence of neuronal death rather than its cause. If true, the observed accumulation of these deposits in TBI is indicative of neuronal death occurring during the secondary injury phase.

4.4 How Are Brain and Spinal Cord Protected from Trauma?

The containment of the brain and spinal cord within a bony cavity provides outstanding protection to the soft brain tissue, as these bones can withstand rather amazing forces. Take, for example, the head of a boxer, where the skull bones withstand an acceleration of 20–100g force without breaking. A soccer player, heading a 500 g soccer ball moving at 120 km/h, impacts the forehead with about 15–18g acceleration. How does the soft brain tissue within deal with such forces?

One answer rests in an elegant design feature whereby the brain sustains much lower forces than it otherwise would. More specifically, brain and spinal cord are surrounded by, and are actually floating in, CSF (Box 1). This clear fluid occupies about 150 ml of the total cranial volume (1650 ml) and is a cell-free ultra-filtrate of blood serum that is secreted by the choroid plexus in the roof of the lateral ventricles. It surrounds essentially all cells and occupies the minute extracellular spaces in the brain parenchyma. It also surrounds the brain and spinal cord by filling the supra-arachnoid space between the dura mater that covers the cranial bone and the pia mater that covers the brain parenchyma. Being completely surrounded by fluid, the brain experiences the same buoyancy as a diver in a pool, reducing the effective mass by a factor of 30, from 1500 g to only 50 g. Consequently, any acceleration

BOX 3

GLIAL SCARRING

Astrocytes occupy essentially the entire brain and spinal cord, where they serve a number of support roles. These include the removal of waste products, the homeostatic regulation of ion and neurotransmitters, and transport of glucose from the blood to the brain. Astrocytes also release cytokines, both pro- and anti-inflammatory, as well as growth factors. Among the most important roles is the removal of glutamate from the extracellular space near synapses. This assures that the postsynaptic neuron is not overstimulated by the continued presence of the transmitter. In response to trauma, astrocytes change their appearance and their function. This stereotypic response to injury is often called reactive gliosis,[13] and is most prominently observed in traumatic injury. However, reactive gliosis occurs in essentially all neurodegenerative diseases. Reactive astrocytes assume thickened

processes, many of which extend toward the site of injury. Some cells begin to divide and produce additional astrocytes. The intermediate filament protein, glial fibrillary acid protein (GFAP), is often used as a marker for gliosis. Being an important part of the cell's cytoskeleton, it is upregulated as cell processes change their shape. An example of a reactive, GFAP-stained astrocyte is shown in Box 3-Figure 1, and the process of scarring is schematically illustrated in Box 3-Figure 2.

Astrocytic processes, which are normally touching each other, become intertwined and less organized. Astrocytes then release extracellular matrix molecules, including chondroitin sulfate proteoglycans. The entire structure begins to look like a scar, and, akin to scars of the skin, it establishes a barrier that seals off the wounded tissue.

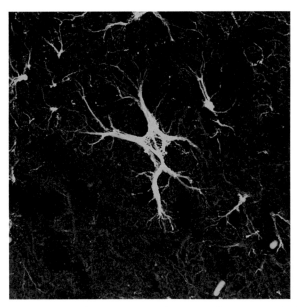

BOX 3-FIGURE 1   Reactive astrocytes from the cortex of a mouse after injury stained for glial fibrillary acidic protein (GFAP). The staining shows a characteristic change in morphology with thickened main processes. *Kindly provided by Dr Stefanie Robel, University of Alabama Birmingham.*

BOX 3 *(cont'd)*

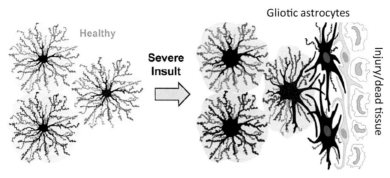

BOX 3-FIGURE 2   The process of reactive gliosis, schematically illustrated. Glial cells turn into tenacious tissue that seals off a wound serving as a physical barrier comparable to scars forming on skin tissue after injury. *Modified from Ref. 13.*

Astrocyte scars can resolve if an injury is mild, but will persist after severe insults. With regard to trauma, where lesions are severe, glial proliferation contributes to the scar, which is largely protective. It contains the spread of inflammatory cells, reduces the lesion size, and assures that blood–brain barrier breakdown and further demyelination are minimal. Although reactive glia also lose some of their support functions, such as regulation of the composition of the extracellular space, the beneficial effects of the scar appear to outweigh these losses in traumatic injury. In other nervous system diseases, debate is ongoing as to whether the scar negatively contributes to disease or whether it helps to reduce disease impact on the brain.

or deceleration exerted on the nervous system now applies to only 1/30 of the effective mass, according to Newton's second law of physics: Force = mass × acceleration.

This reduces the force that brain tissue experiences during acceleration (running/forward motion) or deceleration (hitting the head on a dashboard) by a factor of 30! As long as the head is not penetrated, the CSF also distributes any external force more evenly on the impacted brain or spinal cord. This clever use of a small amount of fluid is an essential and often overlooked feature of evolution. As early vertebrates left the sea, they took a small amount of seawater in the form of CSF with them to land. Prior to that, the same buoyancy applied to the entire submerged body.

4.5 Why Is the CNS Still So Vulnerable to Mechanical Injury?

Lack of protective connective tissue: The brain and spinal cord lack the typical connective tissue that confers structural stability and gives other organs their mechanical rigidity. The tenacity of connective tissues is greatly enhanced by secreted extracellular matrix molecules such as collagen and laminin, which cross-link into a stiff web-like matrix. The lack of these molecules leaves nervous

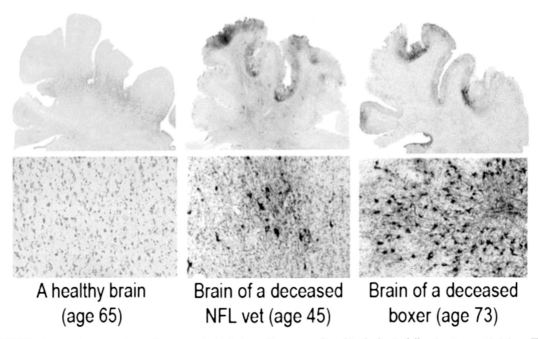

| A healthy brain (age 65) | Brain of a deceased NFL vet (age 45) | Brain of a deceased boxer (age 73) |

FIGURE 10   Tau deposits, akin to those seen in Alzheimer disease, are found in the brain following traumatic injury. These may explain the frequent occurrence of dementia following repeated head injury. The example shown is from the brain of a college football player who died prematurely. Top row: Brain sections showing dense tau protein deposition in multiple areas of the frontal cortex (boxes). Bottom row: Microscopic images showing large numbers of tau-containing neurofibrillary tangles (dark brown spots) in the areas of damage. *Kindly provided by Ann McKee, MD, Boston University School of Medicine.*

tissue in a gelatin-like consistency that is easily deformed. True connective tissue can only be found in the meningeal coverings and the epineurium, a sheath that covers collections of nerve fibers.

Cytotoxic edema: As cells become injured, water has a tendency to enter cells, causing swelling, or cellular edema. In neurons, this challenges the integrity of the cell bodies, processes, and synaptic terminals. Swollen synaptic terminals lose their ability to signal. To make matters worse, astrocytes have an unusual ability to swell,[14] and in doing so, they reduce the already narrow extracellular space. This can push adjacent brain tissue into the ventricular space, or push the brain toward the spinal cord, a process called herniation.

Vascular edema: Because blood vessels are also ruptured by mechanical force, blood serum enters the brain, causing an increase in extracellular water content, or vascular edema. Vascular injury also impairs the energy supply to damaged tissue, causing ischemic neural injury that we discussed for stroke in Chapter 1. Indeed, the same excitotoxic cascade involving aberrant activation of Glu receptors comes to play after head or spinal cord trauma, as well.

White matter injury: An important additional consideration is the vulnerability of white matter to trauma. Myelinated axons are essential for normal motor and sensory signaling between the body and brain, and occupy much of the brain and spinal cord white matter. Some of these fibers are very long, extending from the motor cortex to the lowest segments of the spinal cord. Sensory neurons can span from the midbrain to the foot, up to 5 feet in length. Their processes are very thin, barely 1/100 of the diameter of a human hair, and are therefore very fragile and prone to breakage. In addition to their mechanical fragility, the myelin insulation, provided by oligodendrocytes in the spinal cord and brain and by Schwann cells in the periphery, is also fragile. Myelinating oligodendrocytes can be killed if exposed to high Glu concentrations and, like neurons, can undergo Glu excitotoxicity.[15]

4.6 Why Are Axonal Connections Not Restored After Injury?

Injury that affects a neuronal cell body typically kills the cell almost instantaneously, although, by contrast, cutting off one of its processes does not. Instead, the neuron fuses the damaged membrane and begins to regrow its processes. During development, for example, cutting off the axon causes the next-longest cell process of the cell to develop into an axon.[16] In principle, neurons are quite capable of sprouting new processes and do make every effort to do so after injury. In fact, newly generated cell processes are believed to contribute to certain forms of epilepsy in the hippocampus, probably because they produce abnormal activity (see Chapter 4, Epilepsy). So why do we not simply regrow the axons that are injured in the spinal cord after a car accident?

To understand this, we need to make a brief excursion into the injury responses of the peripheral nervous system (Figure 11). If, for example, a motor axon that innervates a muscle controlling the extension of the index finger is injured by a cut, a number of regenerative processes will engage quickly. First, the Schwann cells that myelinate this nerve will release growth factors such as NGF, BDNF, and CNTF to stimulate axonal sprouting and growth cone motility. As a result, the axon will elongate at a rate of about 1 mm/day and will continue to grow toward its target. Second, Schwann cells will secrete extracellular matrix molecules, including laminin, fibronectin, and collagen, that provide a support substrate

for axons to grow in. They will also maintain their tube-like myelin sheath to actively guide the extending axon toward the muscle fiber it needs to innervate. Schwann cells in the vicinity of muscles may even periodically release acetylcholine, the neurotransmitter used between the axon and the muscle. The resulting twitches will prevent the denervated muscle from atrophying. Schwann cells do everything possible to assure successful regeneration and re-innervation. Due to their length, some peripheral axons continue their journey for over a year.

Unfortunately, the picture is very different in the CNS, where oligodendrocytes are in charge of myelination. In fact, together with astrocytes they make every effort to actively prevent axons from regenerating[18] (Figures 11 and 12). First, as the myelin sheath of an injured CNS axon breaks down, these myelin fragments contain a number of growth-repelling, paralytic molecules. Most well studied is a protein called Nogo-A, which is expressed on oligodendrocytes.[20] It binds to the Nogo-A receptor (NgR1) expressed on neuronal growth cones. There it activates Rho, a small guanosine-triphosphate (GTPase), which signals through the Rho-associated kinase (ROCK). This Rho signaling causes microtubule disassembly and the collapse of the actin cytoskeleton during growth. Another myelin-associated glycoprotein (MAG), similar to Nogo-A, binds either to NgR1 or to a second family member of Nogo receptors, NgR2, causing paralysis of the growth cone. A second way to suppress regeneration involves astrocytes. Not only do they produce a tenacious, impenetrable scar (Box 3), but they also lay down an extracellular matrix rich in growth inhibitory molecules, in particular, the chondroitin sulfate proteoglycans (CSPGs) aggrecan, brevican, neurocan, and phosphacan. Together, these inhibitory forces are hard to overcome. Thus, the difference in the regenerative ability of axons is not due to the neuron or its process. Instead, it is regulated

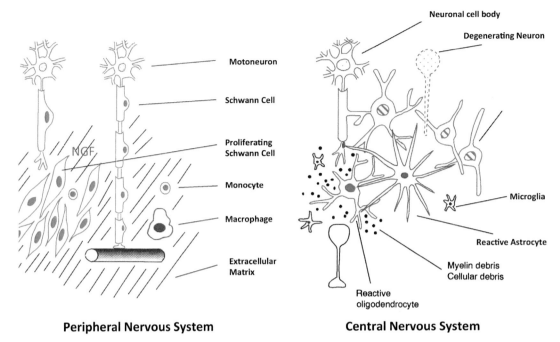

Peripheral Nervous System   **Central Nervous System**

FIGURE 11   Differences in the regenerative response between the peripheral and central nervous system. In the peripheral nervous system, Schwann cells actively support regeneration. In the central nervous system, myelin debris is paralytic for axonal sprouting, the astrocytic scar is a physical barrier, and activated microglial cells release inflammatory cytokines. *Modified after Ref. 17.*

by its glial support cells, with peripheral glia (Schwann cells) being supportive of regeneration while the central glia (oligodendrocytes and astrocytes) are abortive of regeneration.

This difference was elegantly demonstrated over 30 years ago in pioneering studies by Albert Aguayo.[21] He cut the optic nerve of a frog, and then sutured a peripheral nerve graft containing Schwann cells to the cut optic nerve, running it across the brain's surface to the optic tectum, the part of the frog's brain where the optic nerve usually ends. Within weeks, the cut axons regrew through the peripheral nerve graft and successfully innervated neurons in the optic tectum. Not only did this experiment provide evidence for the differential role of central and peripheral glia in regeneration, it also provided the rationale for using peripheral nerve grafts in patients

with SCI. As discussed later in this chapter, a number of efforts are in various stages of development taking advantage of this finding.

4.7 Activity-Dependent Functional Recovery After Injury

The previous paragraphs explained the mechanisms by which CNS evolution has prevented the regrowth and reconnection of axons and target neurons after an injury. One then wonders why the peripheral nervous system is allowed to repair itself constantly while the CNS is prevented from doing so. My personal conjecture is that the peripheral nervous system is much simpler and contains point-to-point wiring, making erroneous connections unlikely. By comparison, the complexity of the brain makes it very error prone. The danger of making just a few wrong

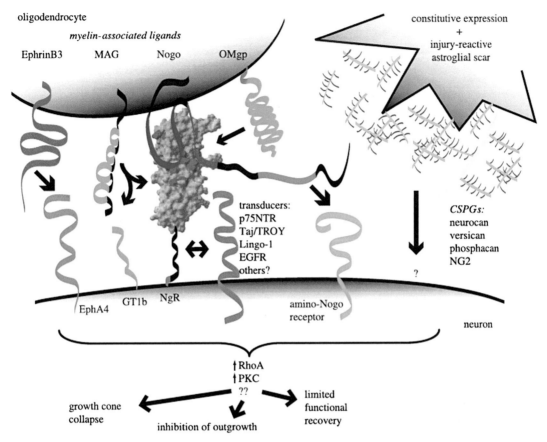

FIGURE 12 A variety of molecules released by oligodendrocytes and astrocytes can actively suppress axonal regeneration after injury. The myelin-associated ligands ephrin-B3, myelin-associated glycoprotein (MAG), Nogo-A, and oligodendrocyte myelin glycoprotein (OMgp) are expressed by the oligodendrocyte. Ephrin-B3 signals through the neuronal ephrin receptors including EphA4. MAG binds NgR1 and specific gangliosides such as GT1b. Nogo has two inhibitory domains: Nogo-66 (dark blue)/Nogo-24 (purple) binds to the neuronal NgR1, while D20 (light blue) binds to a distinct putative amino-Nogo receptor (turquoise). OMgp is expressed by oligodendrocytes and signals through NgR1. NgR1 (crystal structure of the ligand-binding domain depicted) is a GPI-anchored protein that signals through multiple transducers (listed). The NgR1 residues necessary for binding of all the three myelin ligands are shown in red. Chondroitin sulfate proteoglycans (CSPGs) are membrane bound or attached to specific matrix protein such as tenascin (yellow). The core protein (light orange) is covered with inhibitory glycosaminoglycans (GAGs) (burgundy) that limit axon sprouting and outgrowth by an undefined mechanism. These inhibitory molecules activate downstream signaling pathways that prevent neurite outgrowth in vitro and impede axon growth in vivo. *Reproduced with permission from Ref. 19.*

connections would potentially alter who we are, a cost that far exceeds the benefit gained. The work-around to a structural rewiring of the brain to the identical state it had prior to an injury is to exploit the tremendous plasticity that we use throughout development. Here, new circuits form in an activity-dependent fashion. Recovery of function after CNS trauma relies on the same principle, where wires are restored based on activity and experience. Indeed, this forms the basis of constraint-induced therapy, where intensive use of the affected limb is enforced by constraining the unaffected limb in a cast or splint. Of course, the argument that the brain is too complex to just allow axons to resprout does not quite hold true for the spinal cord. Unfortunately, its inability to regrow its axons is simply due to the fact that, as part of the CNS, the spinal cord is myelinated by oligodendrocytes rather than Schwann cells. The meningeal coverings of the CNS define the CNS–PNS boundary. Everything inside uses oligodendrocytes for myelination; everything outside uses Schwann cells. The oligodendrocytes thus impose a restriction on spinal cord regeneration that may be unnecessary. As most scientists know well, not everything in evolution makes sense or works to perfection.

4.8 Inflammation

Inflammation is a natural response to any form of tissue injury and its role is to facilitate resolution of the underlying damage. The inflammatory response is typically regulated by cytokines, a large family of short-lived proteins produced by leukocytes and glial cells. Cytokines that propagate the inflammatory response are called proinflammatory; those that truncate the response are anti-inflammatory. Following tissue damage in the brain and spinal cord, several cytokines, both pro- and anti-inflammatory, are activated. For example, interleukin-1 (IL-1) is a proinflammatory cytokine with neurotoxic function that is released by astrocytes and induces apoptosis, blood–brain barrier disruption, edema, and adhesion of leukocytes to endothelial cells. Tumor necrosis factor is another proinflammatory cytokine produced by microglial cells and astrocytes that induces neuronal apoptosis, but also modulates neuronal Glu receptors, with potential effects on cognition. These cytokines are antagonized by the anti-inflammatory interleukins 6 and 10. Although studies in animal models support a major role for cytokines in the development of brain and spinal cord lesions, changing the cytokine milieu after an injury is a hitherto untapped approach to reduce tissue damage after injury.

4.9 Animal Models of Disease

The ability to model the salient features of disease is essential to gaining a complete understanding of the underlying cellular and molecular processes and, importantly, to developing and testing new treatments. In brain and SCI, rats and mice are the primary model organisms used.

Although readily available and relatively inexpensive, rodents do not show the same trauma pathology as humans. This may in part be responsible for the low predictive value of preclinical studies, which, after showing success of treatments in animals, often fail when used in human clinical trials. Nevertheless, careful experiments in animals are essential to studying mechanisms of disease and to ultimately developing much-needed new therapeutics. The following models of injury are most commonly used to study trauma.

Traumatic Brain Injury/Concussion

A number of approaches and devices have been developed to mimic some of the TBI experienced by humans.[22] All apply a known force on a defined region of the brain, either with the brain exposed or in a closed-head situation. The fluid percussion model is the oldest and still the most widely used. As illustrated in Figure 13, a pendulum is used to strike a piston that transfers this force via a fluid column through a rigid tube to the intact meninges on the brain. The skull bone is removed first, and the injury can be applied to the top or sides of the brain. The contusion of the cortex that results from this injury shows gliosis and ongoing neuronal cell death, and animals show similar cognitive symptoms seen in humans. To simulate closed-head injuries, the weight drop model simply drops a weight from a known height onto the exposed skull or the exposed meninges. Numerous sophisticated permutations of these devices exist that employ different ways to apply the force or modulate the type of force applied. Of all

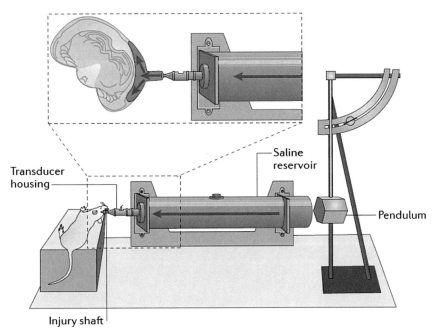

FIGURE 13   Fluid percussion injury model used as a reliable and reproducible way to mimic a human closed-head injury in rodents. A calibrated weight on a pendulum pressurizes fluid in a piston connected via tubing to the intracranial fluid of the animal. *Reproduced with permission from Ref. 22.*

the injuries modeled, the most closely matched to the human exposure is blast injury. Here, a shockwave is generated inside an enclosed compartment, such as a large tube, through explosion of a bullet.

Unfortunately, although many of the drugs developed to reduce secondary injury were successful in animal models, none of these drugs have thus far held up in human clinical trials. This suggests that although these models recapitulate some of the histological changes, they do not completely model human disease. This may in part be inherent to the differences between the rodent and human brain, not least important of which are marked differences in cortical gyration and very different gray-to-white-matter ratios.

Spinal Cord Injury

Multiple methods have been developed to injure the spinal cord, and these include transection (either complete or incomplete), contusion, or compression. After exposing the cord and dissecting the dura, a scalpel is used to dissect the spinal cord, either completely or only partially. A lateral hemisection is advantageous because it preserves bladder and bowel function of the animal, making postoperative care less involved. This approach is frequently used to study axonal regeneration and the effect of implanted devices. However, this type of injury is rarely encountered in human disease.

Spinal cord contusion is a more realistic human injury. In animals, a blunt object is used to contuse the exposed spinal cord, without disruption of the dura. Sophisticated impaction devices are often used that exert a reproducible force. This injury better mimics the lesions seen in patients and is considered clinically most relevant.

Compression is accomplished by using a clip with calibrated force to squeeze the spinal cord. It, too, provides a fairly accurate mimicry of the human disease and recapitulates the early phase of injury, with hemorrhagic necrosis and edema, followed by a late phase of injury characterized by remodeling, scarring, and axonal atrophy.

Following each injury, behavior is typically assessed by an observer who scores according to the Basso, Beattie, Bresnahan scale, which scores motor ability from 0 = no movement to 21 = normal movements. An alternative and less involved motor test uses a horizontal ladder with variable rung spacing in which missteps can be counted. New camera-based systems can assess overall movement of animals using a computer program that assesses biomechanical aspects of movement.

Sensory function is typically assessed with Von Frey filaments and the hot plate test. In the former, nylon fibers of various resistances are placed against the paw until the animal withdraws the paw. The hot plate test looks for evidence of hypersensitivity and early paw withdrawal as temperature is raised. These tests are not unequivocally sensory, because they require motor activity to withdraw the paw.

5. TREATMENT/STANDARD OF CARE/CLINICAL MANAGEMENT

Treatment of all trauma patients begins at the site of injury with stabilization of the patient, assuring clear airways and intubation if necessary. SCI patients must be handled carefully so as to not worsen a potential injury through movement of the head. A protective neck brace should be used and the patient lifted onto a stretcher without moving either head or neck. A quick assessment of sensory loss can be done on site if the patient is conscious by simply touching the trunk and extremities, assessing the patient for sensation. Similarly, asking the patient to move his fingers or wiggle her toes can quickly evaluate the gross extent of motor damage.

5.1 Clinical Management of Traumatic Brain Injury

Clinical management of TBI has been standardized through guidelines jointly developed by the American Association of Neurological Surgeons and the Congress of Neurological Surgeons and issued in 1995 by the Brain Trauma Foundation. Anyone suspected of a mild TBI should be removed from work or play and allowed to recuperate until all symptoms have cleared. Headaches should be treated with nonsteroidal anti-inflammatory drugs. Clearance for work or sport must await resolution of symptoms, such as headache and balance difficulty. Cognitive evaluation may also be necessary due to associated deficits in attention and memory.

For moderate to severe TBI, as with SCI, treatment begins at the site of injury with stabilization of the patient. After successful resuscitation, if necessary, and protection of the airway, the patient is evacuated and transported to a trauma unit capable of neurological surgery. During transport and thereafter, the patient's ventilation must be maintained, through intubation if necessary. Any bleeding must be controlled and blood pressure monitored. In hospital, intracranial pressure should be monitored and lowered by infusion of mannitol or hypertonic saline. Seizures must be controlled and fevers should be avoided at all cost. Imaging should be performed to evaluate structural lesions and search for any sign of bleeding, and evacuation of large hematomas may be required. Following the acute care, extensive rehabilitation of physical and cognitive deficits will be required.

5.2 Clinical Management of Spinal Cord Injury

Tragically, over 65% of individuals suffering an SCI are under the age of 35, with most of their productive life still ahead of them. Currently there is no effective therapy for SCI, and the only approved drug is the corticosteroid methylprednisolone. Although it has shown some beneficial effects in patients when given within 4 h after an accident, several controlled studies are questioning the overall utility of this drug, which has significant side effects. Consequently, corticosteroids are no longer considered standard of care.

This only leaves supportive care and extensive physical and occupational therapy. The major health issues to consider are secondary complications, such as respiratory disorders, neurogenic bowel and bladder management, and frequent bladder infections, sexual dysfunction, skin breakdown, and the risk for deep vein thromboembolism. Accordingly, the bladder may need to be catheterized to allow emptying, and a nasogastric tube may be required to control abdominal distention. Thrombosis can be prevented with low doses of the blood thinner heparin. Spastic, reflexive movement of denervated limbs is common and can be suppressed using GABAergic agonists such as baclofen. Depression and psychological challenges are common after SCI. Finally, chronic and severe pain often develops over time as aberrant synapses form, and must be aggressively treated.

To address the many specific needs of SCI patients, most medical centers in the US have established centers that specialize in SCI and rehabilitation. These encompass specialized injury-scene evacuation to high-level trauma-surgical and emergent care systems with access to intensive care, orthopedic/neurosurgical spine surgery, urology, and acute care hospitalization. Rehabilitation evaluation is initiated early in the intensive or acute care stage, and, when appropriate, includes physical and occupational therapy, psychosocial support, specialized nursing, and other ancillary services as needed. Lifetime follow-up and vocational/independent living opportunities are coordinated once the individual leaves the rehabilitation setting. About 55% of individuals that come to rehabilitation have incomplete spinal cord lesions. Depending on the degree

of injury, recovery can be initially experienced soon after injury or within several months. The amount of neurological return varies with each individual and may occur over many months or years. Effective rehabilitation is particularly concerned with education, retraining, and obtaining maximal functional independence. Mobility, balance, and achieving ability to perform activities of daily living are primary goals, but it is also important to maintain range of motion of the affected limbs and learn how to manage and maintain health of the organ systems that are no longer under voluntary control by the brain, especially lungs, bowel, bladder, and skin.

SCI patient care aims to empower a person with disability to gain as much independence and integration into society as humanly possible. It is difficult to achieve independence without mobility. Fortunately, major advances are being made regarding human–machine interfaces that allow injured patients to control robotic devices. For example, it is possible to steer a motorized chair using a joystick controlled by mouth if the patient has only minimal independent movement left.

6. EXPERIMENTAL APPROACHES/ CLINICAL TRIALS

NMDA-R antagonists: With the rationale that significant primary neuronal injury is due to glutamate excitotoxicity, particularly involving NMDA-R-mediated Ca^{2+} entry, gacyclidine, an NMDA-R antagonist, was examined in a multicenter clinical trial in France. Although gacyclidine treatment was initiated rapidly, within 2 h of injury, there was no functional benefit from treatment when patients were compared to placebo-treated patients 1 year later.

Gangliosides are glycolipids found in the cell membrane. When applied to isolated neurons they promote growth and process extension and, in animals, facilitate recovery after trauma and SCI. Based on laboratory findings, the first human clinical trials started using gangliosides in 1991. The first study on 34 patients showed promise, yet when expanded to almost 800 patients where treatment was compared to placebo, the initial positive findings were not supported.

Masking of repellent molecules on CNS glia: The identification of several molecules that actively suppress the regrowth of axons, illustrated in Figure 12, has generated tremendous interest in developing strategies to overcome this inhibition. This can be done, for example, by masking the released Nogo ligand with antibodies. After successful preclinical testing of anti-Nogo, IgG antibodies were given to 52 patients with an acute SCI by infusion into the lumbar spine. Patients tolerated the treatment well, but the trial is still ongoing and has not reported efficacy data yet.

An alternate strategy used only in animals is an NgR1 mimetic. Such drugs mimic NgR1 and bind to the Nogo receptor, but do not activate it. Another successful preclinical approach targets the repellent properties of the glial scar. The CSPs released by the astrocytic scar can be digested using the bacterial enzyme chondroitinase ABC, and this enzyme has shown promise in animal models with SCI.[18]

Because all the pathways that inhibit axonal regeneration converge on the small GTPase RhoA, an even more effective approach may be to inhibit its activation, which causes the collapse of the axonal growth cone via the Rho-associated kinase ROCK. In the laboratory, this has been done in a number of ways, for example, by using the enzyme C3 isolated from *Clostridium botulinum*, the bacterium that produces Botox. A recombinant version of the C3 enzyme, Cethrin, has been studied in 48 SCI patients. The drug was found to be safe when applied extracellularly, and 66% of the patients who had cervical injuries showed behavioral improvements.[23] It was therefore recommended to further test Cethrin in larger placebo-controlled clinical studies.[24]

The widely used analgesic and anti-inflammatory nonsteroidal drugs such as ibuprofen reduce the levels of activated RhoA in cultured neurons and inhibit the collapse of the growth cone by exposure to Nogo. Since ibuprofen is an anti-inflammatory drug, it may have a number of additional beneficial effects in SCI. Furthermore, ibuprofen crosses the blood–brain barrier and is well tolerated by most patients. In light of this, ibuprofen may be an ideal treatment for SCI. Surprisingly, however, it is not currently pursued as such, possibly since the commercialization of an over-the-counter drug for this indication would be difficult to justify.

Stem cells: Given the regenerative potential that stem cells from various sources have shown in animal models, SCI is an ideal condition to experiment with harnessing their regenerative potential in humans. Several small pilot studies with only 5–10 enrolled patients were conducted to show feasibility and have reported safe outcomes. Obviously, these studies are not able to report on efficacy, and at best provide anecdotal evidence. One larger study reported on 300 patients that received autologous bone marrow-derived stem cells via a lumbar puncture. One-third of the study participants reported neurological improvements.[25]

Another study took advantage of the regenerative potential of the olfactory mucosa. Throughout life the sensory neurons of the nose are constantly replaced. Autologous grafts of the olfactory mucosa were implanted in 20 SCI patients, most of whom showed some neurological benefit.[26] As of January 2014, there are about 30 active clinical trials enrolling patients with different levels of injury and using different stem cell sources. These include, for example, autologous bone marrow-derived stem cells, stem cells derived from adipose tissue, induced pluripotent stem cells derived from somatic cells, and embryonic stem cells.

Estrogen: The idea that estrogen may have an intrinsic neuroprotective effect originated from epidemiological data showing that men are 3–4 times more likely to suffer from SCI, a finding that was not entirely explained with increased risk-taking behavior in men. Laboratory studies using rats show that administration of estrogen 30 min after spinal cord contusion provided significant benefit, with treated animals recovering almost completely from the injury.[27] Histological analysis showed sparing of axons and overall preservation of tissue, suggesting a neuroprotective effect. Interestingly, the doses used were equivalent to those contained in an estrogen patch used for birth control. This raises the possibility that SCI patients could be treated with an estrogen patch during the acute phase of injury, even as they get into an ambulance. To reduce the feminizing effects of estrogen, plant-derived estrogens could be considered as an alternative. Since estrogen receptors are also the target of the widely used anticancer drug tamoxifen, it, too, could be readily explored in human clinical trials with relatively little risk.

Robotic rehabilitation/forced walking: The idea that passive movement of limbs or electrical stimulation of the innervating nerves could improve functional recovery and could be an effective way to augment rehabilitation has been considered for decades. Indeed, well over 200 papers have reported on using such approaches, primarily in paraplegic individuals. However, a recent careful analysis of published work suggests insufficient evidence that such approaches actually work. Surprisingly, in spite of anecdotal evidence that some patients regain function, no conclusive clinical trials have been done.[28] Hopefully, this gap will soon be addressed, as the federal clinical trials registry (clinicaltrials.gov) shows at least 10 clinical trials that are currently examining this important question.

7. CHALLENGES AND OPPORTUNITIES

The vulnerability of the brain and spinal cord to trauma, and the severe deficit resulting from it, has been recognized since ancient times. Surprisingly, until just a couple of decades ago, we have accepted as fact that little could be done to help an individual after such injuries. It took celebrities in film and sports to challenge our complacency and challenge the inevitable outcome of CNS trauma. As a result, hundreds of laboratories are now conducting experiments using valuable animal models of trauma. The National Institutes of Health and Department of Defense are each pouring billions of dollars into research. Although early successes are few, our understanding has grown in leaps and bounds. Without question, sports-related injuries are at a turning point. For the first time in history, we are serious in our admission that young persons are putting their brains and cognitive future at risk when they step onto a ball field. We are recognizing that military personnel, even when at great distance from explosives, can receive life-altering injuries from blasts. We also recognize the compounding nature of subthreshold exposures that add up to chronic disease. It is now clear that early assessment of brain function after injury will have a profound influence on future brain health. Moreover, the recent advances in man–device interfaces have provided astonishingly sophisticated assistance devices that contribute to the independence of disabled individuals. In my expectation, the greatest and quickest immediate advances will come from preventive surveillance, public education, and biomedical engineering. It is hoped that, with the increased investments in research, our understanding of neural injury will quickly catch up. The multiple signaling pathways that prevent regeneration of axons in the CNS provide a significant hurdle. As suggested, this may be by design, in order to prevent wrong connections from forming during the repair process. In my opinion, the use of stem cells holds the greatest potential. First, stem cells produce not only neurons but essentially all cells required in the regenerating spinal cord, including astrocytes and myelinating oligodendrocytes. Since these are immature cells, the inhibitory factors that presented by adult myelin are not yet present. Moreover, the wiring that needs to be restored in the spinal cord is relatively straight-forward, and the potential damage that can be done through human experimentation is minor, given that presently no regeneration takes place at all. Therefore, given the relatively low risk to worsen the condition, and the willingness of patients to participate in studies, SCI may be the ultimate testing ground for stem cell therapy in human health, and, if successful, could be life altering for many patients.

Acknowledgments

This chapter was kindly reviewed by Amie Brown McLain, MD, Chair and Professor, Dept of Physical Medicine and Rehabilitation, and Thomas Novack, PhD, Program Director, Traumatic Brain Injury Model System (UAB-TBIMS), University of Alabama at Birmingham School of Medicine.

References

1. Donovan WH. Donald Munro Lecture. Spinal cord injury–past, present, and future. *J Spinal Cord Med.* 2007;30(2):85–100.
2. Agarwalla PK, Dunn GP, Laws ER. An historical context of modern principles in the management of intracranial injury from projectiles. *Neurosurg Focus.* May 2010;28(5):E23.
3. Nicholls JG, Martin AR, Wallace BG. *From Neuron to Brain*; 1992. Sinauer Associates, Inc., 3rd edition.
4. Yang L, Kress BT, Weber HJ, et al. Evaluating glymphatic pathway function utilizing clinically relevant intrathecal infusion of CSF tracer. *J Transl Med.* May 1, 2013;11(1):107.
5. DeKosky ST, Ikonomovic MD, Gandy S. Traumatic brain injury–football, warfare, and long-term effects. *N Engl J Med.* September 30, 2010;363(14):1293–1296.
6. Ikonomovic MD, Uryu K, Abrahamson EE, et al. Alzheimer's pathology in human temporal cortex surgically excised after severe brain injury. *Exp Neurol.* November 2004;190(1):192–203.
7. Jackson AB, Dijkers M, Devivo MJ, Poczatek RB. A demographic profile of new traumatic spinal cord injuries: change and stability over 30 years. *Arch Phys Med Rehabil.* November 2004;85(11):1740–1748.
8. Blackstone C, O'Kane CJ, Reid E. Hereditary spastic paraplegias: membrane traffic and the motor pathway. *Nat Rev Neurosci.* January 2011;12(1):31–42.
9. Kulkarni MV, Bondurant FJ, Rose SL, Narayana PA, et al. 1.5 tesla magnetic resonance imaging of acute spinal trauma. Radiographics: a review publication of the Radiological Society of North America, Inc. 1988;8(6):1059–1082.
10. Al-Holou WN, O'Hara EA, Cohen-Gadol AA, Maher CO. Nonaccidental head injury in children. Historical vignette. *J Neurosurg Pediatr.* June 2009;3(6):474–483.
11. Hulkower MB, Poliak DB, Rosenbaum SB, Zimmerman ME, Lipton ML. A decade of DTI in traumatic brain injury: 10 years and 100 articles later. *AJNR. Am J Neuroradiol.* November 2013;34(11):2064–2074.
12. Mohamed FB, Hunter LN, Barakat N, et al. Diffusion tensor imaging of the pediatric spinal cord at 1.5T: preliminary results. *AJNR. Am J Neuroradiol.* February 2011;32(2):339–345.
13. Sofroniew MV. Molecular dissection of reactive astrogliosis and glial scar formation. *Trends Neurosci.* 2009;32(12):638–647.
14. Kimelberg HK. Current concepts of brain edema. *J Neurosurg.* December 1995;83(6):1051–1059.
15. Stys PK, Lipton SA. White matter NMDA receptors: an unexpected new therapeutic target? *Trends Pharmacol Sci.* November 2007;28(11):561–566.
16. Goslin K, Schreyer DJ, Skene JHP, Banker G. Development of neuronal polarity: GAP-43 distinguishes axonal from dendritic growth cones. *Nature.* 1988;336:672–674.
17. Bähr M, Bonhoeffer F. Perspectives on axonal regeneration in the mammalian CNS. *Trends Neurosci.* 1994;17(11):473–479.
18. Huebner EA, Strittmatter SM. Axon regeneration in the peripheral and central nervous systems. *Results Probl Cell Differ.* 2009;48:339–351.
19. Liu BP, Cafferty WBJ, Budel SO, Strittmatter SM. Extracellular regulators of axonal growth in the adult central nervous system. *Philos Trans R Soc B: Biol Sci.* September 29, 2006;361(1473):1593–1610.
20. Grandpre T, Strittmatter SM. Nogo: a molecular determinant of axonal growth and regeneration. *Neuroscientist.* October 2001;7(5):377–386.
21. Benfey M, Aguayo AJ. Extensive elongation of axons from rat brain into peripheral nerve grafts. *Nature.* 1982;296:150–152.
22. Xiong Y, Mahmood A, Chopp M. Animal models of traumatic brain injury. *Nat Rev Neurosci.* February 2013;14(2):128–142.
23. Fehlings MG, Theodore N, Harrop J, et al. A phase I/IIa clinical trial of a recombinant Rho protein antagonist in acute spinal cord injury. *J Neurotrauma.* May 2011;28(5):787–796.
24. McKerracher L, Anderson KD. Analysis of recruitment and outcomes in the phase I/IIa Cethrin clinical trial for acute spinal cord injury. *J Neurotrauma.* November 1, 2013;30(21):1795–1804.
25. Kumar AA, Kumar SR, Narayanan R, Arul K, Baskaran M. Autologous bone marrow derived mononuclear cell therapy for spinal cord injury: a phase I/II clinical safety

and primary efficacy data. *Exp Clin transplant*. December 2009;7(4):241–248.

26. Varma AK, Das A, Wallace G, et al. Spinal cord injury: a review of current therapy, future treatments, and basic science frontiers. *Neurochem Res*. May 2013;38(5):895–905.

27. Siriphorn A, Dunham KA, Chompoopong S, Floyd CL. Postinjury administration of 17beta-estradiol induces protection in the gray and white matter with associated functional recovery after cervical spinal cord injury in male rats. *J Comp Neurol*. Aug 15, 2012;520(12):2630–2646.

28. Karimi MT. Robotic rehabilitation of spinal cord injury individual. *Ortop Traumatol Rehabilitacja*. Jan-Feb 2013;15(1):1–7.

General Readings Used as Source

1. Mayer SA, Rowland LP. Head Injury. Chapter 63. In: Merritt, Rowland Lewis P, Rowland Randy, eds. *Textbook of Neurology*. Lippincott; June 2000.

2. Marotta FT. Chapter 64. In: Merritt, Rowland Lewis P, Rowland Randy, eds. *Textbook of Neurology*. Lippincott; June 2000. Spinal Cord Injury Facts and Figures at a Glance, www.uab.edu/nscisc.

3. Greenblatt SH, Dagi TF, Epstein MH (eds). *A History of Neurosurgery*. The American Association of Neurological Surgeons, Williams & Wilkins Publishers; 1997.

4. Center for Disease control and Prevention Blue Book, www.cdc.gov.

5. Silva NA, Sousa N, Reis RL, Salgado AJ. From basics to clinical: a comprehensive review on spinal cord injury. *Prog Neurobiol*. 2014;114:25–57.

6. Donovan WH. Donald Munro Lecture. Spinal cord injury–past, present, and future. *J Spinal Cord Med*. 2007;30(2):85–100.

7. Varma AK, et al. Spinal cord injury: a review of current therapy, future treatments, and basic science frontiers. *Neurochem Res*. 2013;38(5):895–905.

8. Blennow K, et al. The neuropathology and neurobiology of traumatic brain injury. *Neuron*. 2012;76(5):886–899.

9. Agarwalla PK, et al. An historical context of modern principles in the management of intracranial injury from projectiles. *Neurosurg Focus*. 2010;28(5):E23.

10. Ling GSF, et al. Diagnosis and management of traumatic brain injury. *CONTINUUM: Lifelong Learning in Neurology*. 2010;16(6, Traumatic Brain Injury):27–40.

11. Liu BP, et al. Extracellular regulators of axonal growth in the adult central nervous system. *Philos Trans R Soc B: Biol Sci*. 2006;361(1473):1593–1610.

12. Xiong Y, et al. Animal models of traumatic brain injury. *Nat Rev Neurosci*. 2013;14(2):128–142.

13. Jackson AB, et al. A demographic profile of new traumatic spinal cord injuries: change and stability over 30 years. *Arch Phys Med Rehabil*. 2004;85(11):1740–1748.

14. Thuret S, et al. Therapeutic interventions after spinal cord injury. *Nat Rev Neurosci*. 2006;7(8):628–643.

Suggested Papers or Journal Club Assignments

Clinical Papers

1. Fehlings MG, et al. A phase I/IIa clinical trial of a recombinant Rho protein antagonist in acute spinal cord injury. *J Neurotrauma*. 2011;28(5):787–796.

2. McKerracher L, Anderson KD. Analysis of recruitment and outcomes in the phase I/IIa Cethrin clinical trial for acute spinal cord injury. *J Neurotrauma*. 2013;30(21):1795–1804.

3. Bigler ED. Neuropsychological results and neuropathological findings at autopsy in a case of mild traumatic brain injury. *J Int Neuropsychol Soc*. 2004;10(05):794–806 (Interesting clinical case of a TBI resulting in delayed death).

4. Theadom A, Starkey NJ, Dowell T, et al. Sports-related brain injury in the general population: An epidemiological study. *J Sci Med Sport/Sports Med Aust*. 2014;17(6):591–596.

Basic Papers

1. Ikonomovic MD, et al. Alzheimer's pathology in human temporal cortex surgically excised after severe brain injury. *Exp Neurol*. 2004;190(1):192–203.

2. Brody DL, et al. Amyloid-beta dynamics correlate with neurological status in the injured human brain. *Science*. 2008;321(5893):1221–1224.

3. Siriphorn A, et al. Postinjury administration of 17beta-estradiol induces protection in the gray and white matter with associated functional recovery after cervical spinal cord injury in male rats. *J Comp Neurol*. 2012;520(12):2630–2646.

4. Ifft PJ, et al. A brain-machine interface enables bimanual arm movements in monkeys. *Sci Transl Med*. 2013;5(210):210ra154.

5. Rosenzweig ES, Courtine G, Jindrich DL, et al. Extensive spontaneous plasticity of corticospinal projections after primate spinal cord injury. *Nat Neurosci*. 2010;13(12):1505–1510.

Seizure Disorders and Epilepsy

Harald Sontheimer

Diseases of the Nervous System
http://dx.doi.org/10.1016/B978-0-12-800244-5.00003-3

1. CASE STORY

For the past 3 years, Lydia played the saxophone in the high school marching band. Just like any teenager her age, Lydia didn't sleep much during her senior year. She was already accepted into the college of her choice and spent her time out with friends or glued to social media until the wee hours. School days were like a haze and she was often dizzy, but since she had felt like this for years she did not make much of it. Two months earlier, during band warm-ups, she started feeling strange. She could not concentrate and was unable to get her saxophone to play a single note. She turned to her bandmate Heidi with a frightened look on her face and told her something wasn't right. Within seconds, Lydia's hand started shaking, eventually involving her leg and torso as her eyes stared into space, unresponsive. She fell to the floor as Heidi called her name but Lydia, eyes wide open, did not respond. Her body was coiled up and kept contracting rhythmically, with saliva dripping from her mouth. A panic broke out and the band director shouted to call an ambulance. Lydia's contractions seemed to last for ages and did not stop until the paramedics injected her with an anticonvulsant.

Hours later, when Lydia awoke in a hospital bed, she heard her mom and a doctor talking but she could not understand a word. She barely remembers the following week either. It took her several days to be able to dress herself, walk the hall, and have conversations with friends who came to visit. She had suffered a life-threatening generalized seizure called a status epilepticus and was now taking daily doses of Dilantin. Whether it was the medication or the seizures, she felt tired and depressed. Once back in school, she was quiet and withdrawn. She felt humiliated, the girl who fell to the ground before the high school football game. She thought everybody was still talking about her. Most important, however, there was this nagging fear that it could happen again, and indeed it did. Just

2 weeks later, Lydia was shopping at Target, wandering aimlessly through the aisles, no clue what she came to buy. She called her mom to remind her of her shopping list, and shortly after hanging up her vision became blurry and her arms started twitching. She tried to focus on the shelves around her but everything was blurry as her body started shaking and she lost control. When she regained consciousness she was in an ambulance to the hospital, repeating her last ordeal. Now she tells herself that she will not let these seizures run her life, but her self-confidence is shattered and she is overcome with anger. Her doctor tells her that it will get better once the right medication is found. Until then, she lives in fear of having another seizure.

2. HISTORY

Over the centuries epilepsy has engendered a profound fascination and was often called the "falling sickness." There was a commonly held assumption that a supernatural power takes possession of an individual, which would display as a "seizure," a term derived from the Latin *sacire*, which best translates as "taking possession of." A person with recurrent seizures suffers from epilepsy, a term derived from the Greek verb *epilambanein*, meaning "to take hold of" and again implying that the affected person is being manipulated by a higher power or demon.

In old texts epilepsy is called the "sacred disease," and a detailed and rather accurate case description can be found in early scripture (Mark 9: 14–29): "Teacher, I brought you my son, who is possessed by a spirit that has robbed him of speech. Whenever it seizes him, it throws him to the ground. He foams at the mouth, gnashes his teeth, and becomes rigid. I asked your disciples to drive the spirit out, but they could not."

In contrast to the notion that evil spirits are at work, the Greek philosopher Hippocrates (~400 BC) correctly suggested that seizures result from a natural cause: "Its cause lies in the brain, the releasing factors of seizures are cold, sun and winds which change the consistency of the brain. Therefore epilepsy can and must be treated not by magic, but by diet and drugs."[1] Unfortunately, this recognition was lost in medieval Europe, where seizures characterized only witches and justified the killing of thousands of women. During this time period, and indeed into the eighteenth century, seizures were believed to be contagious, and even in church "the possessed needed to be segregated from the faithful."[2] As shocking and inappropriate as this may seem today, the isolation and stigmatization of epilepsy continued well into the twentieth century. Indeed, 17 US states prohibited marriage to individuals with epilepsy until 1980, and 18 states promoted sterilization on eugenic grounds.

The first detailed and authoritative medical account of seizures dates to the English neurologist John Hughlings Jackson, who defined seizures as "an occasional, an excessive, and a disorderly discharge of nerve tissue on muscles" (Jackson JH. A study of convulsions. *Trans St Andrews Med Graduates Assoc* 1870; 3:162–204),[3] quite an accurate description in light of the absence of electrophysiological recordings. Not until several decades later did the first electrical recordings of brain activity during a seizure, captured by Hans Berger in 1929, confirm the unusual electrical discharge that we now record by the commonly used electroencephalography (EEG).

In the nineteenth and early twentieth centuries a popular misconception held that seizures are the result of excessive masturbation, and the earliest widespread epilepsy treatment, potassium bromide, was believed to reduce sexual libido, thereby indirectly reducing seizures. By the end of the nineteenth century, in spite of significant side effects, bromide salt was used widely and in excessive quantities by hospitals. Still in 1928, one of five prescriptions in the United States was for bromides. While bromide may have provided relief to some patients, it also had severe hypnotic and sedative side

effects and was said to "stupefy" many patients. Nevertheless, bromide remained available as an over-the-counter remedy (Bromo-Selzer) until 1975 and is still available to treat epilepsy in dogs. (For a more comprehensive historic account see Ref. 4.)

The use of bromide in humans was gradually replaced by phenobarbital (Luminal), which was synthesized in 1912 and was a true turning point in the effective treatment of epilepsy. Phenobarbital remains the oldest antiepileptic drug (AED) still in clinical use to date. Since the discovery of phenobarbital the interest in epilepsy remedies has surged, and a number of drugs were introduced into clinical practice, most importantly phenytoin (1939, Merritt and Putnam), carbamazepine (1953, Schindler), ethosuximide (1958), and the anticonvulsant sodium valproate (1963). The latter proved to be the most versatile drug for the clinical management of seizure disorders.

Over the centuries, many famous individuals who were tremendous achievers have been reported to suffer from epilepsy. These include, for example, the painters Vincent van Gogh (1853–1890), Michelangelo (1474–1564), and Leonardo Da Vinci (1452–1519); the physicist Sir Isaac Newton (1643–1727); the French leader Napoleon Bonaparte (1769–1821) and the Roman emperor Julius Cesar (100–44 BC); and the writers Agatha Christie (1890–1976) and Charles Dickens (1812–1870). Of course, in many instances we can only infer that these individuals suffered from epilepsy based on historical accounts because their lives predated the recognition of epilepsy as a neurological syndrome and in many cases the disease may have been misdiagnosed.

3. CLINICAL PRESENTATION/ DIAGNOSIS/EPIDEMIOLOGY

This chapter covers a heterogeneous group of conditions that is characterized by a defined behavioral phenotype referred to as seizures.

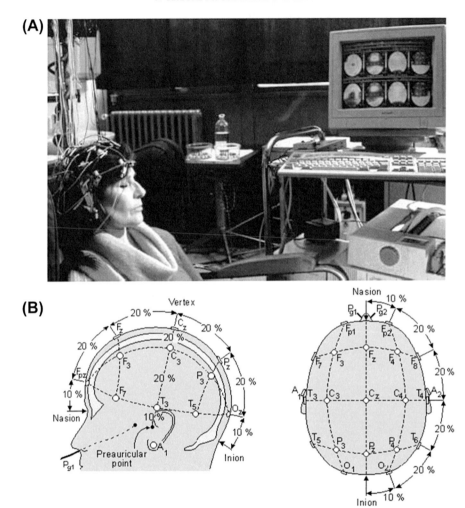

FIGURE 1   In hospital electroencephalography (EEG) recording (A). Placement of EEG electrodes according to the 10–20 international system is the standard naming and positioning scheme for EEG applications. It is based on an iterative subdivision of arcs on the scalp starting from craniometric reference points: nasion (Ns), inion (In), left preauricular (PAL) and right preauricular (PAR) points (B). The intersection of the longitudinal (Ns–In) and lateral (PAL–PAR) is called the vertex. http://www.bci2000.org/wiki/index.php/User_Tutorial:EEG_Measurement_Setup.

Seizures are the result of abnormal, highly synchronous electrical activity of a restricted brain region, in many cases encompassing the entire brain. This activity can often be measured by EEG using a series of small-surface electrodes applied to the scalp (Figure 1). The recorded signals (Figure 2) have a small amplitude and are not always detectable; therefore the presence of abnormal EEG activity can confirm a seizure diagnosis, but an absence of encephalographic activity is not sufficient to rule out seizures. Seizures are almost always associated with behavioral abnormalities, which can range from very subtle, such as a blank gaze staring into space, to severe convulsions involving the entire body.

Stressing the difference between seizures and epilepsy is important. Seizures define an acute condition that is typically associated with

Focal, right temporal Seizure

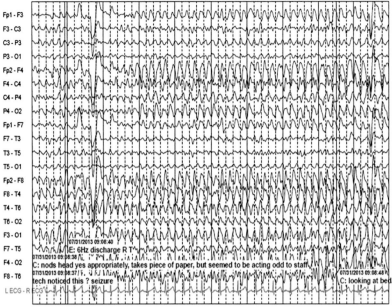

Generalized Seizure

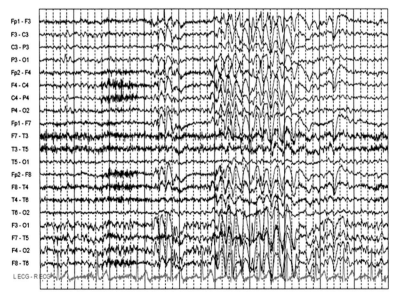

FIGURE 2   Example electroencephalography (EEG) recording of a focal seizure originating from the right temporal lobe (top) and a generalized absence seizure (bottom). Typical 3-Hz spike and wave discharges are seen. *Courtesy of Jerzy P. Szaflarski, MD, PhD, University of Alabama Epilepsy Center.*

abnormal EEG and a behavioral abnormality. Numerous conditions can provoke seizures, ranging from infection to trauma, and many are obscure and not immediately recognized. Epilepsy, by contrast, is a clinical syndrome (as opposed to a disease) characterized by the spontaneous recurrence of seizures and assumes the existence of an underlying abnormal process. This distinction is important because some people may have suffered a seizure at some point in life, mostly likely as infants, yet never developed epilepsy. In the case of infants, an infection presenting with a high temperature may have resulted in "febrile seizures." Once the underlying cause is recognized and corrected, in this instance by controlling the fever, seizures do not recur. In some clinical cases of epilepsy unequivocally identifying the underlying cause is impossible, and clinical management reduces the symptoms without altering the root cause. It is the heterogeneity of both the suspected underlying causes and clinical presentations that warrant the designation of epilepsy as a neurological syndrome as opposed to a disease.

3.1 Epidemiology

Epilepsy is one of the most common neurological syndromes, affecting approximately 50 million people worldwide, or 1% of the population. At age 74 the lifetime risk of developing epilepsy is about 3%, that is, 3 in 100 persons. There is a slight predominance (1.1–1.5 times) of epilepsy in men compared to women. Age is highly correlated to disease incidence, with an early childhood peak and a steep, progressive increase in diagnoses in adults older than 60. This "age peak" is predominated by epilepsy resulting from cerebrovascular or neurodegenerative disease. By comparison, children and young adults predominantly present with developmental, genetic, or idiopathic (unknown) forms of epilepsy. Loss of productivity constitutes an immense societal burden that exceeds $15 billion annually. Quality of life is significantly compromised, particularly for the many patients who do not respond to currently available drugs.

3.2 Seizure Definition and Characterization

Various forms of epilepsy can be defined on the basis of their association with a particular seizure phenotype. An accurate determination and classification is desirable because it aids in identifying potential underlying causes, long-term prognosis, and selection of the most promising therapy. To ensure that clinicians and scientists use a common language to describe and classify seizures, the International League against Epilepsy commissioned a classification that defines and classifies seizures by both characteristic behavioral changes in the patient and unique EEG signatures, if available (Figures 1 and 2); these are discussed in greater detail below.

EEG recordings involve the placement of 21 small electrodes on the scalp in a defined pattern called a 10–20 arrangement (Figure 1(B)). Electrodes are placed so that their relative distance is 10% of the total front-to-back distance and 20% of the total left-to-right distance. An additional reference electrode is placed on either side of the head. This arrangement compensates for differences in head size. Each scalp electrode samples the combined electrical activity originating from a small cortical brain region. Since the electrical discharge of a single neuron is far too small to be detected on the surface of the scalp, the EEG signal is the summation of many thousands of neurons discharging synchronously. The amplified signals are recorded on a mechanical or digital chart recorder; example recordings are shown in Figure 2. A normal EEG is not a flat line but contains a variety of rhythmic deviations, many of which reflect the resting activity of the brain or are related to certain brain states, for example, sleep. Seizures, however, show a characteristic deviation from normal brain activity, often in a defined region of the cortex and with a defined electrical signature, which a trained epileptologist can recognize by visually inspecting the chart. The typical waveform of a seizure contains a combination of spikes and oscillations, that is, a spike and wave pattern with a periodicity in the 3- to 16-Hz range (Figure 2). These discharges coincide with but may precede behavioral abnormalities by several seconds and persist throughout an attack. In patients with relatively infrequent seizures inducing seizures while the patient is being observed may be desirable. This is often possible through sleep deprivation, hyperventilation, or photic stimulation with a strobe light.

Focal Seizures

As the name implies, focal seizures originate from a relatively distinct part of the brain and always involve only one hemisphere. In the past these were called partial seizures, a term no longer recommended. Focal seizures are often associated with a well-delineated structural abnormality of the affected brain region from which the synchronized electrical activity originates. A common structural abnormality is mesial-temporal lobe sclerosis, further discussed below. Focal seizures are additionally subclassified based on whether the patient suffers from cognitive impairment, called dyscognia, as focal seizures with or without dyscognia. The older clinical literature refers to these as complex partial seizure (with cognitive impairment) and simple partial seizure (without cognitive impairment), respectively. Depending on the brain region where the seizure originates, a focal seizure can be associated with a variety of behavioral manifestations. For example, a seizure originating in the motor cortex would be expected to present with abnormal involuntary repetitive (clonic) muscle contractions. Given the proximity in the motor cortex of neurons controlling facial muscles to those controlling the hand, the concurrent contractions and twitching of facial muscles with involuntary hand movements are readily explained.

In the early twentieth century Hughlings Jackson described patients in whom seizure activity started in a finger or hand and spread progressively to both upper and lower extremities and even the face over several minutes. This phenomenon is now often called "Jacksonian march" and can be explained by a spread of electrical activity from a discrete part of the motor cortex to neighboring areas that innervate different muscle groups and body parts.

In addition to motor symptoms, focal seizures may also present with sensory changes such as numbness, tingling, loss of balance or vertigo, or altered autonomic function such as enhanced sweating. Unusual perception of strange or strong odors or intense sounds and lights are also common.

Dyscognitive focal seizures are associated with an altered state of mind that leaves a person unable to communicate or respond to visual and verbal commands. These often start with an aura, which many patients recognize as a unique signature feature that precedes their seizures by several seconds and serves as a warning to get into a safe place, for example, pulling their car to the side of the road. At the time of seizure onset, the individual remains largely motionless, with the gradual onset of repetitive involuntary movements called automatisms, which may resemble wringing hand movements, lip smacking, or swallowing.

EEG recordings obtained during a focal seizure, which is known as the "ictal" state, can often localize the onset of the abnormal activity to a well-circumscribed area of the cerebral cortex; in the case of Figure 2, the focal seizure localized to the right temporal lobe, namely the electrodes referenced by F_2–F_8 (Figure 1). During the so-called interictal state, the time between seizures, EEG recordings are either normal or show brief spikes or sharp waves. It is important to remember that seizures that originate from deeper brain structures that are out of reach of the scalp electrodes and may remain invisible.

Generalized seizures

Generalized seizures always involve networks of neurons across both cerebral hemispheres. While these, too, may originate from a more discrete focal point within the brain, they quickly involve large interconnected neural networks and on EEG involve all electrodes on the scalp (Figure 2).

Generalized seizures vary widely with regard to their behavioral manifestation, from the subtle to the extreme. For example, absence seizures often last only a few seconds, during which affected individuals show a complete absence of attention and sensation and appear to be staring into space, incognizant of their surroundings. A patient regains consciousness just as quickly as it was lost and is often unaware of any preceding lapse in consciousness. An absence seizure is at best associated with subtle motor signs such as rapid eye blinking or chewing movements of the mouth.

Absence seizures are most commonly diagnosed in young children and often noticed by schoolteachers as repeated periods of absences or daydreaming during class. They account for ~20% of all childhood seizures. While these seem relatively benign, they can occur hundreds of times during the day and severely handicap an individual, making normal attention to schoolwork nearly impossible. Importantly, the underlying abnormal electrical activity in the brain, which can occur almost continuously, leads to profound memory and cognitive impairments, which, if untreated, may result in a lifelong disability.

On EEG, absence seizures present with a unique signature characterized by short, 3-Hz spike-wave discharges that start and stop abruptly (Figure 2). Abnormal EEG activity is often even more frequent than behavioral absence episodes.

Tonic–clonic seizures: The other extreme of behavioral manifestations are on display during generalized tonic–clonic seizures, which account for the majority of seizures diagnosed in adults. These begin spontaneously and are unpredictable. As the name implies, seizure onset is defined by tonic muscle contraction causing a clasping of the upper and lower extremities that may also involve the jaw muscles, which may cause a patient to bite their tongue. Concurrently, the patient is likely to show an increase in heart rate and blood pressure and dilated pupils. Within tens of seconds, the tonic phase gives way to the clonic phase, characterized by repeated jerking and muscle contractions lasting up to a minute. Ultimately, the patient is unresponsive, fatigued, and sweating, salivates excessively, and may show signs of incontinence. The patient gradually regains consciousness, although complete recovery may take hours, during which time headaches and muscle pains are not uncommon. On EEG, the ictal phase shows a progressive worsening of electrical activity that begins with low-voltage spikes that increase in amplitude and frequency (Figure 2). The end of the clonic phase is frequently characterized by a marked suppression of electrical activity, that is, a flat-voltage signal called postictal suppression.

While the above description gives a stereotypic view of this class of seizures, clinical variants that present very differently exist. For example, patients may briefly lose muscle tone and collapse (atonic seizure) or, conversely, the sudden contraction of muscle groups may occur, causing a jerk or fall (myoclonic seizure).

Moreover, the distinction between focal and generalized seizures becomes a bit blurred; seizures can begin as focal and evolve into generalized seizures as activity spreads from one hemisphere to the other. Such a development occurs when the initial seizure focus is localized to the frontal lobes of the brain. The behavioral presentation is essentially indistinguishable from that of a generalized seizure, and the evolution from focal to generalized seizure typically happens so fast that EEG analysis is required to establish it. Finally, not all seizures can be classified into one of the above groups, leaving a "grab-bag" of cases diagnosed as *unclassifiable seizures*.

Status Epilepticus

Although in most instances seizures end spontaneously after less than 5 min, this is not always the case. Should seizures persist for 5 min or longer, a life-threatening emergency called status epilepticus occurs, and acute medical termination of seizures is necessary. This typically requires intravenous administration of diazepam, a drug that augments GABAergic inhibition. Status epilepticus affects approximately 150,000 patients each year in the United States and results in 42,000 deaths. Approximately 12% of patients present with status epilepticus on first presentation, whereas patients already treated for epilepsy have a 1–4% lifetime risk of entering status epilepticus. Patients rescued from status epilepticus still carry a 20–25% chance of death in the ensuing 30 days.[5] It is worth mentioning that many rodent models used to study epilepsy chemically induce status epilepticus followed by termination of the seizure with diazepam shortly thereafter. Most of these animals develop spontaneous seizures later in life and show significant structural changes, particularly in the wiring of the hippocampus.

3.3 Epilepsy Classification

As already mentioned, epilepsy is a syndrome that is defined by spontaneous, repeated, unprovoked seizures of any of the phenotypes described above. Generally speaking, a patient who has experienced more than three unprovoked seizures is suspected to suffer from epilepsy. We differentiate idiopathic epilepsy, without an unknown precipitating cause, from acquired epilepsy, which is suspected to result from an underlying condition that has altered the brain, for example, trauma, infection, stroke, or a brain tumor. In addition, rare forms of familial epilepsy provide unique clues to heritability and genetic causes of the disease. While these distinctions are useful to clinicians and scientists, it is important to emphasize that we must assume that whenever epilepsy is present, the brain displays an electrical abnormality. Even if an obvious cause has not yet been identified, future research and medical advances are likely to change this. For this reason the term *idiopathic epilepsy* is less commonly used to date, and even those forms of epilepsy for which no clear causative factor can be identified are now called genetic epilepsy. (The name *idiopathic* was recently abandoned by the International League against Epilepsy and replaced with *genetic epilepsy*. Since the term *idiopathic* is used much more widely for other nervous system diseases, it is used concurrently here for consistency.)

Genetic/idiopathic epilepsy presents spontaneously, and a precipitating event or cause is often not known, yet the patient's brain does have a genetic predisposition to generate seizures—hence the term *genetic epilepsy*. In truth, only for a small number of patients (1%) have genetic changes that cause epilepsy been defined, and many of these affect ion channels. These are diseases often called "channelopathies" and are discussed in more detail later. Genetic/idiopathic epilepsies account for the majority of epilepsies (~60%) and can present at almost any age, although typical onset occurs between early childhood and adolescence.

Symptomatic or acquired epilepsy has a known or suspected cause. Although these can occur at any age, they account for the majority of presentations diagnosed in adults aged 18–45 years. The single largest cause of acquired epilepsy is trauma to the head, such as closed or penetrating head wounds or concussions. The likelihood of developing epilepsy following severe head trauma is estimated to be as high as 50%, and the traumatic event may precede the epilepsy syndrome by many months or even years.

In older adults epilepsy is more commonly associated with cerebrovascular disease, brain tumors, or neurodegenerative diseases. Again, the precipitating event, for example, a stoke, may precede the development of chronic seizures by many months.

By far the most common cause of seizures in adults who have not suffered brain trauma or do not have an underlying illness such as a tumor is exposure to drugs. Alcohol is a γ-aminobutyric acid (GABA) agonist and enhances the inhibitory action of GABA. Chronic alcohol consumption causes compensatory upregulation of the excitatory glutamate (Glu) receptors (specifically

NMDA); abrupt cessation of alcohol intake through withdrawal then tilts the balance toward enhanced excitation, thereby causing seizures.

While the term *acquired epilepsy* implies that the patient was somewhat actively involved in the acquisition of disease, that is often not the case. Hence the term *symptomatic* better describes epilepsies in which there is a well-described underlying cause, such as a cortical heterotopia or mesial-temporal lobe sclerosis, or when epilepsy occurs as a result of a separate neurological syndrome, as is the case in tuberous sclerosis, a rare autosomal dominant disorder that presents with benign tumors and cortical abnormalities called tubers that give rise to intractable epilepsy.

4. DISEASE MECHANISM/CAUSE/ BASIC SCIENCE

To understand some of the advances in basic science research on epilepsy gained from model systems research, it is essential to understand the relative contribution of individual neurons versus networks of neurons to hyperexcitability. Therefore neuronal excitability is briefly reviewed in Box 1 and synaptic transmission in Box 2.

The generation and propagation of the neuronal AP involves a well-orchestrated machinery where a number of proteins must function on task to ensure that a properly encoded signal reaches its target and that the signal is properly terminated. A change in any one of the proteins—be it a presynaptic Ca^{2+} channel required for transmitter release or a postsynaptic receptor involved in generating the receptor potential—could each result in markedly altered neuronal signaling. Even changes in a single population of channels, for example, one that may alter the duration of the Na^+ channel closing or the delay with which K^+ channels repolarize, change the timing and possibly the frequency of the signal sufficiently to alter the information encoded.

Not surprisingly, there does exist evidence for each of the mentioned proteins and signals to function aberrantly in certain forms of epilepsies; these are discussed further under genetic changes below (see 4.6 and 4.7).

4.1 Hyperexcitability, Network Excitability and Synchronization

A burst of abnormal neuronal discharge, as shown in Box 1-Figure 1, which is often called "epileptiform activity," is not sufficient to explain the development of epilepsy. Instead, a neuronal network must spontaneously develop synchronous discharges that occur sporadically. The question then becomes whether groups of neurons that are each individually hyperexcitable give rise to seizure or whether the abnormal interaction between groups of such neurons forms an epileptic unit. In all likelihood it is a combination of both. A good example of the first possibility is an increase in extracellular K^+, which would lead to the synchronous depolarization that leads to a wave of epileptiform discharge spreading across the tissue in otherwise completely unaltered neurons. On the other hand, mutations in ion channels, which are described more extensively below, cause the excitability of individual cells to change and produce altered network activity.

The concept of synchrony is central to our understanding of epilepsy.[6] As EEG recordings suggest, numerous neurons discharge in synchrony, generating an oscillatory change in voltage that lasts many seconds to minutes. How can such synchrony be explained? Pyramidal cells, the principle cells in the cortex, are aligned in a highly orderly fashion (Figure 3). They are interconnected through glutamatergic synapses; in addition, they contain electrical synapses called gap junctions, albeit at a low density. These allow the direct spread of electrical activity from one cell to another. This then permits a single cell to engage its neighbors to discharge in synchrony. Many of these

BOX 1

EXCITABILITY OF SINGLE NEURONS

The principle neuronal signal is the action potential (AP), an all-or-nothing event during which the transmembrane voltage abruptly changes from −80 to +50 mV and back. A typical recording is shown in Box 1-Figure 1. The upstroke of the AP engages Na⁺ channels, which briefly open and flux Na⁺ and then spontaneously close within milliseconds. The return to the resting potential is accelerated by a delayed opening of K⁺ channels that efflux K⁺. This often results in an undershoot, whereby the membrane potential becomes briefly even more negative than the resting potential that is maintained between periods of electrical activity.

This in turn can activate a number of K⁺ channels that collectively dampen activity by delaying a renewed AP from being fired. Some neurons fire a burst of APs, during which the repolarizing phase is incomplete, resulting in many action potentials being fired in quick succession. Others do so only under abnormal conditions, for example, after the extracellular accumulation of K⁺ around neurons, which can slightly depolarize the cell and decrease the threshold for activation. This can be sufficient to cause hyperexcitability. Whether it is a single AP or a train of many, the activity is transmitted via axons to adjacent cells via synapses (see Box 2).

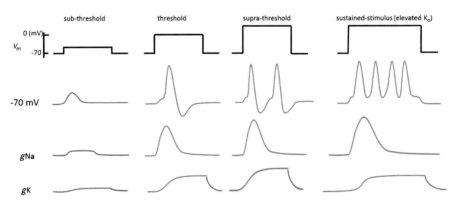

BOX 1-FIGURE 1   Neuronal discharge is the consequence of the activity of Na⁺ and K⁺ channels. Their conductance gNa and gK is schematically illustrated for different stimulus conditions. Short subthreshold stimuli fail to elicit an action potential. Threshold stimulation causes single action potentials, whereas suprathreshold stimulation causes multiple action potentials. Sustained, epileptiform discharge can be caused by aberrant K⁺ clearance or mutations in the underlying ion channels.

pyramidal cells are also innervated by basket cells, which are GABAergic interneurons. Each cell innervates and suppresses the activity of many pyramidal neurons. If only a few of these interneurons release their inhibition simultaneously, they cause the synchronous excitation of a large pyramidal cell network. This arrangement is schematically illustrated in Figure 3.

Yet another way to coerce networks of neurons to act in synchrony is by establishing new connections between individual cells. This occurs in animal models of mesial temporal lobe epilepsy (MTLE) as well as patient biopsies where granule cells in the dentate gyrus of the hippocampus begin to sprout new processes, so-called mossy fibers, and make new connections with cells over a considerable distance (Figure 4).

BOX 2

SYNAPTIC TRANSMISSION

The arrival of an action potential in the presynaptic terminal causes the opening of Ca^{2+} channels. The influx of Ca^{2+} causes synaptic vesicles to fuse, which releases their content into the synaptic cleft to be sensed by the adjacent neuron. Here, the neurotransmitter binds and activates ligand-gated ion channels that flux either Na^+ or Cl^- and sometimes Ca^{2+} into the synaptic terminal. The Na^+ or Cl^- flux generates either a depolarizing or hyperpolarizing receptor potential that propagates passively along the dendrite toward the cell body and axon initial segment (Box 2-Figure 1). Excess neurotransmitter is removed by the surrounding astrocytes through transporters. This ensures the rapid termination of synaptic activity, allowing for the cell to be ready for a new cycle of activation. It is important to remember that APs transmitted along the axon are frequency coded; hence information that they carry is encoded by the timing of each action potential and their frequency. This is in contrast to the receptor potential conducted in the dendrites where the signal is amplitude encoded.

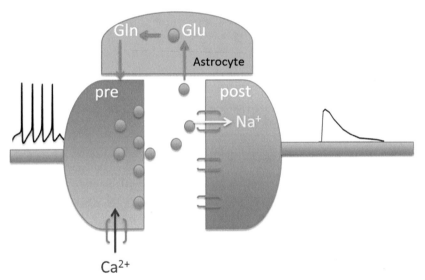

BOX 2-FIGURE 1   Schematic of a synapse. The presynaptic arrival of action potentials causes the terminal to depolarize. The resulting activation of presynaptic Ca^{2+} channels leads to the influx of Ca^{2+}. This leads to presynaptic glutamate (Glu) release which then activates postsynaptic receptors. The resulting postsynaptic depolarization spreads along the dendrite as a graded receptor potential. Excess Glu is being taken up by astrocytes and converted to glutamine (Gln).

Finally, there are synchronous activities in the body that are essential for life. The heartbeat and breathing are excellent examples; their rhythms are controlled by brain nuclei. For example, the reticular neurons in the thalamus provide pacemaker activity that contributes to sleep–wake cycles. Hence the brain contains rhythmic cells that act as pacemakers. If any of these pacemakers become directly or indirectly involved in driving otherwise nonsynchronous

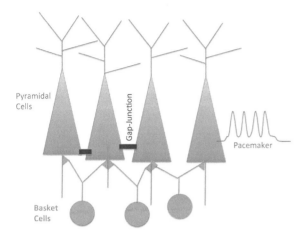

FIGURE 3 Schematized arrangement of excitatory pyramidal cells (blue) and their interactions with inhibitory GABAergic interneurons (orange). A synchronous release of GABAergic inhibition may lead the network to act as pacemaker.

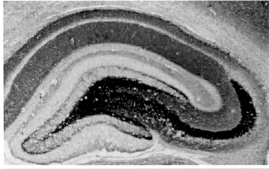
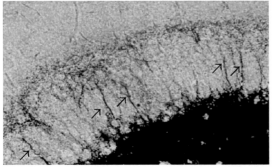

FIGURE 4 Hippocampus showing mossy fibers sprouting, visualized by the dark/black (Timms) staining at low (top) and high magnification (×400; bottom). *From Ref. 7.*

networks, epileptiform neuronal networks could form.

4.2 Emerging Unifying Hypothesis: Excitation–Inhibition Imbalance

Studies of humans and animal models of epilepsy suggest a unifying albeit simple conceptual mechanism that seems to underlie all forms of epilepsy: a change in the excitation–inhibition (E–I) balance. The activity of neural networks in the brain are finely tuned and normally transmit only relevant information. Central to the understanding of brain function is the concept of signal integration, signal processing, and distribution of this information discussed in more detail in Box 3 and illustrated in Box 3-Figure 1.

Aberrant "runaway" excitation of many neurons underlies the epileptic discharge that is recorded by EEG and that leads to convulsive involuntary muscle contractions. This is typically ascribed to an imbalance in glutamatergic excitation and GABA inhibition often simply referred to as E–I balance. Much of our understanding of seizures and epilepsy centers on the concept of an impaired E–I balance. The root cause of this can be multifold and in theory can affect each transmitter system individually or together. Moreover, as described below, pharmacological treatments typically aim to rectify an impaired E–I balance by reducing aberrant excitation or enhancing inhibition. Many factors can influence this delicate balance, including changes in brain chemistry such as (1) altered production or release of a transmitter; (2) changes in the actual network due to structural brain changes, for example, those resulting from a congenitally malformed cortex, trauma, stoke, or infection; and (3) genetic changes that alter the expression and function of receptors or ion channels or cause aberrant networks to form in the first place. These changes can work singly or in combination, and the actual change present in a given patient is often obscure.

BOX 3

NEURONS AS INTEGRATORS OF EXCITATORY AND INHIBITORY SIGNALS

Specialized parts of each nerve cell assume different signaling functions and these are illustrated in Box 3-Figure 1. An elaborate web of cell processes called the dendritic tree contains finely branched processes that are studded with synaptic contact points called spines. Here a neuron can receive a signal from an adjacent neuron, which, for example, may convey a sensory stimulus from the environment. Conservative estimates suggest that cortical neurons receive inputs from at least

10,000 synapses that are spread over this vast tree of dendrites, and therefore in theory these neurons may listen to the input from hundreds of other cells simultaneously. A majority of these dendritic inputs are excitatory and use the neurotransmitter glutamate (Glu). In response to Glu binding to one of its varied receptors, the receiving cell propagates a small, depolarizing (positive) voltage signal along its dendrites toward the cell body. As additional dendritic braches receive similar voltage signals,

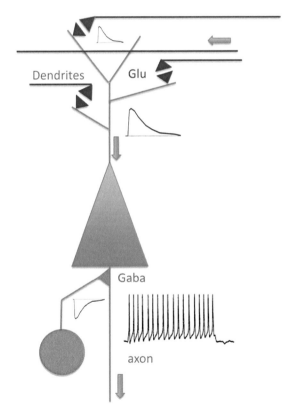

BOX 3-FIGURE 1   Schematized propagation of an electrical signal. Summazion of excitatory glutamatergic inputs are antagonized by inhibitory GABAergic input. If sufficient activity sums in the axon initial segment, an action potential propagates down the axon.

<hr>

BOX 3 *(cont'd)*

<hr>

these add up to a larger signal, thereby enhancing the chance that the combined signal eventually reaches a specialized area near the cell body called the axon initial segment. Here the excitatory input of all the dendrites is combined. If at any given time the signal reaches a preset activation threshold, an action potential or spike of positive-going electricity is triggered that travels down the axon, the principle output structure of a neuron toward an effector cell. This could be a muscle, a gland, or another neuron. Competing information arriving from the periphery or from other computational sites in the brain needs to be able to modulate this excitatory signal. This is achieved through a second class of inhibitory neurotransmitter, typically γ-aminobutyric acid (GABA). Like Glu, GABA binds to specific receptors, where the binding of the transmitter results in a hyperpolarizing, or negative, voltage change. Because this is the opposite of the excitatory Glu signal, these two signals can compete with each other and one can cancel the other. Importantly, the placement of these inhibitory synapses on a given neuron greatly influences the effectiveness with which GABAergic inhibition influences the information conveyed toward the axon initial segment. More specifically, if an inhibitory synapse is placed close to the axon initial segment, activation of just a few inhibitory synapses will be sufficient to modulate or even cancel the excitatory activity received from thousands of excitatory synapses along the vast dendritic tree. Hence, rather than placing GABAergic synapses next to glutamatergic synapses throughout the dendritic tree, the brain evolved in such a way that inhibitor synapses are localized much more closely to the site of action potential generation, that is, near the cell body or axon initial segment. As a result, even small changes in GABAergic inhibition have a profound effect on whether a neuron fires an action potential. Therefore, enhancing GABA receptor activity can readily stop seizures, a strategy used to stop an acute seizure event such as status epilepticus using diazepam or carbamazepine.

<hr>

4.3 Epileptogenesis

Although changes in the E–I balance readily explain the electrical activity underlying a single seizure, how this imbalance only periodically causes seizure is less clear. Most of these occur sporadically, often without a clear trigger or as a result of a seemingly innocuous insult. For example, elevated temperature can trigger febrile seizures in infants, yet only a small number of babies are affected and many outgrow the condition. Some of the features conferring susceptibility are genetic since a family history of seizures increases the likelihood of seizures among family members, yet others may be environmental. Many epileptic patients can go for months or even years without suffering a single seizure. What then provokes the recurrence and serves as a trigger is typically unknown, but the presentation is highly suggestive that such a trigger or dynamic change in susceptibility exists. These could be physiological changes or exposure to environmental stimuli or substances. The best current explanation is that the delicate E–I balance that acts on the normal brain to prevent seizures becomes fragile and more susceptible to distortion in affected individuals, possibly as a result of endogenous and exogenous precipitating factors. Such factors include changes in gene expression, drug or environmental exposure, and aging, to name just a few.

It is believed that early in the disease, during a process called epileptogenesis, when seizures are still absent, progressive changes in the interconnectivity of neural networks set up clusters of cells that tend to fire in synchrony (see Figure 3). They wire together and fire together. Accordingly, epilepsy could best be regarded as a neuronal circuit dysfunction.[8] However, such seizure-prone networks need to be provoked by additional intrinsic or extrinsic factors—many of which we do not yet know—which then trigger a seizure. Factors to consider probably differ from patient to patient but may include diet, drugs, temperature, light, sounds, smells, stress, or sleep deprivation. Unfortunately, little about this period of time preceding presentation of disease is known, and many of the model systems used to study epilepsy (discussed below) examine abnormalities once seizures are present rather than the changes that converge to induce epilepsy. For many acquired epilepsies, for example, tumors or mesiotemporal lobe lesions, the trigger event is known, permitting the scientist to study the process of epileptogenesis. In others, susceptibility as a result of a mutation in an ion channels is known but the trigger event to induce seizures may not. Together, these have yielded valuable animal models in which epilepsy and its treatments can be studied in the laboratory.

4.4 Laboratory Work/Animal Models[a]

Decades of active research have significantly propelled the current knowledge regarding abnormal brain activity that underlies epilepsy. As is the case for other neurological diseases, to gain insight into disease mechanisms reproducing the clinical phenotype of the disease in animal models as accurately as possible is essential. Although there are elegant studies of fruit flies and worms, by far the most widely adopted model organism used to study epilepsy is the rodent brain, where seizures can be induced chemically, electrically, by creating lesions in the brain, or by genetically manipulating ion channels or

transporters; each method mimics a different form of human epilepsy. Note that many of these models reproduce seizures akin to those observed in human patients, yet few provide insight into the period of epileptogenesis that precedes the first seizure.

Chemical Induction

Epilepsy can be reliably induced in mice and rats by injection or topical application of a number of drugs. For example, systemic injection of the muscarinic receptor agonist pilocarpine induces generalized tonic–clonic seizures in rats and mice, which spontaneously recur over the animal's lifetime. Similarly, injection of the Glu receptor agonist kainic acid causes recurrent focal seizures that mimic temporal lobe epilepsy and that are associated with neuronal loss in the CA3 region of the hippocampus similar to that observed in patients with temporal lobe epilepsy. In cats, intramuscular injection of high doses of the antibiotic penicillin induces absence seizures. The cardiostimulant GABA-A receptor antagonist pentylenetetrazol induces acute clonic seizures in mice, rats, monkeys, and humans and is widely used to screen for anticonvulsive properties of newly developed drugs.

Kindling

A nonchemical method of inducing secondary generalized seizures involves implantation of electrodes into a brain structure where focal seizures can be induced, for example, the hippocampus, through repeated high-frequency electrical stimulation. Initially the animal presents with only brief epileptiform discharges that last a few seconds and mimic a focal seizure. When stimulation sessions are repeated over several days, however, the seizure severity increases, they generalize to both hemispheres, and they are accompanied by loss of consciousness. Ultimately, after many days of stimulation, seizures develop spontaneously without further

[a] Extensively covered in Models of Seizures and Epilepsy, Pitkaenen, Schwartzkroin and Moshe, Elsevier 2006.

stimulation and the animal presents with epilepsy for the remainder of its life. This approach is generally called kindling, akin to using a small flame to start a large fire. The kindling model has been used extensively to examine the latent period that precedes the first occurrence of spontaneous seizures. It allows a side-by-side comparison of animals in which a specific signaling cascade is genetically eliminated to wild-type controls. Similarly, it permits the time course to epilepsy to be mapped, comparing animals treated with inhibitors or activators of defined signaling pathways. These studies, for example, elucidated the absolute requirement for brain-derived neurotropic factor (BDNF) signaling via tropomyosin-related kinase (Trk) B receptors for the development of seizures, discussed further below.

Lesions

Yet another approach re-creates the cortical malformations or heterotopias observed in some patients. Here, placing a freezing probe on the exposed skull of a neonatal mouse or rat induces a cortical malformation. As adults, these animals present with spontaneous, recurring focal seizures.

The advantage shared by all these model systems is that living slices of brain tissue can be isolated for more detailed biophysical, biochemical, and anatomic studies. Furthermore, mice can be genetically altered to harbor gene mutations suspected to increase seizure susceptibility in humans. In spite of the seemingly diverse approaches, together these animal models have provided numerous important insights that form the bedrock of our current understanding of the disease. Some of the salient findings are summarized below.

1. EEG abnormalities in rodent models of epilepsy parallel those observed in humans.
2. Seizures in rodents are accompanied by abnormal neural network activity in the cortex and hippocampus.
3. Manipulation of the transmitter systems hypothesized to be involved in the E–I balance is sufficient to induce epileptiform activity.
4. E–I balance is altered after seizure induction either due to a loss of GABAergic cells or altered GABA or Glu signaling.
5. Changes in the expression or function of ion channels involved in neuronal discharge are common, and mutations in these genes often mimic disease in mice.
6. Ion and neurotransmitter homeostasis is compromised near a seizure focus.
7. Structural changes can be observed following successful induction of seizures.
8. Trauma, infections, and vascular changes present with seizures in rodents.

4.5 Transmitter Systems and Epilepsy

As alluded to in Box 3, two transmitter systems, Glu and GABA, are principally responsible for excitation and inhibition and oppose each other to maintain a normal state of brain activity. Increasing excitation or reducing inhibition each has the potential to generate runaway activity that leads to seizures. Most models of epilepsy indeed present with a change in this E–I balance. However, the way that this change comes about can differ widely. For example, in MTLE the number of GABAergic interneurons, often identified by the expression of a Ca^{2+} binding protein called parvalbumin, is reduced compared to normal brain. Thus fewer inhibitory neurons are available to dampen the excitatory activity. In neonatal seizures, by contrast, the number of GABAergic cells is unaltered. However, the effect of GABA is reversed. While in the normal adult brain GABA is inhibitory, in the immature brain it is excitatory.

Astrocytes

In addition to changes in neuronal transmitter release and receptors, there also seem to be changes in the ability of glial support cells, specifically astrocytes to clean up excess transmitter after it has been released. Glu is typically maintained below 1 μM in the small extracellular space that surrounds neurons. Synaptic vesicles contain millimolar concentrations of Glu that are released into the synaptic cleft. After binding to neuronal Glu receptors, Glu must be removed so that it does not continue to activate postsynaptic receptors. This is accomplished by Glu transporters on astrocytes, which envelope glutamatergic synapses (Box 2-Figure 1 and Figure 5). These transporters couple the inward-directed gradient for Na^+ ions to transport Glu from the synaptic space into the astrocyte. Once trapped inside, Glu is converted to glutamine (Gln) by the enzyme glutamine synthetase, and Gln is recycled back to the neuron for the de novo synthesis of Glu. This process is very important and, not surprisingly, these glial Glu transporters make up 1% of total brain protein. Any loss of expression or function of the astrocytic Glu transporters can quickly increase Glu in the extracellular space and, as a consequence, shift the E–I balance toward too much excitation. Changes in the expression or function of these transporters lead to too much Glu in the extracellular space that can shift the E–I balance as well as cause neuronal cell death.

4.6 Ion Channels

As we discussed in Box 1, the conversion from a single AP to a train of epileptiform activity can occur without the direct involvement of neurotransmitters and simply as a result of changes in the ion channels underlying the various phases of an AP, and these do indeed occur in certain forms of epilepsy. For example, mutations in *SCN1A*, encoding for a voltage-gated Na^+ channel, cause myoclonic epilepsy of infancy, characterized by early onset and progressive worsening. These mutations keep the channel active for a longer time than normal. Depending on whether this mutation affects GABAergic or glutamatergic cells, the result could be either decreased or increased excitability.

Other mutations occur in K^+ channels that are responsible for repolarization and cause a

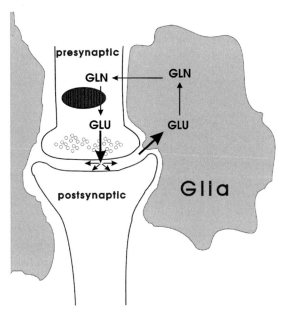

FIGURE 5   Astrocytic recycling of glutamate (Glu) via active transport and subsequent conversion to glutamine (Gln) protect neurons from excessive excitation.

broadening of the AP. As a result, more positive charge enters the cell, increasing the likelihood that the cell will stay near the activation threshold and fire another AP, and another, resulting in a train of activity.

To ensure that a neuron has appropriate rest periods between firing APs, the membrane has special K^+ channels that activate only upon termination of an AP. These hyperpolarization-activated channels, called HCN channels, thereby dampen the neuronal discharge. In generalized idiopathic epilepsy, sequence variations in HCN channels lead to channels that are less efficient in preventing the cell from frequent discharge, creating a scenario where hyperexcitability occurs.

A final aspect to consider is that each action potential extrudes almost 1 mM K^+ into the extracellular space. Trains of APs therefore would quickly increase K^+ from 3 mM at rest to 10 or even 20 mM during activity. This in turn depolarizes the neuron, causing it to fire uncontrollably. Again, astrocytes come in as safeguards here. In addition to clearing Glu, discussed above, they are also responsible for clearing neuronally released K^+. They do so by diffusional uptake via the inwardly rectifying K^+ channel Kir4.1, as well as by energy-dependent transport via the Na^+K^+-ATPase. Once inside the astrocyte, K^+ no longer affects the neuronal membrane potential.

As these examples illustrate, there are ample opportunities for things to go wrong, resulting in neuronal hyperexcitability. This also explains, as described below, why we must assume that epilepsy has polygenetic causes. A mutation in the HCN channel gene, the SCN Na^+ channel gene, or astrocytic K^+ channels may each present with the same outcome: increased neuronal excitability.

4.7 Genetics of Epilepsy

A powerful approach to the study of disease is to search for genetic mutations that either cause disease, and through inheritance pass on the disease trait to offspring, or alternatively, to identify genetic variations that correlate with an enhanced likelihood that an individual presents with disease. As already indicated above, there seems to be an increased epilepsy risk in offspring from affected mothers and a high concordance in monozygotic twins, suggesting that some heritable disease-causing mutations exist.

While the genetic understanding of epilepsy is still evolving, recent advances in gene sequencing technology and cost are beginning to accelerate the discovery process. There are excellent candidate genes to explain some forms of "monogenetic" epilepsy, that, single gene mutations that seem to explain the development of disease, as well as the emergence of candidate genes to explain complex "multigenetic" epilepsy, combinations of genetic changes that alter the likelihood of an individual presenting with epilepsy. The affected mutations are primarily found in receptors and ion channels that are fundamentally important in the generation and propagation of neuronal signals. Table 1 lists some of the most well-understood channel mutations in epilepsy, and a few examples are discussed in greater detail here.

The *SCN1A* gene provides an excellent example of a monogenetic cause of epilepsy. It encodes for the alpha subunit of the neuronal voltage-gated Na^+ channel that is responsible for the AP. More than 150 newly occurring de novo mutations in this gene have been associated with various epilepsies, ranging from febrile seizures to tonic–clonic generalized seizures.[9] Single amino acid substitutions that change the structure of the channels lead to milder forms of epilepsy, whereas more severe forms of epilepsy involve multiple mutations of the same gene. Approximately 80% of patients with Dravet syndrome, a severe form of epilepsy, present with a mutation in the *SCN1A* gene, providing confidence that these mutations cause disease.

While most mutations of the *SCN1A* gene are de novo, autosomal dominant heritable mutations in the alpha subunit of the Na^+ channel

TABLE 1   Known Mutations in Ion Channels and Receptors in Epilepsy

Gene	Protein	Inheritance	Type of seizure	Presumed mode of action
SCN1A	Navl.l Na^+ channel, alpha subunit	>150 De novo mutations	Febrile, generalized tonic–clonic, Dravet syndrome	Slow inactivation, broadening of the action potential, persistent current
SCN2A	Nav1.2 Na^+ channel alpha subunit	Autosomal dominant	Benign familiar epilepsy syndrome	Slow inactivation, broadening of the action potential, persistent current
KCNA1	Kvl.l Shaker related K^+ channel	Autosomal dominant	Focal epilepsy	Slowed repolarization of the action potential noninactivating currents that normally raise the threshold for excitability, loss of function increases excitability
KCNQ2	Kv7.2, M-current cation current	Autosomal dominant	Benign familiar epilepsy syndrome with tonic–clonic seizures	
KCNQ3	Kv7.3, M-current, cation current	Autosomal dominant	Benign familiar epilepsy syndrome	Noninactivating currents that normally raise the threshold for excitability, loss of function increases excitability
HCN1	Hyperpolarization-activated K^+ channel	De novo mutations	Temporal lobe epilepsy	Shortened refractory period, enhanced excitability
KCNMA1	KCal.l, MaxiK Ca^{2+}-activated K^+ channel	De novo mutations	Generalized epilepsy	Contributes to repolarization, stabilize resting membrane potential, loss of function causes hyperexcitability
KCNT1	KCa4.1, Ca^{2+}-activated K^+ channel	De novo mutations	Generalized epilepsy, nonconvulsing malignant migrating partial seizures of infancy is a rare epileptic encephalopathy	Stabilize resting membrane potential, loss of function causes hyperexcitability
CACNA1A	P/Q type Ca^{2+} channel	Autosomal dominant	Absence seizure	Required for neurotransmitter release, which, if mutated, is altered
CACNA1H	T-type Ca^{2+} channels	Autosomal dominant	Childhood absence seizure	Pacemaker, enhances pyramidal cell bursting

TABLE 1   Known Mutations in Ion Channels and Receptors in Epilepsy—cont'd

Gene	Protein	Inheritance	Type of seizure	Presumed mode of action
CHRNA4	nACH receptor alpha-4 subunit	Autosomal dominant	Autosomal dominant nocturnal frontal lobe epilepsy	Regulates excitability
CHRNB2	nACH receptor beta-2 subunit	Autosomal dominant	Autosomal dominant nocturnal frontal lobe epilepsy	Regulates excitability
CHRNA2	nACH receptor, alpha-2 subunit	De novo mutations	Autosomal dominant nocturnal frontal lobe epilepsy	Regulates excitability
GABRG2	GABA-A receptor, gamma-2 subunit	De novo mutations	Febrile seizure, absence seizure	Decreases tonic inhibition

Table established based on data provided in: Lerche et al. J Physiol. 2013; 591.4: 753–764, & Jasper's Basic Mechanisms of the Epilepsies (Internet). 4th ed.

gene *SCN2A* have also been identified and cause a benign familial epilepsy syndrome that presents with generalized partial seizures.

Another example is the *KCNT1* gene that encodes for KCa4.1, a calcium-activated potassium channel, and has been linked to early onset infantile epilepsies. Over 60 mutations have been reported for another voltage-gated K^+ channel, Kv7.2, that is encoded by the *KCNQ2* gene. These are associated with benign familiar neonatal seizures, which are characterized by recurrent seizures in newborns that spontaneously disappear at age 1–4months. These channels are responsible for dampening neuronal activity, and even a small decrease in their function seems sufficient to destabilize the cells' resting potential just enough to cause hyperexcitability and ultimately seizures.

Mutations in calcium channels involved in the presynaptic release of neurotransmitters or burst firing of neurons have been linked to a number of familial epilepsies. These include the P/Q-type Ca^{2+} channel gene *CACNA1A* and the T-type Ca^{2+} channel gene *CACNA1H*, both of which cause absence seizures.

Nicotinic acetylcholine receptors are an example for neurotransmitter-gated ion channels

mutated in epilepsy. Both the alpha-4 (CHRNA4) and beta-2 (CHRNB2) subunit of the receptor can be mutated, causing a loss of function of these receptors that leads to an autosomal dominant inherited form of epilepsy that manifests with episodes of nocturnal motor seizures, called autosomal dominant nocturnal frontal lobe epilepsy.

Mutations in the γ2 subunit of the GABAA receptor (*GABARG2* R43Q) are another important example of a monogenetic cause of epilepsy. The mutated channel reduces tonic GABAergic inhibition and gives rise to a number of different epilepsy phenotypes, most importantly febrile and absence.[10]

An important test to confirm the monogenetic nature of disease is to question whether the introduction of the same mutation into a mouse causes the animal to suffer disease. This has been accomplished for mutations in the *SCN1A*, *HCN1*, and *GABARG2* genes, which each present as epilepsy in mice. Prospective gene targeting in mice by genetic manipulation is laborious and not always successful. Considering the age-dependence of epilepsy and species differences between humans and mice is important. The inability to reproduce a seizure phenotype by introducing a disease-causing mutation in a mouse does not prove lack of causality. It is also important to realize that collectively these mutations account for only a very small percentage of epilepsy cases and that for many affected patients gene changes that would predict disease remain elusive.

With regard to complex mutigenetic epilepsies, we are just beginning to see positive results from large studies conducted by consortia of investigators such as the European EPICURE Consortium.[11] Such studies look for common genetic variations or single-nucleotide polymorphisms. (For a better understanding of these genetic screens the reader is directed to Chapter 14, "Shared Mechanisms of Disease," where genome-wide association studies are explained in further detail). To be successful, such studies must include data from large patient cohorts. Thus far, increased susceptibility has been shown to correlate with variations in the *SCN1A* and *CHRM3* genes, a Na$^+$ channel and muscarinic receptor, respectively, and for three genes (*VRK2*, *ZEB2*, and *PNPO*) for which the link to epilepsy is less clear at the moment. With regard to multigenetic epilepsies, it is important to stress that a combination of small changes in affected genes gives rise to a phenotype that may confer heightened susceptibility that only manifests as seizures under the appropriate environmental conditions, such as stress or hypoxia.

4.8 Emerging Targets and Mechanisms Suspected to Play a Role in Epilepsy

KCC2/NKCC1

The importance of GABA as an inhibitory transmitter in the E–I balance has already been discussed. Being primarily a ligand-gated Cl$^-$ channel, the inhibitory action of GABA is due to the electrochemical gradient for Cl$^-$ across the plasma membrane. Cl$^-$ accumulates via the sodium–potassium–chloride transporter NKCC1 that imports Cl$^-$ together with Na$^+$ into the cell. In mature neurons, NKCC1-mediated Cl$^-$ uptake is counteracted by a second transporter, the potassium chloride cotransporter KCC2, which effluxes Cl$^-$ (Figure 6). From this balance of import and export, most mature neurons have a very low intracellular Cl$^-$ concentration of only 7–10 mM, whereas extracellular (Cl$^-$) is much higher, around 140 mM. As this gradient favors the influx of Cl$^-$ upon opening of the GABA-gated channel, a negative charge enters the cell, thereby hyperpolarizing the membrane. During brain development, however, KCC2 expression is delayed and is not fully expressed until well after birth. Therefore, in the newborn (human or mouse), intracellular Cl$^-$ is considerably higher and GABA receptor activation leads to Cl$^-$ efflux rather than influx, thereby actually depolarizing the membrane and contributing to excitability.[12] This developmental shift is graphically illustrated in Figure 6. This shift in E–I balance is at least one reason why newborns are generally more susceptible to seizures, a condition that rectifies itself as KCC2 expression increases in the ensuing weeks of postnatal life, during which seizures spontaneously disappear. While there currently is no feasible way to induce a premature expression of the KCC2 transporter, there is a potent drug to inhibit the NKCC1-mediated Cl$^-$ uptake. The loop diuretic bumetanide (Bumex), used to treat edema associated with heart failure or liver disease, is being tested in a clinical trial for severe infantile seizures with the rationale that it alters the intracellular Cl$^-$ concentrations (clinicaltrials.gov identifier NCT00830531). A change in KCC2 expression and function also characterizes adult forms of human epilepsy,[10] and a recent clinical study successfully used bumetanide to treat patients with MTLE.[14]

It is worth mentioning that bromide, which was used quite effectively to treat epilepsy in the past, may have done so by affecting the transmembrane Cl$^-$ gradient as well. Bromide, like Cl$^-$, is a monovalent halide anion, and it

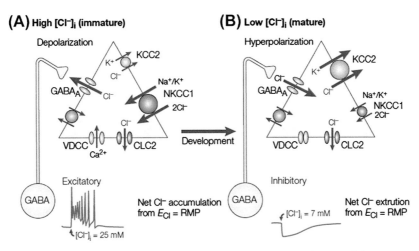

FIGURE 6 Cl⁻ transport affects the action of γ-aminobutyric acid (GABA) as either an inhibitory or excitatory transmitter. In the developing, immature brain, intracellular Cl⁻ is accumulated to a high concentration via the activity of the highly expressed NKCC1 transporter. As a result, GABA acts as an excitatory neurotransmitter (A). Upon maturation, NKCC1 expression is reduced and KCC2 expression now predominates shunting Cl⁻ out of the cell. The resulting low intracellular Cl⁻ makes GABA an inhibitory neurotransmitter. *Reproduced with permission from Ref. 13.*

behaves much like Cl⁻ in biological systems. It can therefore substitute as a charge carrier for the GABA receptor and is about 50% more permeable than Cl⁻. At an extracellular concentration of 10–20 mM, bromide enhances the flux of a negative, inhibitory charge into the cell compared with Cl⁻ alone by about 28–35%. So without any other changes, the same amount of GABA increases inhibition by about one-third.[15]

Mammalian Target of Rapamycin

Mammalian target of rapamycin (mTOR) is a protein kinase involved in regulating cell growth, proliferation, and survival and has been extensively studied in the cancer literature. The mTOR inhibitor rapamycin is used in clinical trials for the treatment of endometrial cancer and soft-tissue and bone sarcomas. Tuberous sclerosis is an inherited disorder with benign tumors and cortical thickenings called tubers. These are caused by the aberrant migration of neurons and cellular growth that presumably occur as a result of defective mTOR signaling. Recordings

in patients determined that some of these tubers are intrinsically epileptogenic. Interestingly, by harboring the mutations that cause tuberous sclerosis the mTOR inhibitor rapamycin suppressed seizures. It was also recently discovered that mTOR activity is regulated by pathways that are involved in glutamatergic, excitatory transmission and seems to be strongly associated with other forms of epilepsy. Because of the pleiotropic effects of mTOR signaling, its mechanistic involvement in epileptogenesis, that is, the development of epilepsy, is not yet known. However, in a number of preclinical epilepsy models, such as kainate-induced status epilepticus, rapamycin was able to suppress seizure development.

Synaptic Vesicle Protein 2A

Synaptic vesicle protein 2A (SV2A) is an abundant membrane-bound glycoprotein that is found on secretory vesicles including synaptic vesicles. It is an essential protein; mice with homozygote deletion die within 3 weeks after birth. SV2A has been identified as the binding site for the AED levetiracetam, although the mechanism whereby the drug reduces seizures is not understood. Interestingly, patients suffering from MTLE show a 32% decrease in SV2A expression,[16] and a similar decrease in SV2A was observed in rodents following experimental induction of seizures. Moreover, the genetic deletion of the *SV2A* gene in neurons results in a decrease in GABA release in the hippocampus. This renders these animals more susceptible to seizures, although it is insufficient to induce epilepsy. In spite of its unclear role in epilepsy, SV2A is an attractive target for the development of future AEDs.

Neurotrophins

Neurotrophins are important modulators of cell survival and development; some scientists consider them the "multivitamin of the brain." The biology of BDNF is extensively discussed in Chapter 14. Four neurotrophins are expressed in the mammalian brain, including nerve growth factor, BDNF, and the neurotrophins 3 and 4. They each bind to distinct receptors called TrK receptors, tyrosine kinases that exhibit distinct affinity for select neurotrophins. TrKA binds nerve growth factor, TrKB is specific for BDNF and neurotrophin 4, whereas TrKC is activated by neurotrophin 3. Upon ligand binding, TrK receptors dimerize to become catalytically active. Numerous research studies support an important role for TrKB receptors activated via BDNF in epileptogenesis, although no single common mechanism has been elucidated.[17] Elimination of TrkB in mice eliminates epileptogenesis in the kindling model of epilepsy, and overexpression of BDNF causes spontaneous seizures in mice and in brain slices. BDNF also encourages the sprouting of dendrites in the hippocampus, which contributes to aberrant connectivity. Finally, TrKB activation causes a reduction in KCC2 expression, which, as discussed above, leads to reduced GABAergic inhibition.

Ca²⁺-Regulated Enzymes

Neuronal activity almost always causes increases in intracellular Ca^{2+}. This in turn can activate a number of Ca^{2+}-activated enzymes. Two of these, CaMKII and calcineurin, have received prominent attention in the epilepsy literature. Ca^{2+} causes a conformational change in CaMKII, activating its catalytic activity and causing phosphorylation. Mice with a mutated and nonfunctional CaMKII show limbic epilepsy, arising from the limbic structure of the brain for example the frontal lobe, and kindled mice show a reduction in CaMKII activity. Currently available data[17] suggest that reduced CaMKII activity alone is sufficient to cause cellular and molecular changes that make the brains of mice epileptic.

Calcineurin is a phosphatase; when activated by Ca^{2+} binding, calcineurin dephosphorylates its substrates. Calcineurin can be pharmacologically inhibited by the immunosuppressant drugs cyclosporine A or FK506, and treatment of mice with these drugs makes it much more difficult to induce seizures in mice through kindling. Increased activity of calcineurin, on the other hand, accompanies febrile seizures in infants. Both CaMKII and calcineurin are also important in learning and memory; their activity regulates Glu and GABA receptor trafficking and thereby alters the number of receptors in the membrane.

Astrocytes

While it is undisputed that seizures are generated by the synchronous discharge of neuronal networks, nonneuronal cells may be complicit elements that support the genesis of seizures. Astrocytes have received most interest in this regard; these cells serve a multitude of functions to support normal neuronal activity. Astrocytes clear K^+ and Glu released from neurons, and the loss of either Kir4.1, the channel responsible for K^+ clearance, or EAAT2, the transporter that removes Glu, causes seizures in mice. Astrocytes catabolize Glu to Gln, which can serve as a precursor for the neuronal synthesis of GABA via glutamic acid decarboxylase. Astrocytes are also highly interconnected through gap junctions, allowing them to quickly shuttle molecules removed from the extracellular space to neighboring cells. Astrocytes show structural and functional changes when injured or when associated with disease, which makes them reactive. Such reactive astrocytes may no longer be able to perform their expected housekeeping roles and may indirectly contribute to disease. For example, reactive astrocytes may no longer import Glu from the extracellular space and convert it to Gln. The consequences may be dramatic as the buildup of Glu near synapses chronically activates neuronal Glu receptors. Moreover, since astrocytes no longer supply neurons with a needed substrate for GABA synthesis, they starve neurons of GABA. Both increased extracellular Glu and decreased neuronal GABA upset the normal E–I balance.

Reactive astrocytes also have been shown to produce and release proinflammatory cytokines including interleukin (IL)-1b and tumor necrosis factor-α. These molecules increase neuronal excitability and activate the adaptive immune system, both of which are undesirable effects in the setting of epilepsy.

Stem Cells

The formation of aberrant projections that is observed, for example, in the hippocampus following seizures suggests that structural plasticity contributes to a rewiring of neural networks. The discovery that the brain harbors neural stem cells throughout life prompted inquiry as to their participation in nervous system diseases and injury. Interestingly, the hippocampus, a highly seizure-prone brain region, also contains one of the niches where stem cells reside. The dentate gyrus produces neural stem cells that migrate into the hippocampus in a process called neurogenesis, and many of these adult-born neurons differentiate into GABAergic interneurons. Their contribution to abnormal networks has been documented in experimental models

of epilepsy where the induction of status epilepticus was followed by marked neuronal cell death in the hippocampus and then by robust proliferation of progenitor cells. In the ensuing 1–3 weeks, these newborn neurons contributed to aberrant networks in the hippocampus that showed abnormal processes and were placed in the wrong parts of the hippocampus, suggesting that they contribute to epileptogenesis.[18]

Epigenetics

One of the most puzzling features of epilepsy is the unpredictable sporadic manifestation and the long delays that can exist between seizures. Moreover, a suspected brain insult such as trauma or infection is often followed by a long latent period until the first seizure presents. We generally believe that this epileptogenesis period wires the brain to have a heightened intrinsic susceptibility to trigger epileptic events. The emerging field of epigenetics describes a plausible way in which environmental factors control gene expression. Chapter 14 contains a much more elaborate discussion of epigenetics as it pertains to neurological disorders. In a nutshell, environmental factors can alter the accessibility of groups of genes on the chromosomes or of individual genes for transcription into proteins. A few proteins that are under environmental control thereby become the master keys to parts of the genome. These proteins are enzymes that attach or detach methyl or acetyl groups to cytosine or adenine nucleotides on DNA. Methylated DNA is typically silenced and not transcribed. Methylation occurs at so-called CpG dinucleotides or CpG islands.

Following status epilepticus, over 300 genes are hypomethylated (reduced methylation), and consequently gene expression is expected to be enhanced.[19] By contrast, patients with MTLE show enhanced expression of the DNA methyltransferase gene, which would be expected to lead to enhanced transcriptional silencing of DNA.

If the histones, which coil stretches of DNA, are methylated or demethylated, the transcription of many genes contained on a larger stretch of DNA is regulated. The master keys that do so include histone acetyltransferase, deacetylases, and histone methyltransferases. These are emerging pharmacological targets in epilepsy and other neurological disorders, particularly since two drugs (Vidaza and Dacogen) that target all DNA methyltransferases and inhibit DNA methylation are already clinically approved.[20] Interestingly, one of the oldest AEDs, sodium valproate, is also an inhibitor of one of these (histone deacetylase), which when inhibited leads to uncoiling of DNA from histones and therefore generally enhances transcription. Hence valproate may act not only as a regulator of neuronal ion channels but also as a master key regulating transcription of proteins that contribute to a number of chemical and structural changes, including neurogenesis, that occur during epileptogenesis.

5. TREATMENT/STANDARD OF CARE/CLINICAL MANAGEMENT

5.1 Diagnosis

During a routine neurological exam ination, the treating neurologist gathers the patient's history and any evidence of recent lifestyle changes, use of medication or drugs, injuries, diseases, or exposure to drugs or stress factors such as divorce or loss of a loved one. EEG evaluation is advisable. Given the infrequent occurrence of seizures in most patients, ambulatory, wearable EEG systems can be used on an outpatient basis. The EEG cannot always establish or refute the diagnosis of epilepsy, but if abnormalities are detected, it can aid in classifying the disease and may aid in determining the most appropriate drug regimen. Most patients will also be imaged by computed tomography or magnetic resonance imaging (MRI), if possible, to establish whether a structural abnormality, such as a tumor or cortical malformation, may contribute to the disease.

5.2 Pharmacological Treatment

There are currently over 20 drugs in use as primary treatments for epilepsy (Table 2). These are typically referred to as AEDs. In general they act to raise the threshold for neuronal network excitability. In light of what we learned above, there is a diverse group of proteins involved in neuronal activity, ranging from ion channels to neurotransmitter receptors and their transporters. Not surprisingly, different AEDs target different proteins within this network and the drugs often have pleiotropic effects. The latter may be surprising, but as we learn in Chapter 15 of this book, few clinically used drugs were developed specifically to target known receptors but instead resulted from serendipitous discovery, with the drug existing long before its application was known. Table 2 summarizes the presumed mode of action and preferred use of current AEDs. As AEDs are supposed to enhance inhibition, it is not surprising that many drugs work by augmenting the GABA response (benzodiazepines and barbiturates) or the availability of GABA through reuptake inhibitors (gabapentin and tiagabine) or enhancing GABA release from synaptic vesicles (levitracetam).

An overall reduction in Na^+ channel activity that drives APs is another target of some widely prescribed drugs such as phenytoin (Dilantin), carbamazepine (TEGretol), lamotrigine, topiramate, and lacosamide. Finally, the release of neurotransmitters can be regulated by inhibition of certain Ca^{2+} channels (phenytoin, gabapentin, and pengabalin); ethusuximide and valproic acid presumably are effective in absence epilepsy by means of inhibiting T-type Ca^{2+} channels. Given their different modes of action, some drugs can be used in combination with synergistic effects.

The first line of treatment for most patients with focal epilepsy is carbamazepine,

TABLE 2   Some Currently Available Antiepileptic Drugs (in Alphabetical Order)

Drug name	Trade names	Target	Presumed mechanism of action	Indication	Comments
Bromide salt	Bromo-Selzer	Unknown	May replace Cl⁻ as permeable ion at GABA-R; has higher permeability than Cl⁻	Tonic–clonic seizures	No longer used in humans, still available for veterinary use
Carbamazepine	Tegretol	Voltage-gated Na^+ channels	Reduces repetitive neuronal discharge, anticonvulsant	Focal and generalized seizures of all types	
Ethosuximide	Zarontin	T-type Ca^{2+} channels	Targets Ca^{2+} channels in thalamic neurons involved in generation of aberrant spike-wave discharges	Absence epilepsy	
Ezogabine	Ezogabine	K^+ channels, KCNQ Kv7.2	Opens K^+ channels, stabilizes membrane potential, reduces excitability	Focal seizures	No known interactions with other AEDs
Felbamate	Felbatol	NR2B subunit of NMDA-R, Na^+ channels, GABA-R	Open channel blocker, inhibits NMDA, thereby reducing glutamatergic excitatory activity	Focal seizures, Lennox–Gastaut syndrome	Enzyme inducer that interacts with several other AEDs, rare cause of liver failure
Gabapentin/ Pregabalin	Neurontin	Ca^{2+} channels, GABA transporters?	Unclear, increases synaptic GABA concentrations, thereby enhancing inhibition	Focal seizure	Originally designed to mimic GABA but does not work through GABA-R, pregabalin is the successor and is more potent, has better bioavailability
Lacosamide	Vimpat	Na^+ channels	Slows recovery from inactivation	Focal seizures	
Lamotrigine	Lamictal	Ca^{2+} channels, Na^+ channels	Blocks voltage-gated Na^+ channels, reduces the release of glutamate	Tonic–clonic focal and generalized seizures; Lennox–Gastaut syndrome	
Levetiracetam	Keppra	Synaptic vesicle protein 2A	Changes release of neurotransmitters	Focal seizures, myoclonic, tonic–clonic	
Oxcarbazepine	Trileptal	Na^+ channels, N-type Ca^{2+} channels	Stabilizes resting membrane potential	Focal seizures	Analogue of carbamazepine

Continued

TABLE 2   Some Currently Available Antiepileptic Drugs (in Alphabetical Order)—cont'd

Drug name	Trade names	Target	Presumed mechanism of action	Indication	Comments
Perampanel	Fycompa	AMPA-R	Noncompetitive agonist of AMPA Glu receptors, reduces excitability	Refractory, focal seizures	High potency, long systemic bioavailability
Phenobarbital	Luminal	GABA-A receptor	Enhances GABAergic inhibition		Enzyme inducer
Phenytoin	Dilantin	Voltage-gated Na$^+$ channels	Use-dependent inhibitor, becomes more effective with repetitive neuronal activity		
Rufinamide	Banzel	Na$^+$ channels	Maintains Na$^+$ channel in closed, inactive state	Adjunct for Lennox–Gastaut syndrome	
Tiagabine	Gabitril	GABA transporter	Inhibit reuptake of GABA, therefore prolonging inhibitory action of GABA	focal seizures, adjunct treatment	
Topiramate	Topamax	AMPA-R, GABA-R, Na$^+$ channel	Multiple effects proposed, inhibitor of AMPA Glu receptors and blocker of Na$^+$ channels reduce excitatory drive, also enhanced GABA-R-mediated inhibition	Lennox–Gastaut syndrome, focal seizures	Inhibits carbonic anhydrase, increases risk for kidney stones, enzyme inducer, interacts with other AEDs
Valproate	Depakote	multiple, GABA-R, GABA transaminase, NMDA-R, HDACs	Reduces glutamatergic excitation via NMDA-R, enhances GABA inhibition, changes transcription of multiple genes via HDAC inhibition	Absence seizures, tonic–clonic, focal seizures with dyscognia, posttraumatic epilepsy, Lennox–Gastaut syndrome	Introduced in 1967, now the most widely prescribed AED worldwide
Vigabatrin	Sabril	GABA transaminase	Irreversible inhibitor of the enzyme that catabolizes GABA; drug increases available GABA and enhances inhibition	Focal and secondary generalized seizures, infantile spasms, Lennox–Gastaut syndrome	Negative interactions with carbamazepine and phenytoin reported
Zonisamide	Zonegran	Na$^+$ and Ca^{2+} channels	Unknown mechanism of action, primary target is carbonic anhydrase but likely reduces excitability by interacting with ion channels	Focal and generalized tonic–clonic seizures	

AED, antiepileptic drug; AMPA-R, AMPA receptor; GABA, γ-aminobutyric acid; GABA-R, γ-aminobutyric acid receptor; Glu, glutamate; HDAC, histone deacetylase; NMDA-R, N-methyl-D-aspartate receptor.
Compiled using data from: French, Gazzola. in Continuum, Lifelong learning in Neurology – Epilepsy. June 2013; 19(3).

lamotrigene, phenytoin, or topiramate. Valproate is the primary choice and "Swiss army knife" for a number of generalized seizures, including absence seizures as well as myoclonic and atonic seizures. It is the most widely used drug worldwide to treat epilepsy. For one-third of patients the first drug choice provides freedom from seizures. However, if even after careful increase in drug dose a therapeutic response cannot be achieved, a second drug is added. Some patients must take three or four drugs simultaneously, and considering how they might interact is important.

The goal of effective treatment is freedom from seizures. It is important to note that freedom from seizures does not necessarily equate to a completely normal EEG, and there is continued debate between scientists and clinicians regarding the value of considering EEG changes in the assessment of the efficacy of a certain treatment approach. From a purely clinical point of view, "we are not treating the EEG but the seizure" is a frequently voiced opinion. However, since recent research suggests that the masking of behavioral seizures in the continued presence of underlying EEG abnormalities may signify a continued deterioration of brain function, including cognitive decline, this opinion may warrant being revisited.

With the currently available drugs, approximately two-thirds of patients are effectively managed, with the majority of patients achieving freedom from seizures for most of their life. For the remaining one-third of drug-resistant patients alternative strategies must be considered. Foremost, surgical approaches are particularly effective in a subgroup of epilepsy syndromes.

5.3 Surgical Treatment

Patients who do not respond to any of the AEDs often have a poor quality of life and an inability to work or even live independently. For some of these patients, particularly those with "lesional epilepsy" harboring structural brain abnormalities that give rise to seizures, surgery can be an effective alternative. Before surgery a meticulous neurological evaluation including EEG and MRI must be performed. Structural changes, if present, are often subtle and require high-field-strength (3 T) images using a combination of T_1 and T_2 and possibly fluid-attenuated inversion recovery sequences to visualize abnormalities. It is essential that multiple tests converge on the same seizure focus to ensure that the offending part of the brain can be surgically removed while sparing as much of the surrounding brain as possible. During surgery, focal cortical stimulation while the patient is awake and naming objects displayed on a computer screen allow the surgeon to further map the "eloquent cortex," those areas of the cortex essential for language and motor and sensory function.

The most common surgically curative lesional epilepsy is MTLE, for which complete or partial removal of the anterior temporal lobe, along with partial resection of the hippocampus and amygdala on the affected hemisphere, provides excellent outcome in many patients. An example of a patient before and after surgery is illustrated in Figure 7. Dissection of the corpus callosum provides benefit to some of the severe, medically intractable forms of epilepsy, for example, Lennox–Gastaut syndrome.

A number of acquired epilepsy syndromes benefit from surgical intervention. Most notable are benign or low-grade brain tumors, including gangliogliomas and oligodendrogliomas, which can give rise to chronic epilepsy. Similarly, traumatic brain injury can leave patients with chronic seizures, and removal of scar tissue surrounding the lesion has the potential to ameliorate the disease. Another surgical indication is seizures resulting from vascular malformations that can in some instances be surgically corrected.

The most drastic surgery that is routinely performed involves the removal of large sections of a brain hemisphere in children suffering from Rasmussen encephalitis, a rare, debilitating, and medically intractable epilepsy presumably caused by an autoimmune response to brain Glu receptors. As drastic as this intervention sounds, it carries a high chance of controlling seizures while maintaining overall acceptable neurological function.

Overall, surgical successes have improved considerably in recent years, particularly with the introduction of preclinical mapping of eloquent brain regions and careful intraoperative stimulation mapping of the seizure focus. Almost 70% of patients with MTLE and 50% of patients with focal neocortical resections report freedom from seizures after surgery.[22] However, because of the understandable reluctance of

patients to undergo surgery, it is pursued in far fewer patients than are likely to benefit and is often only considered after a patient has already suffered uncontrolled seizures for many years.

5.4 Immunological Approaches, Diet, and Alternative Treatments

It is abundantly clear that seizures can trigger various immune responses in the brain. Although typically assumed to be immune-privileged, the brain does have some semblance of both activated and innate immunity. Specifically, microglia, astrocytes, and endothelial cells can become activated and release

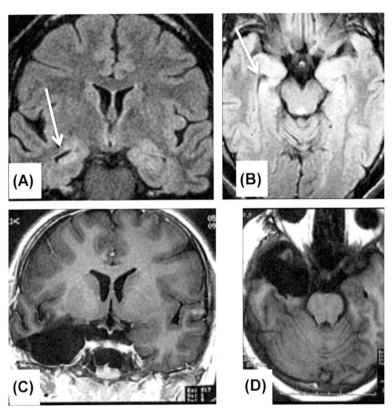

FIGURE 7   Surgical treatment of epilepsy. Coronal magnetic resonance image (MRI; fluid-attenuated inversion recovery) of brain before surgery in coronal (A) and axial view before surgery showing right mesial temporal sclerosis (arrows) with compensatory dilatation of the temporal horn before surgery. (C) and (D) show the same views, respectively, of T_1-weighted MRI scans images after surgical removal of the amygdala, hippocampus and anterior temporal lobe. *From Ref. 21.*

a number of inflammatory molecules including IL-1β, TNF, IL-6, prostaglandin E2, and complement. Some of these cytokines are released as a consequence of fevers. IL-1β, for example, is released into the brain by elevating temperature alone, and it is possible that IL-1β plays a role in febrile, temperature-induced seizures in infants. However, currently there is not sufficient evidence to suggest causality for cytokines in the generation of seizures, let alone epilepsy. Autoantibodies to specific nervous system proteins, for example, GluR3, are present in Rasmussen encephalitis, a catastrophic form of childhood epilepsy that presents with brain atrophy. These brains also contain reactive astrocytes and activated microglial cells. Antibody removal by plasmapheresis or intravenous immunoglobulin as well as reducing inflammation with corticosteroids or corticosteroid-releasing hormones has been explored for the treatment of some of these epilepsies, yet with inconclusive effectiveness. This leaves us to wonder whether the immune response was a consequence of the seizure or whether it may have been the cause. At present these treatments are not generally recommended.[23]

By contrast, an effective alternative treatment pursued by some patients is the ketogenic diet, a diet rich in fat and low in carbohydrates. It is probably the most well documented and longest established. It is based on the serendipitous finding some decades ago that prolonged periods of fasting reduce the frequency of seizures in some epileptic patients. The body typically uses glucose as its main source of energy. Glucose is stored in the form of glycogen in the liver and other organs. The total stored glucose sustains the body for about 24 h. After that, the body begins using fat as an energy substrate, and the liver converts fat to fatty acids or ketone bodies that give this diet its name. Ketone bodies are used as an energy substrate in place of glucose in the brain. The classic ketogenic diet consists of 80% fat, the balance being glucose and protein.

The efficacy of the ketogenic diet has been confirmed in children through four randomized, controlled clinical studies reporting effects comparable to those of standard AEDs. However, long-term compliance with this diet is difficult, particularly in children. Encouraging results have, however, also been reported in children using a modified Atkins diet, which is easier to tolerate by permitting more carbohydrates and proteins. How either of these diets reduces seizures is not known, although animal models suggest that both glucose restriction and ketone bodies per se can affect neuronal hyperexicitability.[24]

5.5 Comorbidities and Psychological Issues to Consider

Given the fact that seizures represent an unusual discharge of electrical activity in relatively large areas of the brain, frequently including brain regions associated with higher cognitive functions (cortex) or memory (hippocampus), it is no surprise that uncontrolled seizures present with a significant decline in cognitive and memory function. Indeed, approximately 75% of newly diagnosed and untreated adults with epilepsy have deficits in attention, executive function, and memory. Significantly, cognitive impairment in patients with newly diagnosed epilepsy surpasses that seen in the early stages of other cerebral diseases such as Parkinson disease or multiple sclerosis.[25] Memory decline can be associated with a loss of hippocampal volume that is visible on MRI.

An important consideration with regard to patients treated with AEDs is the potential direct effect that these drugs may have on cognitive and memory function as well as mood and behavior. Since AEDs interfere with both excitatory and inhibitory neurotransmitter systems, which are fundamentally involved in information processing, it is not a surprise that chronic AED treatments can alter cognitive function.

Of the available drugs, barbiturates, benzodiaz-epines, and topiramate seem to cause the most negative cognitive side effects.

Up to 50% of patients with medically refrac-tory epilepsy and about 10% of patients with well-controlled seizures present with anxiety and/or depression. This is not surprising given that patients report a feeling of helplessness and loss of control as they become increasingly dependent on others and are often unable to live independently, hold a steady job, or even drive. Yet depression is the most underdiagnosed and untreated condition in patients with epilepsy and must be adequately addressed in a compre-hensive treatment plan by, for example, offering selective serotonin reuptake inhibitors in con-junction with AEDs.

Another important consideration regarding the medical control of seizures is how AEDs may affect fetal development if an expecting mother with seizures uses AEDs. Since significant brain development occurs before a prospective mother knows of her pregnancy, care must be taken to avoid potentially harmful exposure of the fetus. This is particularly relevant for valproate, which, although highly effective in a number of seizure disorders, also is a teratogen, that is, a drug that can lead to significant birth defects. Valproate has been reported to significantly lower a child's mean intelligence quotient from 100 to 92 by the age of 3 years, with follow-up showing impaired verbal, nonverbal, and memory function by the age of 6 years. As a consequence, the American Academy of Neurology recommends avoiding valproate in women during pregnancy or women likely to become pregnant.

6. EXPERIMENTAL APPROACHES/ CLINICAL TRIALS

Until we can provide effective relief and achieve seizure control for all patients, we must continue to explore new treatment strat-egies. Following laboratory discoveries and preclinical testing in animal models of disease, new approaches are subject to controlled exper-imentation in human clinical trials. Active clinical trials are published for public access on clinicaltrials.gov, a searchable database that informs patients about their potential eligibil-ity to participate in a clinical trial. The process of clinical experimentation is expanded upon in Chapter 15 of this book. Typically, early stage trials (phase I/II) are exploratory and examine safety, whereas later-stage clinical tri-als (phase III/IV) are more mature and seek to show enhanced efficacy over other avail-able treatment options. There are currently about 200 open clinical epilepsy trials, of which one-half are taking place in the United States. Of the latter, 33 are early (phase 0–II) studies and 32 are phase III or IV. Not all of these are focused on drugs or surgery; some examine exercise, diet, or even stress management. Oth-ers look at comorbidities such as vision effects or decline in cognitive function. By and large, pharmacological studies are examining the efficacy of some second-generation AEDs that have already passed previous clinical trials, including lacosamide, brivaracetam (an ana-logue of levetiracetam), lamotrigine, and reti-gabine for different seizure indications or in different combination therapies. For example, ezogabine is a K^+ channel activator that had significant reduced seizure frequency in a pre-vious phase II study of 399 participants with refractory partial seizures[26] and is now being studied in combination with other established AEDs. Lacosamide was first introduced for clinical use in 2009. It acts on voltage-gated Na^+ channels and enhances slow inactivation of the channel. In a placebo-controlled study, lacosamide significantly reduced the number of partial seizures. It is now being studied in intravenous form as opposed to oral dosing, and another study examines the drug's effect on sleep–wake cycles.

In addition to trials exploring the efficacy of drugs for established epilepsy, there are new trials examining whether prophylactic drug treatment may prevent epilepsy following a known insult, notably trauma. One such study uses biperiden, an atropine-like drug that works on muscarinic acetylcholine receptors and has previously been tested in patients with Parkinson disease. Similarly, levetiracetam (Keppra) is being examined as an epilepsy prophylactic in patients who suffered hemorrhagic stokes.

7. CHALLENGES AND OPPORTUNITIES

One hundred fifty years into the pharmacological management of epilepsy and the statistics are startlingly sobering. Even today, roughly one-third of all patients cannot be effectively treated and are forced to live with a significant disability and a reduced quality of life. Does that mean that we have not made progress? Of course not! The clinical management of epilepsy is vastly superior today than it was even 20 years ago. Extensive research has uncovered many pathways and gene candidates through which epilepsy can manifest. Outstanding animal models of disease now offer the opportunity to test new compounds in clinically relevant model systems. However, as is the case with other nervous system disorders, translating findings in animal models into effective treatment in humans is often a long and tedious process. Without question, we have learned that epilepsies are a large collection of conditions that jointly manifest with similar electrophysiological and behavioral abnormalities, but they certainly do not define a single disease. The underlying root causes may be as varied as the conditions that cause a patient to run a fever. The principal tenet, however, that epilepsy results from an impaired E–I balance, is supported by an abundance of evidence. Given the many pathways through which this balance can be altered, there are tremendous opportunities to attempt to restore this balance through specific drugs.

The greatest challenge remains the identification of new pathways that could yield unexploited drug targets to help those individuals who remain refractory to current pharmacological and surgical treatment. The emergence of second-generation drugs such as ezogabine, which reduce excitability through completely different mechanisms as the first-generation AEDs, are promising steps in this direction. Given the diverse AEDs available, it is possible—even likely—that personalized medicine holds the greatest promise in providing individual patients with a cocktail of drugs tailored to their specific epilepsy. This approach could be aided by further identification of genes associated with disease and the emergence of whole-genome sequencing, which will allow physicians to consider a person's genetic makeup when selecting a drug cocktail best tailored to that patient. It could be argued that seizure disorders may offer an ideal testing ground to explore the implementation of personalized medicine. We have effective tools to diagnose the disease as well as accurately assess a patient's drug response, without the need to wait for years. We also have numerous patient-specific presentations on which to try variants of different treatment protocols. Finally, and probably most important, there exists a significant market and unmet medical need for drug companies to serve. Drug discovery is ultimately driven by market forces, and companies will develop personalized drugs only if significant financial rewards exist; millions of patients worldwide who are suffering from pharmacoresistant epilepsy provide a viable market opportunity.

Acknowledgments

This chapter was kindly reviewed by Christopher B. Ransom, MD, PhD, Director, Epilepsy Center of Excellence and Neurology Service, VA Puget Sound & Department of Neurology, University of Washington.

References

1. Hunt WA, Temkin O. *The Falling Sickness: A history of Epilepsy from the Greeks to the Beginnings of Modern Neurology Baltimore*. MD: Johns Hopkins University Press; 1945, ISBN: 0-8018-1211-9.
2. The history and stigma of epilepsy. *Epilepsia*. 2003; 44:12–14.
3. Jackson J. A study of convulsions. *Arch Neurol*. 1970;22(2):184–188.
4. Friedlander WJ. The rise and fall of bromide therapy in epilepsy. *Arch Neurol*. 2000;57(12):1782–1785.
5. Sirven JI, Waterhouse E. Management of status epilepticus. *Am Fam Physician*. 2003;68(3):469–476.
6. Scharfman HE. The neurobiology of epilepsy. *Curr Neurol Neurosci Rep*. 2007;7(4):348–354.
7. Liu CH, Lin YW, Tang NY, Liu HJ, Hsieh CL. Neuroprotective effect of Uncaria rhynchophylla in kainic acid-induced epileptic seizures by modulating hippocampal mossy fiber sprouting, neuron survival, astrocyte proliferation, and S100B expression. *Evid Based Complementary Altern Med*. 2012;2012:194790.
8. Goldberg EM, Coulter DA. Mechanisms of epileptogenesis: a convergence on neural circuit dysfunction. *Nat Rev Neurosci*. 2013;14(5):337–349.
9. Meisler MH, Kearney JA. Sodium channel mutations in epilepsy and other neurological disorders. *J Clin Invest*. 2005;115(8):2010–2017.
10. Miles R, Blaesse P, Huberfeld G, Wittner L, Kaila K. Chloride homeostasis and GABA signaling in temporal lobe epilepsy. In: Noebels JL, Avoli M, Rogawski MA, Olsen RW, Delgado-Escueta AV, eds. *Jasper's Basic Mechanisms of the Epilepsies*. Bethesda, MD: National Center for Biotechnology Information (US); 2012.
11. Allen AS, Berkovic SF, Cossette P, et al. De novo mutations in epileptic encephalopathies. *Nature*. 2013;501(7466):217–221.
12. Ben-Ari Y, Khalilov I, Kahle KT, Cherubini E. The GABA excitatory/inhibitory shift in brain maturation and neurological disorders. *Neuroscientist*. 2012;18(5):467–486.
13. Ben-Ari Y. Excitatory actions of gaba during development: the nature of the nurture. *Nat Rev Neurosci*. 2002;3(9):728–739.
14. Eftekhari S, Mehvari Habibabadi J, Najafi Ziarani M, et al. Bumetanide reduces seizure frequency in patients with temporal lobe epilepsy. *Epilepsia*. 2013;54(1):e9–12.
15. Suzuki S, Kawakami K, Nakamura F, Nishimura S, Yagi K, Seino M. Bromide, in the therapeutic concentration, enhances GABA-activated currents in cultured neurons of rat cerebral cortex. *Epilepsy Res*. 1994;19(2):89–97.
16. de Groot M, Aronica E, Heimans JJ, Reijneveld JC. Synaptic vesicle protein 2A predicts response to levetiracetam in patients with glioma. *Neurology*. 2011;77(6):532–539.
17. McNamara JO, Huang YZ, Leonard AS. Molecular signaling mechanisms underlying epileptogenesis. *Sci STKE*. 2006;2006(356):re12.
18. Parent JM, Lowenstein DH. Seizure-induced neurogenesis: are more new neurons good for an adult brain? *Prog Brain Res*. 2002;135:121–131.
19. Miller-Delaney SF, Das S, Sano T, et al. Differential DNA methylation patterns define status epilepticus and epileptic tolerance. *J Neurosci*. 2012;32(5):1577–1588.
20. Kobow K, Blumcke I. The emerging role of DNA methylation in epileptogenesis. *Epilepsia*. 2012;53(suppl 9): 11–20.
21. Chowdhury FH, Haque MR, Islam MS, Sarker M, Kawsar K, Sarker A. Microneurosurgical management of temporal lobe epilepsy by amygdalohippocampectomy (AH) plus standard anterior temporal lobectomy (ATL): a report of our initial five cases in Bangladesh. *Asian J Neurosurg*. 2010;5(2):10–18.
22. Spencer SS, Berg AT, Vickrey BG, et al. Predicting long-term seizure outcome after resective epilepsy surgery: the multicenter study. *Neurology*. 2005;65(6):912–918.
23. Vezzani A, French J, Bartfai T, Baram TZ. The role of inflammation in epilepsy. *Nat Rev Neurol*. 2011;7(1):31–40.
24. Danial NN, Hartman AL, Stafstrom CE, Thio LL. How does the ketogenic diet work? Four potential mechanisms. *J Child Neurol*. 2013;28(8):1027–1033.
25. Witt JA, Helmstaedter C. Should cognition be screened in new-onset epilepsies? A study in 247 untreated patients. *J Neurol*. 2012;259(8):1727–1731.
26. Porter RJ, Partiot A, Sachdeo R, Nohria V, Alves WM. Randomized, multicenter, dose-ranging trial of retigabine for partial-onset seizures. *Neurology*. 2007;68(15):1197–1204.

General Readings Used as Source

1. Lowenstein DH. Seizures and Epilepsy. Chapter 369. In: Longo DL, Fauci AS, Kasper DL, Hauser SL, Jameson J, Loscalzo J, eds. *Harrison's Principles of Internal Medicine*, 18e. New York, NY: McGraw-Hill; 2012.
2. Pedley TA, Bazil CW, Morrell MJ. *Meritt's Textbook for Neurology*. 10th ed. Lippincott Williams & Wilkins; 2000. [chapter 140].
3. *Continuum: Lifelong Learning in Neurology – Epilepsy*. June 2013;19(3):551–879 (Comprehensive medical text targeted at physicians and health care providers).
4. Pitkraenen, Schwarzkroin, Moshe. *Models of Seizures and Epilepsy*. Elsevier; 2006 (The definitive text on currently available epilepsy models, primarily of interest to scientists engaging in epilepsy studies).
5. Noebels JL, Avoli Massimo, Rogawski Michael A, Olsen Richard W, Delgado-Escueta Antonio V, eds. *Jasper's Basic Mechanisms of the Epilepsies (Internet)*. 4th ed.

Bethesda, MD: National Center for Biotechnology Information (US); 2012 (very detailed, multi-author book, is based on a series of workshops held for four days at Yosemite National Park in March 2009, freely available online).

6. McNamara JO, et al. Molecular signaling mechanisms underlying epileptogenesis. *Sci STKE*. 2006;2006(356):re12.

7. A public guide to understanding genetic mutations can be found at: www.ghr.nlm.nih.gov. (valuable resource to search for genetic mutations by gene or protein name).

Suggested Papers or Journal Club Assignments

Basic

1. Viitanen T, et al. The K^+–Cl cotransporter KCC_2 promotes GABAergic excitation in the mature rat hippocampus. *J Physiol*. 2010;588(Pt 9):1527–1540.

2. Kearney JA, et al. A gain-of-function mutation in the sodium channel gene Scn2a results in seizures and behavioral abnormalities. *Neuroscience*. 2001;102(2):307–317.

3. Salmi M, et al. Tubacin prevents neuronal migration defects and epileptic activity caused by rat Srpx2 silencing in utero. *Brain*. 2013;136(Pt 8):2457–2473.

4. Hunt RF, Girskis KM, Rubenstein JL, Alvarez-Buylla A, Baraban SC. GABA progenitors grafted into the adult epileptic brain control seizures and abnormal behavior. *Nat Neurosci*. 2013;16(6):692–697.

Clinical

1. Allen AS, et al. De novo mutations in epileptic encephalopathies. *Nature*. 2013;501(7466):217–221.

2. de Groot M, et al. Synaptic vesicle protein 2A predicts response to levetiracetam in patients with glioma. *Neurology*. 2011;77(6):532–539.

3. Spencer SS, et al. Predicting long-term seizure outcome after resective epilepsy surgery: the multicenter study. *Neurology*. 2005;65(6):912–918.

4. Berg AT, Rychlik K, Levy SR, Testa FM. Complete remission of childhood-onset epilepsy: stability and prediction over two decades. *Brain*. 2014;137(12):3213–3222.

PROGRESSIVE NEURODEGENERATIVE DISEASES

Aging, Dementia, and Alzheimer Disease

Harald Sontheimer

Diseases of the Nervous System
http://dx.doi.org/10.1016/B978-0-12-800244-5.00004-5

1. CASE STORY

It is Easter Sunday. The lilies are finally breaking through the ground, which just weeks earlier bore a solid snow cover. I am packing the car to make the drive upstate to visit mother at Saddlebrook Plantation home. I am sad and angry at the same time, having moved her into assisted living just 9 months ago. But caring for her around the clock had become impossible. Should I have tried to keep her at home longer? Should I have hired in-home care? I needed to work to keep up with bills and for my own sanity. I feel fortunate that the elementary school had an opening for me. Teaching distracts me from the daily guilt I feel for abandoning the wonderful woman who raised me.

Mother is sitting in a reclining chair. Alicia, her nurse, is holding her hand while reading her a story. "Hello, Mom," I announce myself. She looks up, bewildered. She has not spoken my name in many months. Slowly, a slight smile develops in her face, suggesting that maybe, just maybe, somewhere deep in her brain a memory of me remains. She slowly rises out of her chair and walks toward me. We hug, I hold her tight. She has become thin and frail, weighing barely 90 pounds. She reaches for the flowers I brought and slowly walks to retrieve a vase.

She takes my hand and leads me to a craft table in her room with a number of small cardboard boxes spread out evenly. Some are colored; others are adorned with ribbons and bows. "Still need color…pretty." She is searching for words. "These are pretty, Mom!" She looks up, her eyes fixated on the flowers. "Flowers?" Yes, Mom, I brought you flowers. Her hand shakes as she points and I am worried she may develop Parkinson's as well. Isn't she burdened enough?

Mother is wearing at least three layers of sweaters although her room feels hot, almost uncomfortably so. "Let's go visit your friends!" She walks slowly, holding on to my arm. We wander past the common area where a group of

residents is playing board games. Others sit in wheelchairs, hunched over. Some are napping. I tell her about her grandchildren, Sam and Susi. She seems to listen without saying a word. I don't believe she remembers them. "Cuddles?" she asks. "Mom, Cuddles died 25 years ago. We have not had a dog since." We walk in silence, taking comfort in each other's company. She tires quickly, and she walks unsteadily. "I brought you a wheelchair," I tell her. "No, no." She was noticeably agitated by the thought of being confined to a wheelchair. She slowly labors along, holding on to my arm. She sits down in her recliner. It is quiet. She closes her eyes and drifts away.

Her birthday is coming up. She does not know it. My next visit is two weeks away. I will bring her a cake. Alicia steps into the room to deliver afternoon tea. As Mom awakes, she startles, clearly surprised about my presence. Her memory of the past hours had already faded. Once again, she doesn't remember me. It pains me. "She has been doing well the past week, but growing frail. Did you bring the chair?" I get the chair from the car and leave it with Alicia outside the room. "Hold off until it is absolutely essential. She hates the thought of being in a wheelchair." I sit with Mom for what seems to be hours, remembering better days. At times, I cannot hold off the tears. She still sleeps. I kiss her on the cheek and steal myself out of the room. Sad, very sad, seeing her confined and robbed of all her past and present and unable to help.

I still have not accepted her disease. I drive home to hug my two children and my husband, afraid that sooner or later I may suffer the same fate.

2. HISTORY

Throughout the ages, the elderly grew forgetful, yet this cognitive decline was called "senile dementia" and considered a normal part of aging rather than a disease. Given that people rarely lived to old age, actual dementia was much less common. When it did occur at an early age, it was presumed to be the result of infections such as syphilis. In 1906 the German psychiatrist Alois Alzheimer described a case of severe dementia in a 51-year-old patient, Auguste D. (Figure 1), who was suffering from severe language deficits, delusions, hallucinations, paranoia, and aggression. She had been admitted to the state asylum in Frankfurt, Germany, where she remained Alzheimer's patient until her death in 1906. Being a trained pathologist, Alzheimer performed an autopsy to examine her brain and found extensive atrophy (shrinkage) of the cortical gray matter not observed in normal individuals at that age. Taking advantage of new staining techniques to visualize microscopic cellular changes, he described abnormal neural fibrillary bundles and plaque-like extracellular deposits, the histopathological hallmarks that have become the defining features of Alzheimer disease (AD). In his original description he wrote that

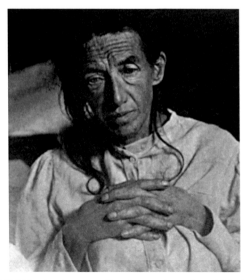

FIGURE 1 Auguste D., the patient that Alzheimer described in 1906 and who served as the name-defining case.

"scattered throughout the entire cortex...one found military foci that were caused by the deposition of a peculiar substance..." that would become known as amyloid in 1984.[1] For many years, this condition was considered an exceptionally rare form of early-onset dementia that became known as Alzheimer disease (AD) through a psychiatry textbook published by Alzheimer's mentor Emil Kraepelin in 1910. Not until 1976 was AD broadly considered to cause senile dementia in the elderly, when Robert Katzmann examined countless autopsies from patients across the age spectrum and found that the pathological changes described by Alzheimer were indeed a common feature in all of them.

The physical nature of plaques and tangles remained a mystery until the 1980s when both amyloid and tau were identified as the principal molecules constituting plaques and tangles, respectively. Since then, countless studies have examined the relative importance of neurofibrillary tangles and amyloid deposits as disease-causing agents in AD, leading to the current view that abnormal amyloid accumulation is the most likely primary cause of the progressive cognitive decline in AD.

This hypothesis is supported by genetic studies of rare familial forms of AD. The initial evidence that AD may have a genetic component came from a study that examined the relatives of 125 individuals with autopsy-confirmed AD. These families had an unusual increase in the number of individuals with dementia, and also a surprisingly greater incidence of Down syndrome. This mysterious connection to Down syndrome, trisomy 21, became clear as the first disease-causing mutation in a Dutch family mapped to chromosome 21. In turn, this led to the identification of the amyloid precursor protein (APP) that gives rise to the disease-causing toxic forms of β-amyloid. Individuals with Down syndrome carry an extra copy of the APP gene, explaining the relatively common occurrence of dementia in Down syndrome.

Since the positional cloning of the APP in 1987, over 50 mutations have been identified in the APP gene and provide the framework for our current genetic understanding of AD.[2] Additional mutations in the presenilin 1 gene were discovered in 1995 and the ε4 allele of the apolipoprotein (APOE) gene was discovered as a major risk factor for AD in 1997. Although AD is only one of many causes of dementia, it vastly outnumbers all other causes and consequently has become synonymous with dementia in the popular press.

3. CLINICAL PRESENTATION/DIAGNOSIS/EPIDEMIOLOGY

Aging is a normal and natural process. Although it carries a negative connotation, this is not warranted, since healthy or successful aging is a positive life experience that is typically associated with positive changes in judgment and personality. For example, the ability to make thoughtful decisions that are not overtly influenced by the immediate emotional state is a skill acquired with age and experience. Similarly, the ability to solve seemingly overwhelming problems and place them into proper context improves with age. In most individuals aging it is accompanied by a measurable but tolerable loss in cognitive ability that does not significantly affect a person's quality of life or ability to function. As the brain ages, opportunities for subtle neuronal changes compound over time and can give rise to pathological changes in mental capacity. Neurologically, older individuals tend to be a little more forgetful. Misplaced car keys and wallets and difficulty remembering names are typical examples that frighten people to consider the possibility that they may suffer from a pathological loss of memory. Typically, none of these are of clinical concern, but all are the source of many jokes. Clinically, this is labeled "benign forgetfulness of the elderly." If memory loss progresses further, and is associated with signs of personality changes, fatigue, restlessness, and irritability, these signify unhealthy brain aging and may be early symptoms of dementia.

Dementia is a serious loss of global cognitive ability in a previously unimpaired individual and is significantly beyond what would be expected from normal aging. Dementia is a syndrome rather than a disease, as it can arise from numerous different causes. Some dementias are static, most notably if the underlying cause was trauma or a stroke; some are reversible, as is the case with drug abuse, where dementias resolve after cessation of drug use. The majority of dementias, however, are progressive and are typically associated with neurodegenerative diseases. The most important forms of neurodegenerative diseases associated with dementia include Creutzfeldt-Jakob disease (CJD), dementia with Lewy bodies (DLB), frontotemporal dementia (FTD), and AD. As elaborated further below, these diseases typically present with abnormal aggregations of specific proteins; for example, $A\beta_{42}$ and tau in AD, α-synuclein in DLB, tau or TDP-43 in FTD, and misfolded prion protein (PrP^{sc}) in CJD. Among the elderly, AD is by far the most common cause of dementia, and we will devote most of this chapter to it. Note, however, that FTD is the largest cause of dementias in people under 65, where AD is uncommon. Vascular disease, stroke, trauma, and brain tumors or infections with human immunodeficiency virus (HIV) or syphilis are additional causes of dementia. More rare causes include vitamin B1 or B12 deficiencies.

3.1 Epidemiology

Worldwide, over 35 million people are affected by dementia, with a new case developing every 4 s. Five percent of the world population over 65 years of age, and 20–40% of those over 85, suffer from dementia.[3] As graphically illustrated in Figure 2, the disease burden is disproportionately high in the industrialized world, in large part due to the significantly longer life expectancy afforded by superior nutrition and health care. AD is by far the leading cause for dementia and affects 5.4 million people in the US currently. These numbers are expected to double by 2030 and quadruple by 2050 with the potential to overwhelm our health care system. Treating and caring for individuals with dementia already consumes over $600 billion worldwide.

Women are almost twice as likely to develop dementia as men, and African Americans have a 2.5-fold increased risk compared to Caucasians, possibly due to increased rates of diabetes, which alone increases the risk for AD by 3-fold. The Framingham study followed nearly 2800 people who at age 65 were healthy and free of dementia over a 29-year period to document the development of dementia. It concluded that on average, a 65-year-old woman has a 21.7% chance of developing dementia and a 17.2% chance of developing AD in her lifetime. The numbers for 65-year-old men were 14.3% lifetime risk for dementia and 9.1% chance for AD.[4]

Numerous environmental factors have been proposed to cause AD, but none have been confirmed as actual risk factors to date. By far the greatest known risk factor for all dementias, including AD, is age, with an exponential increase in risk as a function of age. Some scientists have speculated that dementia is an inevitable consequence of aging. However, many centenarians (individuals over 100 years of age) have normal memory function. A family history of dementia increases the risk to develop AD, with first-degree relatives having a 2-fold increased risk to develop AD. Expression of the apolipoprotein E4 (APOE-ε4) is significantly correlated with late-onset AD, increasing AD risk up to 15-fold.

In an effort to stem this epidemic, many studies have been conducted to explore whether dietary supplements, nutrition, or lifestyle may reduce the likelihood to develop dementia. Thus far, the data are incomplete or inconclusive.

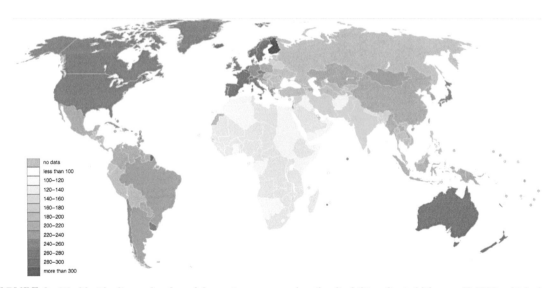

FIGURE 2   Worldwide disease burden of dementias, expressed as the disability-adjusted life year (DALY), which shows the number of years lost due to ill health, disability, or early death per 100,000 inhabitants in 2004. *Wiki Commons, based on 2004 World Health Organization Data.*

Some credible studies suggest that the Mediterranean diet may reduce the risk of AD,[5] and individuals with higher levels of education are less likely to develop AD. This may be due to a greater "cognitive reserve" capacity, although one must consider that education often correlates with income, nutrition, and access to health care, all of which may directly or indirectly affect the likelihood to develop disease. Interestingly, recent imaging studies found that cognitive engagement in midlife slows the build-up of toxic β-amyloid deposits,[6] the pathological hallmark for AD, and suggest that brain activity may indeed modify disease etiology.

3.2 Patient Presentation and Diagnosis

The salient deficit that characterizes all forms of dementia is impaired memory and cognitive function. Unfortunately, we currently have no biomarkers or noninvasive tests to unequivocally diagnose an individual with dementia. Hence the initial diagnosis of suspected dementia relies on cognitive tests, family history, and a comprehensive neurological examination.

Whenever memory deficits are suspected, memory tests should be conducted by a neurologist or neuropsychologist. The easiest way to do so in the clinic or at the bedside involves the Mini-Mental Status Examination (Table 1), which is a 30-point test of both working and episodic memory. Tasks include knowledge of the patient's current whereabouts, naming objects, and recalling a series of objects and words. An easy way to test working memory is to ask a patient to repeat a series of numbers either forward or backward. Most normal adults can repeat six digits forward and five digits backward. Early signs of forgetfulness must be calibrated to age, recognizing that an 85-year-old remembers only half as many words in a list as an 18-year-old.

The smallest measurable deviation in memory function derived from memory tests is called *mild cognitive impairment* (MCI), which is a measurable deficit in memory, but one that does not interfere with daily living. Individuals who fall in this category typically score at least a 23 or higher in the Mini-Mental exam. MCI may be a precursor to AD and may progress to disease, particularly in patients who bear an increased risk due to family history of AD or carrying the APOE-ε4 allele, discussed in greater detail later.

3.3 Alzheimer Disease

Alzheimer disease[7] is often used synonymously with dementia because it is the single most prominent cause of dementia, accounting for ~80% of all dementias worldwide. Ten percent of all individuals over 70 years old have mild memory impairments, but half of these will progress to AD within 4 years. Three-quarters of all AD patients initially present with memory problems, including misplacing objects, being unable to manage money, getting lost, and not following instructions. If these problems progress slowly, the patient is likely to have AD. Motor functions are typically spared until much later in the disease. Mild depression and social withdrawal are common. Unfortunately, most clinical tests, except for amyloid imaging, which is used experimentally, can only rule out other conditions, such as vitamin deficiencies, but cannot unequivocally diagnose AD. As we will discuss later, this may soon change through recently developed noninvasive ways to measure amyloid deposits in the brain. However, currently the pathological hallmarks of AD can only be assessed on autopsy after a patient's death or on imaging for amyloid (discussed later).

AD presents with a similar pattern in most patients and is often divided into stages. During the early stage, lasting 2–4 years, patients have frequent memory losses affecting recent memories but preserving old memories. For example, patients may forget entire conversations, cannot follow directions, and routinely forget where they have placed things, yet still remember the names of family members, friends, and relatives. Although conversational language is initially seemingly intact, language is less fluid and language comprehension declines gradually. A person with

TABLE 1   Memory Assessment using the Mini-Mental Status Exam

	Points
ORIENTATION	
Name: Season/date/day/month/year	5 (1 for each name)
Name: Hospital/floor/town/state/country	5 (1 for each name)
REGISTRATION	
Identify three objects by name and ask patient to repeat	3 (1 for each object)
ATTENTION AND CALCULATION	
Subtract serial 7s from 100 (93-86-79-72-65)	5 (1 for each subtraction)
RECALL	
Recall the three objects presented earlier	3 (1 for each object)
LANGUAGE	
Name pencil and watch	2 (1 for each object)
Repeat: "No ifs, ands, or buts"	1
Follow three-step command (take paper/fold/place on table)	3 (1 for each command)
Write "close your eyes" and have patient follow written command)	1
Ask patient to write a sentence	1
Ask patient to copy a design (pentagon, triangle…)	1
Total	30

The Mini-Mental Status Exam allows for a rapid assessment of a persons memory mental and memory function. A total of 30 points are available for correctly answering 30 questions in 5 categories. 27 or more points are considered normal; 19–24 points are indicative of mild cognitive impairment; 10–18 moderate and fewer than 10 points suggest severe cognitive impairment.

AD may place objects in odd places, for example, leaving his or her keys in the refrigerator. Mood swings occur frequently and patients may be apathetic, with signs of depression. They may need to be motivated and told what to do on a daily basis.

In the second stage of the disease, lasting between 2 and 10 years, patients are no longer able to hide their handicaps and problems. Memory loss becomes much more pervasive and begins to involve older memories as well. Patients may no longer remember the names of their friends and family. Patients may claim that a caregiver is an imposter (Capgras syndrome). Their speech is often nonsensical. They become easily disoriented and get lost in formerly familiar places, and may no longer find their way home if left alone. Patients withdraw and isolate themselves in social situations. They forget social behavioral norms, and a loss of inhibition can lead to awkward behaviors. Memory problems severely interfere with daily activities, such as driving, shopping, or balancing accounts. Patients show difficulties dressing, eating, and even walking. Patients look and walk "Parkinsonian," and are stiff and rigid.

At the final stage of disease, lasting from 1 to 3 years, patients are confused and generally unable to carry out any conversation. They have major mobility difficulties. Problems swallowing increase the risk of aspiration (lodging food or drink in the airways) and are also the most common cause of death in AD. During this stage of the disease, individuals require constant supervision, often in a skilled nursing facility. Overall disease duration ranges from 1 to 25 years, but the typical life expectancy from disease onset ranges between 8 and 10 years.

The constant care required places a tremendous burden on family and caregivers, and may isolate spouses or family members, who themselves are at risk to develop apathy and depression due to the stress of being on call around the clock.

3.4 Frontotemporal Dementia

FTD is the most common dementia diagnosis in individuals less than 65 years of age, and disease onset can be as early as age 45. It currently affects approximately 50,000 individuals in the US. Although FTD is about equally as common as AD in the population under 60, AD is far more common (100-fold) across all age groups. The deterioration of the frontal lobe (Figure 3) causes changes in social and personal behavior. Apathy is a common signature, leading to social withdrawal, with patients frequently staying in bed all day. A lack of inhibition leads to neglect of social norms and to inappropriate behaviors that may include stealing, speeding, and

disinhibited sexual drives. Compulsive behaviors develop along with binge eating. As disease progresses, patients lose fluency in their speech, primarily because they are unable to articulate language, yet they are without loss of language comprehension. The loss of executive function compromises daily planning and organization. Structural magnetic resonance imaging (MRI) often reveals the selective atrophy of the frontal lobe, easily distinguishing the disease from AD. The same differentiation can also be seen on fluorodeoxyglucose positron emission tomography (FDG-PET) (Figure 14), where glucose utilization is selectively reduced in the frontal lobe in FTD, whereas in AD the decrease in glucose utilization is more posterior. The most defining clinical features of FTD include disinhibition, apathy, loss of empathy, compulsive behavior, hyperorality, and loss of executive function. A number of genetic mutations have been identified and include the gene encoding for the microtubule-associated tau protein. There is no known cure and treatment aims at management of behavioral symptoms. Patients typically die within 2–10 years after diagnosis, with continuous supervision required in end stages of disease. Given its early onset, often in the prime

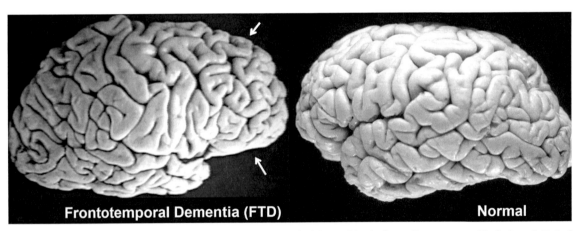

Frontotemporal Dementia (FTD)   **Normal**

FIGURE 3   Images courtesy Dr Peter Anderson, University of Alabama Birmingham, Department of Pathology & Pathology Education Informational Resource (PEIR) Digital Library (http://peir.path.uab.edu).

productive years of a person's life, the socioeconomic effects are particularly devastating.

3.5 Other Causes of Dementia

Numerous other conditions can cause dementia, including vascular disease, tumors, brain trauma, and viral and bacterial infection, as well as drug or alcohol abuse. Many of these conditions produce a static disease that may be reversible if the initiating insult is removed. For example, drug-induced dementia may disappear upon drug withdrawal. Acquired immunodeficiency syndrome (AIDS)-associated dementia has gained increasing importance, since patients infected with the HIV virus now live an almost normal life span, yet frequently develop dementia as a consequence of viral damage to the brain. This condition , discussed more extensively in Chapter 10, is typically associated with low CD4 T cell levels and widespread brain inflammation. In the developed world, the widespread use of highly effective antiviral therapy has made AIDS-associated dementia less common,

yet this form of dementia still poses a significant challenge in the developing world.

4. DISEASE MECHANISM/CAUSE/ BASIC SCIENCE

From a functional point of view, the changes that most dramatically distinguish unhealthy aging and disease from normal healthy aging are altered memory function, personality changes, and compromised executive function. Central to these impairments are the loss of short- and long-term memory and an inability to form new memories. Structural changes, often only accessible in autopsy tissue, include vascular changes with reduced cortical blood flow and breaches in the blood–brain barrier (BBB), extracellular amyloid deposits, intracellular accumulation of neurofibrillary tangles, reduced neuronal numbers, reactive gliosis, and microgliosis. We will extensively discuss each of these below. Since loss of memory is the functional hallmark of all dementias, it is

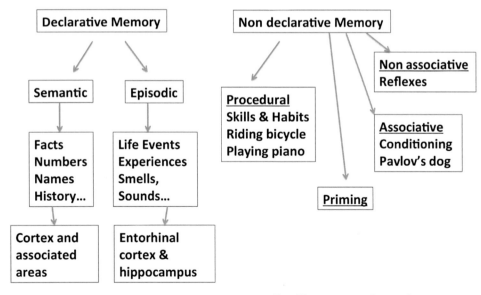

FIGURE 4 Forms of memories. Declarative memories involve the hippocampus and are often termed the "where" memories. Nondeclarative or procedural memories are the "how" memories, and do not involve the hippocampus.

useful to briefly review the various forms of memory and their anatomical and physiological substrates, as this helps to explain the pattern by which lesions in the brain selectively rob certain memories while sparing others.

4.1 Memory and Distribution in the Brain

Based on the duration with which memories are being retained, we typically distinguish short-term and long-term memory, also called proximal and distal memories. As the name implies, short-term memory is fleeting, lasting only seconds to minutes, and is highly sensitive to distractions, whereas long-term memories, by contrast, can last up to a lifetime. *Working memory* has been suggested as a more suitable substitute for short-term memory, since it captures the concept that those memories require attention and are subject to alteration. A sensitive test for working memory requires a subject to spell a word backward or to repeat a series of numbers in reverse order. These tasks are not a matter of simple recall; rather, they require significant processing. Imaging studies and evaluation of patients suffering from brain lesions each suggest that the prefrontal cortex plays an important role in working memory. Working memory declines with age and is one of the early deficits in AD.

Our long-term memory is often divided into *declarative memory*, which is the ability to remember facts, events, and names, and *nondeclarative memory*, which are skills that we perform without awareness, for example, riding a bicycle (Figure 4). As we will soon see, these reside in separate areas of the brain and, accordingly, are differentially affected in AD.

Declarative memories that capture personal experiences, such as a fishing trip with your dad, a recent birthday, or your high school graduation, are examples of *episodic memory*. Memories that encompass general knowledge, such as the use of language, the number system, the purpose and use of tools, classes of animals and plants, and the like, are called *semantic memories*. Semantic memory is

typically retained for a very long time, as are most episodic memories, although some are kept only for a few minutes or hours and then discarded.

Both semantic and episodic memories are required in tandem to generate narrative, relating "what" happened to "whom," "when," and "where." Declarative memories are believed to be stored in the cortical regions where the experiences associated with these memories, such as sounds, smells, and faces, were perceived and processed at the time. Semantic memories are associated with the brain region responsible for the particular semantic task. For example, the concept of language and numbers is stored in the parietal cortex associated with speech and language, while the use and concept of tools resides in the motor cortex. Recognition of colors, faces, objects, animals, and the like are stored in the somatosensory cortex. Spatial memories reside in the posterior hippocampus, making the hippocampus essential for spatial navigation.

Taken together, various individual facts underlying a story, including its content, faces, locations, smells, and sounds, are all stored in a distributed cortical and hippocampal network. In order to recall an experience, this information needs to be brought together in an organized narrative. This occurs through the cortical–medial temporal lobe (MTL) network illustrated in Figure 5, which encompasses the hippocampus, entorhinal cortex, and the perirhinal cortex with its bidirectional projections to the cortex.

The cortical memory traces are retrieved via the entorhinal cortex and modulated within the circular projections of the entorhinal cortex to the dentate gyrus, to the CA1 and CA3 neurons of the hippocampus, and back to the entorhinal cortex. With every passage of information through the cortex–MTL network, memories are strengthened or weakened, presumably within the circuitry of the hippocampus. The more often such a memory is recalled, the stronger and more long-lasting it becomes.

It is important to point out that the cortex–MTL pathway is also modulated by input from the amygdala, providing an emotional input to our

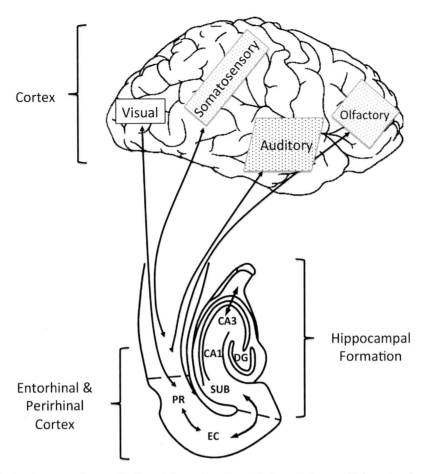

FIGURE 5 Declarative memories are distributed throughout the cortical–mediotemporal lobe network and require information flow through the hippocampus and the surrounding entorhinal and perirhinal cortices.

memories. Not surprisingly, we tend to form longer-lasting memories more quickly if they occur in a state of either intense fear or intense pleasure. Interestingly, this memory pathway is involved not only in recalling past memories but also in envisioning future events. Functional MRI studies show activity in the hippocampus when individuals are asked to project into the future, and persons with bilateral hippocampal lesions are unable to visualize future events.

Loss of the hippocampus and its associated entorhinal and perirhinal cortices causes anterograde amnesia, or the inability to form new memories, yet it preserves very old memories (>20 years). Damage to any part of the cortical–MTL circuit

presents with memory loss, and AD affects essentially all aspects of the cortical–MTL memory pathway at some point in the disease. Indeed, the sequential development of lesions beginning in the hippocampus and surrounding entorhinal cortex, gradually spreading into the cortical gray matter, illustrated in Figure 6, explains the sequential loss of short-term then long-term episodic memory.

The importance of the hippocampus for the acquisition of new memories is most effectively demonstrated by the famous story of H.M., who had both hippocampi surgically removed to treat his intractable epilepsy. This left him unable to remember anything at all for longer than a few minutes, yet he was able to recall old memories

Preclinical AD   Mild to Moderate AD

Severe AD

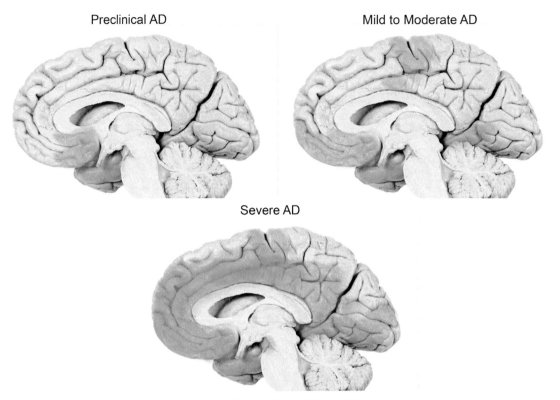

FIGURE 6   Schematic depiction of the spread of AD throughout the brain. During preclinical phases the primary affected areas are the hippocampus and entorhinal cortex. This spreads to include the frontal lobe and ultimately the entire cortex and even cerebellum. *Adapted from a picture from National Institute for Aging (NIA).*

acquired prior to his surgery, and maintained intact procedural memories. The early loss of the hippocampus in AD explains why patients have difficulty acquiring new memories early in the disease process, while sparing recall of old memories.

Nondeclarative memory, or the ability to perform tasks in a nonconscious way, such as using silverware to eat, riding a bicycle, or playing an instrument, is anatomically and functionally entirely different from other forms of memory.

This form of memory can also be called procedural memory and is typically preserved even if declarative memory is lost. In other words, you may not remember practicing or taking piano lessons, but can still be able to play the instrument. Examples of procedural memories are playing sports and riding a bicycle, which typically take a long time to acquire but stay with you for life.

Grandma still remembers how to ride a bicycle even if she has not used one in ages and may no longer remember where she stored it. It is the "knowing how" of nondeclarative memories that separates them from the "knowing that," which characterizes declarative memories. Nondeclarative memories are not hippocampal-dependent, and instead involve the basal ganglia, amygdala, cerebellum, and sensory and motor cortices, and are locally stored in those areas that are normally part of our motor system. The hand and finger representations of musicians occupy larger areas on the somatosensory cortex than in nonmusicians. Motor procedural memory tasks are stored in the basal ganglia and cerebellum and are therefore, not unexpectedly, compromised in Parkinson disease but are largely intact in AD, which spares these structures.

BOX: NEUROCHEMICAL BASIS OF MEMORY (LTP AND LTD)

Our cellular and molecular understanding of memory is very incomplete and is largely based on the study of model systems. Vertebrate memory studies have largely focused on the hippocampus, where a commonly used conditional learning paradigm involves stimulating a group of neurons, followed by a second train of stimulation shortly thereafter. The second train causes the postsynaptic neurons to show a much larger response that persists for minutes to hours. This learning of coincident activity is called long-term potentiation (LTP) and involves both the NMDA-type Glu receptor, which is designed to act as a coincidence detector, and the AMPA receptor, which mediates fast synaptic transmission. A single action potential arriving at a presynaptic terminal would give rise to only a short postsynaptic response mediated

by the AMPA-R, as the NMDA-R is blocked by intracellular Mg^{2+} ions. If the cell receives a strong input, either through repeated activation from the same synapse or from some of its other dendrites, the cell depolarizes just enough to allow a Mg^{2+} ion to dislodge from the pore of the NMDA-R, unblocking it. Any subsequent stimulus arriving within about 50 ms will permit the influx of Ca^{2+} into the postsynaptic cell through the now unblocked NMDA-R. This increase in Ca^{2+} triggers long-term changes in the postsynaptic cell involving activation of CaMK kinase. This leads to increases in the number of AMPA-Rs inserted into the membrane, making the response to subsequent stimuli larger. Central to the idea that this process is a cellular form of memory is the fact that the synchronous activity of two neurons that

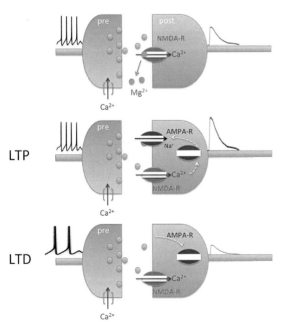

BOX FIGURE 1 Neurochemical basis of memory (LTP and LTD). High-frequency stimulation dislodges Mg^{2+} from NMDA receptors, allowing enhanced influx of Ca^{2+}, which leads to the insertion of AMPA receptors into the postsynaptic membrane, causing an increase in postsynaptic current (LTP). Low-frequency stimulation, on the other hand, leads to much lower postsynaptic Ca^{2+} and removal of postsynaptic AMPA receptors, leading to a reduced postsynaptic current (LTD).

BOX   *(cont'd)*

---

fire together enhances the response between them and, over time, strengthens their connectivity, while cells that are asynchronous in their activity show no enhancement.

Just as LTP enhances the connection between cells, and therefore "learns" a connection, an opposite cellular equivalent of forgetting exists as well. This is called long-term depression or LTD. Here, long-term stimulation of one cell at a low frequency, about once per second over 10–15 min, reduces the postsynaptic activity through a reduction in the number of functional AMPA-Rs. Whether LTP or LTD occurs depends on the size of the Ca^{2+} change in the postsynaptic terminal. Small, long-lasting changes in Ca^{2+} cause LTD; brief, large changes cause LTP.

LTP and LTD are still short-term associative learning phenomena that result in relatively short-lasting changes in the physiology and are the result of phosphorylation of AMPA-R that changes their trafficking into and out of the membrane. These can, however, lead to long-lasting memory traces that require changes in the biosynthesis of these receptors, and probably changes in the neural network. For purposes of AD, it is important to remember that both NMDA- and AMPA-type Glu receptors, particularly those in the hippocampus and entorhinal cortex, are essential for learning and memory, and that NMDA-Rs are the target of memantine, a use-dependent inhibitor of NMDA and one of the few drugs approved to treat AD.

Memory at the cellular level: To understand the pathological changes that occur to memory function in aging, dementia, and AD, it is also important to briefly review our knowledge of learning at the cellular level. Two important synaptic changes that are called long-term potentiation (LTP) and long-term depression (LTD) are now widely accepted as representing the cellular and neurochemical basis for learning and forgetting, respectively. These are briefly reviewed in Box.

4.2 Structural Changes Underlying Alzheimer Disease and Other Forms of Dementia

Anatomical Changes

In AD, pathological changes begin in the transentorhinal region, spreading to the hippocampus, and then move to the lateral and posterior temporal and parietal neocortex (Figure 6). Eventually AD presents with widespread and more diffuse degeneration and a global thinning of the cortical gray matter, readily visible across the entire brain on autopsy (Figure 7). The time course of anatomic lesions explains the gradual appearance of symptoms, from early memory deficits to aphasia (loss of language ability) and later navigational issues.

This pattern of functional loss distinguishes AD from other dementias, most notably FTD (Figure 3), where selective lesions in the frontal lobe cause problems in executive function, such as loss of judgment, as initial symptoms.

Upon autopsy, two characteristic abnormalities can be observed that have become the hallmarks for AD and other dementias. These are neurofibrillary tangles and amyloid plaques (Figure 8). Their gradual appearance in the affected brain regions has been suggested to be

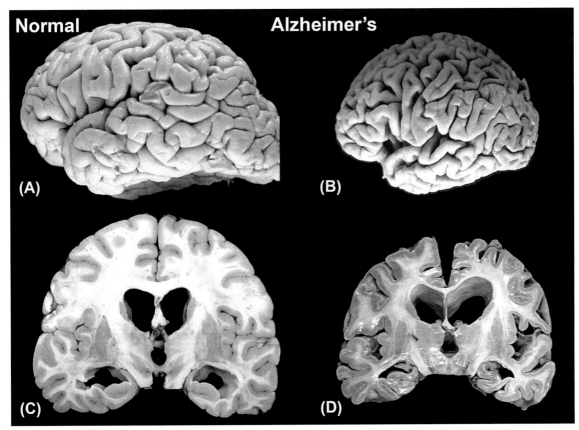

FIGURE 7 Autopsy shows extensive brain atrophy in a patient who died from AD (right) compared to a similar-aged individual who died from natural causes (left). Images in A nd B show overall brain atrophy, while C and D show marked thinning of the cortical gray matter and enlargement of the ventricles in the AD patient. Images courtesy of Dr Peter Anderson, University of Alabama Birmingham, Department of Pathology & Pathology Education Informational Resource (PEIR) Digital Library (http://peir.path.uab.edu).

either the cause of or consequence of neuronal death.

NEUROFIBRILLARY TANGLES

Neurofibrillary tangles (NFTs) are always present in AD autopsy specimens. They are entirely made up of the microtubule-associated protein tau, which, when hyperphosphorylated, forms insoluble aggregates that can fill the entire intracellular space of a neuron.[8] Six isoforms of tau derive from a single gene on chromosome 17. The longest tau isoform contains 441 amino acids and 79 possible phosphorylation sites. In normal neurons, only three residues are phosphorylated, as this promotes tubulin assembly into microtubules. Through its interaction with microtubules, tau serves both a structural and dynamic role. Microtubules run the length of the axon, which can be up to 1 m long, and are an important cytoskeletal element. In addition, microtubules serve as tracks along which cargo is moved from the cell body to the axon terminal. Such cargo includes proteins, lipids, synaptic vesicles, and even mitochondria (Figure 9). The motor proteins that move cargo along microtubules include kinesin and dynein. Some cargo

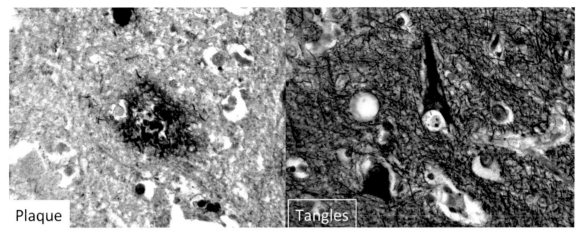

FIGURE 8   Neuropathological hallmarks of Alzheimer disease, neuritic plaques and neurofibrillary tangles visualized with modified Bielschowsky stain. These are from the autopsy of an 83-year-old woman with AD. *Courtesy of Dr Steven L. Carroll, Medical University of South Carolina.*

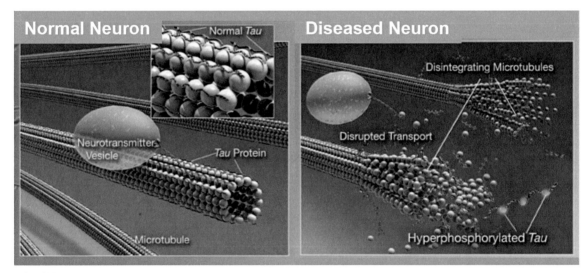

FIGURE 9   Schematized role of tau protein in healthy (left) and diseased (right) brain. Tau stabilizes the microtubules along which cargo such as neurotransmitter vesicles are moved. The hyperphosphorylation of tau causes a loss of microtubule stability, disrupting transport of cargo vesicles. *From National Institute for Aging (NIA).*

molecules move fast, 50–400 mm/day; others move slowly, 8 mm/day. Hyperphosphorylation of tau prevents its binding to microtubules, which in turn become unstable and disintegrate, hence compromising both the axonal cytoskeleton and the tracks needed for axonal transport (Figure 9), which both may hasten neuronal cell death.

In addition to the disruption of microtubules, hyperphosphorylated tau self-aggregates into insoluble inclusions, the NFTs, which no longer bind tubulin. Tau pathology is an early feature of AD, and appearance of tangles correlates with neuronal loss; however, in animal models of AD, considerable neuronal death can precede NFTs,

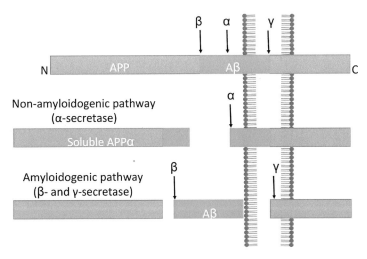

FIGURE 10 Sequential cleavage of Amyloid Precursor Protein (APP) by β- and γ-secretase produce oligomeric Ab40 and 42 fragments. N and C designate the N- and C-terminal of the APP protein. Note that γ-secretase cleaves within the membrane, whereas β-secretase cleaves on the extracellular plasma membrane site. Cleavage by α-secretase produces a soluble longer form of amyloid that is not toxic.

suggesting that NFTs may be a consequence of disease rather than a cause. The neurotoxicity of NFTs has also been called into question. Once assembled into NFTs, tau becomes resistant to proteolysis by calcium-activated proteases, and it has therefore been suggested that tau aggregation serves to protect neuronal tau in stressed neurons. Indeed, even in the presence of NFTs, neurons can survive for years. Although a hallmark for the diagnosis of AD, tau aggregates are also found in FTD and several "Parkinsonian" diseases, such as progressive supranuclear palsy. These diseases are often called "tauopathies."

AMYLOID

Plaque deposits can be found in the brains of patients with a variety of neurodegenerative diseases. These plaques are generally insoluble aggregates containing a number of misfolded proteins that occur naturally in the body. Being misfolded, these proteins stick to each other, forming hydrophobic fibrils. The specific plaques found in the brains of patients with AD contain an overabundance of beta amyloid (Aβ), a 36–43 amino acid peptide. It is generated from the APP through a series of cleavages by enzymes called secretases. The APP is a large transmembrane glycoprotein of unknown function expressed in neurons and non-neuronal cells. As illustrated in schematic form in Figure 10, the bulk of APP is cleaved by the enzyme α-secretase, yielding a soluble APPα and, after further cleavage by γ-secretase, the intracellular domain produces a signaling domain that acts as transcriptional regulator. An alternative cleavage pathway involves the sequential cleavage by β- and γ-secretase, and produces soluble amyloids ranging from 36 to 43 amino acids in length, with the most common variants being Aβ40 and Aβ42. Plaques contain both Aβ40 and Aβ42, with Aβ40 being the more abundant of the two. However, Aβ42 is more hydrophobic and therefore the most fibrillogenic, and is considered the more toxic of the two. In neurons, unlike in non-neuronal cells, the production of Aβ40 and Aβ42 can also occur intracellularly,[9] and intracellular Aβ can be released from neurons at synapses in an activity-dependent manner.

Autosomal dominant mutations in the APP gene cause an increase in the relative production of the plaque-forming Aβ42 and characterize rare familiar forms of AD. Moreover, it is now widely accepted that abnormal expression or processing of APP causes an imbalance between Aβ production and clearance, and that this causes the gradual build-up of Aβ, leading

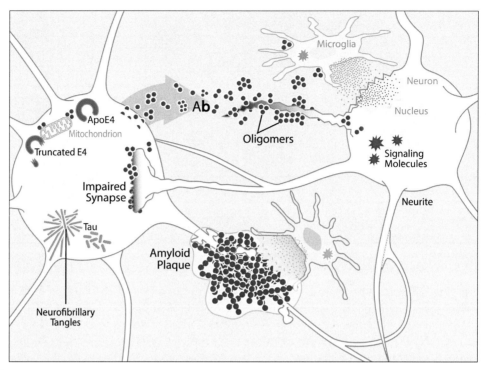

FIGURE 11   The dynamic relationship between plaques and oligomers with differential effects on brain physiology and pathophysiology. Oligomers may affect signaling functions of neurons, while plaques either act as storage sites or directly attract and activate microglia, eliciting an inflammatory response. *From Ref. 11.*

to the formation of plaques in both sporadic and familiar forms of AD. While the disease association of plaques and AD is generally accepted, it is less clear whether the plaques in and of themselves are neurotoxic and sufficient to explain the gradual cognitive decline in AD.[10]

Plaques are typically surrounded by soluble Aβ oligomers, which are in equilibrium, with the plaques acting as storage sites for Aβ oligomers (Figure 11). These oligomers may play a major pathophysiological role in disease. Simply applying soluble Aβ oligomers on hippocampal slices impairs synaptic function, and specifically blocks LTP while enhancing LTD, the cellular substrates of learning and unlearning, respectively (see Box Figure 1). Moreover, neuronal activity increases the production and release of soluble Aβ, which then acts as a feedback regulator on synaptic transmission by reducing neurotransmitter release. Pathologically elevated levels of Aβ would be expected to put this feedback loop into overdrive, depressing excitatory transmission. Some studies even suggest that the relative concentration of soluble Aβ42 at the synapse determines whether synapses are depressed or potentiated. These findings provide a potential normal physiological role for Aβ, acting as a feedback regulator at synapses, while also explaining the abnormal pathological role when too much Aβ causes profound changes in synaptic activity and buildup of plaques. The idea that soluble Aβ may cause aberrant synaptic activity would explain how cognitive changes and memory loss could present long before significant plaque burden is present. Precisely how Aβ regulates synaptic transmission in normal brain and in AD remains to be clarified, and the cellular receptor(s) for Aβ have yet to be identified.

Secretases

As stated above, the default pathway for APP cleavage is via α-secretase, which releases a soluble APPα protein and, after further cleavage by γ-secretase, an intracellular transcriptional regulator P3. β-secretases attack the APP protein at a different site, leaving a longer segment of APP membrane bound, which, upon cleavage by γ-secretase, forms the various toxic forms of Aβ. Our understanding of secretases is still emerging. This group of molecules is functionally defined by their proteolytic activity whereby they liberate a substrate from the membrane in a secreted form. γ-secretase is not a single protein, but instead it is a complex of four proteins that include nicastrin, APH-1, PEN-2, and presenilin. The proteolytic activity is conferred by presenilin, yet the other three proteins are essential regulators of APP cleavage. Presenilins are 8-transmembrane domain proteins that derive from two genes (PSEN1 and 2). Knockout of PSEN1 kills mice at an embryonic stage. This is because γ-secretases have other important roles, particularly during development. Most importantly, a transmembrane receptor called Notch is expressed on the surface of many cells, including neurons. Notch mediates cell–cell signaling required for cell differentiation and overall establishment of polarity in a multicellular organism. Adjacent cells each express both the Notch receptor and a membrane anchored ligand, which in mammals is either Jagged or Delta-like. Upon binding to Notch, the intracellular domain of Notch is cleaved by γ-secretase and then trafficked to the cell nucleus to regulate gene transcription. As illustrated in Figure 12, γ-secretases serve a dual role, enabling Notch signaling for orderly development while also producing Aβ.[13] While this dual role may have evolved by design, it is possible that only the Notch receptor was the intended substrate and that APP cleavage is an erroneous process that has the potential to be pathogenic.

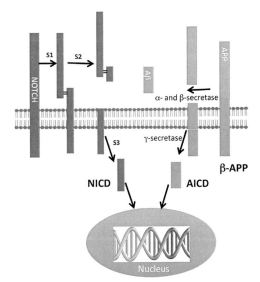

FIGURE 12 Dual role of γ-secretase cleaving both Notch and APP. The same enzyme that produces toxic forms of amyloid serves to produce Notch, an important protein that acts as a nuclear gene regulator required for normal development and cell polarity. Notch signaling depends on three endoproteolytic cleavages (S1–S3). Notch matures in the Golgi by furin-mediated cleavage at site 1 (S1). At the cell surface, Notch is cleaved at S2 (after binding to its ligands Delta/Serrate/Lag-2). Finally, cleavage at S3 liberates the notch intracellular domain (NICD), which translocates to the nucleus, thereby regulating the transcription of target genes by binding to transcription factors. β-APP is processed by a similar pathway. Initial cleavages of β-APP by α- or β-secretase lead to the generation of a membrane-bound complex. γ-secretase cleavage then liberates Aβ and the APP intracellular domain (AICD). The biological function of AICD remains to be determined. *Figure generated from Ref. 12.*

Notch signaling remains important in the adult nervous system, where it has been implicated in learning and memory, and the importance for γ-secretase in normal Notch signaling provides a challenge to clinical trials that target Aβ production, further discussed later.

Clearance of Deposits

In addition to overproduction of Aβ and change in the ratio of Aβ40/42, impaired Aβ clearance has emerged as a potential contributor to disease. Several clearance mechanisms for Aβ have been identified. For example, the

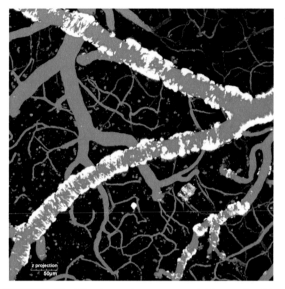

FIGURE 13 Blood vessel association of Aβ visualized in a hAPP mutant mouse in vivo using a fluorescent dye (benzothiazole) that binds to amyloid plaques (white). *Provided by Ian Kimbrough, Department of Neurobiology, University of Alabama Birmingham.*

low-density lipoprotein receptor-related protein (LRP) can form a complex with Aβ on the abluminal side of the blood vessel endothelial cells. These complexes are then either internalized and degraded by the lysosome or released by transcytosis across the BBB into the bloodstream. Clusters of Aβ can be visualized in association with blood vessels in live mice that harbor mutations in the hAPP gene (Figure 13). Other pathways for degradation involve proteolysis via plasminogen or insulin degrading enzyme.

Cell-to-Cell Transmission of Disease

The gradual spread of plaques from the mesiotemporal lobe and entorhinal cortex to surrounding cortical regions, ultimately involving essentially all gray matter, has been a puzzle. Recent experiments now suggest that Aβ toxicity and plaque formation may spread from cell to cell in a prion-like fashion. Prions are misfolded, self-propagating proteins that have been identified in Scapie and Creutzfeldt-Jacob disease. A prion

entering a healthy cell acts as a template to cause normally folded proteins to misfold in the same way as the template. If Aβ behaved as a prion, it could indeed self-propagate in the brain, spreading from one region to the next. In transgenic mice that overexpress a mutant form of APP, the intracranial injection of Aβ from brain homogenates into only one brain hemisphere indeed caused widespread Aβ deposits in both hemispheres, consistent with a prion-like mode of propagation.[14]

4.3 Vascular Changes in Alzheimer Disease

Next to age and family history, any health condition with vascular disease presents a significant risk factor for AD. This includes diabetes, hypertension, obesity, and stroke. It has been well documented that AD presents with vascular pathology, particularly affecting small vessels in the cortex, causing general reduction in cerebral blood flow or hypoperfusion.[15] This alone could negatively affect not only the delivery of glucose and oxygen as energy substrates but also the clearance of Aβ from the cerebrospinal fluid. Lipoprotein receptor-related protein 1 (LRP1) is normally expressed on the abluminal side of cerebral blood vessels, and serves to bind Aβ for its clearance from the brain into the blood across an intact BBB. However, the extensive amyloid deposits along cerebral vessels (Figure 13) compromise vessel integrity, causing a breach in the BBB. This, in turn, will compromise Aβ clearance and indeed may cause further Aβ influx from peripheral blood, along with other harmful reagents, including albumin and glutamate. These molecules may contribute to seizures and excitotoxicity.

4.4 The Genetics of Alzheimer Disease

The two greatest risk factors for AD are age and a family history of dementia. The heritance of AD appears to differ absolutely between early-onset and late-onset disease. Early-onset AD accounts for 5% of all AD cases, and 1%

of cases are caused by rare mutations in APP, PSEN1, or PSEN2.[2] These mutations essentially guarantee disease onset before age 60. Mutations are largely autosomal dominant and follow Mendelian inheritance. By contrast, late-onset or sporadic AD has no consistent mode of transmission, occurs after the age of 60, and accounts for the majority of all AD cases. Nevertheless, genetic predispositions must play a role in sporadic AD as well, since the disease risk is significantly elevated if first-degree relatives suffer from dementia. The gene variant most strongly linked to an increase in disease risk for late-onset AD is the ε4 allele of the APOE gene. APOE is a lipid-binding lipoprotein that is synthesized in the liver, although brain astrocytes and microglial cells produce APOE as well. APOE binds and transports lipids, including cholesterol. Neurons express the low-density lipoprotein receptors that bind APOE. The three isoforms of APOE differ by only a few amino acids, but these are sufficient to alter their relative affinity for cell surface receptors, with ε2 binding most poorly and ε4 most well. ε4 is expressed in 14% of the population and increases disease risk between 4- and 15-fold. By contrast, the ε2 allele, present in 7% of people, confers protection, while the ε3 allele, expressed in 79% of the population, is neutral.

A number of large genome-wide association studies have searched for additional genes that increase disease risk, and these have identified ~15 other candidates. However, none of these candidates changes risk by more than 10%, minute compared to the 400–1500% of APOE4.

4.5 Animal Models of Alzheimer Disease

In order to study disease pathology and develop drugs for treatment, animal models that recapitulate salient aspects of the disease are essential. The vast majority of AD cases are sporadic and do not have a known cause. Only a small number of

patients carry heritable mutation in genes linked to the accumulation of Aβ. The introduction of these genetic mutations has allowed the generation of mice that recapitulate some, yet not all, aspects of the disease. These have been essential to provide support for the amyloid cascade hypothesis. Mice overproducing mutant APP develop extracellular plaques like those found in human autopsies. Increased Aβ42 expression worsens the disease, while increased Aβ40 reduces disease severity. AD mice develop cognitive impairments; however, unlike in humans, these precede rather than trail amyloid deposits. This was the foundation for considering soluble oligomeric amyloid as causative agent in cognitive decline. A disease element not well replicated is the development of NFTs. As discussed above, tangles are intraneuronal aggregates of hyperphosphorylated tau and are a hallmark of human AD. However, none of the APP-overexpressing mice develop tau pathology. Finally, the inflammatory response observed in human disease differs markedly from that observed in mouse models. Several reasons have been given for these discrepancies between mouse and human AD, not least important of which is the very short life span of a mouse compared to that of humans. In spite of these shortcomings, these mouse models have provided detailed insight into many aspects of disease, such as the role of γ-secretases in Aβ42 production.[16] Also, the cell-to-cell propagation of disease, whereby Aβ acts in a prion-like fashion, could not have been disclosed without a suitable mouse model. Unfortunately, none of the drug targets that showed promise in mouse models held up in clinical trials. Hence there remains a pressing need to develop more refined disease models that more accurately mimic the salient elements of disease.

4.6 Changes in Transmitter Systems in Alzheimer Disease

Many of the cortical neurons affected in AD, and particularly those involved in learning and memory, are glutamatergic. However, cholinergic neurons are important modulators of attention and memory formation. Interestingly, choline acetyltransferase, the enzyme responsible for the neuronal synthesis of acetylcholine (ACh), is reduced by 50–90% in the cortex and hippocampus of AD patients, and the enzymatic loss is roughly proportional to the severity of cognitive loss. Autopsies from AD patients have shown a loss of neurons in the basal nucleus of Meynert, which extends cholinergic projections to the cerebral cortex. In light of these findings, drugs that increase the levels of acetylcholine through reducing its breakdown via cholinesterases are frequently used to treat at least early, modest deficits in AD. Furthermore, clinical trials discussed at the end of this chapter suggest that protection of cholinergic neurons from cell death hold promise to slow memory decline in patients. Finally, all of the major modulators of mood, anxiety, and behavior (namely, noradrenaline, serotonin, and dopamine) are also significantly reduced in tissue from AD patients, explaining some of the changes in mood and behavior that characterize late stages of AD.

4.7 Alzheimer Disease and Inflammation

Inflammation is common among acute and chronic nervous system diseases, and AD is no exception. Neuroinflammation encompasses the activation of microglia, the resident immune cells of the brain that derive from macrophages, as well as activation of astrocytes. Each can release proinflammatory molecules and cytokines such as IL-1β, IL-6, and TNFα, as well as complement components. Amyloid deposits are sufficient to activate both microglia and astrocytes in culture and both appear to engage in the removal of Aβ. It has been suggested that early in disease, inflammation may be beneficial and facilitate Aβ clearance, whereas late in disease neuroinflammation may be detrimental. Here, these

same molecules begin to interfere with neuronal function, negatively affecting synaptic transmission and the stability of cell processes. The production of reactive oxygen species (ROS) by microglial cells is sufficient to cause process retraction, and chronic increases in ROS can induce apoptosis. It is possible that a significant aspect of the progressive disease pathology is mediated by reactive astrocytes and activated microglial cells, which, rather than supporting neuronal function, turn into deadly vices. Unfortunately, the immune responses seen in AD patients are poorly replicated in animal models of disease and hence difficult to study. However, one observation that has held up in both animal models and human disease is the positive effect of nonsteroidal anti-inflammatory drugs (NSAIDs) such as ibuprofen, which reduce Aβ burden and, in humans, reduce the likelihood of disease. The beneficial effect appears to be limited to those NSAIDs that inhibit γ-secretase while also attenuating neuroinflammation.[17] Note, however, that meta-analysis of currently available clinical trials, in spite of some reported successes, does not support the use of currently available NSAIDs to treat AD.[18]

5. TREATMENT/STANDARD OF CARE/CLINICAL MANAGEMENT

Most dementias, including AD, are progressive and untreatable, leaving the clinician primarily to manage disease symptoms. However, some dementias that result from vitamin deficiencies or drug exposure can be effectively treated and even cured, and hence it is essential to reach the most accurate diagnosis for any given patient to assure the most adequate support of patient and caregiver.

Patient history and behavioral presentation are often of highest diagnostic value. For example, an elderly patient who has

TABLE 2   Distinguishing Features for the Most Common Forms of Dementia

Disease	Initial symptoms	Cognitive & psychiatric symptoms	Motor symptom	Imaging hallmarks
AD	Loss of memory	Loss of episodic memories	Initially unimpaired	Atrophy of hippocampus & entorhinal cortex
FTD	Apathy; impaired judgment and/or speech; hyperorality	Frontal & executive impairments, euphoria, depression	Palsy, rigidity, dystonia	Atrophy of frontal, insular, and temporal lobes, but spares posterior parietal lobe
DLB	Hallucinations and/or Capgras' syndrome, sleep impairments	Frontal/executive impairments, prone to delirium, but no impairment to memory	Parkinsonism	Atrophy of posterior parietal atrophy, lacks severity of hippocampal atrophy seen in AD
CJD	Dementia, mood/anxiety, movement impairments	Variable frontal & executive impairments, memory, depression, anxiety	Rigidity, Parkinsonism	Ribboning of cortex, hyperintensity of basal ganglia/thalamus
Vascular	Typically sudden on onset, but precise symptoms vary; falls, weakness	Slowing of frontal/executive and cognitive function, delusions, anxiety, but may spare memory	Slowing of movement, spasticity	Infarctions of cortical or subcortical areas, white matter changes

experienced a slow and progressive decline in episodic memory function is most likely to suffer from AD. By comparison, a younger person who has become compulsive, suffering poor judgment and disinhibition, yet without apparent memory loss is more likely to suffer from FTD. Sudden onset and a history of stroke would be highly suggestive of vascular dementia. The neurological evaluation will also search for any evidence of motor involvement, which would be uncommon in AD, yet a frequent feature in FTD and DLB. The Mini-Mental Status examination, already discussed above, is a convenient way to assess early memory deficits. Difficulties recalling names or numbers from a list are early signs of AD, and yet are not affected by FTD. These and other distinguishing features for the most common forms of dementia are summarized in Table 2.

5.1 Imaging, Diagnosis, and Disease Prediction

Imaging plays a major role in the diagnosis of dementias and is now considered standard of care.[19] Firstly, MRI or even computed tomography imaging is useful in establishing whether dementia may be the consequence of a brain insult, such as an infarct, tumor, or infection. Secondly, different imaging modalities can provide complementary information regarding diagnosis, differential diagnosis, and monitoring of disease progression.

Most readily available are standard, structural MRI scans, which permit assessment of global or focal atrophy. Examples comparing two 75-year-old subjects, one with AD and one without, show a striking atrophy of the hippocampus and surrounding mesiotemporal lobe in the AD patient (Figure 14). Longitudinal

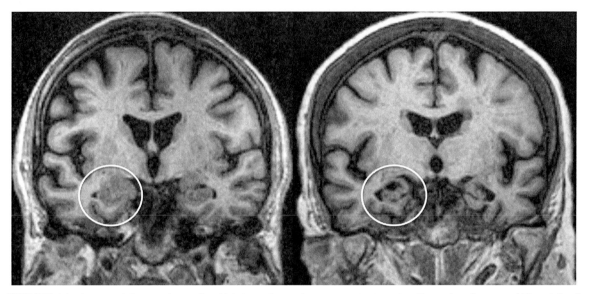

FIGURE 14   Structural changes detected in Alzheimer disease by noninvasive imaging. Examples of structural T1-weighted MRI comparing two 75-year-old individuals, one control (left) to one with AD (right) showing marked mesiotemporal lobe atrophy (circle). *From Ref. 20.*

imaging studies suggest that even mildly affected individuals have a 20–30% decrease in entorhinal and 15–20% decrease in hippocampal volume. The rate of decrease is estimated to be 3–5% per year, and a decrease in hippocampal volume of approximately 10% can already be detected 3 years prior to disease symptoms. The major advantages of structural MRI over other imaging techniques are availability, cost, reproducibility, and quantitative nature of the readout.

A second imaging modality that shows brain activity more directly is FDG-PET. In this modality, fluordeoxyglucose (FDG), a glucose analogue, can be detected by positron emission tomography (PET) when labeled with radioactive isotopes such as fluorine-18. Since the brain's energy use depends almost exclusively on glucose, the consumption of this tracer reports the resting metabolic activity of the brain, primarily attributable to synaptic activity. The drawbacks of FDG-PET are cost and availability, and it is sensitive to potential erroneous contribution of other metabolic changes. However,

it is also a powerful tool to distinguish AD from FTD. As shown in Figure 15, reduced glucose consumption in the temporal and occipital lobe characterizes AD, whereas selective loss in the frontal lobe signifies FTD.

The most recent addition to the imaging arsenal, often considered a "game changer," is amyloid PET. Here, the patient receives a PET tracer that specifically binds to the beta sheet structure of fibrillary, insoluble Aβ. The most commonly used compound is called PiB or "Pittsburg compound." It is a radioactive derivative of thioflavin T (basic yellow 1) that has long been used to label beta sheet structures such as amyloid in histological studies. Of note, these tracers do not detect oligomeric Aβ. Their predictive value for amyloid deposits or cerebral amyloidosis is very high. Fifteen studies that independently used this approach found positive amyloid signal in 96% of all patients (Figure 16). Moreover, when applied to patients with mild cognitive decline, who were followed longitudinally for 3 years, 57 of 155 patients progressed to AD, and an impressive

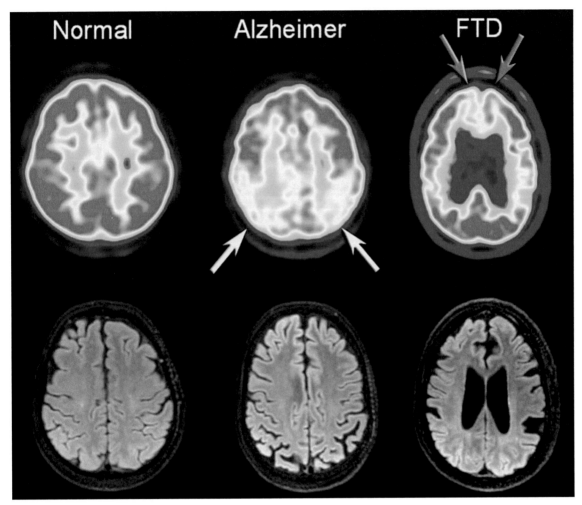

FIGURE 15   FDG-PET used to noninvasively distinguish patients suffering from AD and FTLD. In AD, glucose utilization determined by FDG-PET is lowest in the occipital and parietal lobes of the cortex (arrows), while in FTD the frontal lobe shows the greatest reduction in glucose utilization. *Image courtesy of Drs Frederik Barkhof, Marieke Hazewinkel, Maja Binnewijzend, and Robin Smithuis, Alzheimer Center and Image Analysis Center, Vrije University Medical Center, Amsterdam and the Rijnland Hospital, Leiderdorp, The Netherlands.* http://www.radiologyassistant.nl/data/bin/a509797720938b_FDG-pet.jpg.

53 of the 57 were amyloid positive. Only 7% of the amyloid-negative patients developed AD.[19] Unfortunately, amyloid PET is even less widely available than FDG-PET and is very expensive. This may change with the use of labels with a longer half-life, such as fluorine-18, which does not require an on-site cyclotron for production.

5.2 Drug Treatment

Since we currently lack drugs that prevent or cure dementia, treatments are symptomatic and attempt to reduce the burden on the patient and caregiver. Commonly used drugs include acetylcholinesterase inhibitors such as tacrine or donepezil, which prolong the activity of ACh at

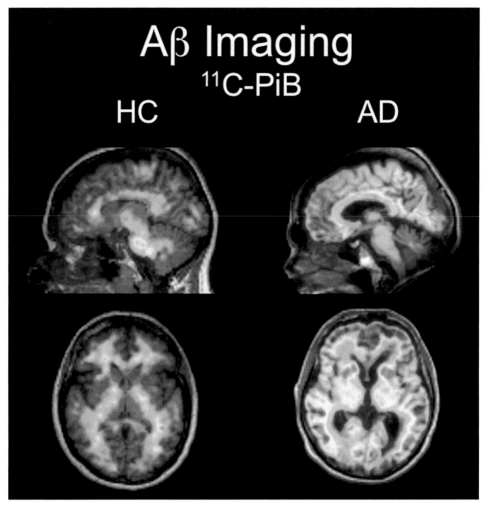

FIGURE 16   Patient diagnosis using the Pittsburg compound which directly binds to amyloid deposits in the brain of AD patients. The example illustrates abundant amyloid labeling in the AD pateint not seen in the control. *Image courtesy of Prof. Rowe and Villemagne, Austin Health, Australia.*

the synaptic cleft, thereby mitigating the overall loss of ACh in cholinergic synapses in AD. This class of drugs has shown benefit in the cognitive performance of some patients. A second approved drug is memantine, a use-dependent blocker of NMDA receptors. In clinical studies it has shown limited benefits to a select group of patients. Since depression is a frequent comorbidity of dementia, antidepressants, particularly selective serotonin reuptake inhibitors such as escitalopram (Lexapro), are useful. Similarly,

psychotropic drugs such as quetiapine (Seroquel) can help reduce delusions and psychosis in late stages of dementia.

6. EXPERIMENTAL APPROACHES/ CLINICAL TRIALS

Given the gravity of this disease, and the tremendous number of patients affected worldwide, it is not surprising that there are

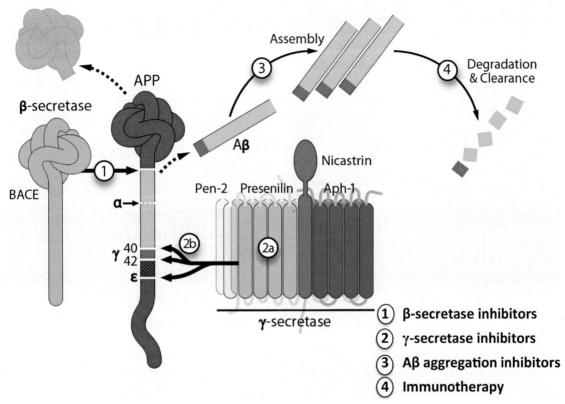

FIGURE 17 Future treatments for Alzheimer disease are exploring four major targets. (1) the cleavage of APP into plaque-prone amyloid by inhibition of the b-or (2) y- secretases (b and y symbol). (3) Interference of plaque formation using inhibitors of Ab. (4) Enhancing the clearance of Ab using immunotherapy. *From Ref. 11.*

numerous efforts under way to explore new treatment strategies. There are currently over 300 active clinical trials in the US, of which ~40 are early safety (phase 1) trials; 60 are early efficacy (phase 2) trials; and 60 are later-stage (phase 3–4) trials. These numbers are somewhat misleading in that many of the trials study early detection as opposed to treatments, or the use of symptomatic treatment of comorbidity. Also included are studies on nutrition, supplements, exercise, and lifestyle change. Hence only a limited number of trials actually study truly novel molecules for treatment, and we will highlight a few examples below. Given what we just learned about the etiology

of disease, any strategy to reduce Aβ or tau burden would appear to be the most likely to succeed. Indeed, the broad consensus that toxic Aβ deposits are central to the development of AD has led to several approaches aimed at reducing the Aβ burden in AD patients, with the various angles taken toward this problem schematically illustrated in Figure 17. A first attempt involved a vaccine that detects and destroys Aβ. This clinical trial came on the heels of very promising studies in animal models and positive safety data in a phase 1 human study. However, a phase 2 vaccine trial using a synthetic Aβ42 (AN1792, Eli Lilly) had to be terminated because five enrolled patients suffered severe brain inflammation.[21] Follow-up examination of brain tissue revealed significant microglial activation, reactive gliosis, and amyloid angiopathy, all consistent with widespread inflammation. Instead of inducing the body to produce antibodies through a vaccine, the studies that followed used humanized monoclonal antibodies, such as bapineuzumab and solanezumab, that directly bind Aβ42 instead. Although these were well tolerated by patients, the results from large phase 3 clinical trials were disappointing for both drugs,[22] as neither demonstrated any improvement of cognitive outcome or daily living.

The main reason that these trials lacked effectiveness may be timing. The clearance of Aβ through antibodies occurs much too late in the disease, at a time when much of the neuronal and cognitive damage is already done. The research we discussed above suggests, as illustrated in Figure 18, that the pathogenic cascade of AD begins at least one or even two decades before cognitive impairments manifest, during which accumulations of Aβ42 and phosphorylated tau cause progressive brain atrophy. If one could introduce these antibodies or other measures to reduce the Aβ42 burden prior to disease onset, one may be able to delay or even prevent dementia. Hence a necessary first step toward an early disease intervention is the reliable identification of persons at risk of developing AD. As we discussed above, this is now within reach using a combination of biomarkers. The currently established markers include structural MRI, which examines regions of global brain shrinkage; FDG-PET to measure a decline in glucose consumption; measurement of fibrillary Aβ using an amyloid PET ligand; and cerebrospinal fluid (CSF) levels of Aβ42 and phosphorylated tau. Although none of these is 100% predictive, in combination they provide a high level of confidence.[23] Clinical studies that evaluated the ability to detect AD risk early using these biomarkers have focused on

two groups: those carrying the APOE ε4 allele and that hence have up to 15-fold increased AD risk, and those with disease-causing mutations in either PSEN1/2 or APP, who have a near 100% likelihood to develop the disease. The latter group in particular provides a powerful sample as, for each individual, given sufficient time, AD symptoms will develop for sure. Studies to date suggest that mutation carriers present with a reduction in hippocampal volume ~15 years before symptom onset, changes in glucose utilization precede symptoms by ~10 years, and PET studies detecting fibrillary Aβ similarly show deposits ~10–25 years prior to clinical disease onset. Tau pathology, judged from changes in CSF samples, also significantly precedes disease. The picture that emerges, graphically illustrated in Figure 18, is a sequential acquisition of pathology during a presymptomatic stage of disease.[24]

Taking advantage of early detection, current efforts are exploring disease intervention strategies where at-risk individuals, biologically defined, are enrolled in trials for the very Aβ monoclonal antibodies that had failed in patients with already established dementia. A large proof-of-concept study has just been launched through a corporate–government joint venture. The National Institutes of Health is participating in a $100 million disease prevention trial in which 200 subjects belonging to a Colombian kindred with the heritable PSEN1 E280A mutation will be enrolled. This mutation reliably causes cognitive impairments by age 44. The trial will evaluate cognitive status in 100 subjects treated with the antibody crenezumab (Genentech) prior to disease onset and throughout its course. These patients will be compared to 100 subjects receiving a placebo.[22] If successful, this approach could be expanded to include other at-risk patients, for example, those presenting with APOE4 with evidence of amyloid deposits detected by amyloid PET imaging.

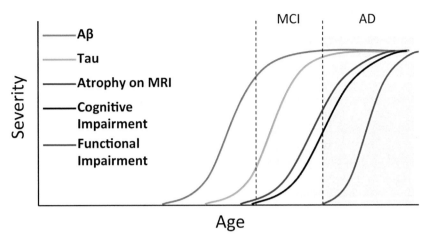

FIGURE 18 Schematic depiction of disease progression. Ab aggregation into plaques precedes disease symptoms by as many as 15 years, while tau hyper-phosphorylation may occur up to 5 years prior to symptoms. As pathology develops, mild cognitive impairment (MCI) eventually gives way to AD-dementia, a stage at which pathological signs tissue changes are abundant.

While the above approaches seek to remove already synthesized Aβ42 from the brain and blood, an alternative approach to disease prevention would be the inhibition of Aβ42 production. In light of the pivotal role of γ-secretases in the production of toxic Aβ42, a number of companies are actively developing such inhibitors. As alluded to before, one complication with this approach is that γ-secretases also cleave other membrane-embedded proteins such as the Notch receptor, which acts as an important signaling molecule. In the skin, Notch acts as a tumor suppressor gene, and blockade of γ-secretase would be expected to also present with a dysfunction of Notch signaling, possibly causing skin cancer. Indeed, early clinical trials with the secretase inhibitor drug LY450139 (Eli Lilly) failed for this very reason. Since then, new drugs have been synthesized that, rather than inhibiting γ-secretase activity, modify its action such that the cleavage produces the more soluble Aβ38 instead of the toxic insoluble Aβ42. As mentioned before, some NSAIDs, for example, ibuprofen, change the γ-secretase cleavage of Aβ, and derivatives

of these γ-secretase modulators are now in various stages of clinical testing as well. Interestingly, long-term chronic users of ibuprofen are 44% less likely to develop AD,[25] which is consistent with the hypothesized reduction in toxic Aβ42. Yet another class of drugs was developed that inhibits APP cleavage selectively while largely sparing Notch. The best of these have a several hundred-fold higher activity on Notch cleavage versus APP. How effective these γ-secretase modulators will be has to await the conclusion of ongoing clinical trials.[26]

Unless any of these disease prevention trials succeeds, the ability to detect disease early leaves us with an ethical dilemma. If applied to the general population, at significant cost to our health care system, a combination of biomarkers will allow us to identify persons at high risk to develop AD a decade or two prior to disease onset. What will we do with this knowledge? If we had treatments to offer that could delay or even prevent disease, such tests would be well justified. However, in the absence of therapy, these evaluations may become entirely academic and indeed may even be psychologically harmful to the patient.

A different approach that does not target amyloid, but instead tries to retain cognitive function by boosting the survival of cholinergic neurons, is beginning to show promise as well.[27] It has long been recognized that over the course of AD, patients lose up to 90% of their cholinergic neurons in the basal forebrain. These neurons project to all areas of the cerebral cortex and hippocampus and modulate synaptic activity involved in learning and memory. Nerve growth factor (NGF) prevents cholinergic neuronal death and enhances memory function in mouse models of AD. Unfortunately, NGF is difficult to deliver as it does not pass the BBB, and intracranial injection causes severe side effects by acting on neurons not involved in the memory circuits. An elegant work-around was to harvest skin fibroblasts from AD patients and genetically modify them to produce and release NGF. These were then stereotaxically implanted into the nucleus basalis, the main site of cholinergic neurons, where these cells took hold and began to release NGF. In a small phase 1 clinical trial involving eight patients, this approach not only was safe but produced marked and long-lasting improvements in cognition and significantly enhanced metabolic activity judged by FDG-PET imaging (Figure 19). The ultimate utility of this exciting approach will have to await further clinical testing in larger patient cohorts.

7. CHALLENGES AND OPPORTUNITIES

Dementia may be among the most frightening conditions a human can experience. Memories are the substrate of learning and cognition, the very fabric of humanity. Improved health care, nutrition, and lifestyle changes now allow us to live decades longer than our ancestors, and our offspring will soon push life expectancy beyond 100 years. Yet, if we cannot find a cure for dementia, this may not be desirable at all. Given the near-exponential rise in disease incidence with age, most individuals would be guaranteed to suffer from AD in their last decade(s) of life. Some argue that dementia is inevitably linked to aging. If true, we must find ways to slow down aging itself rather than be concerned about individual molecular processes of age-related disease. In the near term, however, dementia is a health epidemic and is bound to break our health care system if success does not come soon. The cost to society and the emotional burden on caregivers will be unsustainable. Fortunately, a large number of researchers have devoted their lives to the study of dementia. As illustrated throughout this chapter, our understanding has improved in leaps and bounds and has done so within a relatively short amount of time. However, significant gaps remain in our understanding. For example, we still do not know whether it is the amyloid deposits that are toxic to neurons, or whether instead oligomeric amyloid is the culprit. Similarly, whether tau inclusions are harmful or protective is unclear. Does the prion hypothesis for disease spread hold up to future testing? What is the contribution of cell-autonomous versus non-cell-autonomous processes to disease? Is there a role of astrocytes, microglial cells, and complement components? How about the immune system's role in disease? Breaches in the BBB may significantly contribute to disease, allowing entry of harmful serum components, lymphocytes, and macrophages. The peculiar time course of disease suggests that it begins at least 10, if not 20, years prior to the first signs of dementia. Does that represent the time window of opportunity to interfere? If so, are there modifiable risk factors that one could address to mitigate or delay disease onset? Neurogenetics is still in its infancy, and the ability to sequence whole genomes and search through genomic data from thousands of patients is on the horizon. Will such large genomic screening identify the still missing genes to explain sporadic late-onset AD? One can only hope that answers will come soon!

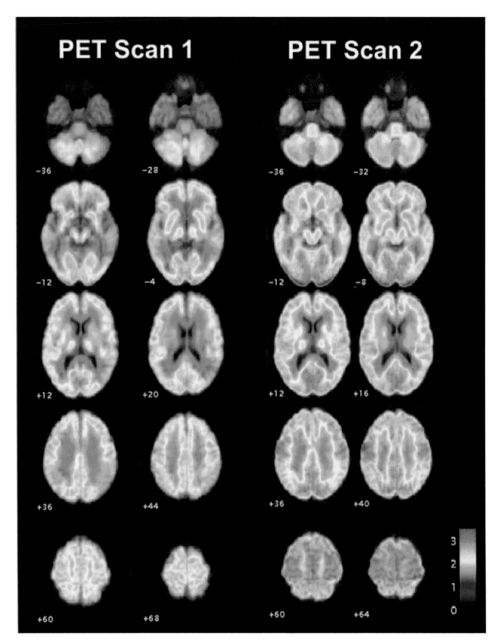

FIGURE 19 Averaged FDG PET scans in four subjects treated with NGF, overlaid on standardized MRI templates. Representative axial sections, with 6–8 months between first and second scan, showing widespread interval increases in brain metabolism. Flame scale indicates FDG use/100 g tissue/min; red color indicates more FDG use than blue. *From Ref. 27.*

Acknowledgments

This chapter was graciously reviewed by Erik Roberson, MD, PhD, who is Associate Professor of Neurology and Virginia B. Spencer Scholar in Neuroscience and directs the UAB Center for Neurodegeneration and Experimental Therapeutics.

References

1. Glenner GG, Wong CW. Alzheimer's disease: initial report of the purification and characterization of a novel cerebrovascular amyloid protein. *Biochem Biophys Res Commun*. 1984;120(3):885–890.
2. Tanzi RE. The genetics of Alzheimer disease. *Cold Spring Harb Perspect Med*. 2012;2:a006296.
3. Wortmann M. Dementia: a global health priority - highlights from an ADI and World Health Organization report. *Alzheimer's Res Ther*. 2012;4(5):40.
4. Seshadri S, Beiser A, Kelly-Hayes M, et al. The lifetime risk of stroke: estimates from the Framingham Study. *Stroke*. 2006;37(2):345–350.
5. Lourida I, Soni M, Thompson-Coon J, et al. Mediterranean diet, cognitive function, and dementia: a systematic review. *Epidemiology (Cambridge, Mass.)*. 2013;24(4):479–489.
6. Landau SM, Marks SM, Mormino EC, et al. Association of lifetime cognitive engagement and low beta-amyloid deposition. *Arch Neurol*. 2012;69(5):623–629.
7. Tarawneh R, Holtzman DM. The clinical problem of symptomatic Alzheimer disease and mild cognitive impairment. *Cold Spring Harb Perspect Med*. 2012;2(5): a006148.
8. Iqbal K, Liu F, Gong CX, Grundke-Iqbal I. Tau in Alzheimer disease and related tauopathies. *Curr Alzheimer Res*. 2010;7(8):656–664.
9. Hartmann T, Bieger SC, Bruhl B, et al. Distinct sites of intracellular production for Alzheimer's disease A beta40/42 amyloid peptides. *Nat Med*. 1997;3(9): 1016–1020.
10. Mucke L, Selkoe DJ. Neurotoxicity of amyloid beta-protein: synaptic and network dysfunction. *Cold Spring Harb Perspect Med*. 2012;2(7):a006338.
11. Roberson ED, Mucke L. 100 years and counting: prospects for defeating Alzheimer's disease. *Science*. 2006;314(5800):781–784.
12. Walter J, Kaether C, Steiner H, Haass C. The cell biology of Alzheimer's disease: uncovering the secrets of secretases. *Curr Opin Neurobiol*. 2001;11(5):585–590.
13. Mattson MP. Neurobiology: ballads of a protein quartet. *Nature*. 2003;422(6930):385–387.
14. Stohr J, Watts JC, Mensinger ZL, et al. Purified and synthetic Alzheimer's amyloid beta (abeta) prions. *Proc Natl Acad Sci USA*. 2012;109(27):11025–11030.
15. Sagare AP, Bell RD, Zlokovic BV. Neurovascular dysfunction and faulty amyloid β-peptide clearance in Alzheimer disease. *Cold Spring Harb Perspect Med*. 2012;2: a011452.
16. LaFerla FM, Green KN. Animal models of Alzheimer disease. *Cold Spring Harb Perspect Med*. 2012;2:a006320.
17. Saura CA. Presenilin/gamma-secretase and inflammation. *Front Aging Neurosci*. 2010;2:16.
18. Jaturapatporn D, Isaac MG, McCleery J, Tabet N. Aspirin, steroidal and non-steroidal anti-inflammatory drugs for the treatment of Alzheimer's disease. *The Cochrane Database System Rev*. 2012;2:Cd006378.
19. Johnson KA, Fox NC, Sperling RA, Klunk WE. Brain imaging in Alzheimer disease. *Cold Spring Harb Perspect Med*. 2012;2:a006213.
20. Scheltens P. Imaging in Alzheimer's disease. *Dialogues in clinical neuroscience*. 2009;11(2):191–199.
21. Birmingham K, Frantz S. Set back to Alzheimer vaccine studies. *Nat Med*. 2002;8(3):199–200.
22. Mullard A. Sting of Alzheimer's failures offset by upcoming prevention trials. *Nat Rev Drug Discovery*. 2012;11(9):657–660.
23. Langbaum JB, Fleisher AS, Chen K, et al. Ushering in the study and treatment of preclinical Alzheimer disease. *Nat Rev Neurol*. 2013;9(7):371–381.
24. Trojanowski JQ, Vandeerstichele H, Korecka M, et al. Update on the biomarker core of the Alzheimer's disease neuroimaging initiative subjects. *Alzheimer's Dementia: J Alzheimer's Assoc*. 2010;6(3):230–238.
25. Vlad SC, Miller DR, Kowall NW, Felson DT. Protective effects of NSAIDs on the development of Alzheimer disease. *Neurology*. 2008;70(19):1672–1677.
26. De Strooper B, Iwatsubo T, Wolfe MS. Presenilins and gamma-secretase: structure, function, and role in Alzheimer disease. *Cold Spring Harb Perspect Med*. 2012;2(1):a006304.
27. Tuszynski MH, Thal L, Pay M, et al. A phase 1 clinical trial of nerve growth factor gene therapy for Alzheimer disease. *Nat Med*. 2005;11(5):551–555.

General Readings Used as Source

1. Harrison's Online, Chapter e9. Memory Loss by Miller BL, & Viskontas V.
2. Harrison's Online, Chapter 371: Dementia by William W. Seeley, Bruce L. Miller (Harrison's online requires a subscription. Many medical libraries make this available for the students and faculty.)
3. The Biology of Alzheimer's Disease: Dennis J. Selkoe, E. Mandelkow, and David M. Holtzman. A collection of articles written by leading experts in the field, and freely available online: http://perspectivesinmedicine.org/cgi/collection/the_biology_of_Alzheimer_disease.

4. Ushering in the study and treatment of preclinical Alzheimer disease, Nature Reviews in Neuroscience. Jessica B. Langbaum, Adam S. Fleisher, Kewei Chen, Napatkamon Ayutyanont, Francisco Lopera, Yakeel T. Quiroz, Richard J. Caselli, Pierre N. Tariot & Eric M. Reiman: http://www.nature.com/nrneurol/journal/v9/n7/full/nrneurol.2013.107.html.

5. Understanding Alzheimer's Disease, I. Zerr: This is an open access online multi-author book: http://www.intechopen.com/books/understanding-alzheimer-s-disease.

6. Alzheimer's Disease, Unraveling the Mystery. National Institute on Aging, NIH: http://www.nia.nih.gov/alzheimers/publication/alzheimers-disease-unraveling-mystery.

7. Alzheimer's disease, facts and figures. Alzheimer's Association; 2013. http://www.alz.org/downloads/facts_figures_2013.pdf.

8. David Sweatt J, ed. *Mechanisms of memory*. 2nd ed. Academic Press; 2010.

Suggested Papers or Journal Club Assignments

Clinical Paper

1. Bateman RJ, et al. Clinical and biomarker changes in dominantly inherited Alzheimer's disease. *N Engl J Med*. 2012;367(9):795–804.

2. Mayeux R. Clinical practice. Early Alzheimer's disease. *N Engl J Med*. 2010;362(23):2194–2201 (case study).

3. Karch CM, et al. Expression of novel Alzheimer's disease risk genes in control and Alzheimer's disease brains. *PLoS One*. 2012;7(11):e50976.

4. Naj AC, et al. Common variants at MS4A4/MS4A6E, CD2AP, CD33 and EPHA1 are associated with late-onset Alzheimer's disease. *Nat Genet*. 2011;43(5):436–441.

5. Tuszynski MH, et al. A phase 1 clinical trial of nerve growth factor gene therapy for Alzheimer disease. *Nat Med*. 2005;11(5):551–555.

6. Brickman AM, Khan UA, Provenzano FA, et al. Enhancing dentate gyrus function with dietary flavanols improves cognition in older adults. *Nat Neurosci*. 2014;17(12):1798–1803.

Basic Science Paper

1. Verret L, et al. Inhibitory interneuron deficit links altered network activity and cognitive dysfunction in Alzheimer model. *Cell*. 2012;149(3):708–721.

2. Stohr J, et al. Purified and synthetic Alzheimer's amyloid beta (Abeta) prions. *Proc Natl Acad Sci USA*. 2012;109(27):11025–11030.

3. Griciuc A, et al. Alzheimer's disease risk gene CD33 inhibits microglial uptake of amyloid beta. *Neuron*. 2013;78(4):631–643.

4. De Jager PL, Srivastava G, Lunnon K, et al. Alzheimer's disease: early alterations in brain DNA methylation at ANK1, BIN1, RHBDF2 and other loci. *Nat Neurosci*. 2014;17(9):1156–1163.

Parkinson Disease

Harald Sontheimer

1. CASE STORY

Anna's 24th birthday was coming up and she hated even the thought of it. She remembered a time when she loved parties, having friends at her house, being the center of attention. No longer! That all changed a year earlier, when her hand started shaking. She first noticed it Christmas evening when she helped grandma set the table. Holding just a fork, her right hand trembled. She was worried. Unable to stop the movement, she let go of the fork and pushed her hand forcefully on the table. "What was that?" her mom asked. "I have no idea," Anna replied. Afraid it would happen again, she only used her left hand to eat. As everyone settled in for the after-dinner Christmas carols, her mind eased as she had a glass of wine. The next morning, picking up her toothbrush, the shaking started again. In fact, it was out of control. She could not stop the hand from trembling until she held on to the towel rod. Something was terribly wrong.

The following day she called her internist, Dr Helmstedt, who was willing to work her into his schedule that same day. She had not seen him since going off to college. He was kind and caring, yet his questions really irritated her. Had she been using drugs or drinking excessively? He should know her better! As a member of the same church, he knows that she has been teaching Bible school since her junior year in high school. She was not that kind of girl.

However, he did not appear to have an answer and told her to keep an eye on it and see if the symptoms wore off. Wear off they did not! The tremors became a daily routine without any sign of abating. Two days later, she passed out in the kitchen without warning while preparing a salad. Yet as quickly as it happened, she regained her consciousness. Her mother, visibly upset, drove her straight to the emergency room at St. Vincent's Hospital. The waiting room was filled with patients, many in pain, some bandaged from work accidents. She seemed a misfit in their company. After sharing her experience with the triage nurse, she was sent to Neurology for a physical exam, which revealed nothing abnormal. "We need to scan your brain primarily to rule out the obvious," she was told. What was the obvious? "A tumor or multiple sclerosis," replied Dr Nicholas, the neurologist on service. A tumor? What a frightening thought. The MRI machine was making pounding noises and the procedure appeared to last forever. The results of the exam, however, would not be in for a couple of days. Every time the phone rang Anna was startled, afraid to learn she had a brain tumor. When the call finally came, the answer was a great relief. Everything looks normal, with no evidence of any visible brain abnormality. However, Dr Nicholas explained that he had called in a prescription for a drug called Sinemet, just as an experiment, he said, to rule out Parkinson disease (PD). "Parkinson's? Isn't that what old people get? I am 23 years old!" "Yes, Anna, it is very unlikely," the neurologist replied. "That is why we are trying the drug only for 10 days as an experiment." Fortunately, or unfortunately, the drug did the trick. The tremors went away and stayed away. A DNA test since confirmed that Anna carries a rare mutation in a gene called Parkin. As bad luck would have it, Anna was among the very youngest patients to develop this rare genetic form of young-onset Parkinson. Why her? Just the thought of it made her cry. She had just accepted a new job at a publishing firm for children's books. Would she need to let her new employer know? What was life going to be like? Would she be able to eventually start a family, have children? She will turn 24 tomorrow. She should be living a carefree life, going out with friends, dancing, having a good time. Yet instead, every time she sees a clock, she sees her life unfold, imagining her body unable to follow her command, a healthy brain trapped in an immobile body. She was scared.

2. HISTORY

In 1817 the British physician James Parkinson (Figure 1) published an article entitled "An Essay on the Shaking Palsy," in which he described the course of disease of six patients who suffered from resting tremors (shaking palsy), abnormal posture, gait difficulties, and reduced muscle strength. He observed three of these patients in the streets of London and the other three in his medical practice. In 1872, a more comprehensive description was compiled by the French

FIGURE 1 James Parkinson (1755–1824) published an article on the "shaking palsy," a condition that was later named Parkinson disease in his honor (public domain).

physician Jean-Martin Charcot, who studied hundreds of patients to develop a detailed description of the disease. Charcot's description distinguished the symptoms from the tremors that are often found in patients suffering from multiple sclerosis. He also advocated that the shaking palsy be named PD in honor of James Parkinson's original description.

Charcot already noted that the "muscle weakness" in PD was not caused by a defect in muscle function, but instead was centered in the brain. This notion was confirmed through subsequent autopsy studies by Tetiakoff (1919) and Brissaud (1925), who identified lesions in the substantia nigra (SN) of the midbrain. Their relationship to movement control was proposed in 1938 by Hassler. The proteinaceous inclusion bodies found abundantly in PD brains were independently shown much earlier, by the pathologist Frederick Lewy in 1912. These are now the pathological hallmark of PD and are largely composed of misfolded alpha-synuclein, discovered 80 years later. Interestingly, we still do not know whether these inclusions are the cause or consequence of neuronal cell death.

The major turning point in our understanding of PD came in the 1950s. First, it was recognized that dopamine is not merely a precursor for the synthesis of other catecholamine neurotransmitters, such as adrenaline and noradrenaline, but that it was itself a neurotransmitter that was present in significantly high concentrations in the midbrain. Next, the Swedish physician-scientist, Arvid Carlsson, showed that blocking dopamine uptake with the drug reserpine produced Parkinson-like features in rabbits. He was able to reverse the loss of movement in these animals by giving them the dopamine precursor L-DOPA. In 1960, Oleh Hornkiewicz showed a depletion of dopamine in the SN of PD patients. These studies together suggested that a loss of dopamine may be responsible for the motor symptoms, and paved the way for the highly effective levodopa therapy used to treat PD patients today. The first human clinical trial was followed soon thereafter and was remarkably successful: "Bed-ridden patients who were unable to sit up, patients who could not stand up when seated, and patients who, when standing, could not start walking performed all these activities with ease after L-DOPA."[1] This treatment was approved under the name levodopa for clinical use to treat PD in 1967. However, given the systemic conversion of L-DOPA to dopamine by aromatic L-amino-acid decarboxylase, L-DOPA had significant side effects along with short-lived positive results. A more refined delivery regimen was developed by George Cotzias that led to a gradual administration of therapeutic doses of L-DOPA that stabilized the patient on an acceptable dose with remarkable remission of symptoms. However, this was quickly superseded by the discovery that the co-administration of decarboxylase inhibitor carbidopa prevented the systemic conversion of L-DOPA to dopamine, thereby reducing peripheral dopamine side effects and ensuring effective delivery to the brain. The combination of L-DOPA and carbidopa, now in a combination tablet, has since become the standard treatment, approved under the name Sinemet. A slow-release version, Sinemet-CR, is also available. Many of the contributors to this remarkable discovery were honored with prestigious science prizes. Carlsson was awarded the Nobel Prize in Medicine in 2000, Hornkiewicz received the Wolf Prize in Medicine, and Cotzias was awarded the Lasker Prize.

In a historical context, it is interesting to consider the evolution of treatments, and the accidental successes achieved along the way. James Parkinson remained committed to the humoral theory prevalent in the day, and therefore suggested blood-letting at the neck as treatment, combined with deliberately placed infections under the skin to divert blood flow away from the brain in order to decompress the brain. By contrast, just a few decades later, Charcot was already experimenting with plant-derived anticholinergic drugs and ergot-contaminated rye grass. The ergot fungi produce ergotamine, which is a synthetic precursor for the synthesis of the modern dopamine agonists pergolide and cabergoline, used to treat PD. So unbeknownst to Charcot, some of his patients benefitted from the same dopamine agonist frequently used as a first line of treatment today. This history is quite remarkable, since this was long before dopamine was even discovered, let alone considered as a neurotransmitter. Equally coincidental, and even more stunning, L-DOPA replacement therapy may have been in use since ancient times in traditional East Indian medicine. L-DOPA is abundantly found in the beans of the cowitch, a tropical legume that is a major ingredient in the traditional Indian medicine Masabaldi Pacana.[2] Unfortunately, no specific reports on the use of this remedy for PD can be found in the literature.

The dopamine re-uptake inhibitor amantadine may also be added to the list of accidentally discovered drugs. This antiviral agent was used widely in nursing homes to treat infections in the 1960s. It reduced tremors, balance-related issues, and akinesia in Parkinson patients, thereby revealing an unexpected dopaminergic effect.[3]

While society is actively discussing the potential benefit of medical marijuana for the treatment of various diseases, it may be of interest that cannabis, either alone or in combination with opium, was explored as treatment for PD well over a century ago by the British physician Gowers, who reported to have witnessed in his patients a "very distinct improvement for a considerable time under their use" (Gowers 1899, as discussed in Ref. 2).

As is often the case, progress is often propelled by a public fascination with celebrities suffering from disease and who become advocates for the cause. Among the notable personalities who have been living with PD are boxer Muhammad Ali, Olympic cyclist Davis Phinney, and famous actor Michael J. Fox. They were recently joined by rock star Linda Ronstadt and Texas A&M basketball coach Billy Kennedy. The Michael J. Fox Foundation has become a major funding agency for PD research.

3. CLINICAL PRESENTATION, DIAGNOSIS AND EPIDEMIOLOGY

3.1 Epidemiology and Risk Factors

Next to Alzheimer dementia, PD is the second-leading neurodegenerative disorder, affecting one million people in the US and over five million worldwide. Disease incidence increases with age from about 1% of the population at age 65–3% at age 85. PD is characterized by a group of "cardinal" features that include rigidity, resting tremor, slowness in movement (bradykinesia), and balance problems. These symptoms are usually asymmetric, affecting one side of the body more than the other. PD is more common in men than in women, and whites are more likely to be diagnosed with PD than are African Americans or Hispanics. Typical disease onset is around 65 years of age, but in rare cases disease can occur in the 20s or even earlier. The vast majority, >99%, of cases are idiopathic and therefore without a known cause. Rare familial forms are characterized by mutations in a group of genes involved with protein biosynthesis and degradation that include α-synuclein, LRRK2, Parkin, Pink1, and others.

The largest risk factor for PD is age; however, environmental factors are suspected to play an important role in the disease etiology. People growing up in rural areas are more likely to develop PD, and exposure to pesticides such as rotenone, or herbicides such as Agent Orange, are suspected risk factors. Surprisingly, smoking cigarettes or using nicotine or caffeine reduces the risk to develop PD. The discovery that MPTP (1-methyl-4-phenyl-1,2,3,6-tetrahydropyridine), a byproduct of trying to illicitly manufacture heroin, produces a remarkably similar disease phenotype has spurred interest in identifying related environmental toxins that may be causal for the disease. However, while MPTP provides one of the most used animal models to study PD, no related chemicals have yet surfaced that may cause idiopathic PD. Nevertheless, current evidence strongly suggests that environmental factors play a major contributing role in the disease, probably acting synergistically with an inherent genetic risk.

3.2 Disease Presentation and Diagnosis

Tremor is an early common symptom that is easily recognizable even by a layperson. It initially affects only one body side and is most noticeable when the affected limb is at rest, and disappears with purposeful movement. There is a smooth quality to the tremor, as if a person were rolling a marble between thumb and index finger. Hands and arms are more affected by the tremor than the lower extremities, and tremors can become quite bothersome for the patient, preventing skilled fine motor tasks, holding a glass without spilling its contents, or placing a key into a lock.

There is also a pronounced overall slowness (or poverty) of motion, called bradykinesia. This impairs the initiation of movement and makes individuals appear frozen. This goes along with shortened stride length, decreased arm swings, and lower rate of eye blinks. The stoic appearance is further enhanced by overall rigidity and stiffness. Once walking, a person with PD is hunched forward and walks with a slow, narrow, shuffling gate (Figure 2). As the disease progresses the patient becomes unsteady and falls frequently.

The major defect that explains the impaired motor symptoms is a loss of dopamine-producing neurons in the SN pars compacta, a nucleus in the midbrain that is part of the basal ganglia. This structure communicates with the motor cortex and the spinal cord and is involved in the modulation of movement. Restoration of dopamine by administration of the dopamine precursor levodopa transiently restores normal motor function and serves as an important tool to unequivocally diagnose PD and to differentiate it from other, related

FIGURE 2 Front and side views of a man with a festinating or forward-leaning gait characteristic of Parkinson disease. *Drawing after St. Leger, first published in Wm. Richard Gowers' Diseases of the Nervous System, in London, 1886 (public domain).*

movement disorders. In addition to neuronal loss in the basal ganglia, nondopaminergic neurons are lost in other parts of the brain. Recent studies actually suggest that the disease may begin in the peripheral autonomic nervous system, spreading centrally over time. This may explain the early loss of olfaction, which is a telltale sign that often precedes disease onset by several years.

A diffuse loss of nondopaminergic cells may also underlie the progressive development of dementia, which eventually affects 80% of patients with PD. Given that dopamine plays a role in the reward system that controls mood and sexual and pleasurable behavior, many PD patients show behavioral changes that may include depression and alterations in sex drive. Other rare psychiatric problems in PD include compulsive gambling, particularly when PD patients are treated with dopaminergic drugs.

3.3 Disease Stages

It is common to categorize four stages of PD, which differ among patients in severity of symptoms and the degree of impairment.[4]

In the **premotor stages** of the disease, constipation, loss of olfaction, and abnormal sleep, particularly the acting out of dreams, may predict the future onset of disease. Ninety percent of PD patients have a measurably decreased sense of smell at time of diagnosis, and a recent prospective study showed that a poor performance on olfactory tests was highly predictive of developing PD years later.[5] Another study of 6790 men without PD showed that those with infrequent bowl movements had a four-fold increased risk to develop PD.[6] Other symptoms that may precede diagnosis by many years include restless leg syndrome, anxiety, and changes in cognitive function. The latter may include subtle attention problems and changes in executive function, including problems with planning, abstract thinking, and cognitive flexibility.

Early stages of PD are dominated by the cardinal motor symptoms, that is, resting tremor, postural instability, bradykinesia, and rigidity. At this stage, levodopa is maximally effective, and in many patients the drug controls motor symptoms so entirely that the patient has little impairment at work or in social life. The "off" states, which are the times that the medication has worn off, are short, and repeated dosing of levodopa maintains the patient in a well-functioning and "on" state all the time.

Over time, the period during which complete dopamine control can be achieved grows shorter. As patients enter a period of **moderate PD**, they spend more time in the disabled "off" state. They may also begin to experience short periods of dyskinesia, although the impairments are not at the point that the patient would wish to discontinue levodopa treatment. Constipation becomes progressively worse, and depression is seen in close to 50% of patients at this stage.

In **advanced stages of disease**, motor and nonmotor symptoms are quite disabling as the levodopa effects are short lived and incomplete. Gait problems worsen, resulting in instability and frequent falls. Dementia and behavioral problems become significant and even in cases where the use of a deep brain stimulator ameliorates motor symptoms, nonmotor symptoms are a major disability, and the burden on caregivers and family members is tremendous.

Histopathologically, the neuronal loss in the susbtantia nigra can readily be seen on autopsy, where the black melanin-pigmented neurons that give the SN its name are missing in PD patients (Figure 3). In addition, protein aggregates called Lewy bodies that contain α-synuclein are found diffusely throughout the brain (Figure 4). New imaging techniques allow visualization of the early dopaminergic neuronal loss by positron emission tomography (PET) (Figure 5), taking advantage of the fact that these neurons bind the 18F labeled dopamine precursor L-DOPA (fluorodopa). This approach, however, is not commonly used clinically.

There is currently no cure for PD, but a number of disease-modifying treatments are

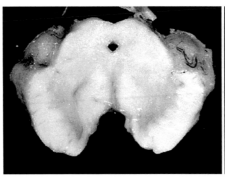

FIGURE 3 Loss of pigmented dopamine neurons in the substantia nigra from a patient with PD (left) compared to a normal sustantia nigra (right). The distinct loss of the black-appearing dopaminergic cells that contain the black pigment neuromelanin, serving as a marker for dopaminergic neurons is prominently seen in the substantia nigra from the patient with PD. *Image was kindly provided by Dr Dimitri Agamanolis, Director, Anatomical Pathology and Neuropathology, Akron Children's Hospital.*

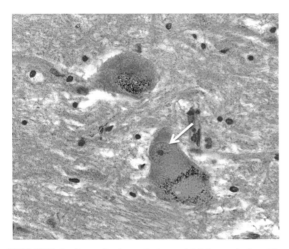

FIGURE 4 Lewy body pathology. A high-resolution histological image of the substantia nigra shows neurons containing brown melatonin granules along with characteristic Lewy bodies (arrow). *Image was kindly provided by Dr Thomas Caceci.*

available that potentially slow the progression of disease and greatly improve the quality of life for most patients. These are largely targeted at the motor symptoms and seek to restore normal dopamine levels in the brain. This is primarily accomplished through dopamine replacement therapy, which involves the systemic administration of dopamine precursor levodopa

in conjunction with enzyme inhibitors that reduce peripheral degradation of this pro-drug. Unfortunately, the effectiveness of this strategy wears off over a period of 5–10 years and yields somewhat abnormal hyperkinetic, jerky movements known as "chorea," akin to that seen in Huntington disease. Recent successes using implanted electrodes that stimulate motor control pathways in the midbrain have provided significant benefit to many patients, and this deep brain stimulation (DBS) is now widely used in advanced stages of PD.

Many of the nonmotor effects of PD cannot be effectively treated. Of these, a somewhat unique form of dementia, which primarily affects executive functions and planning of daily activities, but without memory loss, is a major impairment for which no effective treatment exists. In addition, depression is very common, affecting between 50% and 80% of PD patients, and often goes untreated. With disease progression, the strain on the caregiver becomes significant, and round-the-clock residential nursing may be advisable, if possible.

Most PD patients live to an almost-normal life expectancy. The early phase of disease, during which a patient is able to go about his or her daily life with minimal motor deficits, lasts for

Control   PD Patient

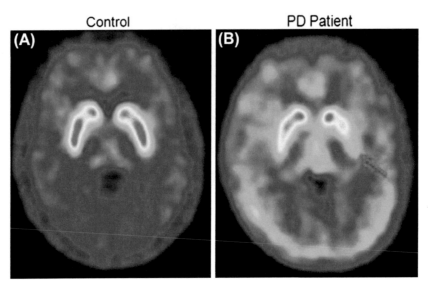

FIGURE 5   Presence of dopamine neurons assessed by 18-fluorodopa positron emission tomography (PET). (A) PET scan from a control subjects showing high striatal uptake of the precursor used to synthesize dopamine, indicative of the presence of dopaminergic neurons (highest value in red). (B) Example of a patient with Parkinson disease with motor signs mainly confined to the left limbs. Uptake of the dopamine precursor is markedly reduced in the right posterior putamen (area indicated by arrow is 70% below normal) and to a lesser extent in the anterior putamen and caudate of the left hemisphere. *Reproduced with permission in a modified form from Ref. 7.*

about 5 years, during which time period most patients can maintain employment. This is followed by a 5–10 year period during which dopamine replacement therapy provides sufficient relief to manage the disease, albeit with impairments. An additional 10 years of quality life may be gained in some patients through implantation of a deep brain stimulator.

Although PD is typically considered a nongenetic disorder, about 15% of patients have a first-degree family member who also suffers from PD. In about 5% of these, de novo mutations in the LRRK2 gene are found, but causality to disease for these mutations is rarely proven.

3.4 Related Movement Disorders

Essential Tremor (ET): Although tremor is one of the hallmarks of PD, tremors also occur in other movement disorders and, indeed, we all have physiological tremors that are quite

normal. Normal tremors are typically invisible, but can show up in periods of stress or after consumption of coffee or nicotine. One of the diseases often misdiagnosed as PD is called ET or benign tremor. It affects primarily the arms and hands, but can include the head and even voice. These tremors are high-frequency movements (4–12 Hz) that begin unilaterally and, with time, involve both sides of the body. A simple way of diagnosing and monitoring ET is having the patient draw an "Archimedian" spiral (Figure 6), which should look smooth in a normal person but is jagged in a person suffering from ET. Head tremors cause shaking "no–no" or "yes–yes" movements. ET is the most common movement disorder and affects 5% of the general population and up to 9% of the population over age 60. Tremors typically worsen with age, becoming higher in amplitude and therefore more pronounced. These particularly impair the dominant extremities and can make simple tasks

(A)   **(B)**

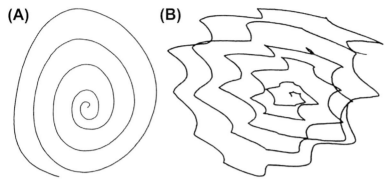

FIGURE 6 "Archimedian" spiral drawn by an individual with tremor (B) and an unaffected individual without tremor (A). The jittery appearance of the spiral indicates the presence of tremor. *Reproduced with permission from Ref. 4.*

such as eating, drinking, or personal hygiene difficult, and operating machinery impossible. Since ET was initially believed to only involve motor function while sparing overall cognitive health, these were originally called benign tremors. However, there is now clear evidence that patients with ET also develop cognitive decline, anxiety, and hearing loss. Although ET does not reduce life expectancy, it can nevertheless be quite disabling. Fortunately, for many patients, tremors are treatable. Arm and hand tremors respond well to simple beta-blockers such as propranolol, with few side effects. Others respond to the phenobarbital analogue primidone. Severe cases may require placement of a deep brain stimulator.

Dystonia: Involuntary sustained muscle contractions that cause the body to twist into an abnormal shape, often with spastic contractions, are the classical symptoms of a heterogeneous collection of conditions called dystonias. These symptoms can start in early childhood, and the earlier their onset, the more severe the condition becomes. Among adults, cervical dystonia is the most common, with 15,000 newly diagnosed patients each year in the US. It causes the neck to be unusually tilted or rotated and is a very painful condition. Dystonias can be the primary presenting feature, or they can be secondary to an injury or disease. The recent identification of over 20 genes involved in dystonia suggests that there is a clear underlying genetic cause for most forms of dystonia. In spite of the advances in identifying disease-causing genes, little is known about its cause, and treatment options are few and largely ineffective. In rare forms patients respond to levodopa; in others, anticholinergic drugs or GABA agonists such as baclofen or clonazepam may help. If the dystonia is focal, botulinum toxin injections can transiently relieve symptoms. DBS is also being used in severe forms of dystonia, with varying success.

4. DISEASE MECHANISM/CAUSE/ BASIC SCIENCE

Since PD is by far the most studied and best understood of the movement disorders, we will focus almost exclusively on its etiology. Because it is primarily a movement disorder, it is useful to briefly review the functional neuroanatomy of normal control of movement, as this will aid our understanding of the defects present in patients with PD.

4.1 Dopamine and the Control of Movement

Willful movements originate from the primary motor cortex, which is located anterior to the central sulcus. Similar to the sensory cortex, the motor cortex is organized somatotopically and this organization is called the motor

FIGURE 7 The motor homunculus based on Wilder Penfield's original drawings from 1940 shows the cortical representation of our body with regard to control of voluntary movement (public domain).

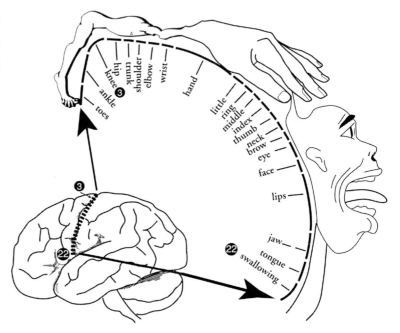

FIGURE 7 The motor homunculus based on Wilder Penfield's original drawings from 1940 shows the cortical representation of our body with regard to control of voluntary movement (public domain).

"homunculus" (Figure 7). The innermost region of the cortex controls movements of the foot, and the outermost regions control hand, face, and tongue. Given the importance of fine movements in our hands, made possible through coordination of 34 distinct muscles and 123 ligaments, a disproportionately large amount of motor cortex is devoted to it.

There are several parallel pathways projecting from the primary motor cortex to the body, with some redundancy between them. Direct projections via the corticospinal tract innervate motor neurons in the spinal cord that control muscles in the extremities. Stimulation of neurons with such direct projections in the primary motor cortex causes discrete movements that involve only a small muscle group. An example would be bending of the index finger. The vast majority of movements, however, arise differently. For these, the primary motor cortex activates adjacent cortical premotor and supplementary motor areas (Figure 8). Stimulation of individual neurons in these associated cortical regions elicit more complex patterns of movement, such as gestures or poses that require coordinated activation of a number of muscle groups. Examples would be a leg moving in a walking stride or an arm folding over the head. These patterned movements involve projections from the primary motor cortex via the basal ganglia back to the premotor and supplementary motor cortex (Figure 8).

4.2 The Basal Ganglia

The basal ganglia, illustrated in Figure 9, are a collection of nuclei at the interface of the telencephalon, diencephalon, and mesencephalon. In the telencephalon, four structures collectively form the corpus striatum, which we will simply call striatum. These are the caudate nucleus, putamen, nucleus accumbens, and globus pallidus. In the diencephalon resides the subthalamic nucleus (STN), and the single mesencephalic nucleus is called the SN. It has two parts, the pars compacta and pars reticulata. Loss of dopaminergic neurons in the pars compacta causes PD, as already illustrated in Figure 3. The major input to the basal ganglia comes from the cortex

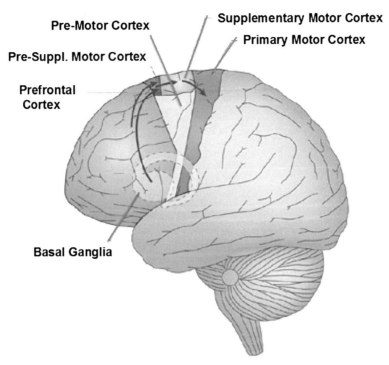

FIGURE 8 Cortical control of movement involves primary motor cortex and premotor and supplementary motor areas. Their activity is modulated by the midbrain, particularly the basal ganglia. *Reproduced with permission from Ref. 8.*

and terminates in the striatum, which runs just underneath the cortex along its entire length. Like the cortex, the input to the striatum is topographically organized, with frontal cortex innervating frontal striatum and posterior cortex innervating posterior striatal neurons. The major output structures from the basal ganglia are the globus pallidus interna and the SN pars reticulata. These signal to the thalamus, which in turn relays this information to the associated motor cortex. The projections from the cortex to the striatum are excitatory and use Glu, while the output from the basal ganglia to the thalamus are GABAergic and therefore inhibitory.

The basal ganglia integrate motor information to filter and integrate incoming and outgoing signals. Overall, their function would be best described as a brake on the motor output; they provide a tonic inhibitory influence on the motor output from the premotor and supplemental motor cortex. For any motor activity to occur, the brake must be transiently removed. This is achieved through dopamine release from neurons in the subtantia nigra, which causes a disinhibition of the output from basal ganglia projecting back to the cortex. High levels of dopamine therefore promote movement (reduced brake action), while low levels promote inactivity (increased brake action). This explains the slowness of movement when dopamine is lost in PD and the chorea-like hyperactive movements when a PD patient has elevated dopamine.

To understand this schema in more detail, we will next examine the pathways and projections involved. There are two distinct pathways at work that balance each other's activity. These are called the direct and indirect pathway, schematically illustrated in Figure 10.

The **direct pathway** sends inhibitory projections from the striatum to the globus pallidus

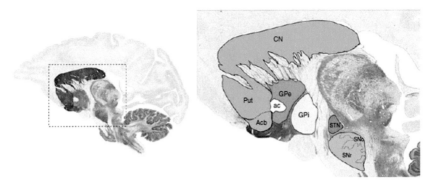

FIGURE 9   The basal ganglia include the caudate nucleus (CN), putamen (Put), nucleus accumbens (Acb), and globus pallidus internal (GPi) and external (GPe). All four are part of the telencephalon. The subthalamic nucleus (STN) and substantia nigra (SN), with its two parts, pars compacta (SNc) and pars reticulata (SNr), modulate the output of the globus pallidus to the thalamus to control the associated motor cortex. The labeled histological sections are parasagittal section through the monkey brain (stained with the acetylcholinesterase method) showing the localization and boundaries of all major components of the basal ganglia system. *Reproduced with permission from Ref. 9.*

internal (GPi). The GPi then sends inhibitory projections to the thalamus, which sends excitatory projections to the associated motor cortex. If we add this all up, activity of the direct pathway causes excitation. One easy way to remember this is to simply multiply the excitatory (+1) and inhibitory (−1) activities. We have two inhibitory and one excitatory:

(−1)*(−1)*(1) = 1 => excitatory action on the cortex.

The **indirect pathway** starts in a different set of neurons in the striatum that sends inhibitory projections to the globus pallidus external (GPe). These neurons make inhibitory projections to the STN, which in turn make excitatory connections to the GPi. This in turn is connected, as we learned above, by inhibitory connections to the thalamus, which then excites the motor cortex. Thus, for the net effect, we have three inhibitory and one excitatory connection:

(−1)*(−1)*(−1)*1 = −1 => inhibitory action on the cortex.

Both of these pathways are under the direct control of the dopaminergic output from the SN that project to the striatum, called the nigrostriatal projections. The striatal neurons that receive synaptic input from the striatum do so via D1 dopamine receptors, which depolarize in

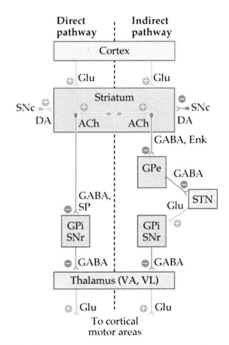

FIGURE 10   Direct and indirect pathway of movement control by the basal ganglia illustrated as a schematic. Abbreviations are as follows: glutamate—Glu; acetylcholine—ACh; enkephaline—Enk; dopamine—DA; substantia nigra pars compacta—SNc; substantia nigra pars reticularis—SNr; globus pallidus external—GP3; glubus pallidus internal—GPi; subthalamic nucleus—STN. *Reproduced with permission from Ref. 10.*

response to dopamine, resulting in excitation. By contrast, the striatal neurons that originate the indirect pathway express D2 dopamine receptors that hyperpolarize the cell in response to the same input. Both act reciprocally, with increases in dopamine activating the "excitatory" direct pathway while inhibiting the "inhibitory" indirect pathway. In PD, the nigrostriatal projections are lost, which causes the excitatory direct pathway to be silent and the inhibitory indirect pathway to be active, resulting in reduced motion.

The above-described parallel loops of direct and indirect pathway that balance each other's activity to assure smooth motor control were described in the 1980s and have served us well toward understanding the pathophysiology of movement disorders. However, although this schema already appears complicated, it does not nearly capture the complexity of the actual wiring of the basal ganglia, and leaves out many projections that contribute to the integrative functions that these nuclei perform. Even today, much of basal ganglia function remains only partially understood.

It must be stressed that the above model only explains one of the cardinal motor features of PD, namely, a poverty of spontaneous movement, or bradykinesia. It does not explain the rigidity and tremor at all. Moreover, it also provides little toward our understanding of the nonmotor symptoms.

4.3 How Might We Explain the Progressive Neuronal Loss Primarily in the SN?

The defining histopathological feature of PD is the selective loss of dopaminergic neurons in the SN. As illustrated from lesion experiments, for example, as seen with exposure of drug addicts to the neurotoxin MPTP, this loss of neurons is sufficient to explain the motor symptoms of PD and it is generally accepted that a loss of these neurons causes PD. That said, this population of neurons is very small when compared to the rest of the

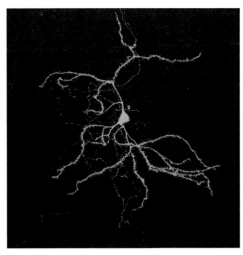

FIGURE 11 A medium spiny neuron in the mouse striatum densely covered with synaptic spines. These are the sites receiving glutamatergic input. The cell was filled with biocytin and labeled with a streptavidin-conjugated fluorophore (green). *Image was kindly provided by Dr Rita Cowell, University of Alabama Birmingham.*

brain. What causes these neurons to die while others are spared? If we knew the answer to this question, we might be able to prevent PD from occurring. There are a number of plausible vulnerabilities emerging, and some may ultimately lead to novel ways of preventing disease.[11]

A starting point to answer this question is to look at the major neuronal types in the basal ganglia and the transmitters they use.

The striatum is by far the largest structure within the basal ganglia. It runs the entire length of the cortex, where it receives input from both ipsilateral and contralateral cortex. The primary striatal neuron is called the medium spiny neuron; this type of neuron makes up 95% of all neurons in the striatum. These cells have elaborate dendritic arbors that are studded with dendritic spines (Figure 11). Spiny synapses are always excitatory and are the sites of glutamatergic input from the cortex. These neurons also receive dopaminergic input from the SN via the nigrostriatal pathway. Neurons originating in the subtantia nigra have extensively branched axons, with each making up to one million synaptic contacts onto striatal neurons, where they primarily release dopamine.

The two pathways involved in control of movement not only differ with regard to their output, but also receive their dopaminergic input via different dopamine receptors. The direct pathway responds via D1 receptors; the indirect pathway, via D2 receptors. Both are G-protein-coupled receptors (GPCRs). D1 receptors couple via Gs type proteins to activate adenyl cyclase to increase the production of cAMP. The D2 receptors do the opposite. They couple via Gi proteins to inhibit adenyl cyclase and therefore inhibit production of cAMP (Figure 12). The concentration of cAMP, in turn, changes the excitability of striatal neurons through complex interactions with ion channels and transmitter receptors.[12] D1 activation causes depolarization and enhances excitability; D2 activation is inhibitory. There are equal numbers of D1-expressing and D2-expressing medium spiny neurons in the striatum.

The release of dopamine occurs from GABAergic neurons in the SN. These neurons have intrinsic rhythmic electrical activity and function as "pacemakers" that drive tonic dopamine release in the striatum. Rhythmic activity entails frequent depolarization, during which Ca^{2+} enters cells through voltage-activated Ca^{2+} channels. This Ca^{2+} entry is an integral part of the pacemaker activity, as it activates Ca^{2+}-activated K^+ channels required for the repolarization following each Na^+-dependent depolarization. This constant influx of Ca^{2+} is one vulnerability believed to play an important role in the pathogenicity in PD.

For the pacemaker neurons to fire action potentials (APs) at a rate of 5–10 per second, they require a lot of energy. Each AP causes a Na^+ influx that needs to be removed via the Na^+/K^+-ATPase. Each membrane depolarization also activates Ca^{2+} channels, causing a Ca^{2+} influx. Unlike many other pacemaker neurons, SN neurons do not have significant Ca^{2+} buffering capacity in the form of proteins like parvalbumin that bind Ca^{2+}. Therefore, Ca^{2+} must be constantly shuttled out of the cell or into

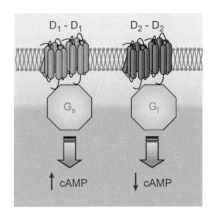

FIGURE 12   Stimulation of the D1 receptor (left) couples via the stimulatory G protein G_s to increase cAMP, causing the striatal neuron to be excited, whereas activation of the D2 receptor (right) couples via the inhibitory G protein G_i to reduce cAMP levels and inhibits striatal neurons. *National Institutes on Drug Abuse.*

organelles against a steep (10,000-fold) concentration gradient, again taxing cellular energy. As illustrated in Figure 13, a number of pump systems that are either directly or indirectly fueled by ATP are constantly at work to maintain this balance. The Na^+/K^+ ATPase returns the Na^+ and K^+ concentrations to normal. Multiple Ca^{2+} ATPases remove the Ca^{2+} that enters as cells depolarize. The synaptic vesicles are loaded using a H^+ gradient across the vesicle that is established by a vacuolar H^+-ATPase.

To meet this incredible energy demand, SN neurons must use every ounce of glucose to produce ATP in the mitochondria by oxidative metabolism, which generates 36 moles of ATP per mole glucose. This conversion occurs in the cellular energy generators, the mitochondria. Not surprisingly, SN neurons are rich in mitochondria. To be particularly effective in providing energy where it is most needed, mitochondria travel close to the sites of highest energy demand: the axon initial segment, where the rhythmic APs are generated, and synaptic terminals, where energy is needed to sustain synaptic activity.

The emerging picture is that SN neurons requires very high mitochondrial ATP production

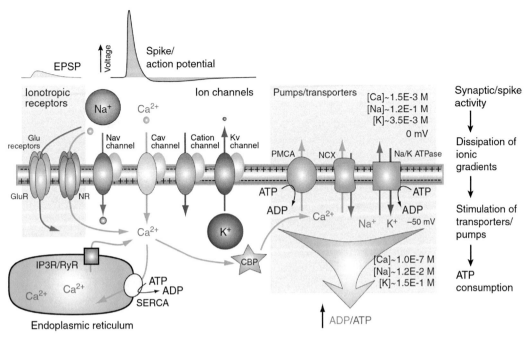

FIGURE 13   Ion flux via numerous ion channels and transporters across the neuronal membrane is essential for neuronal activity. Fluxes are particularly pronounced in pacemaking neurons, which persistently fire action potentials at a high frequency. Ionic movement is directly or indirectly fueled by ATP, putting cells at a high risk for failure should energy supplies deplete. *Reproduced with permission from Ref. 11.*

and, as such, are dependent on superior mitochondrial function and health. Of course, we already learned that essentially all neurons are vulnerable to cell death. This is perhaps best exemplified by ischemic stroke, where energy substrates are lost. Yet the SN neurons have an additional vulnerability with regard to mitochondrial function.

Several chemicals that were shown to kill SN neurons and produce Parkinson-like symptoms turned out to be mitochondrial poisons. The pesticide rotenone, the defoliating chemical Agent Orange, and the synthetic heroin byproduct MPTP are prime examples discussed in more detail later.

It is well known that during normal oxidative ATP production, mitochondria are somewhat sloppy in their handling of oxygen, and produce some reactive oxygen species (ROS) (Figure 14). These are highly reactive molecules with unpaired electrons, and include superoxide O_2^-, hydrogen peroxide H_2O_2, and hydroxyl

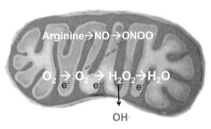

FIGURE 14   Mitochondrial production of energy by oxidative phosphorylation produces reactive oxygen (ROS) and nitrogen species (NOS), particularly O_2^-, OH^-, and H_2O_2. These radicals with unpaired electrons are highly reactive and are destructive to DNA, RNA, proteins, and lipids. In PD, the high energy consumption of pacemaking neurons generates an overabundance of ROS and NOS.

radicals $-OH^-$. These ROS molecules can react with lipid membranes, causing lipid peroxidation, but can also damage RNA and DNA. ROS damage to mitochondria induces apoptotic cell death and, consequently, enzymes such as superoxide dismutase or cellular antioxidants such as glutathione attempt to contain toxic concentrations of ROS in healthy mitochondria.

The mitochondria are unique organelles that contain their own mitochondrial DNA (mDNA), through which they produce many of the enzymes required for oxidative metabolism. Most important among these are proteins that form complex I–IV, the sites at which the proton gradient across the mitochondrial membrane is converted to chemical energy in the form of ATP. This mDNA is at constant risk of being damaged or mutated through mitochondrially produced ROS. This in turn can result in further production of ROS, causing a "perfect storm" scenario whereby mitochondrial ROS cause runaway damage. To add insult to injury, the oxidation of dopamine and its metabolites can produce additional ROS,[13] and levels of the protective antioxidant glutathione are reduced in PD patients.[14] As we will see further below in the section on genetics of PD, mutations that occur in familial forms of PD affect two proteins, Parkin and PINK1. These proteins normally recognize damaged mitochondria and assure that they are disposed of rather than producing further damaging ROS; but, after mutation, these proteins now keep faulty mitochondria hanging around SN cells.

Interestingly, a relationship between metabolism, energy use, and aging had already been proposed as long ago as the early 1900s, based on the finding that organisms with low baseline metabolism live longer than those with high energetic demand. Subsequently, Harman proposed that degenerative diseases may be related to the damaging effects of ROS (Harman 1956, as discussed in Ref. 11). The oxygen free radical theory of aging then morphed into the mitochondrial theory of aging, with the recognition that most of the ROS were produced in conjunction with the mitochondrial electron chain.

Why Then Does It Take 60 Years or Longer for PD to Develop?

As with all neurodegenerative diseases, the time course of disease is a big question mark. Cumulative toxicity, combined with environmental exposure risks that can hasten the onset of disease, is certainly an attractive explanation for the progressive late-onset nature of PD. While we do not know if any of the above-discussed vulnerabilities occur in dopaminergic SN neurons, one can make the argument that from an evolutionary point of view it is acceptable for these neurons to be vulnerable and "live on the edge." Note that the safety margin for dopamine in the brain is extraordinary. Approximately 80% of all dopaminergic SN neurons must die before motor symptoms manifest clinically.[15] Obviously, in most of us, an 80% loss does not occur over our lifetime and we do not develop PD. In those who unfortunately do develop PD, it typically will have taken 60 years to develop. This is well past the usual years of procreation for most people, and therefore the continuation of the human race is not at risk because of this neuronal loss. Because of this, there is no evolutionary pressure to select against this phenotype.

The **cellular inclusions called Lewy bodies** are the pathological hallmark of PD. These were already recognized in 1912 by Friedrich Lewy and have since served as a pathological signature for PD (Figure 4). They are formed predominantly by accumulations of α-synuclein and fill the cell bodies (perikarya) and processes of neurons. They are sometimes found in oligodendrocytes as well, albeit much less frequently. Lewy bodies contain dense globular material as well as 10–15 nm straight filaments,[16] which can be formed from recombinant α-synuclein in the test-tube. **α-Synuclein** is an abundant protein highly expressed in neurons of all types. It accounts for 1% of total cytosolic protein in the nervous system; however, its function is not entirely understood. It is found in close proximity to synaptic vesicles and appears to modulate the SNARE complex of proteins (SNARE—Soluble NSF Attachment Protein, NSF—N-ethylmaleimide-sensitive factor) involved in vesicle binding, having a dampening effect on synaptic vesicle release. Cytosolic α-synuclein does not have a tertiary structure, yet when it associates with phospholipid membranes it assumes an α-helical structure. When it is overexpressed, however, it changes to a β-sheet

BOX 1: THE UBIQUITIN PROTEASOME SYSTEM (UPS)

Most nervous system proteins are short lived (hours to days) and are being constantly replaced. We refer to this as protein turnover. Also, misfolded or otherwise damaged proteins are rapidly cleared from the cell. Protein degradation and recycling occurs via the ubiquitin proteasome system (Box Figure 1). At the heart of this machinery is the proteasome, a large intracellular complex that binds and enzymatically cleaves proteins through proteolysis. It works in conjunction with a family of enzymes that targets proteins for degradation by attaching ubiquitin labels to their surface. Ubiquitin is a small, abundant 76-amino-acid polypeptide that becomes

activated by three enzymes called E1/E2 and E3 ubiquitin ligases. By using ATP, these activate and then covalently bind ubiquitin to a protein substrate. Multiple ubiquitins are typically attached. The result is polyubiquitinated protein recognized by the 19S proteasome cap, which removes the ubiquitin for future use. It then shuttles the protein to be degraded into the hollow core of the proteasome, where the protein is cleaved into its individual amino acids. Here a number of enzymes work together, resembling the activity of caspase, trypsin, and chemotrypsin, to cleave the protein. The identification of which proteins are to be targeted for degradation rests entirely

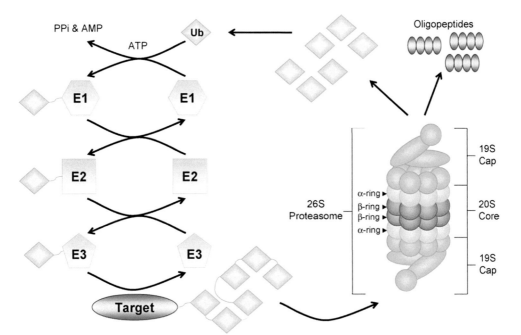

BOX 1-FIGURE 1 The ubiquitin-proteasome pathway. Target proteins of the proteasome are tagged with polyubiquitin molecules in an ATP-dependent process through E1, E2, and E3 ligases. Polyubiquitinated proteins are then recognized by the 19S regulatory complex of the 26S proteasome and fed into the 20S catalytic core for degradation and the ubiquitin molecules recycled. *Reproduced with permission from Ref. 17.*

Continued

BOX 1: THE UBIQUITIN PROTEASOME SYSTEM (UPS) *(cont'd)*

on the ubiquitin ligases, particularly the E3 ligase, the last one in the cascade. A number of proteins have been shown to harbor this activity. Notably, Parkin, which is one of the mutated proteins in familial

PD, has been identified to harbor E3-ligase activity. Mutated Parkin may therefore cause aberrant protein turnover that may contribute to the formation of Lewy bodies.

---

formation, which, much as is the case with amyloid in Alzheimer disease, has a tendency to assemble into sticky aggregates. In PD, mutations in the α-synuclein gene cause a doubling or even tripling of the encoded gene, leading the cell to make an overabundance of the protein. Moreover, abnormal phosphorylation and ROS modifications contribute to the filamentous aggregation of α-synuclein into Lewy bodies.

In addition to enhanced production, inadequate protein degradation via the ubiquitin proteasome pathway (Box, UPS) likely contributes to Lewy body formation as well. Lewy bodies are immune-reactive for ubiquitin and ubiquitin binding proteins. It is debatable whether Lewy bodies are contributing to disease pathology or are simply an inevitable aggregation of diseased proteins, possibly even as a protective measure.

4.4 Genetics of PD

The vast majority of PD cases show no evidence for heritance and are considered idiopathic or with unknown cause. However, a small but significant number of genetic causes of the disease have been reported and provide a starting point toward a molecular understanding of the underlying disease mechanisms that are likely shared in PD.

The first disease-causing genetic mutation in PD was reported in 1996.[18] Today there are 28 known chromosomal regions that harbor disease-causing mutations, yet only six contain genes that, when mutated, actually cause monogenetic PD[19] (Table 1). All six are autosomal, linked to chromosomes other than sex

chromosomes. Two are dominant and four are recessive genes. In dominant inheritance, one mutated allele is sufficient to cause disease, and inheritance usually shows affected individuals in every generation. By contrast, recessive genes only become symptomatic if both alleles are mutated, and these typically skip generations.

The two autosomal dominant mutations unequivocally linked to the cause of monogenetic PD are α-synuclein and LRRK2.

Carriers of **mutated α-synuclein** present with early-onset PD (<50 years) and show rapid disease progression. Dementia and cognitive decline are common, and Lewy bodies are always prominent on histopathological examination. The normal function of α-synuclein has long been a mystery. Mutations in the gene lead to gene amplification and abnormal aggregation into Lewy bodies. Transgenic mice with α-synuclein deletion exhibit increased neurotransmitter release, suggesting that α-synuclein dampens synaptic function.[20] Recent studies show a direct interaction of α-synuclein with molecules that form the SNARE complex responsible for orchestrating presynaptic vesicle docking, fusion, and transmitter release.[21] Synuclein binds directly to synaptobrevin, also known as vesicle-associated membrane protein (VAMP), and stimulates the assembly of a synaptic vesicle complex (Figure 15). Synuclein serves as a molecular chaperone that regulates the availability of synaptic vesicles for release, a function that is impaired when α-synuclein is mutated or aggregated into Lewy bodies.

Mutations in the **leucine-rich repeat kinase 2 (LRRK2) gene** are the most frequent cause of

TABLE 1   Six genes have been identified as causes of familial PD. These encode proteins of different function and characterize different variants of disease as indicated

Protein	Gene	Heritance	Presentation	Function
α-Synuclein	SCNA	Dominant	Early onset	Involved in synaptic vesicle release
LRRK2 (leucine-rich repeat kinase 2)	PARK8	Dominant	Mid-late onset	Large multifunctional signaling protein; synaptic transmission
Parkin	PARK2	Recessive	Juvenile Parkinson's disease (PD)	E3 ubiquitin ligase
PINK1 (PTEN-induced putative kinase 1)	PARK6	Recessive	Juvenile & early onset	Targets malfunctioning mitochondria
DJ-1	PARK7	Recessive	Early onset	Cytoplasmic sensor for oxidative stress
ATP13A, probable cation-transporting ATPase	ATP13A2	Recessive	Early onset	Lysosomal membrane protein

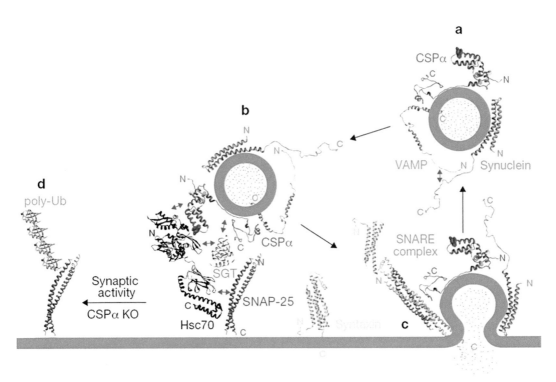

FIGURE 15   Role of synuclein in synaptic function. Synuclein acts as a molecular chaperone, directly interacting with the vesicle associated membrane protein (VAMP) that stimulates the assembly of vesicles containing neurotransmitter. *Reproduced with permission from Ref. 22.*

autosomal dominant PD. The gene is mutated in up to 40% of North African Arabs and 20% of Ashkenazi Jews. The disease is mid-to-late onset and progresses slowly. LRRK2 is a very large protein with many protein–protein interaction domains. It also carries kinase activity. Yet the role of normal LRRK2 remains enigmatic and, consequently, little is known concerning a mechanistic involvement of mutated LRRK2 in disease.

The four autosomal recessive genes unequivocally shown to cause PD are Parkin, PINK1, DJ-1, and ATP13A2. The disease phenotype of all four is indistinguishable, which is likely because they all operate along the same biological pathway related to mitochondrial health. All four genes cause early-onset PD, starting between 30 and 40 years of age. Parkin mutations are the most frequent cause of juvenile PD, starting prior to age 21, and account for over 75% of PD cases under age 30.

The **Parkin** protein encodes for an E3 ubiquitin ligase involved in protein degradation and recycling along the ubiquitin–proteasome pathway (Box). Not surprisingly, mutations in Parkin cause reduced protein degradation; however, unlike with α-synuclein mutations, aggregations of Lewy bodies are usually absent in patients with mutated Parkin.

Similarly, **PTEN-induced putative kinase 1 (PINK1)** is an abundant protein kinase, and recent studies show that PINK1 and Parkin work together to eliminate failing mitochondria from cells. Specifically, PINK1 detects and binds to mitochondria that have a compromised transmembrane potential, which is a sign of functional impairment. PINK1 association recruits Parkin, which labels the mitochondria for autophagic clearance by lysosomes[23] (Figure 16).

The other two recessive genes are relatively rare causes of PD and only a few families have been

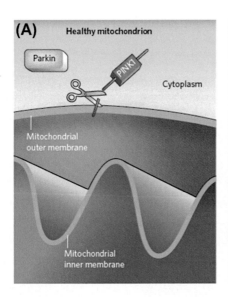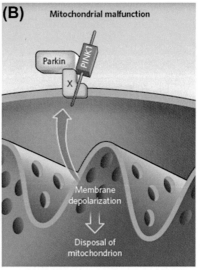

FIGURE 16   Parkin and PINK synergistically act as surveyors of mitochondrial health. (A) In healthy mitochondria, PINK1 is maintained at low levels because it is cleaved by an unidentified protease. (B) When a mitochondrion malfunctions, with associated depolarization of the membrane electrical-potential gradient, PINK1 is stabilized at the outer mitochondrial membrane with its kinase domain facing the cytoplasm. Directly or indirectly through an unknown protein, X, PINK1 then recruits Parkin to the mitochondrial surface. A loss of membrane potential in damaged mitochondria causes PINK1 to stay associated with the outer membrane, where it is then recognized by Parkin. Parkin is an E3 ligase that marks damaged mitochondria for degradation. *Reproduced with permission from Ref. 24.*

reported in the literature. Consequently much less is known about their mechanistic disease contribution. **DJ-1** is a small cytoplasmic molecule believed to be a sensor for oxidative stress, and **ATP13A2** is a lysosomal membrane protein.

Finally, while monogenetic disease causing mutations are the "holy grail" in disease genetics, it is just as important to identify genes that increase disease susceptibility. Unfortunately, in spite of the many suspected environmental risk factors, genes that unequivocally predict increased risk are few. One that does stand out is beta-glucocerebrosidase (GBA), a lysosomal enzyme that is involved in the metabolism of glycolipids. Its loss causes accumulations of glycolipids throughout the body, affecting lung, liver, blood, and nervous system function in a disease called Gaucher disease. Mutations in GBA are found in 8–14% of PD patients, and first-degree relatives of patients with Gaucher disease are at significantly increased risk to develop PD.

Taken together, and graphically depicted in Figure 17, our current genetic understanding suggests that PD is a multifactorial disease that involves many currently unrecognized factors. Genetics establish a certain susceptibility, which, under the wrong environmental influences, such as exposure to toxins, establishes a progressive neuronal loss. This multifactorial picture can be shifted dramatically by the presence of disease-causing mutations, which hasten disease onset, or aberrant exposure to toxins, as illustrated by the effects of rotenone or MPTP.

4.5 Animal Models of PD

In light of the known monogenetic causes of familial PD, where mutations in α-synuclein, LRRK2, Parkin, and PINK1 are sufficient to cause disease, PD appeared to be an ideal disease to model in transgenic mice. Indeed, over the past decade a majority of the mutations identified in

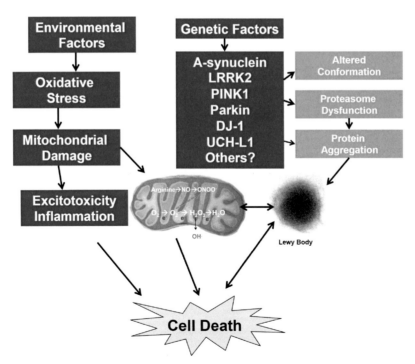

FIGURE 17   Putting the various disease-causing events into relationship reveals multiple ways in which single events or combinations of events or abnormalities can readily contribute to neuronal cell death in PD.

patients have been introduced into mice to evaluate the underlying biology. The result was a mixed bag of success and failure. Only some biological aspects of the disease could be replicated, but the most defining aspect of the human disease, the loss of dopaminergic neurons in the SN, was not.[25] For example, transgenic mice harboring mutations in the α-synuclein gene showed, as expected, that these mutations indeed lead to intracellular protein aggregates reminiscent of Lewy bodies. Somewhat unexpectedly, overexpression of wild-type α-synuclein was also sufficient to do so. However, neither mutated nor overexpressed α-synuclein caused dopaminergic neurons to die.

Similarly, multiple approaches to mutate LRRK2 in mice, using different promoters and strains of animals, have been unable to induce an age-dependent loss of dopaminergic neurons. However, such animals do show impairments in dopaminergic signaling in the striatum, suggesting that the LRRK2 plays a specific signaling role in dopaminergic neurons. Studies on the autosomal recessive genes fared no better, as complete knockout of the Parkin or PINK1 genes each failed to show neurodegeneration of the dopaminergic neurons in the SN.

What these studies collectively illustrate is that in spite of the power generally afforded by transgenic animals, allowing us to rigorously study cellular and molecular mechanisms of defined signaling pathways, it is not always possible to model the complexity of human disease in transgenic animals. It is likely that compensatory mechanisms come into play and assure the survival of these animals. These, in and of themselves, may be interesting to study, as we must assume that such mechanisms may also be at work in humans who are affected by mutations in these genes yet do not become symptomatic for many decades.

Chemically Induced Animal Models of PD

In light of the mitochondrial damage observed in PD, particularly the generation of oxidative stress, environmental toxins have long been suspected to contribute to disease.[26] A major breakthrough regarding the environmental toxin hypothesis of PD came as several drug addicts in California presented with Parkinson-like symptoms after intravenously injecting a contaminant in an attempt to develop a synthetic form of heroin. This contaminant was later found to be MPTP.[27] Most interestingly, these patients were successfully treated with sinemet, suggesting that the underlying symptoms were also caused by the loss of dopaminergic neurons in the midbrain,[28] which was indeed later substantiated on autopsy.

This discovery naturally led to a tremendous interest in MPTP and how it may selectively affect the biology of dopaminergic neurons. Being a lipophilic molecule, MPTP can readily enter the brain, bypassing the blood–brain barrier. MPTP itself is not toxic to neurons. However, MPTP enters astrocytes in the midbrain, which metabolize it via monoamine oxidase B to the highly toxic MPP$^+$. Once released from astrocytes, MPP$^+$ poses as a substrate for the DAT dopamine transporter, which carries the toxic molecule into dopaminergic neurons. Once inside, the drug acts as a potent inhibitor of complex I of the mitochondrial electron transport chain, causing ATP depletion and buildup of ROS (Figure 18). MPP$^+$ can also enter synaptic vesicles and is then co-released from neurons during synaptic transmission.

MPTP causes a selective loss of dopaminergic neurons in the SN and mimics the characteristic motor symptoms of PD in most vertebrate animals from mice to monkeys. Not surprisingly, it has become the standard for evaluating the effectiveness of new drugs to treat PD. However, it must be emphasized that the main shortcoming of this model is the sudden loss of almost all dopaminergic cells, failing to replicate the natural slow etiology of this disease. Consistent with this acute presentation, Lewy bodies are usually absent in affected patients and in animals treated with MPTP.

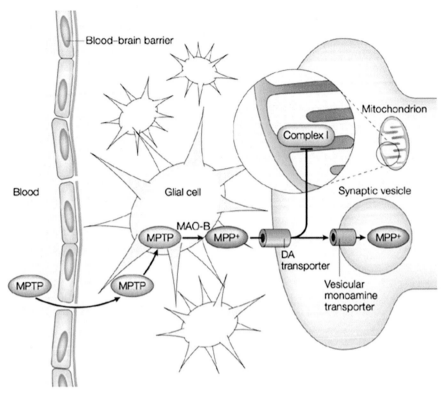

FIGURE 18   The chemical MPTP induces Parkinson-like symptoms by killing dopaminergic neurons. It is converted to the toxic MPP+ molecule that is the substrate for uptake via the dopamine transporter. MPP+ inhibits complex 1 of the mitochondrial oxidative chain, causing a failure to produce ATP. *Reproduced with permission from Ref. 29.*

The findings on MPTP toxicity fueled further research on other environmental toxins. Among the most interesting compounds is the widely used pesticide rotenone. It is structurally similar to MPTP and is naturally found in the seeds of a number of plants, including Mexican yam and kudzu. It was synthesized in 1902 and has been used as an insecticide and fish killer for well over 100 years. Like MPTP, it is highly lipophilic and is taken up by brain cells without any need for active transport. It, too, inhibits complex I of the respiratory mitochondrial chain.

Another chemical that causes less specific but potent neuronal death is the hydroxylated dopamine analogue 6-hydroxy-dopamine. It is taken up into dopaminergic and noradrenergic neurons via DAT, where it produces a high ROS load, killing the affected neuron. The drug does not enter the brain and is typically injected with a stereotaxic needle into the pathway to be lesioned. This gives the advantage that the investigator can lesion only the striatum on one side of the brain, leaving the other side intact to use as a control. This induces hemi-parkinsonism, whereby the animals have the tendency to rotate when walking toward the injected affected side of the body.

Drosophila *as Model to Study PD*

Although mouse models continue to be most widely used in the study of neurological disease, it is increasingly clear that many signaling pathways of the nervous system are highly conserved throughout evolution, with fruit flies expressing analogous proteins. With regard to PD, following some disappointments in the study of genetic mouse models of disease, some remarkable discoveries were made in *Drosophila*. Most important of these was the discovery that Parkin and PINK1 work together to protect cells from nonfunctional mitochondria.[30] Specifically, PINK1 is normally found on the outer mitochondrial wall, but is regularly cleaved unless the mitochondria are compromised in their function, evident from a decreased transmembrane potential. In such cases, Parkin binds to PINK1, inducing the degradation of the nonfunctional mitochondrion (Figure 16). Mutations and loss of function of either Parkin or PINK1 cause impaired mitochondria to stay around and generate toxic ROS, eventually killing the affected neuron. In addition to aiding our understanding of mechanisms of disease, fruit flies may be a great model system in which to search for new drug candidates for PD.

5. TREATMENT/STANDARD OF CARE/CLINICAL MANAGEMENT

Treatment must address both motor and non-motor symptoms of the disease and is staged according to the time-dependent development of symptoms. The initial concern is primarily motor impairment, yet over time, depression, apathy, and cognitive health must be confronted.

5.1 Dopamine and Dopamine Replacement Therapy

Knowing that a deficiency in dopamine release is primarily responsible for the motor symptoms of PD, the majority of treatments are focused on dopamine and its receptors in an attempt to increase dopamine levels in the midbrain.

Dopamine is normally synthesized in the nerve terminals of dopaminergic neurons of the SN. Like other monoamines, dopamine derives from the nonessential amino acid L-tyrosine in a two-step process. Tyrosine hydroxylase produces the intermediary L-DOPA, which is decarboxylated to dopamine by the enzyme aromatic acid decarboxylase. Dopamine is then loaded into synaptic vesicles via the vesicular monoamine transporter VMAT2 for synaptic release. The synaptic action of dopamine is terminated by its removal from the synaptic cleft into presynaptic terminals by the DAT dopamine transporter. DAT is a $2Na^+/1Cl^-$ coupled transporter that harnesses the electrochemical gradient of Na^+ to import dopamine against its concentration gradient back into the cells.

As dopaminergic neurons are gradually lost in PD, replacement of dopamine is vital to ameliorate the motor symptoms, essentially bypassing the tonic dopamine release from neurons in the SN onto striatal D1 and D2 receptors that activate the direct and indirect motor pathways. Unfortunately, dopamine cannot be directly delivered to the brain, since it does not pass the blood–brain barrier. However, its precursor, L-DOPA, is readily taken up from cerebral blood vessels; and once in the brain, L-DOPA is decarboxylated to dopamine. Since this decarboxylation would also occur in the periphery, where the resulting dopamine causes severe adverse effects, including hypotension, impaired kidney function, and cardiac arrhythmias, L-DOPA must be given in conjunction with the dopamine decarboxylase inhibitor carbidopa. Typically these drugs are administered as a single combination tablet. This dopamine replacement therapy using levodopa/carbidopa has become the "gold standard" for treatment of PD, and no other approach has as yet come close in its effectiveness.

Unfortunately the effect of L-DOPA wears off gradually with continued use, providing shorter and shorter "on" times during which the patient finds relief. Consequently, patients require frequent redosing, causing transient overshoots in the delivered dopamine, leading patients to experience dyskinesia, characterized by stereotypic rhythmic movements of the lower extremities. The reason L-DOPA wears off is not fully understood but is very likely related to the progressive loss of neurons that can convert the L-DOPA to dopamine and the nonphysiological concentration spikes of dopamine that may cause changes in receptor expression.

Most patients show the best results from L-DOPA for the first 5–10 years of its use, after which they experience significant side effects and dyskinesia. Therefore many physicians recommend delaying therapy until the patient cannot function without this treatment. To delay the use of dopamine replacement therapy, or to reduce the amount of L-DOPA given, dopamine receptor agonists such as rotigotine can be considered. These are longer-lasting therapies and consequently do not wear off as quickly. Unfortunately, however, as monotherapy, they do not provide the same relief as the levodopa/carbidopa combination. Another way to prolong the action of dopamine is to reduce its metabolism by monoamine oxidase inhibitors. Again, these are particularly effective in conjunction with levodopa.

To fully understand the time course of levodopa therapy, and why it wears off eventually, it is important to learn where and how L-DOPA is converted to dopamine. In the normal brain this obviously occurs in dopaminergic neurons, yet in PD these neurons are gradually lost. The current assumption is that the decarboxylation of L-DOPA also occurs in other monoaminergic neurons, many of which also contain aromatic acid decarboxylase. Interestingly, serotoninergic neurons express the VMAT-2 transporter required to load dopamine into synaptic vesicles. These neurons may in fact release dopamine together with serotonin from their synaptic vesicles. It is also interesting that aromatic acid decarboxylase expression is negatively regulated by dopamine, such that dopamine depletion results in

a compensatory upregulation. This upregulation is not limited only to dopaminergic cells, which may explain why the effect of L-DOPA eventually leads to hyperactivity, requiring patients to discontinue treatment. In dopaminergic neurons, this regulation would calibrate dopamine synthesis to the required amount for release; however, dopamine produced in non-dopaminergic neurons may produce abnormal swings in dopamine concentration. Hence, the regulation of the enzymatic conversion of dopamine may hold a key to unraveling why L-DOPA replacement therapy eventually wears off.

It is also important to emphasize that dopamine is an important transmitter that regulates mood and emotion in other parts of the midbrain, particularly the reward pathway involving the ventral tegmental area (VTA) and nucleus accumbens. Here, drugs of abuse such as cocaine and amphetamine increase the availability of dopamine in the synaptic cleft by inhibiting the DAT transporter either directly (cocaine) or by acting as a competitive substrate (amphetamine).

The interdependence of dopamine as a major transmitter in movement control and in controlling emotions explains why PD patients often show changes in mood and personality, particularly acutely after administration of dopamine agonists. It also explains why excessive use of cocaine and amphetamines can present with Parkinson-like symptoms.

5.2 Deep Brain Stimulation

One of the exciting newer treatment modalities, particularly for patients with advanced PD, is the electrical stimulation of midbrain nuclei via an implanted pacemaker.[31] This approach, called DBS, consists of an implantable stimulus generator similar to a cardiac pacemaker. This device delivers high-frequency (30–150 Hz) electrical currents via a pair of platinum iridium electrodes implanted into the midbrain (Figure 19). A wire connecting the electrode and the stimulator is subcutaneously placed under the scalp and neck (Figure 20). Typical sites of stimulation are the

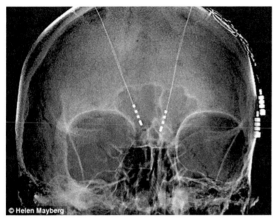

FIGURE 19   Deep brain stimulation (DBS) electrodes and electric leads visualized after implantation through X-rays. *Images courtesy of Dr Helen Mayberg, Emory University.*

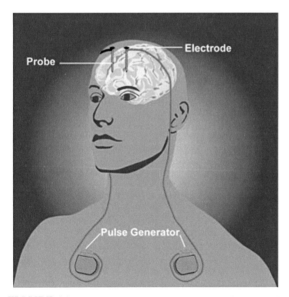

FIGURE 20   Deep brain stimulation set-up schematized. The battery containing pulse generators are implanted subcutaneously near the clavicle. All lead wires are run under the scalp and skin. *National Institute for Mental Health (www. nimh.nih.gov).*

globus pallidus internus or STN. The electrode placement is done under local anesthesia, with the patient being awake and responsive, allowing the surgeon to place the electrode such that stimulation causes an immediate attenuation of tremors or rigidity. After the implantation, further programming of the device tailors the stimulation to achieve maximum relief for the individual patient. DBS is typically used in conjunction with levodopa therapy, and is started 10–12 years after first diagnosis in patients for whom levodopa no longer provides sufficient control of motor symptoms. Over 100,000 patients have now received a DBS device and clinical studies overwhelmingly support the device's safety and efficacy. The US Food and Drug Administration approved DBS for treatment of PD in 2002. Some patients who participated in the early trials have had effective use from a DBS for over 10 years. There is no question that DBS works in most PD patients; however, it is less clear *how* it works.[32] DBS studies in animals and humans suggest that DBS primarily stimulates myelinated nerve fibers rather than neuronal cell bodies. Depending on the frequency and amplitude of stimulation, activity can be induced or inhibited. One theory suggests that the loss of dopaminergic projections to the striatum leads to pathological oscillations of electrical activity within the basal ganglia, and that the frequency of this activity is disrupted by the high-frequency stimulation provided by DBS.[33] It is important to emphasize that like levodopa, DBS treats motor symptoms almost exclusively, leaving dementia and cognitive and emotional symptoms largely untouched. Indeed, since the midbrain structures that are stimulated with DBS are also involved in processing emotions, depression is among the serious side effects encountered with DBS. Other complications are a consequence of the complex setup where electrodes, wires, and batteries periodically fail and require replacement. Frequent surgical interventions, with the possibility of infections, scarring of tissue, and vascular damage, all must be considered in patients who are elderly and often frail. So while effective, this approach is certainly much more involved than taking a pill. At present about 1% of all PD patients take advantage of DBS to manage later stages of disease.

5.3 Neuroprotection

As with other neurodegenerative diseases, an important strategy to consider is a reduction in neurotoxicity, and therefore a slowing of neuronal loss or preservation of existing dopaminergic neurons. Glutamate excitotoxicity was first recognized in stroke but is now much more broadly implicated in neuronal cell death in acute and chronic neurological conditions. Indeed, glutamate toxicity has been implicated as a contributing factor in PD.

A major issue regarding the development of neuroprotective strategies is the inadequacy of our available animal models. As already mentioned, presently available animal models of PD lack the characteristic age-dependent selective loss of dopaminergic neurons, making it essentially impossible to screen for effective compounds in preclinical studies.

Fortunately, one neuroprotective drug was accidentally identified during the treatment of influenza outbreaks in nursing homes. Amantadine was given to treat influenza but, surprisingly, also reduced dyskinesia. Subsequently, it has been shown to block NMDA receptors and may therefore be protective against glutamate toxicity, which typically involves aberrant activation of NMDA receptors, causing an uncontrolled influx of Ca^{2+} that triggers cell death. Amantadine effects are, unfortunately, very short lived but can provide transient symptomatic relief. Another use-dependent blocker of NMDA receptors, memantine, showed symptomatic improvements in some PD patients in early open-label studies; yet, in a larger placebo-controlled study, it was primarily effective in patients suffering from Lewy body dementia.[34]

Consequently amantadine remains the only approved neuroprotective drug for PD.

Intriguingly, epidemiological studies have identified an independent protective effect for coffee, smoking, and use of nonsteroidal anti-inflammatory drugs (NSAIDs). Each reduces the relative risk by about 20–50% and people who smoke, drink coffee, and use NSAIDs have a 62% lower risk to develop PD than people who never used any of these substances. Smoking provided the greatest reduction in risk.[35] This may indeed be one of the few health benefits of tobacco use. It must be stressed, however, that the health risk of lifetime tobacco use by far outweighs the reduction in the likelihood of developing PD. A number of active clinical trials are evaluating the potential benefit of nicotine used as a drug delivered orally or via transdermal patches. Published findings thus far are equivocal, with some showing benefit and others not.[36]

5.4 Treatment of Nonmotor Symptoms

With the effective management of motor symptoms through levodopa/carbidopa and dopamine receptor agonists, the nonmotor symptoms have become the largest challenge clinically. Unfortunately dementia is very common and affects 80% of PD patients. Unlike Alzheimer dementia, which presents with decline in memory and language function, PD dementia primarily affects attention and executive function. Several acetylcholinesterase inhibitors are approved for treatment of dementia, yet their effectiveness is unclear. Up to 50% of PD patients develop depression, and unfortunately levodopa worsens it. Like other forms of depression, it should be treated with selective serotonin reuptake inhibitors. Psychosis and hallucinations can also be a major problem, and are best treated with neuroleptics such as clozapine.

Patients with PD frequently suffer from various sleep problems ranging from insomnia to rapid eye movements (REM) sleep behavior disorder, where an individual acts out dreams during this important phase of the sleep cycle characterized by REM. This condition is best treated with benzodiazepines such as clonazepam that enhance GABAergic inhibition.

6. EXPERIMENTAL APPROACHES/ CLINICAL TRIALS

In spite of the obvious successes in managing many of the disease symptoms, there remains a tremendous unmet need with regard to disease prevention and disease modification, as opposed to disease management. Not surprisingly, there are over 200 interventional clinical studies that are recruiting patients (clinicaltrials.gov); a few emerging new targets and strategies are briefly reviewed here.

6.1 Ca^{2+} Channel Blockers

Our discussion on cellular mechanisms of neuronal death suggested that the use of Ca^{2+} channels in the pacemaker activity of SN neurons, together with a lack of Ca^{2+} binding proteins, put these cells at an elevated risk to sustain a toxic Ca^{2+} overload that could trigger programmed cell death. Therefore, Ca^{2+} channels may be a suitable pharmacological target. The particular Ca^{2+} channel involved belongs to the family of low-voltage gated, L-type channels for which clinically effective drugs (dihydropyridines) are already available. These are typically used to control high blood pressure by blocking peripheral L-type channels in vascular smooth muscle. A clinical study that evaluated risk profiles for the development of PD concluded that hypertensive patients who took Ca^{2+} channel antagonists were indeed at significantly reduced risk to develop PD.[37] Ca^{2+} channel blockers therefore warrant further study, ideally with patients who are at the earliest stages of disease or those carrying familial genetic risk.

6.2 Prevention of Lewy Body Formation

A conformational change in α-synuclein from an α-helical to a beta sheet configuration is pivotal to their aggregation into Lewy bodies. Assuming that misfolded α-synuclein actively contributes to disease pathology and that Lewy bodies are detrimental, preventing their formation would be beneficial. The health benefit of green tea is widely recognized, and is often attributed to antioxidants such as epigallocatechin-3-gallate (EGCH). In animal models of PD, EGCH indeed reduces the formation of Lewy bodies by preventing α-synuclein protein from assuming a beta sheet conformation.[38] This provides proof of principle that molecules that can prevent the formation of Lewy bodies exist, and EGCH may serve as a lead compound to develop a second generation of PD drugs that prevent the formation of Lewy bodies. In the meantime, having a cup of green tea for breakfast may not be a bad idea.

6.3 Cell-Based Therapies

Since the 1970s, scientists have been exploring the use of tissue grafts as a therapeutic strategy in neurological disease. In PD, the early interest focused on the implantation of dopamine-producing cells into the midbrain.[39] The discovery that adrenal chromaffin cells also produce dopamine and survive when grafted into animals led a team of investigators at the Karolinska Institute in Sweden to test this approach in patients. Over several years, beginning in 1982, patients received adrenal cell transplants taken from their own adrenal gland. Although the outcome was largely disappointing, a Mexican research group treated 61 patients, of whom some showed improvements. Many medical centers in the US were exploring this approach throughout the 1980s, but the data were never adequately cataloged or shared and, by and large, the medical community lost interest in this approach.

This was in part because much better outcomes were obtained in preclinical studies utilizing fetal tissue using dopaminergic cells harvested from embryonic human midbrain. However, implementing such an approach in humans was ethically challenging and indeed was prohibited by law in the US in 1988. In Sweden, however, where the attitude toward abortion and the use of fetal tissue was more liberal, clinical studies were initiated in Lund and Stockholm. Here PD patients received fetal tissue implants harvested from the midbrain of aborted fetuses in an open-label, unblinded study, so that patients knew they were indeed receiving a live transplant. The initial results were very encouraging, and patients reported motor improvements. These were associated with increases in dopamine measured by 18F-dopa PET. By 1992, three independent open-label studies reported positive outcomes from this approach in the *New England Journal of Medicine*. With encouragement from US scientists, the newly elected president, Bill Clinton, lifted the ban on federal support for fetal tissue research, paving the way for two placebo-controlled studies in the US funded through the National Institutes of Health. Over a 4-year period, these trials together enrolled 74 patients. Unfortunately, unlike prior studies, their outcome was negative and many patients reported major negative side effects, particularly graft-induced dyskinesia. This stresses the importance of placebo-controlled trials, as open-label studies often report positive findings that may in large measure be attributable to the powerful placebo effect, discussed in greater detail in Chapter 15. No further fetal transplants have since been attempted.

However, scientists continue to debate the principal merit of the approach, fueled in part by reassessment of the original data that revealed that subpopulations of PD patients clearly showed a treatment benefit, and by an overall critical analysis of the study approach identifying poor-quality of engraftment tissue and patient selection as critical flaws. In hindsight, the trials were also criticized for being underpowered, enrolling too few patients to support the conclusions reached. However, unlike with drug treatment, there are significant barriers, including technical, financial, and ethical barriers, to conducting larger-scale placebo-controlled surgical trials. They require the placebo-controlled group to receive an invasive brain surgery without receiving any potential benefit. Importantly, it is unclear what the placebo entails. Should it be dead cells? These could be pathogenic as well.

Under significant political pressure, scientists in the US have now largely abandoned the idea of using cells from aborted fetuses for transplantation. Fortunately, the recent development of techniques to reprogram a person's somatic cells, for example, skin cells, into pluripotent stem cells (so-called induced pluripotent stem cells or IPS cells), has rekindled the interest in autologous implantation of stem cells as therapy. However, these IPS cells have their own issues. Because they are truly multipotent and self-renewing, a major concern is their potential to grow tumors after implantation. Given the relatively long life expectancy of most PD patients, this poses a significant risk. Moreover, successful engraftment with reinnervation of the striatum has yet to be demonstrated for IPS cells in animal models of disease. So at the moment, in spite of much publicity surrounding the promise of IPS cells, the therapeutic use in PD patients is still premature.

For PD, unlike many other neurological diseases, we already have effective treatments available that provide significant relief for patients. Therefore, there is a greater risk of adversity from both the patient and physician perspective. The growing promise of implanted DBS devices is also detracting the attention of clinicians away from cell-based implants.

7. CHALLENGES AND OPPORTUNITIES

Two centuries have passed since James Parkinson observed patients in London experiencing the cardinal symptoms of a disease that would ultimately bear his name and affect millions of people worldwide. The discovery that a discrete lesion in the midbrain is responsible for all the motor symptoms came quickly, yet our understanding of how dopamine-producing cells in the SN die during aging still remains poorly understood. Toxin exposure and rare genetic changes converge on a dysfunction of the neuronal mitochondria being pivotal in the disease etiology. In all likelihood, we are witnessing a normal biological process in this population of neurons that, due to its use of dopamine and its metabolism, is particularly vulnerable. This age-dependent cell loss would not cause disease in most people. Yet, if exposed to stressors such as environmental factors or genetic risks, people develop disease at an earlier age. In the days of James Parkinson, few lived long enough to ever develop disease. Today, a constantly increasing life expectancy and greater opportunity to be exposed to potentially harmful environmental factors heightens the risk for all of us.

It is, however, important to acknowledge that in its own strange ways, PD is among the neurological success stories. Effective treatments are available to manage the disease in most affected individuals for at least 10–20 years, allowing many to live out their normal lifespan with acceptable levels of disability. Unfortunately, current treatment fails to adequately treat the cognitive, emotional, and behavioral symptoms, which are less visible. It is imperative that future efforts focus specifically on understanding the nonmotor symptoms and develop more effective treatment approaches for this largely neglected aspect of PD.

By the time treatment for PD patients begins, most of the dopaminergic neurons are already lost for good. Cell-based implantation strategies may at some point replenish these lost neurons; however, it seems more prudent in the near term to focus research on neuroprotection strategies. We already know that toxins such as MPTP and rotenone kill these neurons by impairing their mitochondrial function. It is likely that additional environmental factors can be identified that, similarly, may impair mitochondrial function and do so in a more gradual age-dependent fashion. The notion that long-term exposure to agents with marginal mitochondrial toxicity would over time "tip the balance" toward disease makes their identification a real challenge. Mice and rats do not live long enough to mimic this cumulative exposure. Therefore we need better animal models. The successful adoption of *Drosophila* as a model system for PD has taught scientists to be more open-minded. *Drosophila* can also be used in large-scale environmental screens. Moreover, cell-based screens, where mitochondrial function is used as a direct readout for mitochondrial toxicity, could be employed to identify disease-causing toxins.

There must be an increased sense of urgency. Unless we identify potential exposure risks and protective strategies soon, the "graying" of the world's population will pose an unacceptable societal burden.

Finally, the recent advances in neuroimaging are beginning to permit the identification of individuals who are at risk to develop disease long before the first symptoms occur. It is hoped that advances in the identification of environmental risk, novel protective strategies, and early detection will converge in a timely manner such that future patients can receive protective treatments long before becoming symptomatic.

Acknowledgments

This chapter was kindly reviewed by Anthony Nicholas, MD, Ph.D., Associate Professor Neurology, University of Alabama Birmingham and Assistant Chief of Neurology, Birmingham VA Medical Center.

References

1. Birkmayer W, Hornykiewicz O. The L-3,4-dioxyphenyl-alanine (DOPA)-effect in Parkinson-akinesia. *Wien Klin Wochenschr*. November 10, 1961;73:787–788.
2. Goetz CG. The history of Parkinson's disease: early clinical descriptions and neurological therapies. *Cold Spring Harb Perspect Med*. September 2011;1(1):a008862.
3. Schwab RS, England Jr AC, Poskanzer DC, Young RR. Amantadine in the treatment of Parkinson's disease. *Jama*. May 19, 1969;208(7):1168–1170.
4. Ostrem JL, Galifianakis NB. Overview of common movement disorders. *Continuum (Minneap Minn)*. February 2010;16:13–48 (1 Movement Disorders).
5. Ross GW, Petrovitch H, Abbott RD, et al. Association of olfactory dysfunction with risk for future Parkinson's disease. *Ann Neurol*. February 2008;63(2):167–173.
6. Abbott RD, Petrovitch H, White LR, et al. Frequency of bowel movements and the future risk of Parkinson's disease. *Neurology*. August 14, 2001;57(3):456–462.
7. Obeso JA, Rodriguez-Oroz MC, Goetz CG, et al. Missing pieces in the Parkinson's disease puzzle. *Nat Med*. June 2010;16(6):653–661.
8. Haggard P. Human volition: towards a neuroscience of will. *Nat Rev Neurosci*. 12//print. 2008;9(12):934–946.
9. Lanciego JL, Luquin N, Obeso JA. Functional neuroanatomy of the basal ganglia. *Cold Spring Harb Perspect Med*. December 2012;2(12):a009621.
10. Leisman G, Melillo R, Carrick FR. Clinical motor and cognitive neurobehavioral relationships. In: Fernando A. Barrios. (Ed.), Basal Ganglia; An Integrative View. Intech, Open Science; 2013. http://dx.doi.org/10.5772/55227.
11. Surmeier DJ, Guzman JN, Sanchez J, Schumacker PT. Physiological phenotype and vulnerability in Parkinson's disease. *Cold Spring Harb Perspect Med*. July 2012;2(7):a009290.
12. Nicola SM, Surmeier J, Malenka RC. Dopaminergic modulation of neuronal excitability in the striatum and nucleus accumbens. *Annu Rev Neurosci*. 2000;23:185–215.
13. Greenamyre JT, Hastings TG. Biomedicine. Parkinson's—divergent causes, convergent mechanisms. *Science*. May 21, 2004;304(5674):1120–1122.
14. Riederer P, Sofic E, Rausch WD, et al. Transition metals, ferritin, glutathione, and ascorbic acid in parkinsonian brains. *J Neurochem*. February 1989;52(2):515–520.
15. Zigmond MJ, Abercrombie ED, Stricker EM. Partial damage to nigrostriatal bundle: compensatory changes and the action of L-DOPA. *J Neural Transm Suppl*. 1990;29:217–232.
16. Dickson DW. Parkinson's disease and parkinsonism: neuropathology. *Cold Spring Harb Perspect Med*. 2012;2:a009258.
17. Landis-Piwowar KR, Milacic V, Chen D, et al. The proteasome as a potential target for novel anticancer drugs and chemosensitizers. *Drug Resist Updates*. 2006;9(6):263–273.
18. Polymeropoulos MH, Higgins JJ, Golbe LI, et al. Mapping of a gene for Parkinson's disease to chromosome 4q21-q23. *Science*. November 15, 1996;274(5290):1197–1199.
19. Klein C, Westenberger A. Genetics of Parkinson's disease. *Cold Spring Harb Perspect Med*. January 2012;2(1):a008888.
20. Abeliovich A, Schmitz Y, Farinas I, et al. Mice lacking α-synuclein display functional deficits in the nigrostriatal dopamine system. *Neuron*. January 2000;25(1):239–252.
21. Burre J, Sharma M, Tsetsenis T, Buchman V, Etherton MR, Sudhof TC. α-Synuclein promotes SNARE-complex assembly in vivo and in vitro. *Science*. September 24, 2010;329(5999):1663–1667.
22. Burgoyne RD, Morgan A. Chaperoning the SNAREs: a role in preventing neurodegeneration? *Nat Cell Biol*. January 2011;13(1):8–9.
23. Youle RJ, Narendra DP. Mechanisms of mitophagy. *Nat Rev Mol Cell Biol*. January 2011;12(1):9–14.
24. Abeliovich A. Parkinson's disease: mitochondrial damage control. *Nature*. February 11, 2010;463(7282):744–745.
25. Lee Y, Dawson VL, Dawson TM. Animal models of Parkinson's disease: vertebrate genetics. *Cold Spring Harb Perspect Med*. October 2012;2:a009324.
26. Tieu K. A guide to neurotoxic animal models of Parkinson's disease. *Cold Spring Harb Perspect Med*. September 2011;1(1):a009316.
27. Langston JW, Ballard P, Tetrud JW, Irwin I. Chronic parkinsonism in humans due to a product of meperidine-analog synthesis. *Science*. February 25, 1983;219(4587):979–980.
28. Langston JW, Langston EB, Irwin I. MPTP-induced parkinsonism in human and non-human primates—clinical and experimental aspects. *Acta Neurol Scand Suppl*. 1984;100:49–54.
29. Vila M, Przedborski S. Targeting programmed cell death in neurodegenerative diseases. *Nat Rev Neurosci*. May 2003;4(5):365–375.
30. Narendra DP, Jin SM, Tanaka A, et al. PINK1 is selectively stabilized on impaired mitochondria to activate Parkin. *PLoS Biol*. January 2010;8(1):e1000298.
31. Kalia SK, Sankar T, Lozano AM. Deep brain stimulation for Parkinson's disease and other movement disorders. *Curr Opin Neurol*. August 2013;26(4):374–380.
32. Vitek JL. Deep brain stimulation: how does it work? *Cleve Clin J Med*. March 2008;75(2):S59–S65.
33. Kringelbach ML, Jenkinson N, Owen SL, Aziz TZ. Translational principles of deep brain stimulation. *Nat Rev Neurosci*. August 2007;8(8):623–635.
34. Emre M, Tsolaki M, Bonuccelli U, et al. Memantine for patients with Parkinson's disease dementia or dementia with Lewy bodies: a randomised, double-blind, placebo-controlled trial. *Lancet Neurol*. October 2010;9(10):969–977.

35. Powers KM, Kay DM, Factor SA, et al. Combined effects of smoking, coffee, and NSAIDs on Parkinson's disease risk. *Mov Disord*. January 2008;23(1):88–95.
36. Quik M, Perez XA, Bordia T. Nicotine as a potential neuroprotective agent for Parkinson's disease. *Mov Disord*. 2012;27(8):947–957.
37. Becker C, Jick SS, Meier CR. Use of antihypertensives and the risk of Parkinson disease. *Neurology*. April 15, 2008;70(16 Pt 2):1438–1444.
38. Bieschke J, Russ J, Friedrich RP, et al. EGCG remodels mature α-synuclein and amyloid-β fibrils and reduces cellular toxicity. *Proc Natl Acad Sci U S A*. April 27, 2010;107(17):7710–7715.
39. Bjorklund A, Kordower JH. Cell therapy for Parkinson's disease: what next? *Mov Disord*. January 2013;28(1): 110–115.

General Readings Used as Source

1. Harrison's online, [chapter 372]: Parkinson's disease and other movement disorders. Warren Olanow C, Anthony HV. Schapira.
2. Przedborski S, ed. *Cold Spring Harbor, Perspectives in Medicine Parkinson's Disease*; 2011. http://perspectivesinmedicine.cshlp.org/.
3. Leisman G, Melillo R, Carrick FR. In: *Clinical Motor and Cognitive Neurobehavioral Relationships in the Basal Ganglia, Basal Ganglia - An Integrative View*. InTech; 2013, ISBN: 978-953-51-0918-1. http://dx.doi.org/10.5772/55227. Available from http://www.intechopen.com/books/basal-ganglia-an-integrative-view/clinical-motor-and-cognitive-neurobehavioral-relationships-in-the-basal-ganglia.
4. Surmeier DJ, et al. Physiological phenotype and vulnerability in Parkinson's disease. *Cold Spring Harb Perspect Med*. 2012;2(7):a009290.
5. Perfeito R, et al. Reprint of: revisiting oxidative stress and mitochondrial dysfunction in the pathogenesis of Parkinson disease-resemblance to the effect of amphetamine drugs of abuse. *Free Radic Biol Med*. 2013;62:186–201.
6. Girault JA. Signaling in striatal neurons: the phosphoproteins of reward, addiction, and dyskinesia. *Prog Mol Biol Transl Sci*. 2012;106:33–62.

7. Tieu K. A guide to neurotoxic animal models of Parkinson's disease. *Cold Spring Harb Perspect Med*. 2011;1(1):a009316.
8. Goetz CG. The history of Parkinson's disease: early clinical descriptions and neurological therapies. *Cold Spring Harb Perspect Med*. 2011;1(1).

Suggested Papers or Journal Club Assignments

Clinical Paper

1. Becker C, et al. Use of antihypertensives and the risk of Parkinson disease. *Neurology*. 2008;70(16 Pt 2):1438–1444.
2. Consortium, I. P. D. G. Imputation of sequence variants for identification of genetic risks for Parkinson's disease: a meta-analysis of genome-wide association studies. *Lancet*. 2011;377(9766):641–649.
3. Powers KM, et al. Combined effects of smoking, coffee, and NSAIDs on Parkinson's disease risk. *Mov Disord*. 2008;23(1):88–95.
4. Freed CR, et al. Transplantation of embryonic dopamine neurons for severe Parkinson's disease. *N Engl J Med*. 2001;344(10):710–719.

Basic

1. Kravitz AV, et al. Regulation of parkinsonian motor behaviours by optogenetic control of basal ganglia circuitry. *Nature*. 2010;466(7306):622–626.
2. Narendra DP, et al. PINK1 is selectively stabilized on impaired mitochondria to activate Parkin. *PLoS Biol*. 2010;8(1):e1000298 (plus see NATURE commentary: Abeliovich A. Parkinson's disease: Mitochondrial damage control. *Nature*. Feb 11, 2010;463(7282):744–745.
3. Burre J, et al. α-synuclein promotes SNARE-complex assembly in vivo and in vitro. *Science*. 2010;329(5999): 1663–1667 (plus commentary: Burgoyne, RD, Morgan A. "Chaperoning the SNAREs: a role in preventing neurodegeneration?" *Nat Cell Biol*. 2011;13(1):8–9).
4. Chan CS, et al. 'Rejuvenation' protects neurons in mouse models of Parkinson's disease. *Nature*. 2007;447(7148):1081–1086.

Diseases of Motor Neurons and Neuromuscular Junctions

Harald Sontheimer

1. CASE STORY

The first unusual symptom Jerry remembers is dropping his pen while taking inventory at the Caremark warehouse he supervises. For no reason, the pen just slipped from his hand. It must have been the chill of winter that left his hands stiffer than usual, since he had not been wearing gloves outside. He did not make much of the episode until it happened again multiple times in the following weeks. One morning he had difficulty starting his car in the garage. He simply could not turn the ignition key. How hard could it be to turn a key, a task he had performed thousands of times before? He was angry and irritated, and he ultimately started the car by using his left hand to turn the key.

As winter gave way to spring, unexplainable episodes like these repeated themselves, and on more than one occasion he experienced embarrassing moments in front of his colleagues or family. One day he dropped a cup of coffee in the office; another day his soup spoon just slipped from his hand, spilling soup all over his shirt. When brushing his teeth, he would occasionally notice annoying muscle twitches. One day Jerry's right foot literally refused to follow his command, failing to push the accelerator pedal as he was driving. He pulled to the side of the road to lift his leg off the pedal with his hands. While the left foot was perfectly fine, the right foot refused to follow his orders. He did not feel any pain, not even numbness. The foot just refused to move at will. Over the following weeks, these episodes kept recurring, and he began to be increasingly impatient and irritated.

Jerry was due for his annual checkup with his internist, Dr Oaks, whom he had been seeing for the past 10 years. When Jerry explained his episodes, Dr Oaks suggested that he might have injured a nerve, possibly in a skiing accident earlier in the year. Just to be sure, however, Dr Oaks ordered Jerry to see a hand specialist. Magnetic resonance imaging (MRI) and an electromyogram (EMG) that tested his arm and hand muscles indicated nothing unusual, so Jerry began to see more specialists. Blood tests ruled out cancer. A spinal exam by his chiropractor was also normal. Eventually, Dr Oaks sent him to see a neurologist who specialized in muscle diseases. The neurologist ran more tests and a week later he called Jerry to his office.

Jerry chose to go without his wife since she had already been overly concerned about his health. The neurologist told him that they had ruled out Parkinson disease and multiple sclerosis, and Jerry felt a sense of relief. Until, that is, the doctor asked, "Have you ever heard of amyotrophic lateral sclerosis (ALS)?"

"Oh my god!" Jerry almost fainted. "That is Lou Gehrig's disease, isn't it?" Of course he had heard of it. As a baseball fanatic he had learned about Lou Gehrig's inspirational story in high school. What to do now? Was there a treatment or a cure? The doctor's recommendation shocked Jerry more than anything else. "Get your affairs in order and write your will and advanced directive. Quit your job and do all the things you ever wanted to do in your life, but do it quickly before it's too late!"

On his way home, Jerry stopped at the baseball field where he used to play in high school. He got out of the car and picked up a glove and ball left by the practicing high school team. He threw the ball as far as he could. He threw another one, and still another one. Then he broke down on the field and started crying inconsolably, until he had no tears left.

2. HISTORY

Muscle weaknesses have been reported since ancient times yet have rarely been attributed to disease, let alone diseases affecting the nervous system. Instead, the Greek philosophers of Aristotle's school believed that the body was controlled by the "psyche," an inner agent and the originator of action that was not a substance but a spirit. Weak people were weak in soul or spirit, rather than suffering from disease.[1] Early suggestions that distinct anatomic entities, namely nerves, communicate with muscles and coordinate their contraction appear in the writings of the Greek philosopher Galen (130–200 ad). In the absence of any knowledge about electrical signals conveyed by nerves, he proposed that nerves originate in the spinal cord and serve as a conduit to propagate the "psychic pneuma" to the muscle. Although Galen recognized the independence of the spinal cord in eliciting voluntary motor movement through reflex pathways, these were discovered in earnest by Thomas Willis (1621–1675). Only when Luigi Galvani (1737–1798) showed that nerves conduct electricity akin to an electric wire and that muscle contraction can be elicited through electric currents was it possible to truly begin to understand the voluntary and involuntary control of movement via the motor cortex and spinal cord. Charles Bell (1774–1842) characterized the motor roots in the anterior spinal cord, and Charles Sherrington (1857–1952) provided the first compellingly complete description of reflex pathways in the spinal cord and the role of the flexion–reflex pathway in the control of limb movement.[2] In his 1910 publication Sherrington also first explained how spinal motor neurons receive their input from the upper motor neurons originating in the motor cortex, which was discovered independently by Fritsch and Hirtzig in 1870. Because we discuss diseases of the motor system here, it is worth emphasizing in this historic account that our understanding of motor control pathways is barely 100 years old. Moreover, the ion channels and neurotransmitter receptors involved in nerve conduction and muscle contraction were not molecularly identified until the 1980s, which illustrates how very young this field of science truly is.

With regard to diseases affecting motor control, the earliest accounts date to the mid-1800s, with a wealth of literature from a number of French and English neurologists who devoted their clinical practices to the study of nerve and muscle disorders. These were classified as either neurogenic or myopathic diseases, respectively. For example, Sir Charles Bell described myotonias, and the French neurologist Guillaume Duchenne described muscular dystrophy.

Around that time the first medical description of ALS by the French neurologist Aran appears; he reported 11 patients who presented with progressive muscle weakness that he termed *progressive spinal muscular atrophy*.[3] In his report he describes a 43-year-old man who initially complained of muscle cramps in his arm. The disease then generalized and eventually affected most muscle groups in his body. The patient died within 2 years. Interestingly, the patient had a remarkable family history; three of his sisters and two uncles died in a similar fashion, clearly suggesting that the disease was hereditary. What was missing in Aran's description, however, was histological confirmation of neuronal death and muscle atrophy. This was provided through the clinicopathological approach first introduced by Jean-Martin Charcot, who published a comprehensive report on progressive motor neuron disease in 1874 and coined the term *amyotrophic lateral sclerosis*, which is still in use today. Surprisingly, Charcot was convinced that, unlike Duchenne's muscular dystrophy, which was discovered around the same time, ALS was not hereditary. This viewpoint persisted until 1993, when 11 mutations in the superoxide dismutase gene were discovered as a cause of ALS in 13 different families and provided the first unequivocal genetic cause of ALS.[4]

In the United States little attention was paid to motor neuron and neuromuscular diseases. This changed in 1938, when the "luckiest man on the face of the Earth" retired from a successful career in professional baseball with the New York Yankees at age 36, having set 7 major league records, including 23 grand slams. It was not his success that prompted Lou Gehrig to leave professional sports, but his beginning battle with ALS, which came to be known more widely in North America as Lou Gehrig's disease. In the latter part of the 1938 season, Gehrig complained to friends, "I tired mid-season. I don't know why, but I just couldn't get going again." As Lou Gehrig's debilitation became steadily worse, he was examined by Charles William at the Mayo Clinic in Rochester, New York, where the diagnosis of ALS was confirmed on June 19, Gehrig's 36th birthday. The prognosis was grim: rapidly increasing paralysis, difficulty swallowing and speaking, and a life expectancy of less than 3 years, although there would be no impairment of mental functions. "Don't think I am depressed or pessimistic about my condition at present," Lou Gehrig wrote following his retirement from baseball. Struggling against his ever-worsening physical condition, he added, "I intend to hold on as long as possible and then if the inevitable comes, I will accept it philosophically and hope for the best. That's all we can do." On June 2, 1941, at 10:10 p.m., 2 years after his retirement from baseball, Lou Gehrig died at his home in the Bronx, New York, and the Yankees retired his no. 4 jersey.

In 1963 21-year-old Stephen Hawking, now regarded as the most brilliant theoretical physicist of our time, was diagnosed with ALS and given just 2 years to live. Through being on a ventilator, Hawking is still alive, and, in spite of losing essentially all motor function except the movement of his eyes, has been able to lecture and write some of the most acclaimed textbooks about the universe and fundamental physical principles, including string theory.

Since Gehrig and Hawking were first diagnosed, relatively little regarding the disease outlook has changed, except for the discovery of mutations in 16 genes that are believed to be causal of both familial and sporadic disease. Unfortunately, none have yet led to better treatments, but the knowledge gained from studying their role in disease biology has been informative.

3. CLINICAL PRESENTATION/DIAGNOSIS/EPIDEMIOLOGY

In this chapter we focus on ALS, one of the most important neurodegenerative disorders in which muscle weakness is the primary and, initially, the only symptom. For completeness, however, we also briefly discuss three related diseases that also present primarily with motor loss: Lambert Eaton myotonia (LEM), myasthenia gravis (MG), and Guillain-Barré syndrome (GBS). The common feature among all four diseases is neurogenic muscle weakness, which is due to pathology of nerve or synapse, as opposed to myopathic diseases, such as myotonias and muscular dystrophies, which affect the muscle per se. In spite of some similarities in their presentation, the four diseases have distinctly different pathology, and therefore treatments and prognosis differ absolutely. In ALS the progressive degeneration of the upper and lower motor neurons, which are responsible for conveying motor commands from the cortex via the spinal cord to the body's voluntary muscles, results in progressive and irreversible paralysis. By contrast, muscle weaknesses associated with LEM and MG are caused by a failure of the neuromuscular junction (NMJ), the synapse delivering signals from the motor nerve to the muscle fiber. Here, insufficient release of the neurotransmitter acetylcholine (ACH) causes LEM, whereas a loss of postsynaptic ACH receptors is responsible for MG. In GBS muscle weakness and paralysis are caused by impaired signal conduction along peripheral myelinated motor nerves connecting the motor neurons with their presynaptic terminals.

3.1 Amyotrophic Lateral Sclerosis

Amyotrophic lateral sclerosis is the most common motor neuron disease, and the term is often used interchangeably with its acronym, ALS. It first was described as a distinct neurological

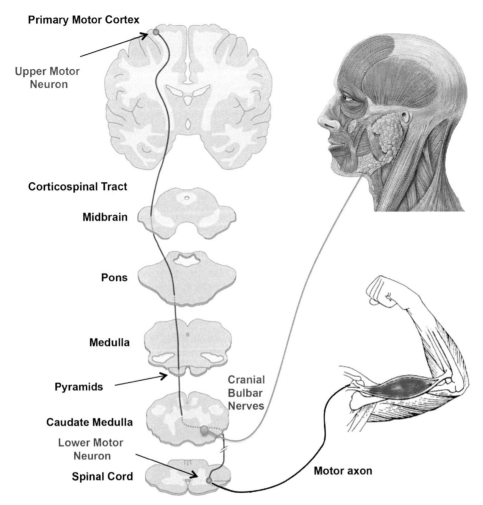

FIGURE 1 Schematic of the human motor system. The corticospinal tract carries motor commands from upper motor neurons to the lower motor neurons in the anterior horn of the spinal cord. These project their axons to the skeletal muscles that are involved in voluntary movement. The lower motor neurons that innervate the bulbar muscles of the face and throat are in the medulla. *Head drawing from Patrick J. Lynch, Medical Illustrator, Creative Commons.*

disease by Marie Charcot in 1879, who also provided a detailed and still accurate description of the disease symptomatology. Charcot first differentiated ALS from MS: the former exclusively affects motor neurons, whereas the latter affects sensory neurons, as well.

ALS is a progressive disease that is caused by the loss of motor neurons in the brain and spinal cord. It affects primarily the upper motor neurons in layer 5 of the motor cortex, whose motor axons comprise the corticospinal tract, which innervates lower motor neurons in the anterior horn of the spinal cord or brain stem (Figure 1). In addition ALS causes progressive loss of interneurons in both spinal cord and motor cortex. The disease is diagnosed almost exclusively by its symptoms, with only confirmatory contributions from MRI and EMG. The cardinal symptom is muscle weakness that progressively involves multiple muscle groups, yet the disease generally spares sensory neurons. ALS typically starts with asymmetric flaccid weakness in one hand or arm, sometimes accompanied by morning cramps. Difficulties with fine motor skills, such as operating the ignition key in a car, may be the first noted symptoms. This may go hand in hand with stiffness of the fingers and overall weakness in one hand. Over the following weeks and months, as the disease progresses, the other hand and arm may be similarly affected. Moreover, arm muscles may show atrophy and spontaneous twitches, often called fasciculations. In some patients the bulbar muscles of the face and throat (Figure 1) that are innervated by cranial nerves IX–XII of the brain stem are the first muscles to be affected, resulting in difficulties chewing, swallowing, and moving tongue and face, which often also are accompanied by muscle twitches. This may produce odd grimaces. In some patients degeneration of the muscles controlling respiration causes premature death.

In most instances, however, weakness progresses continuously, with slightly spastic weakness of the upper extremities and legs. Later in the disease the facial and laryngeal muscles and the phrenic nerves that innervate the diaphragm required for breathing are affected as well. Virtually any muscle group can be the first to show symptoms, and in some patients the legs may be affected before the arms. The disease starts asymmetrically, yet as the disease progresses, weakness is symmetrical and involves essentially all of the body's muscles, except those controlling eye movements and the sphincters of the bowl and bladder.

The disease typically progresses quite rapidly, typically causing death within just 2–5 years, most commonly due to respiratory failure or aspiration pneumonia. With an incidence of 1 in 10,000, ALS is not at all a rare disease and has an incidence comparable with that of multiple sclerosis. The lifetime chance of an individual acquiring ALS is 1 in 400 in men and 1 in 350 in women, with no racial bias. The disease risk increases to 1 in 100 if one parent also has ALS.[5] It can strike as early as 18 years of age or as late as age 90, yet the predominant age at onset is in the late 50s and early 60s, making ALS the most common neurodegenerative disorder of middle age and, arguably, the most devastating neurodegenerative disorder.

While it was long assumed that the disease affects only motor neurons and therefore spares cognitive function, it is now known that up to 50% of patients experience some changes in cognition over the course of their illness, and at least mild deficiencies in executive function, as well as language and memory function, are commonly observed. As we explore further below, this may be explained by a gene defect shared with frontotemporal dementia (FTD; Picks disease), a disease that presents with a profound impairment in executive function, as discussed in Chapter 4.

Risk factors: A number of studies suggest that exposure to pesticides, insecticides, and smoking enhances risk for ALS. Surprisingly, serving in the military, particularly on deployment, is associated with a significant increase in disease incidence, and the US Veterans Administration makes specific accommodations for patients with ALS. The highest documented risk is among professional soccer players, who show a sevenfold increase in incidence compared to the general population.

Genetic risk: By and large, the cause of ALS remains an enigma, but a strong genetic association and at least moderate heritability is well established.[5,6] Population studies suggest that up to 5% of individuals with ALS have an affected family member, as well. Among twins, heritability ranges between 60 and 75%.

Approximately 10% of ALS cases are caused by congenital mutations in one of 16 genes, most of which are inherited in an autosomal dominant fashion. These cases are considered the familial ALS (fALS). Clinically, fALS and sporadic ALS (sALS) are indistinguishable and, interestingly, many mutations are shared between the sporadic and familial forms of the disease, suggesting a common disease mechanism. The first mutations identified were in the superoxide dismutase-1 gene (SOD1) that encodes for a cytosolic enzyme required to detoxify free radicals generated during cell metabolism. This led to the assumption that free radical injury may underlie the progressive neuronal loss. Even mutations that leave the enzymatic activity intact cause neuronal cell death, however, requiring an alternative explanation. A second important gene mutation affects the RNA-binding protein TDP43. This mutation is shared between FTD and ALS, suggesting that these two diseases may also share a common pathology. The recent discovery of an expansion of chromosome 9 open reading frame 72 (C9ORF72) that was found in 38% of fALS

and 7% of sALS cases may well be the crucial turning point in our molecular understanding of the disease.[7]

Prognosis: ALS is always fatal and progresses rapidly in most patients. The only specific treatment is riluzole, a drug that limits glutamate release and inhibits Na^+ channels. It increases survival by only 3 months, however, and does little to improve quality of life. All other treatments are palliative and ideally are administered by a clinical team that specializes in ALS care and includes a neurologist, dietician, physical and occupational therapists, speech therapist, psychologist, and social worker. By proactively managing patient care and educating both the patient and caregivers in the proper management of disease-associated disabilities, this team significantly improves the quality of life of patients dying from ALS.

ALS is used synonymously with motor neuron disease. However, several less common motor neuron disorders, which are typically less severe and progress much more slowly than ALS, deserve mention (Table 1). The three most

TABLE 1   Most Common Adult-Onset Motor Neuron Disorders (Data for the Table in Parts from Ref. 8.)

	Amyotrophic Lateral Sclerosis (ALS)	Progressive Muscular Atrophy (PMA)	Spinal Muscular Atrophy (SMA)	Kennedy Disease
Age at diagnosis	30–60	30–60	20–50	30–60
Duration	12–60 month	Many years	Decades	Decades
5-year survival	14%	56%	100%	100%
Upper motor neurons	Affected	Not affected	Not affected	Not affected
Lower motor neurons	Affected	Affected	Affected	Affected
Weakness	Symmetric distal before proximal	Symmetric predominantly distal	Symmetric proximal	Symmetric proximal
Inheritance	10% familial 90% sporadic	Sporadic	Autosomal recessive SMN gene	X-linked recessive (CAG repeats)
Incidence	1:10,000	1:100,000	1:50,000	1:40,000 males

important are progressive muscular atrophy, spinal muscular atrophy (SMA), and Kennedy disease.

3.2 Progressive Muscular Atrophy

Progressive muscular atrophy accounts for about 10% of all cases of motor neuron disease. Patients with progressive muscular atrophy only show weakness attributed to lower motor neuron dysfunction. While little is known about its genetic causes, the disease seems to be primarily sporadic. It progresses much more slowly than ALS and has a much better long-term prognosis, with 56% of patients alive 5 years after diagnosis compared with only 14% of patients with ALS.

3.3 Spinal Muscular Atrophy

SMA is an autosomal recessive disease that also presents with muscle weakness. An infantile form of the disease is severe with a prognosis comparable to that of ALS. The adult-onset form of SMA, however, is much less debilitating. Weakness is most pronounced in the legs, trunk, shoulders, and upper back. In adults the disease progresses very slowly; only a small percentage of patients eventually require a wheelchair. Both forms are caused by a mutation in the *SMN* (survival of motor neuron) gene family. In the absence of the proteins encoded by these genes spinal motor neurons die and muscles atrophy. SMA affects only the lower motor neurons in the anterior horn of the spinal cord and spares the cranial nerves; therefore the disease spares the bulbar muscles of the face and throat.

3.4 Kennedy Disease

Kennedy disease is an X-linked inherited disease that affects only males in their 30s and 40s. Most patients have facial muscle twitches and hand tremors. Like SMA, the upper motor neurons of the cortex are spared. The genetic cause of the illness is a trinucleotide (CAG)

expansion in the androgen receptor gene on the X chromosome.

Next we briefly discuss three diseases or syndromes that share muscle weakness with ALS as their major symptom. Unlike ALS, these all share an autoimmune disease etiology.

3.5 Myasthenia Gravis

MG is an autoimmune disease that causes fluctuating muscle weakness as the primary symptom. The disease typically begins with weakness of facial muscles that control facial expressions, eye movements, chewing, talking, and swallowing. Slurred speech may be among the early signs, as are drooping eyelids (ptosis) or double vision (diplopia). Weakness in the hands, fingers, arms, and legs can vary as muscles tire with extended periods of use. The symptoms are intermittent, however, and a patient recovers after periods of rest. Symptoms are sometimes less severe in cold weather. The disease is caused by antibodies attacking the nicotinic ACH receptors found on the postsynaptic muscle side of the NMJ, the synapse that connects the lower motor neuron with the muscle (Figure 2). As with many other autoimmune diseases, why the body generates antibodies to its own proteins is unclear. These antibodies are aberrantly produced by circulating B-lymphocytes after they are presented with antigen epitopes from activated T cells. Autoactive T cells that recognize self-antigens are normally removed by the thymus. The antibodies produced can recognize different sites on the ACH receptor. Some interfere with the binding of ACH to the receptor, whereas others destroy internalize the receptor. In either case the released neurotransmitter is unable to activate a sufficient number of postsynaptic receptors. Therefore the muscle action potential is reduced in size, and muscle contraction is consequently weak or absent (Figures 3 and 4). Motor neuron function, as well as presynaptic function, is intact, but the postsynaptic muscle endplate is unable to respond properly to the transmitter released. Removal of

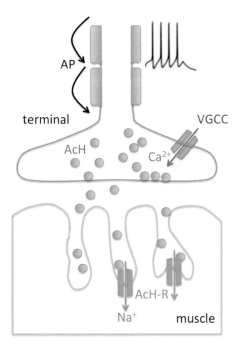

FIGURE 2 Simplified schematic of the neuromuscular junction. Action potentials (APs) arriving from lower motor neurons depolarize the presynaptic terminal, resulting in a Ca^{2+} influx via voltage-gated Ca^{2+} channels (VGCCs). These are the target of autoantibodies in Lambert Eaton myotonia. The released acetylcholine (AcH) transmitter binds to postsynaptic AcH receptors (AcH-Rs) on the motor endplate. Autoantibodies to these receptors cause myasthenia gravis. AcH receptors mediate the influx of Na^+, which gives rise to the postsynaptic muscle AP. This in turn causes muscle contraction via release of Ca^{2+} from the sarcoplasmic reticulum.

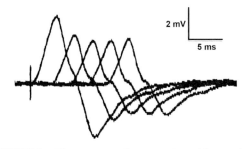

FIGURE 3 Electromyography in a patient with myasthenia gravis shows a characteristic decrease in the amplitude of the compound muscle action potential of the abductor digiti quinti, with repeated ulnar nerve stimulation, reaching the lowest amplitude after three or four stimuli, when the action potentials plateaus. *Courtesy of Dr Peter King, University of Alabama Birmingham.*

antibodies through plasmapheresis transiently reduces symptoms, and drugs that inhibit acetylcholinesterase, the enzyme that breaks down ACH at the synapse, improve muscle contraction by allowing the transmitter to exert its function over a longer period of time.

The disease can be diagnosed by asking patients to repeat muscle contractions such as sit-ups or knee bends or by simply keeping their arms stretched out for 60 s, tasks that they will find difficult to do. EMG is the most sensitive test; it shows a decrease in the muscle action potential with repeat nerve stimulation (Figure 3). Administering edrophonium chloride (Tensilon), often called the Tensilon test, temporarily increases the ACH concentration, thereby increasing muscle strength and mitigating the decrease in the muscle action potential. Patients with MG are typically treated with immunosuppressant drugs to reduce a further decrease in antibody attacks on their ACH receptors. Removing the thymus can stabilize the disease in some patients. Plasmapheresis can be used as emergency treatment to temporarily remove as many of the autoreactive antibodies as possible.

Most patients with MG have a normal life expectancy, yet quality of life can be negatively affected depending on the severity of the autoimmune response. The disease presents most commonly in women 20–40 years old, whereas men are typically over 50. With an incidence of 2 in 10,000, MG is slightly more common than ALS.

3.6 Lambert Eaton Myotonia

LEM is characterized by muscle weakness that is entirely caused by a deficit in presynaptic release of ACH. The autoimmune loss of presynaptic P/Q-type Ca^{2+} channels deprives the presynaptic terminal of the ability to release vesicles containing ACH (Figure 2). LEM often is associated with systemic cancers such as small-cell lung carcinomas, and the disease belongs to a group often called paraneoplastic syndromes, which are diseases that result indirectly from an

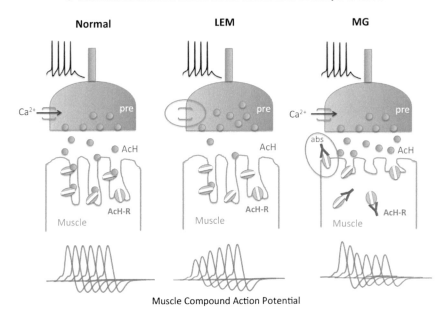

Muscle Compound Action Potential

FIGURE 4 Schematized sites of pathology comparing Lambert Eaton myotonia (LEM) and myasthenia gravis (MG) with the neuromuscular junction in an unaffected normal individual. In LEM presynaptic (pre) Ca^{2+} channels are lost because of an immune attack, causing fewer transmitter-containing vesicles to be released, whereas in MG it is the postsynaptic acetylcholine (AcH) receptors that are reduced in number by the immune attack. The compound muscle action potential normally has a constant amplitude yet shows a marked reduction in LEM that increases gradually as more transmitter is released. In MG the action potential decreases over time since not enough AcH receptors (AcH-Rs) are available for activation with repeated stimulation.

association with cancer. It is believed that small-cell carcinomas express similar Ca^{2+} channels, and as the immune system mounts an attack on these cancer cells, antibodies to epitopes of these Ca^{2+} channels are generated. These in turn erroneously attack the presynaptic Ca^{2+} channels on the NMJ. LEM is a relatively rare disorder, with an incidence of 3 or 4 in 1,000,000. It can occur at any age but is most frequent in people over 40 who smoke. Interestingly, it is not uncommon for neurological symptoms to present first, leading to a subsequent cancer diagnoses. LEM primarily affects proximal muscles, that is, those closest to the trunk, like the arms and legs. Weakness of eye muscles is uncommon. The muscle weakness in LEM is distinctly different from that in MG in that repeated use of the muscles or stimulation of the nerve causes a gradual strengthening of the muscle as the residual Ca^{2+} channels allow more and

more Ca^{2+} to enter, thereby allowing more and more vesicles to be released into an otherwise normal NMJ. These differences are illustrated graphically in Figure 4, comparing LEM and MG with a normal NMJ. Since there is relatively little ACH present, the use of the acetylcholinesterase inhibitor edrophonium chloride has little effect. Lung radiographs typically reveal a tumor, and thus treatment involves care for the systemic cancer, primarily through radiation and chemotherapy.

3.7 Guillain-Barré Syndrome

GBS affects the peripheral nerves that convey motor signals from the spinal cord or brain stem to muscles. GBS is a syndrome as opposed to a disease because it has a spectrum of presentations. Symptoms develop days to weeks after a gastrointestinal viral infection or, in rare cases, after vaccination. Unlike ALS, MG, or LEM, GBS resolves spontaneously in most patients after the infection causing the disease clears. Progressive weakness, typically symmetrically affecting arms and/or legs, develops over 2–3 weeks and then resolves. In severe cases the person is entirely paralyzed and requires mechanical ventilation. GBS is a relatively rare disorder, with an incidence of 1 in 100,000. The muscle weakness is the consequence of an autoimmune response whereby antibodies recognize and degrade proteins in the myelin sheath of peripheral nerves. The NMJ is entirely normal and signal conduction of the lower motor axon is impaired. This can be readily diagnosed through conduction velocity studies of the nerve. How the infection leads to the generation of antibodies against the body's own proteins is not known, but it is hypothesized that the infection sensitizes T lymphocytes that cooperate with B lymphocytes to produce antibodies. While symptoms resolve spontaneously in most patients, recovery can be accelerated and disease severity reduced by plasma exchange (plasmapheresis) and/or treatment with high doses of immunoglobulins. Around 70% of patients recover completely in 6–36 months, whereas 30% retain some residual muscle weakness.

3.8 Differential Diagnoses for Motor Neuron Diseases

We have already discussed several types of muscle weakness with distinctly different underlying pathologies and causes. A number of other disorders, however, some of which are also discussed in this book, can produce remarkably similar motor symptoms and must be considered as alternative possibilities. These include, for example, poliomyelitis, postpolio syndrome, multiple sclerosis, peripheral neuropathy, neurosyphillis, Lyme disease, heavy metal poisoning, or vitamin B_{12} deficiency. Tumors that compress the cervical region of the spinal cord can produce weakness in the upper limbs and spasticity in the legs that can be mistaken for ALS or lymphoma;

multiple myeloma can cause neuropathy in the lower motor axons. Given the poor prognosis of ALS, ruling out any of these less severe and often treatable alternative explanations before rendering an ALS diagnosis is imperative.

4. DISEASE MECHANISM/CAUSE/ BASIC SCIENCE

Of the neurological motor weaknesses, ALS is by far the best understood and arguably the most important disorder, and we therefore restrict our discussion of disease mechanisms to ALS.

4.1 Motor Pathways

To fully understand the muscle weakness encountered in ALS, a brief review of the anatomy and function of the primary motor pathways is useful. Movements of voluntary muscles originate with signals from motor neurons in the motor cortex. These send axon bundles toward the brain stem and spinal cord. These axon bundles are called pyramidal tracts because they pass through the "pyramids" of the medulla (Figure 1). They include both corticospinal and corticobulbar tracts. The corticobulbar tract carries information to the motor nuclei of the cranial nerves in the brain stem that innervate muscles of the face and neck and are involved in facial expressions, chewing, speaking, and swallowing. These are often called the bulbar muscles. The corticospinal tract carries motor signals to the spinal cord, where they synapse on motor neurons in the anterior (ventral) horn. Note that the corticospinal and corticobulbar tracts each cross to the other side of the body in the medulla (Figure 1); hence the left brain innervates the right part of the body and vice versa. The motor neurons in the cortex are called upper motor neurons; the ones in the anterior horn of the spinal cord are called lower motor neurons. The lower motor neurons extend their axons into the periphery, where they innervate muscles via a specialized synapse, the NMJ.

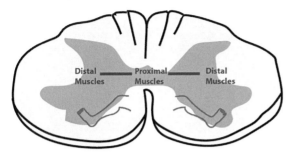

FIGURE 5   Topographic organization of the motor neurons in the anterior horn of the spinal cord.

In the spinal cord motor neurons are organized in a topographic manner (Figure 5), with distal muscle groups innervated by more peripheral motor neurons and proximal muscle groups innervated by motor neurons closest to the center of the cord. Adjacent motor neurons innervate adjacent muscle groups or myotomes, and the progression of ALS progressively affects contiguous myotomes. Groups of neurons that innervate the same muscle aggregate along a column that extends about 2–3 spinal cord segments.

Each muscle cell is innervated by only one lower motor neuron, but each motor neuron branches and innervates up to 100 muscle fibers. These contract together and form a motor unit (Figure 6). Each muscle, therefore, is innervated by many motor neurons, which contract in an orderly fashion based on the size of the motor neuron. Those units that produce the least amount of force fire first and those that produce the maximal force fire last. This allows for a finely tuned muscle contraction that is appropriate for a given motor task. The contraction of muscles is fine-tuned via sensory muscle spindles that monitor contractile force and extension. These signal back to the anterior horn, where interneurons modulate the activity of the lower motor neuron. Muscle spindles are also part of the tendon reflex, often used to examine functionality of the spinal motor segment.

The main visible pathology in ALS is a progressive degeneration of upper motor neurons

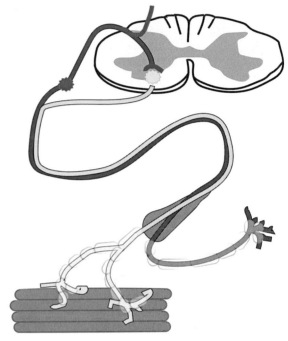

FIGURE 6   Innervation of multiple muscle fibers by a single motor neuron. These multiple fibers constitute a motor unit.

and degeneration of their axons along the corticospinal tracts that innervate the lower motor neurons. The loss of the corticospinal tract is visible during autopsies of affected individuals (Figure 7).

Lower motor neurons innervate muscle cells at the NMJ, a complex, highly folded membrane structure that contains a large number of presynaptic ion channels, vesicles containing neurotransmitters, and, on the postsynaptic side, a high density of ACH receptors (Figure 3). Here, the arriving action potential causes the presynaptic Ca^{2+} channels to open, and loading with Ca^{2+} causes the release of vesicles containing ACH. The released ACH then activates ACH receptors on the postsynaptic muscle. As these receptors flux Na^+ ions into the cells the muscle depolarizes, which excites the muscle to generate an action potential. The muscle action potential causes further Ca^{2+} release from the sarcoplasmic

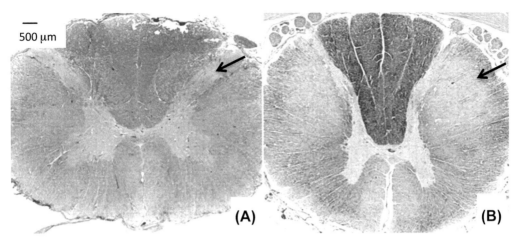

FIGURE 7   Spinal cord pathology in amyotrophic lateral sclerosis showing mild (A) and severe atrophy (B) of corticospinal tracts (arrows). *From Ref. 9.*

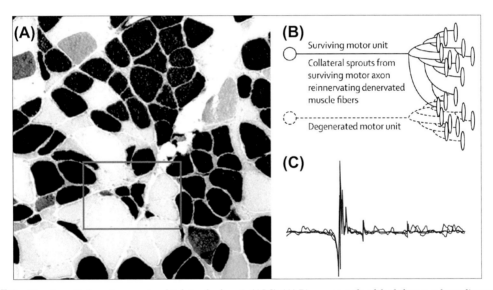

FIGURE 8   Muscle pathology in amyotrophic lateral sclerosis (ALS). (A) Biopsy sample of the left vastus lateralis muscle from a patient with ALS, stained with ATPase (pH 9.4). The biopsy sample highlights grouped atrophic fibers with both type I and type II fibers (mixed-type fibers, encompassed by the red box). Pathophysiology of motor unit degeneration and reinnervation (B), with superimposition of 10 traces (C) demonstrating the typically large, polyphasic, unstable (complex) motor units observed in established ALS (sweep duration, 50 ms), with late components, indicating some reinnervation. *From Ref. 10.*

reticulum in the muscle fiber, which initiates the actin–myosin contraction of the muscle.

As the neuron that innervates a motor unit becomes diseased, its axon goes through a period of unpredictability wherein aberrant action potentials cause twitching or fasciculation of a motor unit. Such twitches are commonly observed in the early stages of ALS, and their electrical activity can be measured through an EMG (Figure 8) that reports voltage changes via fine needles inserted into the muscle. EMG serves as a useful tool to diagnose diseases affecting muscles, motor neurons, and the NMJ. As the axon dies back, the muscle becomes denervated and undergoes atrophy.

Lower motor neurons also receive modulatory input from the basal ganglia and a number of cortical pathways that travel through the brain stem, including the corticorubrospinal, corticovestibulospinal, and corticoreticulospinal pathways. The modulatory neurons that give rise to these tracts also are designated as upper motor neurons. In the end lower motor neurons in the spinal cord receive both excitatory input from the upper motor neurons and a mixture of modulatory, excitatory, and inhibitory inputs from sensory muscle spindles and modulatory descending pathways. The modulation of the lower motor neurons explains why upper and lower motor neuron lesions present with opposite paralysis. When lower motor neurons or their axons cease to function, muscle tone is reduced and the muscle shows a complete absence of contraction, even when a reflex pathway is activated (areflexia). This is often called flaccid paralysis. By contrast, when only the upper motor neurons are lesioned, reflex pathways remain intact, and, indeed, show more brisk reflexes since the descending inhibitor modulatory pathways that normally dampen their activity are lost. Muscles have increased tone but are unable to receive excitatory commands from the upper motor neurons, causing spastic paralysis and hyperreflexia. Neurologists often take advantage of this distinction and use a common reflex, the plantar reflex, to determine whether a patient has upper or lower motor neuron signs. When the sole of the foot is stroked, the toes curl inward in a normal individual. An upward response is called a positive Babinski sign (after the neurologist who first described this test in 1896) and suggests hyperreflexia and, hence, an upper motor neuron problem.

Obviously, lesions at any point along this motor pathway disrupt muscle control in different manners, and the symptomatology allows a trained neurologist to pinpoint the site of the lesion. As already mentioned, flaccid paralysis points toward lower motor neurons, whereas spastic paralysis points toward an upper motor lesion. Muscle fatigue that improves with high-frequency stimulation of the presynaptic terminal or upon administration of drugs that prevent the breakdown of ACH is likely caused by inadequate neurotransmitter release, characteristic of LEM. By contrast, little change to the muscle contraction with repeated stimulation would be expected in MG, though muscle exercise slightly improves the condition. Any lesion of the peripheral nerve, for example, due to demyelination, as is the case in GBS, presents with lower motor neuron symptoms yet typically does not spare the sensory nerves, which would also be blocked. In addition to symptomatology, careful use of EMG and nerve conduction studies can pinpoint the lesion in each given disease (Figure 4).

4.2 ALS Pathology

The cardinal feature of ALS is the progressive death of motor neurons in the cortex and spinal cord (Figure 7), causing their corresponding muscle fibers to atrophy over time (Figure 8(A)). Early in the process, muscle can be innervated through sprouts from adjacent motor nerves (Figure 8(B)), giving rise to aberrant electrical activity (Figure 8(C)) and muscle twitching. Ultimately, complete denervation can be visualized through muscle biopsies and gives rise to the term *amyotrophic*, which simply means progressive muscle wasting. Upper motor neuron loss causes progressive thinning of the corticospinal tracts that is sometimes visualized by MRI (Figure 14). Remarkably, the sensory neurons remain completely unaffected. For unknown reasons, a few motor systems remain unaffected, such as the motor neurons that mediate eye movements and parasympathetic neurons in the sacral spinal cord that control the sphincter muscles of bowel and bladder.

Upon autopsy, motor neurons from patients with ALS show characteristic cytoplasmic

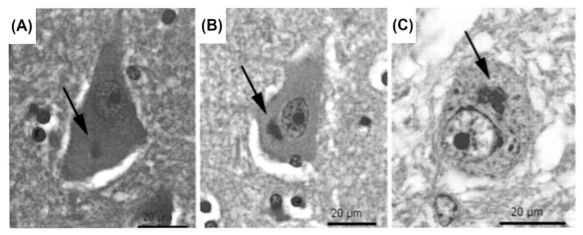

FIGURE 9   Examples of inclusion bodies in three representative examples of motor neurons. The arrows point to compact basophilic neuronal cytoplasmic inclusions. (A) and (B) are upper motor neurons in layer V of the primary motor cortex from two different patients. (C) is an example of a lower motorneuron in the nucleus hypoglossus of yet another ALS patient. *From Ref. 9.*

inclusion bodies, sometimes called stress granules. These appear to contain aggregated proteins (Figure 9). In addition, accumulation of neurofilament in the cell body and in the initial axon segment are pathological hallmarks of ALS.

To better understand the many mechanisms that may explain motor neuron cell death and the progression of the disease, we first examine known genetic causes of ALS and animal models derived from them for the study of disease mechanism(s).

4.3 Gene Mutations Associated with ALS and Heritability

Heritability of ALS is clear from epidemiological studies, which unequivocally support the existence of sALS and fALS.[5,11] These seem to be clinically indistinguishable, and essentially all gene mutations that have been found in fALS are also mutated in sALS. In general, fALS has an earlier onset than sALS and often shows more rapid progression. Many forms of fALS may be misdiagnosed as sporadic because of the incomplete penetrance of many ALS genes, which often leave family members who are gene carriers unaffected. Hence the true prevalence

of fALS is probably underestimated, with many cases being erroneously labeled sALS.

As of 2014, mutations in 16 genes have been unequivocally linked to ALS; some of the most important ones are listed in Table 2. Some of these affect large numbers of patients, others just a small minority. The proteins encoded by these genes span many aspects of cell biology, including the regulation of transcription and translation, mitochondrial function, protein trafficking, and structural support. It seems likely that ALS is polygenetic; subtle changes in multiple proteins confer an increased risk for motor neuronal death, with an environmental component required to trigger disease.

Superoxide Dismutase-1 (SOD1): In 1993 linkage analysis studies showed that 11 missense mutations in the superoxide dismutase (SOD)-1 (*SOD1*) gene are responsible for fALS. These studies provided the first clue that a point mutation in a single gene is sufficient to produce ALS. Today, 166 different mutations that account for over 20% of all fALS cases and up to 7% of sALS cases have been reported.[5] Overexpression of mutated *SOD1* yielded the first transgenic mouse model for ALS, which recapitulated progressive motor weakness and shortened

TABLE 2   Most Common Gene Mutations Identified in Amyotrophic Lateral Sclerosis (ALS)

Gene	Locus	Mutations (n)	Prevalence of fALS/sALS (%)	Inheritance
SOD1	21q22.1	166	20/7	Dominant
FUS	16q1.2	42	6/2	Dominant
TARDBP/TDP-43	1q36.2	44	6/2	Dominant
C9orf72	9p21	?	40–50	Dominant
VAPB	20q13	2	<1	Dominant
ANG	14q1.2	17	<1	Dominant
OPTN	10q15	5	<1	Dominant
ATXN2	12q24	6	<1	Dominant
UBQLN2	X	?	<1	Dominant

fALS, familial ALS; sALS, sporadic ALS; SOD1, superoxide dismutase-1; FUS, fused in sarcoma; TARDBP, Tar DNA-binding protein; C9orf72, chromosome 9 open reading frame 72; VAPB, VAMP-associated protein type B; ANG, angiogenin; OPTN, optineurin; ATXN2, ataxin-2; UBQLN2, ubiquilin 2 or ubiquitin-like protein 2.

lifespan. SOD is a ubiquitous enzyme that is tasked with the conversion and neutralization of highly reactive superoxide (oxygen with an extra electron) to either oxygen or hydrogen peroxide. The enzyme is a dimer of two 153-amino acid polypeptides, each containing a catalytic Cu^+ and a stabilizing Zn^+ ion. Given its role in the degradation of free radicals, it was initially thought that the motor neuron loss in ALS might be caused by the toxicity of free radicals. That view has since been revised, however, given evidence that even catalytically active mutant SODs that effectively dispose of superoxide are still toxic to neurons. Interestingly, expression of mutated *SOD1* in glia rather than neurons is sufficient to induce ALS in mice. Therefore, in spite of unequivocally causing disease, the mechanism(s) whereby *SOD1* mutations cause neurodegeneration and the role of neuronal versus glial expression of the mutated gene remain to be explored.

TDP-43 (transactive-region DNA-binding protein gene, TARDBP, also known as TDP-43) and FUS (fused in sarcoma): The cytoplasmic inclusions (Figure 9), or stress granules, altered in ALS are believed to function as storage sites for silenced messenger RNA to protect RNA during periods of cellular stress. Once stress periods have resolved, these messenger RNAs are released into the cytoplasm for protein biosynthesis. In ALS these granules often contain aggregated or mutated forms of two messenger RNA/DNA processing proteins, namely, TDP-43 and FUS.[12]

TDP-43 is a 414-amino acid protein with two RNA binding motives. It shuttles between the nucleus and the cell cytoplasm, yet most of it is found within the nucleus (Figure 10). TDP-43 was first identified as a binding partner of the transactivation response element of human HIV virus, hence the name TAR DNA-binding protein 43. TDP-43 acts to suppress transcription alongside other related regulators, such as MeCP2, which, when mutated, causes Rett syndrome. TDP-43 has highly diverse targets that include over 6000 RNAs, equal to 30% of a cell's entire transcriptome, making it difficult to elucidate the most important targets in ALS. Among the identified targets, however, are proteins that have been suspected to

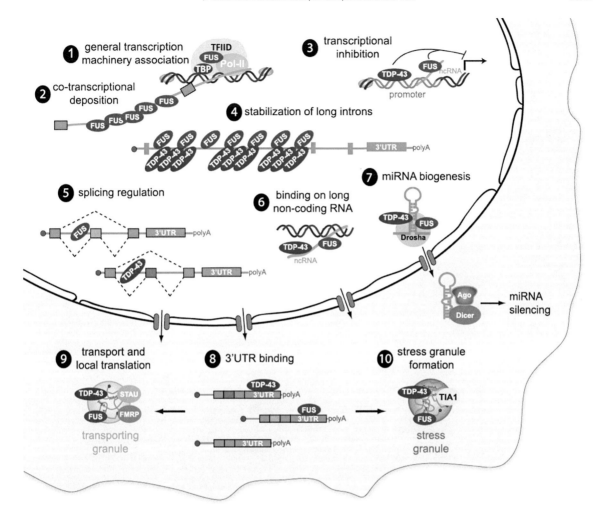

FIGURE 10 Physiological roles for TDP-43 and FUS. Proposed roles for FUS/TLS include general transcriptional regulation (1) and cotranscriptional deposition (2). Both TDP-43 and FUS/TLS associate with promoter regions (3). TDP-43 binds single-stranded, TG-rich elements in promoter regions, thereby blocking transcription of the downstream gene. In response to DNA damage FUS/TLS is recruited in the promoter region of cyclin D1 (CCND1) by sense and antisense noncoding RNAs (ncRNAs) and represses CCND1 transcription. Both TDP-43 and FUS/TLS bind long intron-containing RNAs (4), thereby sustaining their levels. TDP-43 and FUS/TLS control the splicing of >950 or >370 RNAs, respectively, either via direct binding or indirectly (5). TDP-43 and FUS/TLS bind long ncRNAs (6), complex with (7) Drosha (consistent with an involvement in micro RNA [miRNA] processing), and bind 3′ untranslated regions (UTRs) of a large number of messenger RNAs (8). Both TDP-43 and FUS/TLS shuttle between the nucleus and the cytosol and are incorporated into transporting RNA granules (9) and stress granules (10), in which they form complexes with messenger RNAs and other RNA-binding proteins. *Figure and legend reproduced with permission from Ref. 12.*

contribute to disease, including, among others, the glial excitatory amino acid transporter 2 (EAAT2), presenilin, α-synuclein, parkin, neurexin, and neuroligin. TDP-43 need not be mutated to cause disease pathology. Instead, abnormal phosphorylation or ubiquitination is sufficient to cause protein aggregation, and these aggregated proteins no longer participate in RNA processing. In addition, over 40 mutations of *TDP-43* that lead to protein aggregates have been documented. Interestingly, selective expression of mutant *TDP-43* in astrocytes rather than neurons has induced progressive motor neuron loss and paralysis. As already discussed above for *SOD1*, this suggests the contribution of non-neuronal cells to causing so-called non-cell-autonomous toxicity in ALS, an emerging concept discussed further below.

One of several FUS proteins associates with RNA polymerase, affecting the transcription of a diverse set of genes. FUS can also associate with nuclear hormone receptors to cause gene-specific transcription. In addition to DNA, FUS can bind to over 8000 RNAs and alter protein production by RNA splicing; therefore FUS can be described as a multifunctional DNA/RNA regulator that can bind and modify both DNA and RNA. Like *TDP-43, FUS* loses it function when aggregated in its wild-type form or through one of over 50 reported mutations. Loss of function causes abnormal spine morphology, presumably by affecting specific synaptic proteins. *FDP-43/FUS* mutations are found in 4–6% of fALS cases and up to 2% of sALS cases. Surprisingly, *FUS* mutations have thus far failed to be produced in an animal model of disease.

FUS and *TDP-43* are not the only leading gene candidates to cause ALS, but they are also implicated in the pathology of FTD, the second most common dementia after Alzheimer disease. FTD is accompanied by neural atrophy in the frontal and temporal cortex with profound personality changes and impairment of language and executive function. The clinical presentation of these two diseases also overlaps, whereby 15% of patients with FTD also present with typical features of ALS, and as many as 15% of patients with ALS develop the key symptoms of FTD. Therefore it seems that these two proteins contribute to certain forms of ALS that share common cognitive deficits

with FTD. Since both of these proteins are involved in RNA processing, from a disease mechanism point of view, these findings are exciting because they suggest that errors in RNA processing may be the central pathogenic mechanism in both ALS and FTD. Figure 10 illustrates the many physiological roles served by FUS and TDP-43, where dysfunction at each of the illustrated stages may be contributing to disease.

The chromosome 9p21 locus: Among the most exciting recent findings on the genetic causes of ALS is the discovery of an intronic GGGGCC repeat expansion on C9ORF72 that is present in a large number of patients with ALS.[11] Normal individuals have fewer than 24 of these repeats, yet individuals with ALS may have up to 1600 repeats, affecting the expression of a number of downstream genes. These expansions are found in 20–50% of fALS and 20% of sALS cases. In addition, they are present in 80% of familial ALS/FTD cases and 10–30% of FTD cases, making the C9ORF72 expansion the single most common cause of ALS and FTD. When present in ALS, the affected individuals show a severely progressive disease, often with bulbar onset. Half of those patients show signs of cognitive impairment and 30% show dementia. How C9ORF72 causes specific motor neuron toxicity is not fully understood, but essentially all patients with C9ORF72 also show TDP-43 inclusions; hence defective RNA processing as a result of TDP-43 aggregation is one likely mechanism of disease. The fact that in a given family the same C9ORF72 expansion can cause FTD in one family member and ALS in another suggests that these two diseases are different manifestations of the same disease, which exists on a spectrum.

4.4 Animal Models

The discovery that mutant *SOD1* is a cause of fALS allowed the first mouse model to be generated, which reproduced salient aspects of the disease.[4] In this transgenic mouse model overexpression of mutated *SOD1* with a glycine-to-alanine substitution at position 93 (SOD1^{G93A}) was sufficient to cause motor neuron disease; mice became paralyzed in one or more limbs and died prematurely at age 5–6 months. Surprisingly, the enzyme activity of the mutated SOD1 was largely unaffected, suggesting that the mutation causes a toxic gain of function through an unknown mechanism, presumably via protein aggregation. Many studies since have used this model of point mutations at different sites in the *SOD1* gene, each providing important clues about disease biology, such as glutamate excitotoxicity, mitochondrial dysfunction, inflammation, and the role of glia in disease, discussed further below.[13,14] Moreover, this mouse model has become the gold standard for preclinical evaluation of drug candidates, with prolonged life expectancy and motor neuron sparing being the primary readouts. Unfortunately, the initial excitement about this mouse model has given way to increased frustration, in large part because many of the drugs that worked to increase survival in the mouse model, such as minocycline, creatine, celecoxib, sodium phenylbutyrate, ceftriaxone, WHI-P131, and thalidomide, had no therapeutic effect in subsequent human studies. Whether this can be blamed on the inadequacy of the mouse as a disease model or whether poor study design may have contributed to some of these failures remains to be seen. As with many other diseases, notably stroke and Alzheimer disease, animal models of disease are often a sticking point when it comes to advancing ideas and new treatments to the clinic. In the case of ALS the rapid progression from onset of symptoms to death often requires treatment to be given long before animals become symptomatic, which of course does not replicate the treatment course in humans. Hopefully, the discovery of the *TDP-43/FUS* mutations and the C9ORF72 expansions in ALS will yield additional transgenic mice with a better predictive value for treatment responses in human disease.

4.5 When and How Do Motor Neurons Die?

Assessing the exact onset and progression of disease is difficult yet essential to fully understanding disease pathology. From a clinical perspective, ALS is a progressive illness and suggestive of cumulative functional loss over time. Clearly, the progressive weakness suggests that a gradually increasing number of motor neurons die, causing an expanding weakness that eventually encompasses most muscle groups of the body. In other diseases, however, such as schizophrenia, Parkinson disease, and, in particular, Alzheimer disease, recent research suggests that a prodromal phase exists, during which the disease is developing but only subclinical signs are present. Might that be the case in ALS as well? The available data are scant. Electrophysiological assessment of patients who are *SOD1* mutation carriers shows the abrupt, catastrophic loss of motor neuron function immediately preceding the onset of visible symptoms as opposed to a gradual decline in motor neuron health.[15] However, this does not rule out the existence of a prodromal phase during which neurons are slowly choked to death yet maintain axonal signaling throughout.

Also contributing to the debate is the interdependence of upper and lower motor neurons, given that a distinguishing feature of ALS is the loss of both upper and lower motor neurons. Three competing hypotheses address the sequence of events. The "dying forward" hypothesis suggests that upper motor neurons in the cortex kill the lower motor neurons through release of excessive glutamate. This hypothesis would explain why lower motor neurons that lack direct synapses with upper motor neurons, such as the neurons controlling eye movements and the sphincters of the bladder and bowl, are spared in disease.

Alternatively, the "dying-back" hypothesis suggests that the disease begins in the muscle as opposed to the motor nerves. Here, a loss of tropic factors released from the muscle to the innervating lower motor neurons causes its degeneration, which in turn causes the retrograde death of upper motor neurons. This hypothesis is supported by observations that synaptic denervation often precedes lower motor neuron cell death. However, the putative tropic factor released by muscle cells remains elusive.

Finally, an independent death of upper and lower motor neurons is postulated by the "independence hypothesis," which is largely based on histopathological observation, showing simultaneous and seemingly unrelated loss of upper and lower motor neurons.

One may argue that understanding the progression of cell death along the neuraxis is largely academic, but that is clearly not the case. Understanding how pathology at the cellular level is related to symptomatology is ultimately essential to developing more effective interventions. Most such insight, however, cannot be gleaned from human studies; it must rely on animal models. In the SOD1 transgenic mouse motor neurons clearly die via an apoptotic death pathway, often called programmed cell death. Apoptosis requires the sequential activation of caspase-1 and caspase-3, two proteolytic enzymes that initiate and execute cell death. In SOD1 mice chronic activation of caspase-1 occurs well before the first motor symptoms occur, whereas caspase-3 activation coincides with symptomatic disease onset.[16] Interestingly, infusion of caspase inhibitors protects motor neurons from death and delays disease onset in these mice. So what, then, causes the activation of caspases to initiate apoptotic death?

4.6 Cellular Mechanisms of Motor Neuron Death

Cellular mechanisms of motor neuron death were largely identified in SOD1 mutant mice, although compelling supportive evidence has been derived through sampling of cerebrospinal fluid and through studies of patient derived cells. The best hypotheses to date suggest free radical (reactive oxygen species [ROS]) damage, mitochondrial failure, cytokines or other toxins, toxic protein aggregation, and glutamate toxicity as cellular mechanisms of disease, as illustrated in Figure 11. Importantly, these mechanisms are not limited to motor neurons but may also engage astrocytes, microglia, and oligodendrocytes, making the death "non-cell-autonomous."

ROS: The primary role of SOD1 is to neutralize ROS by converting superoxide to oxygen or hydrogen peroxide. The failure of SOD1 to carry out its normal enzymatic function seems not to cause disease in most instances. Nevertheless, oxidative damage has been proposed as a contributor to disease, possibly as a result of structural changes in the active site of SOD1. These changes expose the copper site to aberrant substrates, resulting in the generation of peroxynitrates that can cause lipid peroxidation.[18] Alternatively, the inclusions seen in biopsies contain insoluble SOD1 and ubiquitin, suggesting that mutated SOD1 causes aggregation of presumably misfolded protein. It is, therefore, possible that ROS and reactive nitrogen species accumulate due to SOD1 aggregation of otherwise enzymatically active SOD1. ROS are certainly highly reactive and destructive and have been implicated in essentially all neurodegenerative diseases to date. Their damaging effects include those on DNA and mitochondrial enzymes, as well as peroxidation of membrane lipids. ROS damage is also an attractive mechanism to explain how glial cells or microglial cells may contribute to neuronal death, as these molecules readily diffuse from cell to cell.

Mitochondrial failure: In most mouse models of ALS and in patent autopsy tissue, abnormal mitochondria are found in motor neurons that present with swelling and vacuoles. Such mitochondria are impaired in their ability to generate high-energy substrates and to buffer Ca^{2+}; they

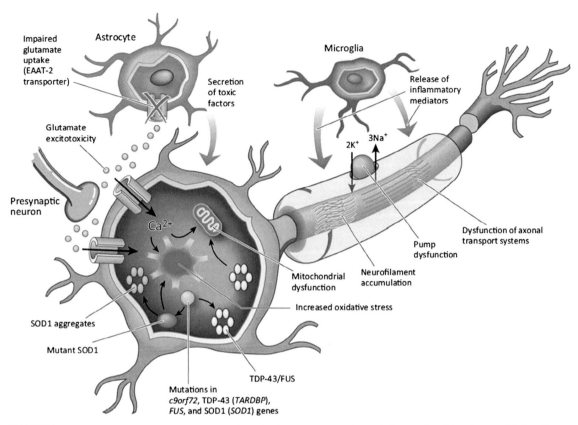

FIGURE 11   Pathophysiology of amyotrophic lateral sclerosis seems to be multifactorial and involves multiple cell types. Mitochondrial dysfunction, aggregation of RNA regulatory proteins, and neurofilament accumulations lead to neuronal and axonal dysfunction. The loss of astrocytic glutamate transporters causes excitotoxic increases in glutamate. The release of inflammatory molecules by microglial cells causes damage to the neuron and axon. *Reproduced with permission from Ref. 17.*

also contain proapoptotic molecules (Figure 12).[19] When mutations in *SOD1* are present, SOD1 aggregates localize to the intramembranous space, as well as to the mitochondrial surface. SOD1 can block an important anion channel in the outer mitochondrial membrane (VDAC-1), which is involved in the production of ATP. Most important, however, mitochondria in ALS show signs of oxidative stress. It may seem counterintuitive that the very organelle that generates the free oxygen radicals, as a natural part of the respiratory chain, is also vulnerable to those same ROS it generated. This vulnerability is due to the large infolded membrane area

where the electron chain occurs, which is itself susceptible to lipid peroxidation. Importantly, mitochondrial damage through oxidative stress can readily become a runaway process whereby initially minor damage causes increased ROS production, which in turn causes further membrane damage, and so forth.

Another important and often overlooked role of mitochondria is the storage of cytosolic Ca^{2+}. Impairments in the mitochondrial membrane potential cause a catastrophic Ca^{2+} overload and inability to buffer Ca^{2+} entry via NMDA receptors. Consequently, motor neurons are more susceptible to glutamate toxicity.

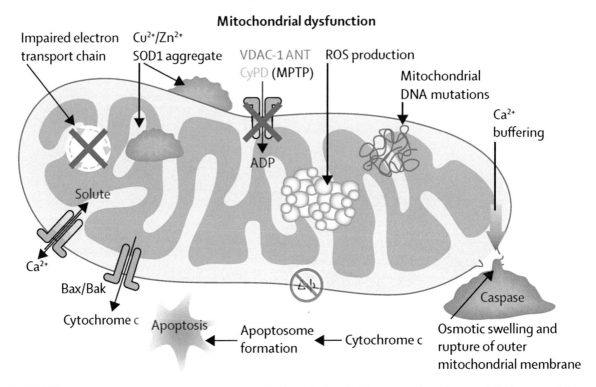

FIGURE 12   Mitochondrial dysfunction in amyotrophic lateral sclerosis. The aggregation of mutant *SOD1* causes a failure in energy production, a breakdown of the mitochondrial membrane potential, and a loss in Ca^{2+} buffering by mitochondria. The release of cytochrome c initiates apoptotic death in the affected neuron. ADP, adenosine diphosphate; ROS, reactive oxygen species. *Reproduced with permission from Ref. 20.*

Finally, release of cytochrome C from the mitochondria into the cytoplasm is an early step in the induction of apoptosis. Mitochondrial stress alone, therefore, is sufficient to induce apoptosis.

Mitochondria are transported via axon transport to the synaptic terminals. Accumulations of phosphorylated neurofilaments impair the successful transport of mitochondria, causing a depletion of these energy-producing organelles from synaptic terminals. Taken together, multiple vulnerabilities exist, whereby even subtle mitochondrial dysfunction may contribute to motor neuron death; these are graphically illustrated in Figure 12.

Glutamate toxicity: Glutamate toxicity[21] is recognized as a mechanism of cell death in stroke, traumatic injury, and glioma, and it is among the leading candidates to cause motor neuron death. Glutamate released from upper motor neurons onto lower motor neurons may not be cleared adequately by surrounding astrocytes, resulting in an overactivation of NMDA or Ca^{2+}-permeable AMPA receptors (Figure 11). The resulting uncontrolled Ca^{2+} influx causes activation of caspases, thereby initiating apoptosis. Evidence in support of glutamate toxicity in ALS initially came from the analysis of cerebrospinal fluid samples taken from patients with ALS; increased glutamate concentrations were found in 40% of patients with sporadic ALS, with glutamate concentration correlating positively with disease severity.[22] Evidence of an impairment of the astrocytic ability to clear glutamate comes from studies of humans and mice. In *SOD1* mutant mice, expression of EAAT2 in the spinal cord is reduced by 50%, and the loss of EAAT2 expression precedes motor symptoms; conversely, overexpression of EAAT2 in astrocytes slows disease progression. Importantly, examination of patient tissue showed abnormal expression of EAAT2 protein in ~80% of ALS cases studied. These findings make a strong case for glutamate dysregulation, particularly deficient glial glutamate reuptake, playing an important role in motor neuron toxicity in ALS. Impairment of glutamate release, however, could be a consequence of the neuronal impairment rather than a cause. What speaks against this is that the RNA for EAAT2 is the target of *TDP-43/FUS*, a leading genetic cause of ALS, which may be one explanation for how this mutation becomes pathological.

Another argument in support of a hypothesized glutamate receptor–mediated toxicity contributing to motor neuron toxicity in ALS comes through a rare related disease called GUAM-ALS. This ALS-like syndrome presents with dementia and sometimes parkinsonian features and was first recognized in the 1940s among tribal natives of the Guam islands in the Pacific. The disease was traced to the dietary intake of β-methylamino-L-alanine (BMAA), a glutamate analog and NMDA receptor agonist that is found in the seeds of the cycad *Cycas micronesica*.[23] The toxin is produced by cyanobacteria living in symbiosis with the plant in its roots. The seeds were a staple food of the Chamorro natives (used as baking flour), yet the accumulation of BMAA to toxic concentrations may have occurred through flying foxes (a species of bats) that also feed on these seeds and then were consumed as a delicacy by the Chamorro. Interestingly, cyanobacteria are ubiquitously found in fresh and marine water throughout the world, suggesting the possibility that exposure to BMAA may not be restricted to Guam islanders eating cycad seeds. Indeed, BMAA can be detected in shrimp and crab and in unfiltered water from Lake Champlain in Vermont. Cyanobacteria are found in many bodies of warm water, as well as cooling towers for buildings; thus it is easy to envision how the bacteria could become aerosolized and inhaled. Along the same lines, BMAA-mediated ALS has been hypothesized as one explanation why Veterans of the 1990–1991 Persian Gulf war are more likely to develop ALS. The desert crust around Qatar contains cyanobacteria and bacterial toxins, including BMAA. Both become aerosolized in desert dust storms and may have been inhaled in excessive quantities by soldiers.

While BMAA initially was primarily assumed to be toxic via activation of glutamate receptors, other toxic mechanisms have now been established. Most interesting among these is the suggestion that during protein biosynthesis BMAA is misincorporated in place of serine into TDP-43. This in turn causes misfolding and aggregation of TDP-43 in stress granules, which we already discussed as being a pathological hallmark of ALS.

Microglially released diffusible toxins: The peculiar spread of disease from one muscle group to an adjacent one over time has led to the hypothesis that a diffusible factor may spread among adjacent motor neurons. Any diffusible factor would have to travel via the cerebrospinal fluid, which can be harvested from patients through a spinal tap. When cerebrospinal fluid from patients with ALS who suffer from either fALS with an *SOD1* mutation, non-*SOD1* fALS, or sALS is administered to motor neuron explants, the CSF causes neuronal toxicity and phosphorylation of their neurofilament cytoskeleton. This seems to be due to microglial release of a toxic mix of cytokines, chemokines, ROS, and reactive nitrogen species. Minocycline attenuates microglial activation in culture and protects neurons from CSF-induced apoptosis. These activated microglial cells seem to release an unknown neurotoxic factor (other than or in addition to glutamate).[24] Over the years, a number of soluble factors have been proposed to be involved in motor neuron death, some released by neurons, others by astrocytes or microglia cells, which are both activated in the context of inflammation and neurodegeneration and have been shown to be the source of free radicals, tumor necrosis factor-α, and glutamate, among others. Their relative role in disease is not well understood, and all may be equal contributors to neuronal death.

4.7 Non-cell-autonomous Toxicity and Neurodegeneration in ALS

The gene mutations in patients with ALS occur in every cell of the body, not just in motor neurons. Similarly, in the *SOD1* mutant mice used for study, mutations were introduced into the germ line to produce an animal model of ALS in which every cell is affected. Therefore it is possible that the mutation may have affected non-neuronal cells, as well.[25] It is even possible that mutations in non-neuronal cells are required to induce disease. For example, the above-described changes in astrocytic glutamate uptake suggest that the ability of glial cells to clear glutamate may be impaired and possibly contributes to neuronal death. Similarly, the release of toxic substances from microglial cells may be the result of mutations affecting them directly.

To address the question of whether mutations in motor neurons are sufficient to induce disease, scientists selectively introduced the *SOD1* mutations only in motor neurons, leaving wild-type SOD in all surrounding cells. Surprisingly, these animals did not show motor neuron loss until the mutation was either also introduced in the microglia cells or introduced in very high copy numbers in the motor neurons. Selective expression only in astrocytes or in microglial cells, too, failed to induce disease.

The converse approach then was used to examine the possible protective role of non-neuronal cells in disease. In this case wild-type SOD was selectively restored only in astrocytes or microglial cells, leaving the mutated form in all other cells. In both instances motor neuron loss began much later and was much less severe. Since microglia derive from the bone marrow, scientists next transplanted bone marrow from wild-type mice into SOD mutant mice and once again saw a neuroprotective effect whereby disease onset was significantly delayed.

The emerging picture from these and other studies, schematically summarized in Figure 13, is that *SOD1* mutations in neurons are required but not sufficient to cause disease. High levels of mutant SOD accelerates disease onset, yet both microglia and astrocytes are contributors that facilitate onset and accelerate the course of disease. By contrast, neither vascular cells nor muscle cells influence disease onset or progression.

How non-neuronal cells contribute to disease pathology is not fully understood, but a number of mechanisms have been proposed.[26] With regard to astrocytes, the reduction of glutamate uptake via a transcriptional reduction of EAAT2 was already mentioned. There is evidence that the mutated *SOD1* directly reduces transcription of the *EAAT2* gene, as does the mutation in *TAR-43/FUS*. In addition, normal but not mutant astrocytes have been shown to upregulate the expression of the GluR2 subunit of the AMPA–glutamate receptor in motor neurons, which, when expressed, makes these receptors impermeable to Ca^{2+}. This in turn would protect the motor neuron from glutamate toxicity due to pathological influx of Ca^{2+}.[27] With regard to microglia, a toxic mixture of inflammatory cytokines and chemokines, as well as nitric oxide and superoxide, is most likely to blame.

These studies make a strong case for ALS being a non-cell-autonomous disease, with contributions from aberrant biology in several affected cell types, whereby the same mutation affects different proteins and pathways in different cells. For example, the *SOD1* mutation reduces transcription of the EAAT2 transporter in astrocytes while activating cytokine release from microglial cells and inducing caspase-3-mediated apoptosis in motor neurons. Hence a single mutation, via different modes of action, creates a "perfect storm" scenario. As discussed

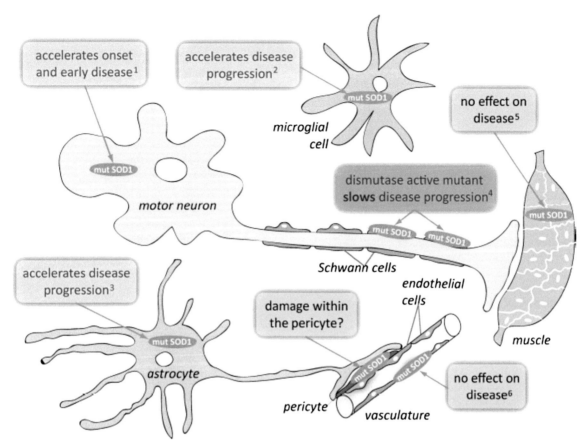

FIGURE 13 Non-cell-autonomous toxicity in amyotrophic lateral sclerosis (ALS). *SOD1* mutations in different cell types contribute in different ways to ALS. Mutations in motor neurons initiate disease, yet microglial and astrocytic changes accelerate disease onset and increase severity. *Reproduced with permission from Ref. 25.*

further in Chapter 14, a similar non-cell-autonomous mode of action must now be considered more broadly in many neurodegenerative diseases. With regard to ALS, it is both exciting and troubling. It is exciting in that it provides opportunities to explain the heterogeneity of disease presentation but troubling in its suggestion that simple strategies for treatment that target a single protein or pathway are likely elusive.

4.8 Inflammation and ALS

Inflammation is a common response to injury and disease throughout the body. In the brain the major contributors to inflammation are microglial cells, astrocytes, and T cells.

Activated microglial cells change their appearance to an ameboid shape and also express CD11b and IBA1 antigens. Activated astrocytes develop hypertrophied bodies and overexpress glial fibrillary acidic protein. Both microglial activation and astrogliosis are commonly seen in patients during autopsy.[28] This is the case in both the primary motor cortex and in the spinal cord. Interestingly, inflammation is not restricted to the areas occupied by motor neurons but also extends into the corticospinal tract and the dorsal horn of the spinal cord.

The temporal sequence in which inflammation develops with regard to disease symptoms has primarily been studied in SOD mutant mice. Here, microglial activation can be seen in the lumbar spinal cord long before disease onset. By contrast, astrocyte activation occurs only concomitant with neuronal cell death. Once activated, the number of microglial cells increases, presumably by local proliferation rather than through myeloid cells contributed from the periphery. As microglial cells become activated and proliferate, CD4+ and CD8+ T cells also infiltrate the central nervous system. Interestingly, the CD4+ helper T cells attenuate inflammation regardless of whether they carry the mutation, and they seem to do so by attenuating microglial activation, shifting them into a resting state. As disease progresses, however, the CD8+ cytotoxic T cells predominate and actively contribute to cell death.

Attenuation of microglial activation seems to be a key component with regard to regulating the inflammatory response. One way to do so is using the broad-spectrum antibiotic minocycline, which indeed improves the lifespan of SOD mutant mice.

A key question remaining is whether the inflammation occurs as a result of neuronal death or whether it causes cell death as result, for example, of gene mutations and whether inflammation is good or bad. This in turn could affect the potential development of disease specific anti-inflammatory treatments. No clear answer exists at the moment.

4.9 Mechanism(s) of Disease Spread

Rapid spread of motor neuron death is the cardinal feature of ALS, so an understanding of how the disease spreads, in the context of the above-discussed pathogenic mechanisms, is imperative. Diffusible substances such as cytokines and glutamate have been suggested. These could gradually affect adjacent motor neurons and spread up and down the corticospinal tract.

Such cell-to-cell spreading would explain the gradually expanding impairment of adjacent muscle groups that are innervated by adjacent motor neurons, which are arranged in a topographic fashion in the anterior horn of the spinal cord.

As an alternative to diffusive toxins, there is increasing interest in the prion-like spread of misfolded proteins, whereby the misfolded proteins may affect adjacent normal proteins to assume their shape by coaxing them into a misfolded stage. This would, therefore, cause aggregates of toxicity to spread and would be consistent with a requirement for misfolded proteins in disease. Similar prion-like disease propagation also has been proposed for Alzheimer and Parkinson diseases (discussed in Chapters 4 and 5) and was validated in animal models. This mode of spread, while exciting from a scientific perspective, would also be frightening from a clinical point of view. As we discuss in Chapter 10, prions are very stable proteins that are resistive to almost any form of treatment. It is therefore difficult to envision how disease progression, once started, could be stopped.

4.10 Putting it All Together: The "Perfect Storm"

At present we can only speculate how the various genetic changes and protein aggregates may cumulatively lead to disease. However, one can assume that protein aggregates that contain RNA and DNA regulatory proteins such as TDP-43 and FUS are likely among those inclusions. Since these are regulators of, for example, EAAT2 and SOD, assuming that loss of EAAT2 causes a glial deficit in glutamate transport, whereas a neuronal deficit in SOD function may cause mitochondrial dysfunction and the resulting ROS may cause lipid peroxidation, is not far-fetched. Energy failure, in turn, causes ATPases in the axon to fail. Mutations in the oligodendrocytes deprive the axon of glycogen as an energy substrate. The toxic aggregates also disrupt axonal transport, depleting the axon of organelles and disrupting the transport of neurotropic factor from the target muscle back to the motor neuron. Over time, the insoluble inclusion bodies spread to adjacent cells in a prion-like fashion, putting an increasing number of neighboring motor neurons at risk. Once the death spiral has started, it continues rapidly as more and more toxic proteins spread almost exponentially to affect an ever-increasing number of neurons.

5. TREATMENT/STANDARD OF CARE/CLINICAL MANAGEMENT

Since there are no unique laboratory markers for ALS, clinical diagnosis relies almost exclusively on symptomatology, and in 95% of cases a trained neurologist will correctly diagnose ALS based on symptoms alone. However, a definitive and unequivocal clinical diagnosis of ALS typically cannot be reached until the disease has progressed and EMG can confirm denervation of at least three limbs. Nerve conduction velocity testing should provide normal results in these affected limbs. Fewer than three involved limbs are defined as probable ALS and monitored further. Imaging can be used to detect a thinning of the corticospinal tracts in some patients. Diffusion tensor MRI is particularly well suited to identifying changes in white matter tracts and therefore can be used to show physical signs of corticospinal tract denervation (Figure 14).

Unfortunately, no disease-modifying treatments are available. The only drug specifically approved for the treatment of ALS is riluzole (Rilutek), a drug with a number of targets that each reduces the amount of glutamate released, thereby limiting glutamate toxicity, which we discussed above as a major contributor to motor neuron death. In a placebo-controlled trial riluzole increased average survival from 15 to 18 months.

Absent specific treatment, palliative care, and emotional support are critical to improving the quality of life of patients with ALS. For care to be most effective, patients with ALS often are managed by multidisciplinary clinics, wherein a neurologist manages a treatment team comprised of a dietician, physical and occupational therapists, a psychologist, a speech therapist, and a social worker. This team can best prepare the patient and his or her family in anticipation of the various disabilities that develop over time. The team addresses issues related to rehabilitative aids that facilitate breathing, ambulation and mobility, language, and alternative communications.

At their most basic, rehabilitative aids such as splints are used to maintain hands and feet in their extended state. Respiratory support will become essential in late stages of the disease, and devices to assist with coughing and clearing the airways are effective in mitigating against aspiration pneumonia. As communication by speech becomes difficult, artificial computer-based communication devices can be used, some of which can translate eye movements, which are always spared even in the late stages of ALS, into letters and words. The progressive weakening of the respiratory muscles as a result of phrenic nerve activity is the leading cause of death, with close to 60% of patients with ALS dying of respiratory distress. While sustaining a patient through mechanical ventilation is possible, this invasive approach severely compromises quality of life and, not surprisingly, is rejected by 95% of patients. Alternative, less invasive approaches to ventilation are possible; for example, an intermittent positive-pressure ventilator that delivers oxygenated air via a face mask when triggered by the patient's inhalation can be used in an ambulatory setting for several hours at a time.

It is imperative for the treatment team to initiate conversations regarding end-of-life planning while the affected individual is still able to actively participate in the process. Clinical

ALS Patient   Healthy Control

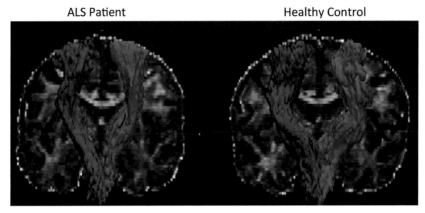

FIGURE 14   Tract-based imaging using diffusion tensor magnetic resonance imaging, an imaging approach that highlight axonal fiber tracts, revealing a thinning in the corticospinal tracts in a patient with ALS (left) compared with a healthy control (right). *Reproduced with permission from Ref. 29.*

studies have evaluated the effectiveness of such a specialized team-based approach to the management of ALS and demonstrated a significant prolongation of life and, importantly, enhancement in the patient's quality of life.

6. EXPERIMENTAL APPROACHES/ CLINICAL TRIALS

Given the grim prognosis, it is not surprising that numerous clinical trials are exploring new treatments targeting many of the pathways identified in laboratory studies. Many of these have failed already, even when preclinical studies of mouse models were successful. As of 2014 there are 72 clinical trials open and actively enrolling patients. The majority of these are early stage trials that examine safety and feasibility rather than efficacy. Some of the most attractive new approaches and their rationales are discussed below.

Modulation of glutamate: Glutamate toxicity remains one of the favored explanations for motor neuron death and is the primary target of riluzole, the only drug specifically approved for the treatment of ALS. Riluzole blocks voltage-gated Na^+ channels, thereby

presumably reducing glutamate release. An alternative way to reduce glutamate toxicity is enhancing its clearance from the synaptic cleft. This is primarily accomplished by the glial glutamate transporter EAAT2, which has significantly reduced expression in patients with ALS. Interestingly, EAAT2 expression can be enhanced two- to threefold with ceftri-axone, a cephalosporin antibiotic used to treat severe meningococcal infections and meningitis. It has been shown to act as transcriptional regulator for EAAT2 and was effective in normalizing average life expectancy to almost control levels in *SOD1* mutant mice. Based on these promising animal studies and the ready availability of this drug, it is currently being examined through a multicenter clinical study for ALS (www.clinicaltrials.gov identifier NCT00349622). Thus far, the collected data suggest that the drug is safe and well tolerated, but efficacy data are not yet available.[30]

Yet another alternative to reduce glutamate toxicity is the use of a noncompetitive NMDA receptor inhibitor such as memantine, which has been explored in the treatment of stroke and Alzheimer disease. Unfortunately, in a placebo-controlled phase III study memantine did not show any improvements over placebo.[31]

Use of antisense oligonucleotides: The evidence that mutations in *SOD1* cause disease is overwhelming, and although the mechanism of its toxicity is not fully understood, the prevailing thought is that mutated *SOD1* itself is toxic. It therefore stands to reason that removal of the mutated *SOD1* from motor neurons would be beneficial, and this is indeed the case in *SOD1* mutant mice. Therefore the removal of mutated or toxic SOD1 proteins from motor neurons or their non-neuronal partners has the potential, in theory, to treat ALS. In mouse studies proteins are removed through generation of transgenic animals in which the gene encoding a given protein is deleted; this can even be done in a cell type–specific manner. Clearly, this approach is not feasible in humans and therefore only provides proof of principle. An alternative approach, however, is commonly used in cell-based laboratory studies. Here, antisense oligonucleotides (ASOs) are used to suppress protein synthesis.[32] These are single-stranded short strings of 8–50 nucleic acids that bind to target RNA by traditional Watson-Crick base pairing. ASOs can be engineered to be stable and are very effective in disrupting protein synthesis through degradation of the encoding RNAs. The end result is a gradual depletion of the targeted protein from the cell.

A major hurdle in applying this approach to human treatment is the specific delivery of ASOs in sufficient quantities to the motor neurons in the cortex and spinal cord. ASO molecules do not pass the blood–brain barrier and therefore cannot be given by systemic infusion. However, invasive injection into the cerebrospinal fluid via a spinal tap, akin to that used for epidural anesthesia or by injection into the ventricle, is possible. Another possibility is the use of a slow-release catheter to allow the continuous infusion of ASOs over many hours, possibly even days. This approach was used in the first phase I clinical study using ASOs to target SOD1 in patients with familial ALS (www.clinicaltrials.gov identifier NCT01041222). The procedure was well tolerated and deemed safe based on results from 32 patients.[33] The next step is to examine efficacy after repeated infusion of ASOs. In principle, a similar benefit should be attainable by removing toxic TAR-43 aggregates or removing the disease-causing GGGGCC expansion due to mutations in C9ORF72. The latter has profound effects in ameliorating disease in mice.[34] Removing SOD1, TAR-43, and C9ORF72 would capture 50% of ALS cases, so that this strategy may have benefits for many patients with ALS, including those with sporadic disease.

The delivery strategy, however, seems somewhat pedestrian and unspecific. There is the potential that ASOs could be delivered with replication-deficient neurotropic viruses, which are discussed more extensively in Chapter 10. For example, adeno-associated virus can specifically infect neurons, sparing most non-neuronal cells. Studies and clinical trials using these strategies are expected to yield interesting results in the years ahead.

Cell replacement therapy: Cell replacement therapy[35] is being used as experimental treatment for a number of neurodegenerative diseases, including ALS. However, unlike Parkinson disease, for example, where neuronal loss is largely confined to the substantia nigra, the motor neuron loss in ALS is diffuse, necessitating the delivery of stem cells along the entire neuraxis. A further challenge is the rapid progression of the disease, providing a relatively short window of opportunity for intervention. These challenges, however, are overcome by a heightened willingness of patients and physicians to accept the risk associated with participation in clinical studies. Different stem cell sources are now used in clinical studies, including fetal stem cells, hematopoietic stem cells, and induced pluripotent mesenchymal cells. A recently completed phase I study reports the feasibility and safety (NS2008-1) of injecting up to 1.5 million neural stem cells intracervically at up to 10 injection sites along the spinal cord.[36] The study also

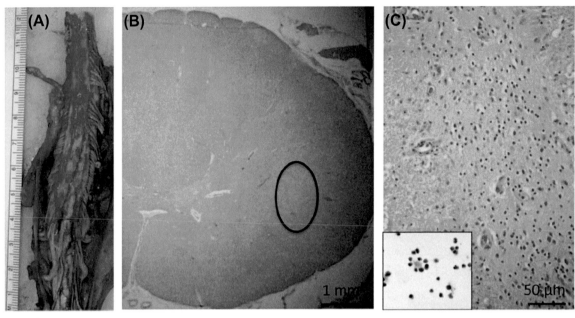

FIGURE 15   Neuropathological findings in a patient after stem cell implantation. (A) Gross image of the cervical spinal cord at the time of autopsy. Serial sections through the region of transplantation did not demonstrate regions of cystic change, hemorrhage, or significant tissue disruption. (B) Representative cross section showing intact cord morphology using hematoxylin and eosin (H&E) staining. There is a nest of cells (circled) that are not intrinsic to the spinal cord and do not stain with glial or neuronal markers (not shown). (C) Magnification of the circled region in B showing the morphology of these cells, which is reminiscent of the morphology of the stem cells before transplantation (inset, H&E stain). *Reproduced with permission from Ref. 36.*

documents the successful grafting of stem cells in the spinal cord (Figure 15). To date there is only limited evidence for efficacy in animal models and no report yet on therapeutic benefit in patients. Note also that all studies to date were open label, not placebo controlled, and typically involved only a few patients (<20). On the positive side, there are no adverse effects reported, the most feared being accelerated disease progression of aberrant tumorous growth of the implanted cells.

Brain–Computer Interface: The loss of bulbar muscles required to speak makes essential alternative nonverbal ways to communicate. Efforts to use brain waves to control devices or generate language are progressing in the laboratory. Typically, these efforts exploit very small changes in electrical brain activity that can be detected by surface electrodes, akin to electroencephalography electrodes. The challenge is to train the affected individual to voluntarily elicit a reproducible change in brain activity that can be interpreted by a computer, which then turns them into appropriate commands, such as a cursor movement. A challenge for patients with ALS is the rapid progression of disease, compressing the time available to be trained to utilize computer-based assistive devices.

7. CHALLENGES AND OPPORTUNITIES

ALS is arguably the most frightening of all the neurological disorders. It traps a person's healthy mind within a failing body, where progressive impairment in the use of skeletal muscles causes immobility, loss of communication, and, ultimately, respiratory distress. While the underlying pathology has been known for over 150 years, the relatively recent revelation of multiple genetic mutations that cause disease promises the development of strategies to intervene. The fact that these genetic mutations characterize both familial and sporadic forms of the disease equally is particularly promising because they point to a shared disease mechanism.

Fifty years of clinical trials using over 150 different drugs or approaches have failed, producing only one drug, rilozole, which has questionable benefit. In 1993, when a mutation in *SOD1* was first discovered as a cause for ALS, it was widely assumed that a loss of enzymatic activity of this enzyme, which contains reactive oxygen radicals, is responsible for oxidative damage of motor neurons. Based on that idea, it was assumed that a cure was imminent. Unfortunately, we know now that this is not the disease mechanism at all, and thus we are no closer to a cure.

Recent research has identified several additional genes that account for a large number of familial and sporadic disease cases. These findings are very exciting, although their actual function and how they cause disease is equally unclear. The single biggest remaining challenge is to elucidate the mechanism(s) whereby motor neurons die. Sadly, while good hypotheses abound, we are largely clueless. One may question whether previous and current clinical trials largely failed because they were premature and not informed by a solid mechanistic understanding of disease. But how do we get such much-needed mechanistic understanding? There may simply not be enough scientists working on the problem, and new ideas are few. There still remains only one rodent model that is held out as the gold standard and with which all preclinical tests are performed. Yet many investigators are questioning the usefulness of

this transgenic mouse that has led many clinical studies astray.

Recent attempts to transplant stem cells of various origins, ranging from fetal to pluripotent mesenchymal stem cells, into small cohorts of patients with ALS have not yet provided tangible benefits to any patient. While safe, there is little evidence that the implanted cells survive or alter the survival and function of existing motor neurons. Even mechanisms by which implanted cells may benefit the diseased spinal cord are not well understood. Is it the release of tropic factors or the attenuation of the immune response? Do any of these cells give rise to glial cells with potentially protective effects? Again, these studies may well be premature because they are based on limited and questionable preclinical success and an overly simplistic rationale.

Our best hope, it seems, rests on careful, deliberate research to mechanistically identify how the most prominent mutations—those in chromosomes 9 and 21—cause disease. This will take time and require patience. Additional animal model systems for study must be consulted, including *Drosophila*, *Caenorhabditis elegans*, and zebrafish. While these may not mimic all aspects of human disease, they may allow us to gain a more refined molecular understanding.

Finally, there is a need for increased awareness and advocacy. ALS is not a rare disease, yet it is rarely seen or talked about. The lifetime chance of developing ALS is comparable to that of multiple sclerosis. Given its rapid progression, however, there is a paucity of prominent individuals or celebrities affected by the disease who could, in spite of losing their physical voice, become spokespersons for raising awareness and much-needed research support.

Acknowledgments

This chapter was kindly reviewed by Jeffrey D. Rothstein MD, PhD, Professor of Neurology and Neuroscience and Director of the Robert Packard Center for ALS Research at Johns Hopkins University School of Medicine.

References

1. Bennett MR, Hacker PMS. The motor system in neuroscience: a history and analysis of conceptual developments. *Progr Neurobiol.* 2002;67(1):1–52.

2. Rowland LP. How amyotrophic lateral sclerosis got its name: the clinical-pathologic genius of jean-martin charcot. *Arch Neurol.* 2001;58(3):512–515.

3. Aran F. Recherches sur une maladie non encore décrite du système musculaire (atrophie musculaire progressive). *Arch Gen Med.* 1850;24(1850):5–35.

4. Gurney ME, Pu H, Chiu AY, et al. Motor neuron degeneration in mice that express a human Cu,Zn superoxide dismutase mutation. *Science.* 1994;264(5166):1772–1775.

5. Andersen PM, Al-Chalabi A. Clinical genetics of amyotrophic lateral sclerosis: what do we really know? *Nat Rev Neurol.* 2011;7(11):603–615.

6. Wingo TS, Cutler DJ, Yarab N, Kelly CM, Glass JD. The heritability of amyotrophic lateral sclerosis in a clinically ascertained United States research registry. *PloS One.* 2011;6(11):e27985.

7. Renton AE, Majounie E, Waite A, et al. A hexanucleotide repeat expansion in C9ORF72 is the cause of chromosome 9p21-linked ALS-FTD. *Neuron.* 2011;72(2):257–268.

8. Barohn RJ. Clinical spectrum of motor neuron disorders. *Continuum (Minneap, Minn).* 2009;15(1):20.

9. Baumer D, Hilton D, Paine SM, et al. Juvenile ALS with basophilic inclusions is a FUS proteinopathy with FUS mutations. *Neurology.* 2010;75(7):611–618.

10. Kiernan MC, Vucic S, Cheah BC, et al. Amyotrophic lateral sclerosis. *The Lancet.* 2011;377(9796):942–955.

11. Rademakers R, van Blitterswijk M. Motor neuron disease in 2012: novel causal genes and disease modifiers. *Nat Rev Neurol.* 2013;9(2):63–64.

12. Ling SC, Polymenidou M, Cleveland DW. Converging mechanisms in ALS and FTD: disrupted RNA and protein homeostasis. *Neuron.* 2013;79(3):416–438.

13. Henriques A, Pitzer C, Schneider A. Characterization of a novel SOD-1(G93A) transgenic mouse line with very decelerated disease development. *PloS One.* 2010;5(11):e15445.

14. Scott S, Kranz JE, Cole J, et al. Design, power, and interpretation of studies in the standard murine model of ALS. *Amyotroph Lateral Scler.* 2008;9(1):4–15.

15. Aggarwal A, Nicholson G. Detection of preclinical motor neurone loss in SOD1 mutation carriers using motor unit number estimation. *J Neurol Neurosurg Psychiatry.* 2002;73(2):199–201.

16. Pasinelli P, Houseweart MK, Brown Jr RH, Cleveland DW. Caspase-1 and -3 are sequentially activated in motor neuron death in Cu,Zn superoxide dismutase-mediated familial amyotrophic lateral sclerosis. *Proc Natl Acad Sci USA.* 2000;97(25):13901–13906.

17. Vucic S, Rothstein JD, Kiernan MC. Advances in treating amyotrophic lateral sclerosis: insights from pathophysiological studies. *Trends Neurosci.* 2014;37(8):433–442.

18. Beckman JS, Estevez AG, Crow JP, Barbeito L. Superoxide dismutase and the death of motoneurons in ALS. *Trends Neurosci.* 2001;24(suppl 5):S15–S20.

19. Cozzolino M, Carrì MT. Mitochondrial dysfunction in ALS. *Prog Neurobiol.* 2012;97(2):54–66.

20. Turner MR, Hardiman O, Benatar M, et al. Controversies and priorities in amyotrophic lateral sclerosis. *Lancet Neurol.* 2013;12(3):310–322.

21. Rothstein JD. Current hypotheses for the underlying biology of amyotrophic lateral sclerosis. *Ann Neurol.* 2009;65(suppl 1):S3–S9.

22. Spreux-Varoquaux O, Bensimon G, Lacomblez L, et al. Glutamate levels in cerebrospinal fluid in amyotrophic lateral sclerosis: a reappraisal using a new HPLC method with coulometric detection in a large cohort of patients. *J Neurol Sci.* 2002;193(2):73–78.

23. Bradley WG, Mash DC. Beyond Guam: the cyanobacteria/ BMAA hypothesis of the cause of ALS and other neurodegenerative diseases. *Amyotroph Lateral Scler.* 2009;10 (Suppl 2):7–20.

24. Tikka TM, Vartiainen NE, Goldsteins G, et al. Minocycline prevents neurotoxicity induced by cerebrospinal fluid from patients with motor neurone disease. *Brain.* 2002;125(Pt 4):722–731.

25. Ilieva H, Polymenidou M, Cleveland DW. Non-cell autonomous toxicity in neurodegenerative disorders: ALS and beyond. *J Cell Biol.* 2009;187(6):761–772.

26. Philips T, Rothstein JD. Glial cells in amyotrophic lateral sclerosis. *Exp Neurol.* 2014.

27. Van Damme P, Bogaert E, Dewil M, et al. Astrocytes regulate GluR2 expression in motor neurons and their vulnerability to excitotoxicity. *Proc Natl Acad Sci USA.* 2007;104(37):14825–14830.

28. Philips T, Robberecht W. Neuroinflammation in amyotrophic lateral sclerosis: role of glial activation in motor neuron disease. *Lancet Neurol.* 2011;10(3):253–263.

29. Sarica A, Cerasa A, Vasta R, et al. Tractography in amyotrophic lateral sclerosis using a novel probabilistic tool: a study with tract-based reconstruction compared to voxel-based approach. *J Neurosci Methods.* 2014;224:79–87.

30. Berry JD, Shefner JM, Conwit R, et al. Design and initial results of a multi-phase randomized trial of ceftriaxone in amyotrophic lateral sclerosis. *PloS One.* 2013;8(4):e61177.

31. de Carvalho M, Pinto S, Costa J, Evangelista T, Ohana B, Pinto A. A randomized, placebo-controlled trial of memantine for functional disability in amyotrophic lateral sclerosis. *Amyotroph Lateral Scler.* 2010;11(5):456–460.

32. DeVos SL, Miller TM. Antisense oligonucleotides: treating neurodegeneration at the level of RNA. *Neurotherapeutics.* 2013;10(3):486–497.

33. Miller TM, Pestronk A, David W, et al. An antisense oligonucleotide against SOD1 delivered intrathecally for patients with SOD1 familial amyotrophic lateral sclerosis: a phase 1, randomised, first-in-man study. *Lancet Neurol.* 2013;12(5):435–442.

34. Lagier-Tourenne C, Baughn M, Rigo F, et al. Targeted degradation of sense and antisense C9orf72 RNA foci as therapy for ALS and frontotemporal degeneration. *Proc Natl Acad Sci USA.* 2013;110(47):E4530–E4539.

35. Thomsen GM, Gowing G, Svendsen S, Svendsen CN. The past, present and future of stem cell clinical trials for ALS. *Exp Neurol.* 2014.

36. Feldman EL, Boulis NM, Hur J, et al. Intraspinal neural stem cell transplantation in amyotrophic lateral sclerosis: phase 1 trial outcomes. *Ann Neurol.* 2014;75(3): 363–373.

General Readings Used as Source

Adams and Victor's, Principles of Neurology, 7th ed.

Barohn RJ. Clinical spectrum of motor neuron disorders. *Continuum (Minneap, Minn).* 2009;15(1):20.

Fundamental Neuroscience, Squire, Berg, Bloom, du Lac, Ghosh & Spitzer 4th ed.

Gordon PH. Amyotrophic lateral sclerosis: an update for 2013 clinical features, pathophysiology, management and therapeutic trials. *Aging Dis.* 2013;4(5):295–310.

Mazzoni P, Rowland LP. *Merritt's Neurology.* 10th ed. Philadelphia, PA: Lippincott Williams & Wilkins; 2001.

Vucic S, Rothstein JD, Kiernan MC. Advances in treating amyotrophic lateral sclerosis: insights from pathophysiological studies. *Trends Neurosci.* 2014;37(8):433–442.

Suggested Papers or Journal Club Assignments

Clinical Papers

Berry JD, Shefner JM, Conwit R, et al. Design and initial results of a multi-phase randomized trial of ceftriaxone in amyotrophic lateral sclerosis. *PloS One.* 2013;(4):8. e61177.

Feldman EL, Boulis NM, Hur J, et al. Intraspinal neural stem cell transplantation in amyotrophic lateral sclerosis: phase 1 trial outcomes. *Ann Neurol.* 2014;75(3): 363–373.

Renton AE, Majounie E, Waite A, et al. A hexanucleotide repeat expansion in C9ORF72 is the cause of chromosome 9p21-linked ALS-FTD. *Neuron.* 2011;72(2):257–268.

Basic Papers

Donnelly CJ, Zhang PW, Pham JT, et al. RNA toxicity from the ALS/FTD C9ORF72 expansion is mitigated by antisense intervention. *Neuron.* 2013;80(2):415–428.

Lagier-Tourenne C, Baughn M, Rigo F, et al. Targeted degradation of sense and antisense C9orf72 RNA foci as therapy for ALS and frontotemporal degeneration. *Proc Natl Acad Sci USA.* 2013;110(47):E4530–E4539.

Lee Y, Morrison BM, Li Y, et al. Oligodendroglia metabolically support axons and contribute to neurodegeneration. *Nature.* 2012;487(7408):443–448.

Teng YD, Benn SC, Kalkanis SN, et al. Multimodal actions of neural stem cells in a mouse model of ALS: a meta-analysis. *Sci Transl Med.* 2012;4(165):165ra164s.

CHAPTER

7

Huntington Disease

Harald Sontheimer

1. CASE STORY

Hanna was only 7 years old when her father, at age 42, was diagnosed with Huntington disease (HD). Neither Hanna nor her older brother Jeremy knew about their dad's disease at the time, nor were they ever told that they would each have a 50% chance of developing it as well. All Hanna remembers from her youth is that her dad had a strange way of moving his body, with gyrating movements involving his arms, legs, and trunk. At first she thought it was funny but she later became embarrassed, particularly when in the company of her friends. Within a few years, his muscles became so jerky that he had to use a wheelchair every time they left the house. Initially, the family tried to maintain as much normalcy as possible, which involved joint family dinners and evening movie hours. But dad soon became unable to participate in conversation. His facial muscles kept twitching and contorting, making his speech slurred and unintelligible. Watching his facial grimaces was particularly difficult, and Hanna never knew whether he was happy or sad. She stopped inviting friends to her house, afraid they would be terrified upon seeing her dad's strange behavior. In high school biology classes Hanna first heard about heritable genetic diseases and learned how little could be done to treat them. It was then that she realized that she and her brother were genetically at risk to develop HD.

Hanna was looking forward to leaving the family for college, hoping that she could finally establish some normal friendships and focus on her own life. Rather than helping her mother deal with her dad's illness, she was now committed to studying medicine to find a cure for his disease. Following high school graduation, which was attended by her entire family, including her dad, Hanna enrolled at the University of Tennessee. Midway through her sophomore year, Hanna noticed that she is often distracted and irritated for no reason. Also, while she liked to keep things neat and organized, she was

clumsy and was often dropping things such as a pencil or even her backpack. It often seemed like her hands just could not hold a grip. She had taken up tennis in her freshman year, and her trainer suspected that she might have developed tendinitis. Her weakness and clumsiness did not improve. On one particularly embarrassing occasion she picked up a pitcher to pour a glass of ice tea at a reception when the pitcher just slipped out of her hand, spilling tea all over the carpet. Her friends jokingly asked her if she was drunk, which really upset her since she never drank alcohol at all. Hanna called her mom that night angry and agitated: "Mom, I think I may have dad's disease!" "I know," her mother replied, "I have noticed subtle symptoms for quite some time. That's how dad started out, too. I just didn't know it at the time." For many weeks, Hanna felt a deep-seated anger, a feeling of helplessness and frustration. At times she was angry with her father for passing on such crappy genes. Her anger more than her disabilities caused her to skip classes and start drinking; she ultimately failed her classes and dropped out of college. After a summer at home, Hanna mustered the courage to return to school, determined to complete her education or, at the very least, earn an associate's degree. Her professors were exceedingly helpful and understanding, but her impairments worsened and her speech gradually began to slur, such that she was often unable to sustain everyday conversations. Giving oral class presentations became entirely impossible. Soon she could no longer sit still and focus on class work. Instead, her body seemed to be constantly moving and her mind was unable to concentrate. Frustrated, she left college again and moved back home. Soon thereafter, her dad died at age 56. To this day, her brother Jeremy is still symptom free. Although he was given the option to have genetic testing, he declined. Seeing his father die of HD and now his sister go through this horrible disease, Jeremy prefers not to know whether he, too, will ultimately succumb to this horrible illness.

2. HISTORY

The origin of HD is not known, and early historical accounts are equivocal. The symptomatic hallmark of the disease, a dance-like, contorted, and involuntary but seemingly choreographed body movement that is often called chorea (derived from the Greek *choros*, for dance), was first reported as "dancing mania" in the middle ages in Europe. Hundreds or even thousands of people reportedly participated in processions, dancing in the streets with chorea-like movements. In all likelihood these were psychogenic illnesses or rituals and most likely had no physical cause. Alternatively, however, these marches may have been initiated by one or more individuals who actually displayed chorea as a result of the very neurological illness we now call Huntington disease, with others copying their movements. In 1500 the Swiss–German physician Paracelsus suggested that the dancing mania, or chorea, was in fact caused by a disorder of the brain; in 1832 physician Sir John Elliston provided the first evidence that chorea was heritable. HD eventually got its name from a young American physician, George Huntington, who published in 1872 the first comprehensive description, entitled "On Chorea." His paper included observations of patients treated by his father and grandfather, who were both country doctors in East Hampton, New York. With clinical observations spanning three generations, Huntington was able to see the disease propagated within extended families, allowing him to outline a clear Mendelian autosomal dominant inheritance pattern. He also described the presence of cognitive and behavioral abnormalities comorbid with the motor chorea.

After Gregor Mendel's work on heredity was rediscovered in 1900, the British geneticist William Bateson unequivocally established HD as a heritable disorder. Unfortunately, this did little to remove the stigma. Much to the contrary, the eugenics movement, prevalent throughout the world in the first half of the 20th century, quickly used this knowledge to suggest compulsory sterilization of affected individuals. It also inspired questionable genealogical research to identify families who may have first introduced HD to the United States, which prompted calls for restrictions on immigration.

In 1944 the famous folk musician and antiwar activist Woody Guthrie recorded what would become one of the most popular American folk songs, "This Land is your Land." While most Americans can readily sing along to the song, few may know that Woody Guthrie also suffered from HD. Indeed, as he recorded this song he was already showing the characteristic symptoms of HD, yet he may not have known about the severity of his illness, which would worsen progressively. Shortly after his death in 1969, his wife Marjorie founded the Committee to Combat Huntington Disease, which is now the Huntington's Disease Society of America, one of the most active societies in the search of a cure for HD.

The contemporary history of HD is a fascinating story of discovery that exemplifies the power of advocacy and how the commitment of a few individuals can have a transformational outcome on research.[1] HD is intricately associated with the Wexler family. Leonore Wexler was a biologist who was diagnosed with HD at age 53. Her father had died of HD in 1926, when she was only 15, and all three of her uncles had developed HD. At the time it was believed that only men were susceptible to the disease, so Leonore's symptoms came as a total shock to her. At the time of her diagnosis, Leonore had two daughters, Alice and Nancy, who obviously shared her genetic risk for HD. Leonore's husband, Milton Wexler, was unwilling to accept his wife's fate and poured his life's energy into creating the Hereditary Disease Foundation. This foundation recruited some of the world's leading geneticists and biochemists, including notables such as Seymour Benzer, William Dreyer, and Julius Axelrod, to develop a strategy for finding a cure. Yet, ultimately, it was Wexler's own daughter Nancy who paved the way to uncovering the genetic cause of HD.

In the 1970s, when the search for the HD gene began, gene cloning was still in its infancy, and scientists relied on identifying chromosomal regions that share similar alterations across multiple generations of affected offspring, a process called positional cloning. To be successful, this approach required large families in which chromosomes from multiple generations of affected individuals could be studied. Through the Venezuelan physician Amerigo Negrette, Nancy Wexler and her colleagues learned about an unusually large number of families affected by HD living along Lake Maracaibo in a remote region of Venezuela. There, hundreds of large families included affected individuals spanning multiple generations. Also, many of their children developed HD at an early age, and in some families as many as 11 of 13 children were affected. To date, this group remains the most well-studied cluster of individuals with HD and now has provided genetic data from over 18,000 individuals spanning 10 generations. Interestingly, almost 15,000 cases share ancestry that traces to a single founder who lived in the early 1800s.

Through positional cloning, the genetic locus of HD was progressively narrowed to the short arm of chromosome 4p16.3, where the huntingtin gene was finally cloned in 1993 and shown to encode a protein with an unknown function. The search for the gene was technologically transformational, generating tools and strategies that laid the foundation for the subsequent sequencing of the entire human genome. While the search for a cure for HD is ongoing, great strides have been made in terms of the molecular understanding of the disease.

Unfortunately, even today the general public remains poorly informed about HD, with little advocacy to remove the stigma associated with the disease. Originating in medieval times, the relationship of HD with witchcraft and the suggestion that those affected were undesirable characters with criminal tendencies was propagated well into the 20th century. Most troubling, this view was even prevalent among physicians. Today, families are reluctant to talk about their experiences and become advocates for research because of both the stigma of the disease and the fear that they may lose access to health insurance.

3. CLINICAL PRESENTATION/ DIAGNOSIS/EPIDEMIOLOGY

HD is a hereditary neurodegenerative disorder that is always fatal and presents with a distinct combination of symptoms in three domains, namely, motor, cognitive, and behavioral. HD is caused by a mutation in the Huntington gene (*htt*), which encodes for a protein of unknown function called the huntingtin protein (HTT). The mutated gene (*mhtt*) contains an unstable expansion of trinucleotide (CAG) repeats; each repeat encodes for the amino acid glutamine. The wild-type protein has perhaps 20–30 glutamines, yet mutated HTT (mHTT) has between 40 and 250 polyglutamine repeats.[2]

As discussed further below, the number of the polyglutamines determines disease onset but not disease severity. A large number of polyglutamines causes juvenile onset, which presents with different motor signs more similar to Parkinson disease and commonly associated with seizures. In these cases chorea is absent, and patients suffer from slowness of motion (bradykinesia) and poor muscle control (ataxia).

The presence of a CAG expansion in the *htt* gene combined with the motor signs (chorea, rigidity, bradykinesia) provides a fairly unequivocal diagnosis, regardless of whether cognitive and/or behavioral abnormalities are present. Typically, the family history alone, even in the absence of sequencing the affected person's *htt* gene, points to HD as the cause. Only 8% of diagnosed patients do not know of other affected family members.

3.1 The Diagnosis of HD

The diagnosis of HD is straightforward, and the disease is hard to miss for a trained neurologist. Among the earliest motor signs are peculiar dance-like, involuntary body movements, called chorea. More subtle motor signs include difficulty sustaining muscle contractions required to lift objects and maintain a grip. For example, the affected person's handshake is often called a "milkmaid's grip," one that starts strong but then wanes. In addition, fine motor skills such as tapping a rhythmic pattern with one finger are impaired. Eye movements are also abnormal, with incomplete saccades being typical. Saccades are quick simultaneous movements of both eyes as a person is scanning an object or scene.

As disease progresses, patients show muscle rigidity and motor incoordination rather than chorea. Behavioral and emotional abnormalities develop gradually and may precede visible motor abnormalities. They include personality changes, mood swings, anger, psychosis, delusions, and hallucinations. An affected individual might complain excessively, be suspicious of others, and show outbursts of temper. Cognitive symptoms are always present and typically develop several years before HD is diagnosed. Such symptoms begin with changes in concentration, planning of activities, and multitasking and memory functions, and some patients develop dementia. This so-called subcortical dementia differs from Alzheimer disease-associated dementia in that it does not cause quick forgetfulness and spares language, although patients often become disengaged from conversations and socially withdrawn.

3.2 Imaging

Magnetic resonance imaging (MRI) is a sensitive method for detecting even small decreases in brain volume that may be indicative of neurodegeneration. The most significant changes seen when comparing MRIs from individuals with HD with those from unaffected control subjects is a decrease in the volume occupied by the striatum, shown for a representative patient with HD compared with an unaffected individual in Figure 1. This volume

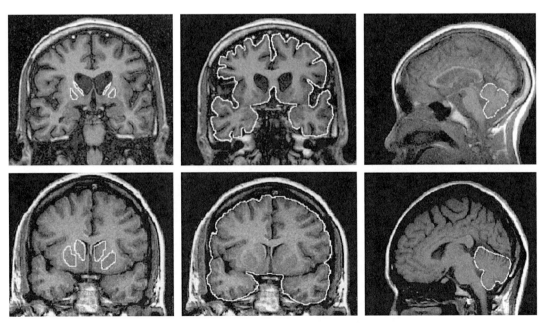

FIGURE 1 Samples of coronal and sagittal magnetic resonance images from a patient with Huntington disease (top row) and a normal control (bottom row) showing the outlines of the caudate and putamen that are part of the striatum (left), cerebral (center), and cerebellar volumes (right). *Reproduced with permission from Ref. 3.*

decrease is already detectable 15 years before symptom onset, and the atrophy continues at a fairly constant rate over the course of the disease (Figure 2). Therefore quantitative MRI can predict the future onset of motor symptoms. Interestingly, the rate of the observed volume changes can also predict the rate of subsequent disease progression. By contrast to the striatal volume, cortical volume decreases much more subtly and only after disease is clearly established.

In addition to the gray matter harboring neuronal cell bodies and dendrites, the volume of myelinated axons that form the white matter decreases as well. This is particularly evident in the frontal lobe, the globus pallidus, and the putamen. The myelin loss in the frontal lobe may explain some of the personality changes and impairments in executive function, which utilize the frontal lobe. Newer imaging modalities such as diffusion tensor MRI are now providing better visualization of white matter tracts. In addition, functional MRI allows the study of changes in connectivity between different brain regions by non-invasively examining the relative consumption of blood oxygen. When applied to HD, functional MRI shows changes in functional connectivity along the cortical and striatal motor pathways. Other experimental imaging approaches used to study chemical changes in HD apply magnetic resonance spectroscopy to accurately determine changes in chemical composition, for example, neurotransmitters such as glutamate or γ-aminobutyric acid, of a given brain region.

3.3 Epidemiology

HD is a relatively rare disease, with an incidence of ~0.4 in 100,000 and a prevalence of ~5 in 100,000. This translates to approximately ~15,000 patients living with HD in the United States and 350,000 worldwide. The disease affects men and women equally, but Caucasians are at a higher risk of developing HD compared with African Americans. For unknown reasons, certain parts of the world have significantly lower rates of HD; for example, in Japan, a person is 10-fold less likely to develop HD than a person in Europe. Similarly, there are several areas with a high prevalence of cases, such as the Lake Maracaibo region in Venezuela, which has an extraordinarily high risk for HD.

3.4 Prognosis

The typical disease onset is around age 35–45, yet it can strike at almost any age between 2 and 90. The disease progresses relentlessly, typically causing death 15–20 years after the initial diagnosis, most frequently from complications such as falls or aspiration. Motor incoordination causes frequent falls, while the failing coordination of facial and laryngeal muscles eventually makes speaking impossible and swallowing food a major challenge. Therefore patients have great difficulty eating and begin to lose weight. The constant movement and random muscle contractions, combined with poor mitochondrial function, creating an ineffective use of glucose, increases a patient's caloric needs to over 5000 calories/day, making sufficient food intake even more challenging.

4. DISEASE MECHANISM/CAUSE/ BASIC SCIENCE

4.1 Neuropathology

HD is characterized by the death of projection neurons throughout the basal ganglia, preferentially affecting the medium spiny neurons in the striatum. These are fast-spiking GABAergic neurons that receive glutamatergic input from the motor cortex and modulatory dopaminergic input from the substantia nigra. As discussed extensively in Chapter 5 (Parkinson disease), the activity of the striatum controls motor coordination via two pathways, called the direct and indirect pathways. HD primarily impairs function of striatal neurons expressing the dopamine

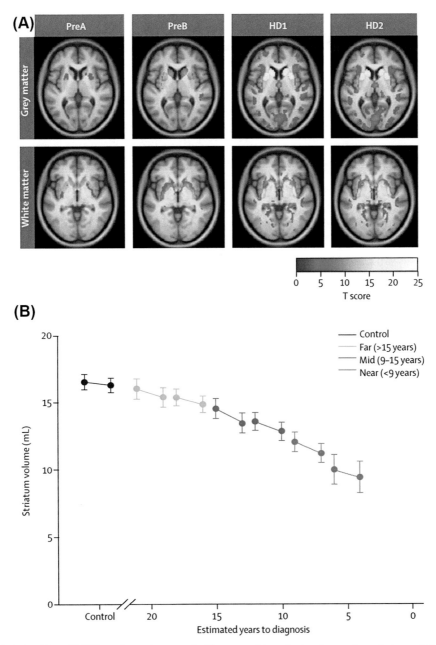

FIGURE 2   (A) Voxel-based MRI based morphometry in 2 human subjects during the prodromal phase of disease (PreA and PreB) indicates early changes in striatum and other brain regions including subcortical white matter compared with controls. As disease progresses (HD1 and HD2), striatal atrophy remains severe, but widespread brain atrophy arises, especially in other subcortical nuclei and subcortical white matter and in cortical grey matter. Red indicates substantial atrophy and yellow the greatest degree of atrophy. (B) Striatal volume derived from magnetic resonance images obtained from asymptomatic individuals who carry the mutated *htt* gene and therefore will develop disease. Striatal volume gradually declined over a 20-year time period preceding disease onset. The three groups of patients (far from onset, mid, and near to predicted onset) were each divided into two subgroups (n=40–50). For all groups, the first point is striatal volume at the time of the first MRI scan, and the second point is volume at the second scan (about 2 years later). Error bars indicate SE. *Reproduced with permission from Ref. 4.*

D2 receptor. These neurons project via the subthalamic nucleus to the globus pallidus and are part of the indirect pathway.

The atrophy of these striatal neurons is visible macroscopically as a gradual decline in the overall striatal volume, particularly in the putamen and caudate nucleus, as detected by MRI (Figure 1). As mentioned above, volumetric MRI measurement has proven to be a sensitive biomarker for HD, often showing a progressive decline in striatal volume up to 15 years before the onset of symptoms (Figure 2).

At the microscopic level, neurons in brains with HD show nuclear and cytoplasmic inclusions that contain the mHTT protein, as well as amyloid and chaperone proteins involved in protein folding and ubiquitin required for protein degradation. Inclusion bodies are universally associated with disease and appear before a patient becomes symptomatic. Inclusions may be a disposal site of nonfunctional proteins, yet the inclusions *per se* are probably not toxic since cells harboring inclusions actually survive longer than those without these inclusions. We explore the disease mechanism and the role of inclusions in more detail below.

4.2 Genetics of HD

Humans have two allelic copies of the *htt* gene on the short arm of chromosome 4. The coding region of the gene includes at least two nucleotide sequences that encode for glutamine (CAG) at the 5′ end, beginning 18 amino acids from the N terminus. These are called polyglutamine repeats (…CAGCAG…) since the nucleotide encodes for glutamine.[5] For unknown reasons, these repeats usually are amplified to a number ranging between 3 and 34, with most people having about 20 repeats. The role of these repeats is not well understood, but they may be involved in binding transcription factors to DNA. They certainly do not cause disease; therefore neither the gene nor the resulting protein is considered mutated. However, the presence of 40 or more repeats is considered a mutated gene since essentially all carriers

develop HD. As discussed below, it is assumed that the resulting mHTT protein has become toxic.

While repeat numbers between 35 and 39 also are considered mutated, they may or may not cause disease late in life. This is called incomplete penetrance since the gene does not necessarily translate to disease. Interestingly, there is a negative correlation between the number of CAG repeats and the age of disease onset; this is graphically illustrated in Figure 3 (red line). The longer the CAG stretch, the earlier the disease begins. Most common repeat lengths are 40–50 and cause disease onset between ages 35 and 45. Longer repeats containing 60 or more CAGs cause juvenile onset of disease, with the youngest known cases presenting at age 2. The highest repeat length reported is 250, but numbers above 80 are very rare. Interestingly, the CAG repeat length does not influence disease severity or how long a person will live with the

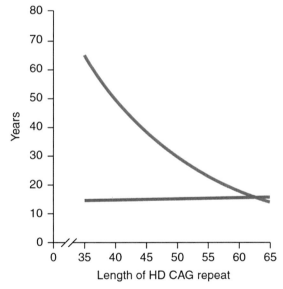

FIGURE 3　CAG repeat length is a predictor of the age at which disease manifests. The red line shows the age at onset as a function of the length of the CAG repeat, indicating that longer repeats cause much earlier disease onset. The blue line shows the duration of disease, from onset to death, as a function of the number of repeats. This line shows that repeat length does not affect disease duration. HD, Huntington disease. *Reproduced with permission from Ref. 7.*

disease (blue line in Figure 3), which seems to be fairly constant and averages about 15 years.

4.3 Heritability and Anticipation

The mutated gene is inherited in an autosomal dominant fashion and, provided at least 40 CAG repeats are present, the mutation has high penetrance, meaning essentially every carrier of the mutated gene will develop the disease. Therefore, the disease follows a "Mendelian" pattern of inheritance, whereby offspring of an affected carrier has a 50% chance of developing disease if they inherit one allele. If both father and mother are carriers, the offspring has at least a 75% chance of developing disease. A peculiar observation among families affected by HD is that the disease can begin earlier and earlier in successive generations if transmitted from an affected father through a phenomenon called genetic anticipation. Not only is the mutated gene inherited, but the number of CAG repeats present is typically at least as long as in the parents but frequently is even longer.

Genetic anticipation occurs if the gene is inherited from an affected father. On average, his offspring will develop HD at an age of 8 years younger than the father's age at onset. Anticipation is the direct result of a lengthening of the CAG expansions in the father's germ line. This occurs only if a long sequence of CAG repeats (at least 28 repeats) are present, which makes the expanded segments of the DNA unstable during replication. Spermatogenesis in the father involves a large number of cell divisions to generate millions of sperm throughout life (300 million/day), and it is during these divisions of still diploid spermatocytes that erroneous, extra CAG repeats are introduced into some of the father's DNA.[8] Different sperm cells have differing repeat lengths, and these can range from slightly shorter than the father's to much longer. Anticipation does not occur when disease is passed on through the mother's germ line because the mother's ~0.5 million oocytes do not divide and are already present in her ovaries at the time of her birth, making germ-line errors due to genetic instability unlikely.

Note that the instability of the male genes is also responsible for the *de novo* occurrence of disease in families that were previously unaffected. Let us assume, for example, that a father carries 33 repeats, and during spermatogenesis he produces an *htt* gene with 35 repeats. Because this number is below the disease threshold, his son is unlikely to develop the disease. But if the *htt* gene expands further in the son's germ line to 40 or more repeats, the son's offspring will develop HD, and they will be the first founders in this family. Since their repeat count is close to the threshold required to cause disease, his children will not develop any disease symptoms until they are well into their 40s. At this time, they probably have already conceived their own children, unaware of the genetic risk that they have passed on.

Such a *de novo* appearance of new disease is probably a rare event, and therefore the vast majority of HD cases develop in families with a long history of disease. In theory, the first *htt* mutation could have occurred in a single individual and spread throughout the world. However, sequencing data suggest that was not the case; rather, *htt* mutations formed independently in several parts of the world. However, as mentioned already, the Venezuela project that studied families along Lake Maracaibo suggests that the largest group of kindred of almost 15,000 HD cases can be traced to a single mother who lived in the early 1900s,[1] while the remaining 4000 cases are not her offspring.

4.4 Animal Models

Genetically modified model organisms that harbor the exact mutations seen in patients are considered the "holy grail" when studying disease in a laboratory. Since HD is a monogenetic disease with complete penetrance, it should be possible, theoretically, to accurately reproduce disease in transgenic mice. A number of different

FIGURE 4 Intranuclear inclusion bodies in neurons in a sample from the autopsy from a patient with Huntington disease (A, arrow) and in a tissue section from a mouse model of Huntington disease (B, arrows). *Reproduced with permission from Ref. 6.*

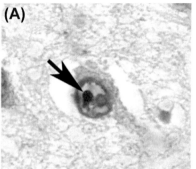 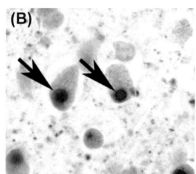

mouse models that express either the full-length *mhtt* gene, with at least 75 CAG repeats, or fragments of the gene with >100 CAG repeats have been produced over the years. Commonly used strains called R6/1, R6/2, YAC128, and BACHD are now commercially available so that investigators can test their hypotheses and drug treatments on the same platform. While these different mice show some phenotypic differences, they generally show aggregation of mutant huntingtin in the form of inclusion bodies (Figure 4(B)); they show atrophy of the striatum and motor cortex, and they have significant motor impairments, all at least qualitatively comparable to human disease. For example, *mhtt* mice show a characteristic hind limb clasping response when suspended by their tail, as illustrated in Figure 5; this clasping response never occurs in wild-type mice. Motor symptoms can be conveniently identified using the rotarod test, which measures the ability of a mouse to balance itself on a rotating metal rod; all of these mouse lines consistently show deficits on this task. There is more variability between these mice when assessing cognitive deficits and anxiety- and depression-related phenotypes, and therefore different researchers prefer different mice for their studies. Mice with symptomatic HD, however, typically score below controls on the Morris water maze, which tests special memory, and show deficits in novel object recognition. They also typically show depression-like behavior during forced swim or tail suspension tests,

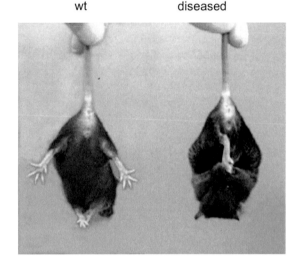

FIGURE 5 Hind limb clasping is an abnormal motor response in a mouse carrying the mutated huntingtin gene (right) as opposed to the splaying of the hind limbs observed in wild-type (wt) mice (left). *Reproduced with permission from Ref. 9.*

where they accept their helpless situation and adopt an immobile position more quickly than do control mice.[10]

While these mouse models have been excellent tools to aid in the study of disease mechanisms, as discussed further below, they have not performed well in the screening and validation of drug candidates for treatment. In spite of well over 5000 articles that used animal models to study HD, none of the treatments that showed promise in animal models have thus far worked

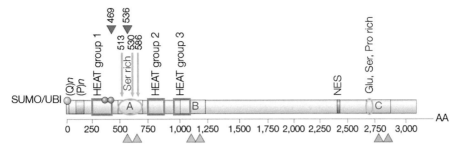

FIGURE 6   Schematic diagram of the huntingtin amino acid sequence. The polyglutamine tract (Q)n is followed by the polyproline sequence (P)n, and the red squares indicate the three main clusters of HEAT repeats. The green arrows indicate the caspase cleavage sites and their amino acid positions, and the blue arrowheads indicate the calpain cleavage sites and their amino acid position. B identifies the regions cleaved preferentially in the cerebral cortex, C indicates those cleaved mainly in the striatum, and A indicates regions cleaved in both. Green and orange arrowheads point to the approximate amino acid regions for protease cleavage. The red and blue circles indicate post-translational modifications: ubiquitination (UBI) and/or sumoylation (SUMO) (red) and phosphorylation at serine 421 and serine 434 (blue). The glutamic acid (Glu)-, serine (Ser)-, and proline (Pro)-rich regions are indicated (Ser-rich regions are encircled in green). NES, nuclear export signal. *Reproduced with permission from Ref. 11.*

in human disease. This is a problem encountered consistently throughout the diseases described this book, and has prompted researchers to question the use of rodent models in the development of new treatments. A more elaborate discussion of the many reasons why mouse models of disease may be poor predictors for treatment is found in Chapters 1 and 15.

4.5 Normal Function of Huntingtin Protein (HTT)

HTT is a large protein that contains 3144 amino acids and does not have any homology to other proteins. In spite of its large size, it is water-soluble and found throughout cell cytoplasm. It associates with the Golgi network, endoplasmic reticulum, cell nucleus, neurites, and synapses and is often associated with vesicles. HTT is primarily found in the nervous system and testes. The primary amino acid sequence, shown schematically in Figure 6, contains few known motifs that could provide information about its normal function. The most conspicuous features are a polyQ region, which presumably binds transcription factors, and a series of HEAT repeats, which are 40-amino acid segments that

mediate protein–protein interactions. There is a nuclear export signal, suggesting that the protein may help transport molecules in and out of the nucleus. HTT has several modulation sites where the HTT activity can be altered through phosphorylation, ubiquitination, sumoylation, and palmitoylation. These reversible chemical changes can alter the biological function of HTT.[11]

HTT must be important during embryonic development since its germ-line deletion causes embryonic lethality, which, surprisingly, can be rescued by expression of mHTT. Reduced expression of HTT (<50%) during embryological development causes impaired neurogenesis and malformations of the cortex and striatum. Studies of cultured striatal neurons suggest that HTT enhances a neuron's resistance to stress, mitochondrial toxins, and proapoptotic stimuli. Indeed, ablation of HTT only in neurons causes neuronal cell death in the hippocampus, cortex, and striatum by 1 year of age.

HD seems to present with a regional deficiency in brain-derived neurotropic factor (BDNF), a trophic factor primarily produced by cortical neurons. BDNF is transported via axonal transport to the corticostriatal synapses, where it is released onto the medium spiny striatal neurons that are the first to atrophy in HD (Figure 7). Wild-type HTT regulates BDNF transcription in a positive way by binding to REST, a transcriptional repressor. When bound by HTT, the REST repressor is retained in the cytoplasm, unable to repress BDNF expression. When mutated, however, HTT is less well able to retain REST, which traffics to the nucleus and binds to repressor element 1/neuron-restrictive silencer element, which is located in the BDNF promoter exon II and prevents BDNF transcription (Figure 7). After its transport to the striatum, BDNF modulates glutamate release at the corticostriatal synapses. This may explain why overexpression of wild-type HTT protects striatal neurons from glutamate excitotoxicity. In addition, HTT regulates vesicular BDNF transport along the axonal microtubules, stimulating its transport toward the synapse.

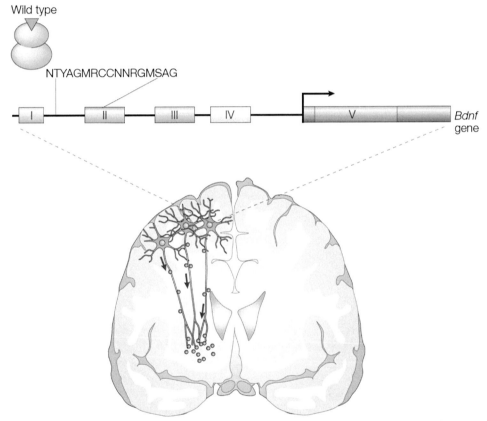

FIGURE 7　Brain-derived neurotrophic factor (BDNF) is produced by cortical neurons and transported to striatal neurons. Wild-type huntingtin contributes to *Bdnf* transcription in the cortical neurons that project to the striatum by inhibiting the repressor element 1/neuron-restrictive silencer element (RE1/NRSE) that is located in the BDNF promoter exon II. I–IV indicate BDNF promoter exons in rodent *Bdnf*; V indicates the coding region. The RE1/NRSE consensus sequence is shown. Inactivation of the RE1/NRSE in *Bdnf* leads to increased messenger RNA transcription and protein production in the cortex. BDNF, which is also produced through translation from exons III and IV, then is made available to the striatal targets via the corticostriatal afferents. Wild-type huntingtin might also facilitate vesicular BDNF transport from the cortex to the striatum. *Reproduced with permission from Ref. 11.*

HTT also interacts with proteins involved in the axonal transport of other substrates, including organelles. Loss of functional HTT reduces the transport of mitochondria along the axon, which in turn may deprive synapses of sufficient adenosine triphosphate (ATP) to perform their function.

At synapses, HTT interacts with the postsynaptic density protein 95, a scaffolding protein that regulates the function of N-methyl-D-aspartate (NMDA) receptors. Inappropriate regulation of this interaction by mHTT likely contributes to NMDA receptor (NMDA-R)-mediated toxicity. Taken together, HTT seems to have numerous, largely beneficial roles during nervous system development and adult nervous system function. This, then, suggests that loss of HTT function in HD may have pleiotropic effects, thereby affecting several characteristics of the disease.

4.6 Mutant Huntingtin

mHTT may cause disease in one of two ways: either the mutated protein has deleterious properties in its own right, referred to as a "negative gain of function," or its aggregation into inclusion bodies prevents it from performing the normal functions just discussed, called a "loss of function." The most prevalent current assumption holds that the disease is primarily caused by a negative gain of function, with mHTT possibly being toxic. As discussed below, however, loss of wild-type function certainly seems to contribute to disease severity, as well.

mHTT aggregates contain amyloid fibers that assume a β-sheet configuration. As we learned in Chapter 4 on Alzheimer's disease, proteins forming β-sheets are highly insoluble. The protein aggregates also contain a number of transcription factors and molecules involved in the quality control of proteins. Their sequestration into aggregates possibly prevents them from participating in normal protein biosynthesis, trafficking, and removal of ill-formed proteins. Consequently, these aggregates negatively affect overall cell health. Note that, similar to Alzheimer's disease, there is evidence that the mHTT oligomers may be the most toxic and that the polymeric protein inclusions represent a strategy for cells to partition off these detrimental proteins.

mHTT oligomers interact with molecules involved in the cell cycle, transcription, DNA/RNA processing, protein transport, and energy metabolism; consequently, each of these cellular processes can be negatively affected by disease. These targets are schematically depicted in Figure 8, and the role that each of these plays in disease is discussed further below. They can be roughly divided into changes that directly affect neuronal signaling, indirect effects on neuronal health and function, such as metabolic effects, and neurotoxicity.

4.7 What Causes Disease: Mutated DNA, RNA, or Protein?

Before investigating disease mechanisms in greater detail, we first need to address whether the CAG repeats contained in the DNA are sufficient to cause disease. Then we must consider whether disease is the result of altered properties of the mutated protein or whether the DNA or its transcribed messenger RNA alone are sufficient to cause disease. For example, changes in the structure of DNA may prevent it from being transcribed into RNA, whereas changes in RNA structure, such as hairpins loops, could bind to other RNAs and block their translation into proteins. Therefore, if CAG repeats are sufficient to cause disease, do they do so at a DNA level, an RNA level, or a protein level?

An elegant approach was used to show that CAG repeats are indeed sufficient to cause disease. Researchers produced a transgenic mouse in which the *HTT* gene was introduced with a long CAG expansion (94 CAG's) under the control of a biological on/off switch. This switch was a doxycycline-dependent promoter, which allowed gene expression to be switched off by adding doxycycline to the drinking water. When

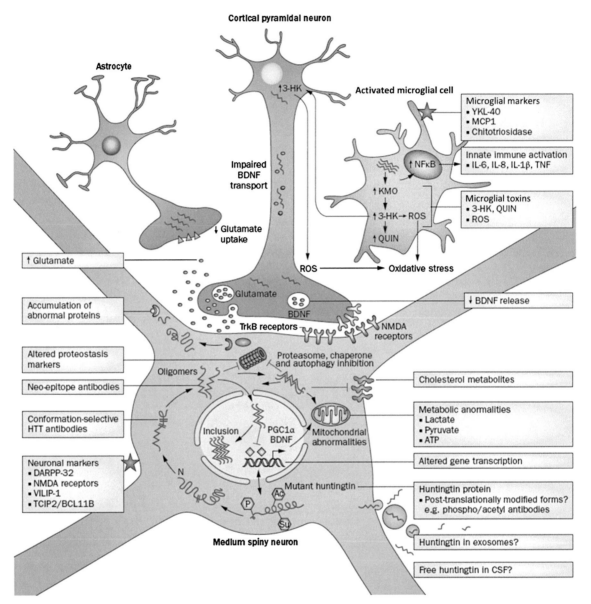

FIGURE 8   Cellular pathogenesis in Huntington disease. Mutated HTT oligomers and polymeric inclusion have been shown to affect a wide variety of cell biological process in the motor neuron, as well as in surrounding astrocytes and microglial cells. Among these, the most important pathways affected are proteostasis, mitochondrial energy production, regulation of gene transcription and translation, axonal transport, and vesicular transmitter and peptide release. ATP, adenosine triphosphate; BDNF, brain-derived neurotrophic factor; CSF, cerebrospinal fluid; IL, interleukin; NMDA, N-methyl-D-aspartate; ROS, reactive oxygen species; TNF, tumor necrosis factor. *Reproduced with permission from Ref. 2.*

the mutated *HTT* gene was active (no doxycycline), animals developed HD symptoms and huntingtin aggregates throughout the striatum. Remarkably, however, when the gene was switched off (with doxycycline) in adult animals that had already developed aggregates and disease symptoms, the aggregates and motor symptoms both rapidly disappeared.[12] This experiment shows that the CAG expansion is both necessary and sufficient to produce disease. But, even more importantly, this experiment also shows that, in principle, HD is reversible, a profound finding suggesting that if one extrapolates findings in mouse models to human disease, turning off the mutated gene may cure the disease.

While these experiments turned off the mutated gene and demonstrated the necessity and sufficiency of CAG repeats in causing disease, they do not identify whether it is the protein, the DNA, or the RNA that causes disease. The answer to this question came from studies of different transgenic mice. When scientists introduced a mutated human HD gene into the germ line of transgenic mice, they were surprised to find that the mice were completely normal and had no evidence of motor symptoms. Upon further analysis they found that the *mHTT* gene accidently contained a stop codon, and therefore its RNA transcripts did not produce any huntingtin protein.[13] This suggests that the CAG repeats alone, absent mHTT, do not produce disease. Another study provided evidence to suggest that mutated RNA is also not toxic. This team of scientists generated a transgenic mouse that harbored a mixture of both CAG and CAA triplets. Their rationale was that both CAG and CAA encode for glutamine, yet repeats that contain heterogeneous triplets are more stable than pure CAG triplets because they give rise to RNA that folds differently. Specifically, the lack of intrastrand pairing prevents the formation of RNA hairpins, which are hypothesized to be toxic. The resulting animals had different DNA and RNA sequences, yet the resulting mutated protein was still the same and was sufficient to cause disease. This suggests that RNA hairpins are also not responsible for the observed neurodegeneration.[14] Taken together, these experiments show that mHTT is indeed necessary and sufficient to cause disease, whereas mutated DNA or RNA that is not turned into mutated protein is not sufficient.

4.8 Are Intranuclear Protein Inclusions Toxic?

As we have seen, intranuclear inclusions are a pathological hallmark of HD and are found in both mouse models of the disease and human autopsies, as shown in Figure 3. They are intricately related to disease, developing before symptom onset in humans and mouse models. Intracellular inclusions contain a number of proteins, including mHTT, ubiquitin, and amyloid and chaperones such as heat shock proteins. Importantly, the previously described studies that used a doxycycline switch to turn on or off the HTT mutation show that turning off the mutated gene after disease is already established causes the inclusions to disappear, along with the motor symptoms.[12] Therefore, a disease-causing and possibly toxic role for the inclusions seems likely.[7]

However, there are several ways in which these inclusions may contribute mechanistically to disease without being themselves toxic. In Chapter 4 we learned that oligomeric Aβ42 is the toxic species in Alzheimer's disease, and the accumulated amyloid plaques simply represent a storage reservoir for toxic oligomers. A similar possibility exists for HD. Alternatively, it is possible that oligomeric HTT is also nontoxic but that the aggregated proteins bind to and sequester many other important proteins, such as transcription factors and chaperone molecules. Unfortunately, there is no unequivocal answer as to which possibility is correct. Indeed, the role of inclusions and their toxicity may be one of the most studied and debated questions in HD research. That said, there is strong support for the notion that trafficking of misfolded proteins is key to understanding the pathology of HD. Central to this idea is a concept called *proteostasis*, a term that describes the maintenance of a constant protein content, that is, equilibrium between production of proteins in and removal of proteins from a cell.

4.9 The Proteostasis Network

Proteins require a defined three-dimensional shape to function optimally. For example, ion channels create a water-filled pore in the membrane only if they are in the right configuration, and enzymes can bind a catalytic substrate only if they are properly folded. Cells use a number of molecular chaperones to accomplish the correct configuration of proteins. Importantly, if things go wrong, these chaperone proteins can also play a role in swiftly removing misfolded proteins under normal circumstances.

This role of chaperone proteins in removing misfolded proteins is also the case for proteins that lose their correct shape and become misfolded as a consequence of disease, as is the case in HD. Moreover, proteins have a limited lifespan and are constantly replenished, with old proteins being removed as new ones take their place. This occurs via two parallel disposal systems. The ubiquitin proteasome system (also discussed in Chapter 5, Box), where proteins are ubiquitinated via E3 ligase and marked for degradation by the proteasome. The second disposal system is the autophagosome system, which picks up large, bulky proteins or protein aggregates that are too big for the ubiquitin proteasome system. Together, these two processes, illustrated in Figure 9, ensure protein homeostasis, also referred to as proteostasis. This is a highly regulated process that involves a number of regulatory proteins, which are collectively called the proteostasis network.[15]

In the case of HD the appearance of inclusion bodies is a visible defect in proteostasis, whereby these two protein degradation processes can no longer keep up with the mutated and misfolded HTT protein that is being synthesized. The accumulation of aggregated proteins may simply be a "cellular trash compartment," and thus the piling up of cellular trash may be a coping strategy. This would explain why the disease can be reversed in transgenic mice when the *mHTT* gene is turned off, even after disease has already started. Whether simply too much mutant huntingtin is being produced or whether there are defects in the proteins that participate in the proteostasis network is not clear. In fact, both may occur in conjunction, since many of the chaperone molecules and proteins involved in protein degradation, such as ubiquitin and heat shock proteins, also are trapped in the intracellular protein aggregates. One may therefore consider HD a disease that presents with a "protein traffic jam." This traffic jam in turn creates collateral damage and toxicity to the cell via defective axonal transport, mitochondrial dysfunction, oxidative stress, impaired glutamate transport, oxidative toxicity, and changes in DNA/RNA transcription and translational regulation. These various pathogenesis pathways are schematically illustrated in Figure 7 and further discussed below.

4.10 Glutamate, Excitotoxicity, and Synaptic Function

HD is associated with cognitive decline, and mice carrying mHTT show poor performance on cognitive tests that are suggestive of changes in synaptic plasticity. Brain slices from mHTT mice show deficits in both long-term potentiation (LTP) and long-term depression (LTD), the cellular correlates of learning and forgetting, respectively. Both forms of learning engage the NMDA subtype of the glutamate receptor. As we learned in Chapter 4 (Alzheimer disease), these receptors are permeable to Ca^{2+}, and Ca^{2+} entry activates many of the downstream kinases required to establish LTP and LTD. We also learned in

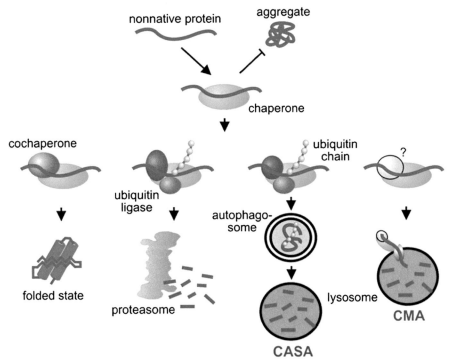

FIGURE 9   The proteostasis network. Molecular chaperones bind nonnative proteins and prevent their aggregation. Folding or degradation of the client protein is initiated in conjunction with regulatory co-chaperones. Association with an ubiquitin ligase leads to the formation of a ubiquitin chain on the chaperone-bound client. This induces client sorting to the proteasome or triggers the autophagic engulfment of the client during chaperone-assisted selective autophagy (CASA). On the latter pathway, the client is eventually degraded in lysosomes. During chaperone-mediated autophagy (CMA), the client is directly translocated across the lysosome membrane. *Reproduced with permission from Ref. 16.*

Chapter 1 (Stroke) that if too much Ca^{2+} fluxes through NMDA-Rs, an excitotoxic cascade develops whereby destructive enzymes, including calpains, phosphatases, and caspases, are activated and ultimately cause cell death.

Changes in the cellular location and subunit composition of NMDA-Rs may explain both the cognitive decline and the striatal cell death that occur in HD. We typically assume that NMDA-Rs are localized only at the synapse, where they receive the presynaptically released neurotransmitter glutamate. However, NMDA-Rs can also be found extrasynaptically, as illustrated in Figure 10. This occurs in the normal brain but seems to be particularly pronounced in HD.[17] The extrasynaptic NMDA-Rs show tonic, sustained activation and are therefore more likely to mediate a sustained Ca^{2+} influx that drives the cell death cascades.

Extrasynaptic receptors are composed of different subunits. While typical synaptic NMDA-Rs are heterotetramers containing an obligatory NR1 subunit combined with NR2A, extrasynaptic receptors instead pair NR1 with NR2B. This change alters the timing of the postsynaptic response and changes cellular learning behavior, whereby NR2B activation favors LTD (the cellular equivalent of forgetting), whereas NR2A subunits favor LTP (the cellular equivalent of learning).[18] Moreover, synaptic NR1/NR2A receptors couple to downstream signaling pathways that promote the survival and release of

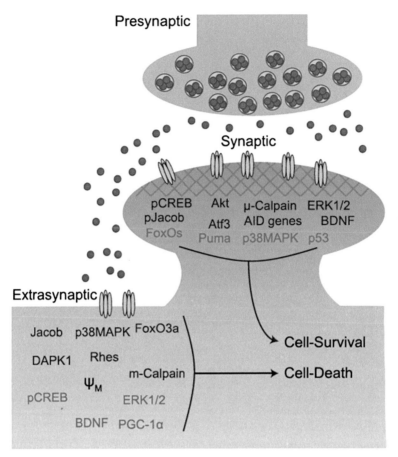

FIGURE 10   Contribution of extrasynaptic N-methyl-D-aspartate (NMDA) receptors to neuronal cell death in Huntington disease. Spillover of glutamate from synapses activates extrasynaptic NMDA receptors that couple via p38MAPK and m-Calpain to promote cell death. By contrast, synaptic NMDA receptors act via a variety of signaling cascades to enhance survival. *Reproduced with permission from Ref. 17.*

tropic factors including BDNF, whereas extrasynaptic NR1/NR2B receptors couple to cell death pathways, illustrated schematically in Figure 10. Taken together, in HD the enhanced expression of NR1/NR2B receptors at extrasynaptic sites changes synaptic plasticity in favor of LTD and suppresses LTP, while also shifting from a pro-survival to a pro-death mode.[17]

Extrasynaptic receptors are the primary target of memantine, a use-dependent, noncompetitive antagonist of NMDA-Rs with relatively low affinity to its receptor. Memantine is clinically approved to treat Alzheimer dementia and

has ameliorated motor symptoms and reduced striatal neuronal loss in *mHTT* mutant mice, suggesting its potential efficacy as a treatment for HD.

Since the extrasynaptic activation of NMDA-Rs occurs by glutamate spilling over from adjacent synapses, there must also be changes in astrocytic function since spillover is usually prevented by astrocytic glutamate uptake via the EAAT2 glutamate transporter (Figure 8). Indeed, selective expression of the *mHTT* gene only in astrocytes causes a loss of EAAT2 expression and reduced glial glutamate uptake, and this is sufficient to cause HD-like motor symptoms in mice even if the neurons contain wild-type HTT.[19] This suggests that extrasynaptic glutamate may contribute to disease by activating NMDA-Rs under conditions in which glial uptake is impaired. Interestingly, pharmacological enhancement of glial reuptake, using ceftriaxone, an antibiotic that causes an increase in EAAT2 transcription, has been shown to be neuroprotective in mHTT mice.[20]

The notion that glutamate toxicity may play a role in HD dates back many decades. Indeed, long before the *htt* gene was cloned and genetic mouse models of disease were established, several animal models of HD were produced through lesioning of the striatum by infusion of the glutamate receptor agonists kainate or quinolinic acid.[10] The ensuing excitotoxic death of striatal neurons presented with HD-like motor symptoms and suggested that overactivation of glutamate receptors may play a role in the natural development of the disease. One caveat, however, is that unlike wild-type mice, the striatal neurons of mice harboring mHTT show a remarkable resistance to glutamate.[21] This may be a compensatory effect because these neurons have significantly elevated basal Ca^{2+} and show enhanced Ca^{2+} buffering capacity, suggesting that they may have adapted to the toxic milieu in which they exist. The role of glutamate toxicity in patients with HD requires much further study via magnetic resonance spectroscopy. In addition, the clinical use of memantine in patients with HD may provide this answer, although a small pilot study has thus far failed to observe any symptomatic improvements with memantine.[22]

4.11 BDNF and Synaptic Function

An alternative way to explain a preferential death of striatal neurons that indirectly involves glutamate and NMDA-Rs implicates a loss of BDNF. We learned above that BDNF is made by layer 5 motor neurons in the cortex and that its transcription and transport to the striatum are regulated by wild-type HTT. In the presence of mHTT BDNF transcript levels are reduced in mouse models of disease, as well as in postmortem cortical tissue of patients with HD.[23] Presynaptically, BDNF facilitates the release of glutamate-containing synaptic vesicles, whereas on the postsynaptic terminal, BDNF activates mitogen-activated protein kinase and CaMKII, both of which are involved in LTP. In Chapter 11 we discuss Rett syndrome, a severe neurodevelopmental disease affecting young girls that is largely caused by suppression of normal BDNF signaling as a result of a mutation in a gene (*MECP2*) that acts as a transcriptional regulator. Moreover, in Chapter 5 (Parkinson Disease) we learned that BDNF plays a particularly important role in the survival of striatal neurons—the very ones lost in HD. Therefore a loss of function of wild-type BDNF can contribute to disease through a variety of changes, including an enhanced susceptibility to glutamate. Loss of BDNF may also explain the early and preferential loss of striatal neurons in disease and may provide a hitherto unexplored avenue for a possible therapeutic intervention.

4.12 Mitochondria and Metabolic Dysfunction

The striatal neurons that die in HD receive synaptic input at an incredibly high rate, probably exceeding 1000 Hz. In addition, their axons fire action potentials at a very high rate. This requires much energy in the form of ATP produced by the mitochondria. Inclusion bodies that contain mHTT are frequently found on the outer membrane of mitochondria and impair electron transport via complexes II and III, resulting in a depletion of cellular ATP and an increase in reactive oxygen species. To make matters worse, the presence of mHTT also impairs the transport of mitochondria from the cell body to the synapse via axonal transport. As a consequence, cells cannot deposit their mitochondria at sites with the highest energy demand. Not surprisingly, studies of glucose metabolism in patients with HD show reduced glucose uptake in the striatum and the cerebral cortex; this decreased metabolism is already present during the presymptomatic stages of disease. In addition, reduced activity of the key components in oxidative phosphorylation and the tricyclic acid cycle has been documented in mitochondria isolated from patients with HD. Finally, poisoning mitochondria chemically with drugs such as 3-nitropropionic acid or malonate, which specifically inhibit Complex II of the respiratory chain, is sufficient to cause HD-like motor symptoms. Hence ample evidence points toward mitochondrial dysfunction contributing to neuronal death in HD.[24]

Importantly, the impairment of mitochondria by mHTT inclusions is not restricted to the brain but occurs throughout the body. Swollen mitochondria with vacuoles and abnormal cristae are found in fibroblasts and muscle tissue in patients with HD. It has long been known that patients with HD experience significant weight loss and are frequently emaciated. In symptomatic patients continuous involuntary muscle contractions are partially to blame for a significantly increased energy need, estimated at about 5000 calories per day. Yet weight loss precedes motor symptoms in mutation carriers. Therefore it is likely that impaired ATP production and ineffective use of glucose throughout the body are major reasons for the increased dietary energy requirement of patients. Poorly functioning mitochondria are also much more likely to generate reactive oxygen molecules, which can cause significant damage to cell membranes and DNA. Therefore impaired mitochondria may contribute in a somewhat nonspecific way to the symptoms of HD.

Mitochondrial health is the target of the Q10 coenzyme, which functions both as an electron carrier and antioxidant in normal mitochondria. Although an initial multicenter clinical trial failed to show the efficacy of coenzyme Q10 (CoQ10) in symptomatic patients,[25] CoQ10 is still being pursued as a potential preventive therapeutic in patients in the prodromal phase

of disease (www.clinicaltrials.gov, identifier NCT00608881).

4.13 Role of Glia in HD

The selective neuronal atrophy in the striatum affects medium spiny projection neurons but spares interneurons and glia, and the exclusive appearance of inclusion bodies in these neurons is highly suggestive of a cell autonomous mechanism of disease, that is, processes that affect only these neurons. However, the mutated protein is found in all cells of the brain and body. It is therefore possible that non-neuronal cells, particularly astrocytes and microglia, may contribute to disease. Such a non-cell-autonomous disease mechanism has been demonstrated in ALS and Rett syndrome. Evidence of a contribution of microglial cells comes from imaging studies that use a positron emission tomography agent called 11C-(R)-PK11195 that selectively attaches only to activated microglial cells, where it binds to a receptor for benzodiazepines. Positron emission tomography imaging of presymptomatic HD mutation carriers was able to detect activated microglial cells in the striatum of patients up to 15 years before the onset of symptoms.[26] In all likelihood these microglia are involved in an inflammatory response that apparently is present long before a patient becomes symptomatic. Astrocytes, too, seem to play an important, albeit incompletely understood, role. Their role in glutamate removal has been abundantly emphasized in the previous chapters, and impairment of astrocytic glutamate transporter function is of key importance in stroke and ALS. Coculture of wild-type striatal neurons with mHTT-expressing astrocytes is surprisingly sufficient to cause neuronal excitotoxicity. Moreover, transgenic mice in which the mutant HD gene is expressed only in astrocytes, leaving neurons with wild-type huntingtin, develop many HD-like features, including motor symptoms, weight loss, and premature death, albeit with a delayed onset compared with mice with neuronal expression of the mutant gene.[19] Also, astrocytes in the striatum of mice carrying mutant huntingtin show a loss of the Kir4.1 K$^+$ channel; this channel plays an important role in controlling extracellular K$^+$ homeostasis, which in turn affects exitability.[27] Virally restoring the channel expression in astrocytes normalizes many of the disease phenotypes, suggesting that astrocytic dysfunction, in conjunction with neuronal mHTT, contributes to disease. Surprisingly, however, mice carrying mHTT that had already developed the full-blown disease could be cured if the defective protein was selectively removed only from striatal and cortical neurons, leaving the mutated protein in all other cells, including astrocytes.[28] Taken together, the best interpretation is that neuronal mHTT expression drives disease, but neuronal loss can be attenuated by normal or supranormal astrocytic function.

4.14 Related Glutamine Repeat Diseases

There are nine known polyglutamine repeat diseases that cause neurodegeneration. In addition to HD, these include six different presentations of spinocerebellar ataxia (SCA 1, 2, 3, 6, 7, and 17), spinobulbar atrophy, and dentatorubral-pallidoluysian atrophy. The only commonality shared by these diseases is a polyglutamine expansion of at least 40 glutamines. In all but two of these diseases the affected proteins are of unknown function. In spinobulbar atrophy the affected protein is a nuclear receptor for androgens (testosterone); in SCA6 it is a necessary subunit of the synaptic P/Q type (Cav2.1) voltage-gated Ca^{2+} channel. The molecular understanding of all of these diseases is still incomplete, but as in HD, the polyglutamine repeats are believed to cause disease. In light of the role of glutamines as binding sites for transcription factors, it is suspected that these diseases arise primarily as a result of altered gene regulation, rather than a loss of function of the polyglutamine-containing protein itself. As in

HD, the affected proteins also share a common tendency to misfold and aggregate; in doing so they also tend to sequester many important regulators of proteostasis. Therefore, many of the previously discussed ideas regarding the cellular pathology of HD probably also apply to these related, albeit much less common, polyglutamine disorders.[29]

5. TREATMENT/STANDARD OF CARE/CLINICAL MANAGEMENT

5.1 Treatment

Currently available treatments cannot modify the disease course and instead exclusively treat disease symptoms. The only drug specifically approved to treat HD is tetrabenazine, which suppresses chorea movements. Tetrabenazine inhibits the vesicular monoamine transporter 2 (VMAT2), which is responsible for loading dopamine and related monoamines into synaptic vesicles. Its suppressive effect on chorea is attributed to an inhibition of the "indirect pathway" of the basal ganglia that regulates motor control in the midbrain, discussed more extensively in Chapter 5 (Parkinson Disease). The drug was approved only recently by the US Food and Drug Administration for the specific treatment of motor syndromes in HD but has been in use as such in Europe since the 1960s. VMAT2 is also the target of amphetamine. Given its somewhat nonspecific effects, the drug can cause mood changes and depression. Other drugs to treat chorea include the neuroleptics clozapine, haloperidol, or fluphenazine, which target dopamine receptors and are used more frequently to treat psychosis in patients with schizophrenia.[30]

Neurobehavioral effects are treated as in other patients who present with anxiety, apathy, or depression—typically, through use of the selective serotonin reuptake inhibitors. Note, however, that their comparative efficacy to placebo is questionable (see Chapters 12 and 15), and this is also the case in HD. Unfortunately, cognitive effects are equally difficult to treat. Acetylcholinesterase inhibitors such as donepezil, used to treat Alzheimer disease, do not yield significant improvements in HD.

In the treatment of patients with HD, much emphasis is on sufficient nutrient intake to meet the high caloric demand and in preventing injury from falls and asphyxiation from poor control of laryngeal muscles.

5.2 Genetic Testing

Since the cloning of the HD gene, accurate predictive genetic testing is readily feasible at any age and requires only the collection of DNA through a saliva swab. Testing offers individuals who are born into an affected family the opportunity to predict whether they will develop disease, but in the United States fewer than 5% of patients take advantage of this possibility, whereas a much higher percentage do so in Canada and Europe. In the United States families may be afraid to be denied health coverage for a preexisting condition. Testing also offers individuals with a family history of disease the opportunity to help with family planning. Given the accuracy with which genetic testing can predict HD onset, the test carries significant psychological risk and the potential for genetic discrimination. It may be surprising that even a negative test result can cause harm; it may upset family dynamics, particularly if one child is positive and one negative. The latter may experience feelings of guilt rather than relief. Interestingly, while individuals who receive a positive diagnosis often initially show increased anxiety and stress, knowledge of their disease status typically has a positive outcome, with reduced anxiety in the long term. Given the complexity of psychological harm and the ethical issues surrounding genetic testing, tests are currently offered only to individuals who have a documented familial risk of HD.[5]

6. EXPERIMENTAL APPROACHES/ CLINICAL TRIALS

A challenge in developing drugs to treat HD is the identification of valid disease targets that have a high likelihood of altering disease.[4] At present, there is only one such target, the mutant huntingtin protein. As we learned, there is little doubt that removing mHTT, ideally in presymptomatic mutation carriers, is likely to cure the disease. While this approach is possible in transgenic mice, however, it is far more challenging in patients. Antisense oligonucleotides, in theory, can be used to interfere with the translation of new mHTT, and feasibility studies of patients with ALS have been done (see Chapter 6). Absent other common pathological pathways, essentially all the mechanisms that could contribute to disease downstream of the mutant protein are equally suitable targets for exploration. Indeed, many strategies have been pursued in the laboratory and showed some promise in animal models, yet few drugs have moved into clinical studies. Table 1 lists the drug candidates that showed greatest promise in mouse models and have undergone clinical trials. Except for tetrabenazine, which is already approved to treat chorea associated with HD, none of these drugs showed significant improvements over placebo.[30] Only a few of the ~150 clinical trials registered at www.clinicaltrials.gov (2014) are actually examining new pathways, strategies, or drugs. Stem cell implants and biomarker studies are among the ongoing studies and are at various stages of development. Some of the newer ideas are briefly discussed below.

6.1 Targeting Mutant HTT

The removal of mutated huntingtin has shown disease arrest in transgenic mice, where this was achieved using small-interfering RNA or antisense oligonucleotides.[4] Both strategies interfere with *mhtt*, albeit through different molecular strategies. ASOs have yielded

TABLE 1 Relative Efficacy of Experimental Drugs That Showed Promise in Animal Models of Disease

Drug	Mode of action	Effect in animal models	Effective in patients	Duration of treatments
Cannabidol	Cannabinoid receptor activation	Reduced lesion size	Not significant	6 Weeks
Creatine	Boosts energy supply	18% Improved survival	Not significant	16 Weeks
Ethyl-EPA, Omega-fatty acid	Reduce oxidative stress	Improved rotarod	Not significant	6 Months
Fluoxetine	Serotonin reuptake inhibitor	Reduced neurodegeneration	Not significant	4 Months
Minocycline	Anti-inflammatory	13% Improved survival	Not significant	8 Weeks
Remacemide	NMDA antagonist	16% Improved survival	Not significant	6 Weeks
Remacemide + CoQ10	NMDA antagonist, mitochondrial stabilizer	32% Improved survival	Not significant	30 Months
Riluzole	Na⁺ channel, NMDA?, AMPA?	10% Improved survival	Not significant	8 Weeks
Tetrabenazine	Inhibits vesicular monoamine transporter 2	60% Improved rotarod results	Significant 3.5-unit improvements on UHDRS	12 Weeks

EPA, eicosapentaeonic acid; NMDA, N-methyl-D-aspartate; UHDRS, unified Huntington disease rating scale.
Table modified from Table 2 in Crook and Housman,[15] with clinical study data from the Huntington's Study Group. A summary analysis suggests that none of these drugs, except for tetrabenazine, provides any symptomatic improvements in HD. Reproduced with permission from Ref. 31

promising results when infused into the ventricles of nonhuman primates.[32] While suppressing the generation of mHTT chronically over the lifetime of a human is much more challenging than in mice, the success of primate studies provide hope that a similar approach may eventually be feasible and translatable to humans. A similar antisense oligonucleotide approach was found to be safe in ALS and will soon be examined in HD as well.

An alternative approach to reducing mHTT would be to increase the degradation of the mutant protein via ubiquitin-specific proteases or through autophagy. The latter can be stimulated using pharmacological activators of the mammalian target of rapamycin pathway using drugs such as calpastatin or minoxidil, neither of which has yet been studied in patients with HD.

6.2 Metabolism

As discussed above, mHTT negatively interacts with mitochondria. In addition, patients demand a higher caloric intake to accommodate their hyperactivity. The mitochondrial abnormalities are suspected to generate toxic byproducts, such as reactive oxygen species and nitric oxide synthase. Antioxidant strategies in culture models of HD were promising and have given way to two large, ongoing, placebo-controlled phase II clinical trials investigating the comparative benefit of dietary supplementation with CoQ10 (www.clinicaltrials.gov identifiers NCT00920699 and NCT00980694). This approach is unlikely to modify disease, but it may slow the time course of disease progression. Since CoQ10 is widely available as a nutritional supplement with few safety concerns, it is likely already used by many patients outside controlled clinical studies.

6.3 Excitotoxicity

The extensive death of medium spiny neurons in the striatum is suspected to be caused in part by glutamate excitotoxicity. Consequently, drugs that target the glutamate system and are currently used to treat ALS and stroke are being evaluated for use in HD as well. Unfortunately, the efficacy of this approach in most neurological diseases has been very difficult to show. In stroke and ALS such approaches have consistently failed. In ALS the antiglutamatergic drug riluzole is the only drug specifically approved to treat ALS, where it extends life expectancy by 3 months. Riluzole has many targets and may provide its benefit in ALS without affecting excitotoxicity. Thus far it has failed to show any benefit in HD.[31] The use-dependent NMDA receptor antagonist memantine, which is approved to treat Alzheimer-associated dementia, should be the ideal drug to treat glutamate excitotoxicity. However, it failed in stroke and Parkinson disease and has yet to be systematically studied in patients with HD. Other glutamatergic drugs that have failed in HD trials include amantadine, remacemide, and ketamine.[31]

6.4 PDE10 Inhibition

Many signaling processes in the nervous system involve the activity of cyclic nucleotide second messengers such as cyclic adenosine monophosphate (cAMP) or cyclic guanosine monophosphate (cGMP). These signals are inactivated through a class of enzymes called phosphodiesterases (PDEs). Among the many such enzymes present in the nervous system, PDE10A is highly expressed in striatal medium spiny neurons. PDE10A enhances signaling of medium spiny neurons, and inhibition of PDE10A protected striatal neurons in laboratory studies. Chronic suppression of PDE10A has been proposed as a potential neuroprotective strategy. The synthetic inhibitor OMS-643762 is now being evaluated in a double-blind, placebo-controlled, phase II clinical study (www.clinicaltrials.gov identifier NCT02074410), but results are not yet available.

6.5 Stem Cells

Given that the neuronal cell loss initially affects only the spiny neurons in the striatum, HD has been argued to be an "ideal" test case to explore the feasibility of stem cell implants to restore function in individuals with neurological disease. As of today, only a few pilot studies have been attempted, and these have treated very few people, typically two to five per trial. These were all open-labeled studies and lacked placebo comparison groups. At such small enrollment numbers it is impossible to evaluate efficacy. However, these trials collectively suggest that fetal implants are relatively safe, and that fetal striatal stem cells survive implantation for many years and may develop processes that interact with midbrain structures. Nevertheless, these studies have also revealed a number of challenges that make their larger-scale application questionable. First, two to four human embryos per patient are required to obtain a sufficient number of stem cells for implantation. Second, significant uncertainty exists with regard to whether patients require permanent immunosuppression for the grafts to survive. Third, for a study of efficacy, more consistent and robust outcome measures must exist. Since the integration of stem cells requires about 18–24 months, patients' symptoms are expected to slowly worsen regardless of treatment. A patient-specific baseline of symptoms must therefore exist, ideally acquired over the 12-month period preceding implantation, to compare their trajectory after treatment.[33]

One must wonder whether fetal stem cell implantation rests on a solid rationale. The challenge in HD is that the implanted cells must restore entire pathways, not just single cells or cell types. They must integrate into multiple places in the cortex and the striatum, and they must replenish white matter tracts as well as different types of neurons. While multipotent stem cells may theoretically yield all the required cell types, it is difficult to envision intracranial placement of stem cells that would spontaneously wire themselves to restore the defective indirect motor control pathway. Of course, only further studies will answer these questions. Given the many challenges, in particular the limited tissue availability, however, fetal stem cell implants are unlikely to provide the much-needed cure for the many thousands of affected patients. These studies clearly illustrate the helplessness and frustration of both patients and doctors.

7. CHALLENGES AND OPPORTUNITIES

HD is arguably one of the most fascinating neurodegenerative diseases. It is the only neurodegenerative disease for which a single disease-causing gene with near complete penetrance has been identified. The disease strikes with unprecedented precision when a stretch of glutamine repeats exceeds 40 multiplications. The disease pathology is remarkably similar to many other neurodegenerative diseases in which protein inclusions are a pathological hallmark. HD is, however, the first example of a disease where it is possible to show that the mutated protein that forms these aggregates is directly responsible for the disease and in which stopping further production of the mutated protein can completely reverse disease symptoms, essentially curing disease. This raises hope that such therapeutic approaches could be developed in the future.

One of the biggest challenges, however, is gaining a better understanding of the affected Huntington protein, both in normal biology and in disease. In spite of many possible roles that have been proposed, none has yet risen above the fray. Only with solid knowledge of the relationship between the mutated protein and the disease will it be possible to strategically intervene with up- and downstream regulatory pathways.

From a societal point of view, HD carries more stigma than any other neurodegenerative disease, probably comparable only to schizophrenia. This is in large part because of the unusual disease presentation, where patients show strange patterns of involuntary movements along with facial grimaces. Combined with strange behavioral and psychiatric presentations it is clear why, in early history, patients with HD were believed to be possessed by demons.

The development of accurate genetic testing with high predictive value has created another societal challenge, namely, how to use these screens productively without creating additional risks or psychological harm. While beneficial to some individuals, for example, for use in family planning, others perceive this knowledge to be psychologically harmful and worry that genetic information may be used to discriminate against affected individuals.

As the history of HD exemplifies, just a few devoted individuals have been able to move the perception from a peculiar behavioral illness to an almost completely understood genetic neurodegenerative disorder. A few more dedicated individuals will hopefully close the gap toward finding direly needed disease-modifying treatments or even a cure. The success of complete disease reversal in animal models of HD should serve as the strongest encouragement that this goal should be achievable. Indeed, I am optimistic that HD may be the first neurodegenerative disease to be cured.

Acknowledgments

This chapter was kindly reviewed by Dr Leon S. Dure, Professor of Neurology and Neurobiology, University of Alabama at Birmingham, and by Dr Steven Finkbeiner, Professor, Departments of Neurology and Physiology,

University of California, San Francisco, Senior Investigator and Associate Director, Gladstone Institute of Neurological Disease, and Director of the Taube/Koret Center for Neurodegenerative Disease Research and Hellman Family Foundation Alzheimer's Disease Research Program.

References

1. Wexler NS. Huntington's disease: advocacy driving science. *Annu Rev Med*. 2012;63:1–22.
2. Ross CA, Aylward EH, Wild EJ, et al. Huntington disease: natural history, biomarkers and prospects for therapeutics. *Nat Rev Neurol*. April 2014;10(4):204–216.
3. Ruocco HH, Lopes-Cendes I, Li LM, Santos-Silva M, Cendes F. Striatal and extrastriatal atrophy in Huntington's disease and its relationship with length of the CAG repeat. *Braz J Med Biol Res*. August 2006;39(8): 1129–1136.
4. Ross CA, Tabrizi SJ. Huntington's disease: from molecular pathogenesis to clinical treatment. *Lancet Neurol*. January 2011;10(1):83–98.
5. Bates GP. History of genetic disease: the molecular genetics of Huntington disease—a history. *Nat Rev Genet*. October 2005;6(10):766–773.
6. Feany MB. New approaches to the pathology and genetics of neurodegeneration. *Am J Pathol*. 2010;176(5): 2058–2066.
7. Finkbeiner S. Huntington's disease. *Cold Spring Harb Perspect Biol*. June 2011;3(6).
8. Pearson CE. Slipping while sleeping? Trinucleotide repeat expansions in germ cells. *Trends Mol Med*. 2003;9(11):490–495.
9. Garriga-Canut M, Agustin-Pavon C, Herrmann F, et al. Synthetic zinc finger repressors reduce mutant huntingtin expression in the brain of R6/2 mice. *Proc Natl Acad Sci USA*. November 6, 2012;109(45): E3136–E3145.
10. Pouladi MA, Morton AJ, Hayden MR. Choosing an animal model for the study of Huntington's disease. *Nat Rev Neurosci*. October 2013;14(10):708–721.
11. Cattaneo E, Zuccato C, Tartari M. Normal huntingtin function: an alternative approach to Huntington's disease. *Nat Rev Neurosci*. 2005;6(12):919–930.
12. Yamamoto A, Lucas JJ, Hen R. Reversal of neuropathology and motor dysfunction in a conditional model of Huntington's disease. *Cell*. March 31, 2000;101(1): 57–66.
13. Goldberg YP, Kalchman MA, Metzler M, et al. Absence of disease phenotype and intergenerational stability of the CAG repeat in transgenic mice expressing the human Huntington disease transcript. *Hum Mol Genet*. February 1996;5(2):177–185.
14. Gray M, Shirasaki DI, Cepeda C, et al. Full-length human mutant huntingtin with a stable polyglutamine repeat can elicit progressive and selective neuropathogenesis in BACHD mice. *J Neurosci*. June 11, 2008;28(24):6182–6195.
15. Powers ET, Balch WE. Diversity in the origins of proteostasis networks—a driver for protein function in evolution. *Nat Rev Mol Cell Biol*. April 2013;14(4): 237–248.
16. Ulbricht A, Arndt V, Hohfeld J. Chaperone-assisted proteostasis is essential for mechanotransduction in mammalian cells. *Commun Integr Biol*. July 1, 2013;6(4): e24925.
17. Parsons MP, Raymond LA. Extrasynaptic NMDA receptor involvement in central nervous system disorders. *Neuron*. April 16, 2014;82(2):279–293.
18. Cui Z, Feng R, Jacobs S, et al. Increased NR2A: NR2B ratio compresses long-term depression range and constrains long-term memory. *Sci Rep*. 2013;3. 01/08/online.
19. Bradford J, Shin JY, Roberts M, Wang CE, Li XJ, Li S. Expression of mutant huntingtin in mouse brain astrocytes causes age-dependent neurological symptoms. *Proc Natl Acad Sci USA*. December 29, 2009;106(52): 22480–22485.
20. Miller BR, Dorner JL, Shou M, et al. Up-regulation of GLT1 expression increases glutamate uptake and attenuates the Huntington's disease phenotype in the R6/2 mouse. *Neuroscience*. April 22, 2008;153(1):329–337.
21. Crook ZR, Housman D. Huntington's disease: can mice lead the way to treatment? *Neuron*. February 10, 2011;69(3):423–435.
22. Ondo WG, Mejia NI, Hunter CB. A pilot study of the clinical efficacy and safety of memantine for Huntington's disease. *Parkinsonism Relat Disord*. October 2007;13(7): 453–454.
23. Zuccato C, Ciammola A, Rigamonti D, et al. Loss of huntingtin-mediated BDNF gene transcription in Huntington's disease. *Science*. July 20, 2001;293(5529):493–498.
24. Chaturvedi RK, Flint Beal M. Mitochondrial diseases of the brain. *Free Radic Biol Med*. 2013;63(0):1–29.
25. A randomized, placebo-controlled trial of coenzyme Q10 and remacemide in Huntington's disease. *Neurology*. August 14, 2001;57(3):397–404.
26. Tai YF, Pavese N, Gerhard A, et al. Microglial activation in presymptomatic Huntington's disease gene carriers. *Brain*. July 2007;130(Pt 7):1759–1766.
27. Tong X, Ao Y, Faas GC, et al. Astrocyte Kir4.1 ion channel deficits contribute to neuronal dysfunction in Huntington's disease model mice. *Nat Neurosci*. May 2014;17(5):694–703.
28. Wang N, Gray M, Lu XH, et al. Neuronal targets for reducing mutant huntingtin expression to ameliorate disease in a mouse model of Huntington's disease. *Nat Med*. May 2014;20(5):536–541.

29. Nelson DL, Orr HT, Warren ST. The unstable repeats—three evolving faces of neurological disease. *Neuron.* March 6, 2013;77(5):825–843.

30. Mestre TA, Ferreira JJ. An evidence-basekd approach in the treatment of Huntington's disease. *Parkinsonism Relat Disord.* May 2012;18(4):316–320.

31. Mestre T, Ferreira J, Coelho MM, Rosa M, Sampaio C. Therapeutic interventions for symptomatic treatment in Huntington's disease. *Cochrane Database Syst Rev.* 2009 (3):Cd006456.

32. Kordasiewicz HB, Stanek LM, Wancewicz EV, et al. Sustained therapeutic reversal of Huntington's disease by transient repression of huntingtin synthesis. *Neuron.* June 21, 2012;74(6):1031–1044.

33. Rosser AE, Bachoud-Lévi A-C. Chapter 17—Clinical trials of neural transplantation in Huntington's disease. In: Stephen BD, Anders B, eds. *Progress in Brain Research.* Vol. 200. Elsevier; 2012:345–371.

General Readings Used as Source

1. Labbadia J, Morimoto RI. Huntington's disease: underlying molecular mechanisms and emerging concepts. *Trends Biochem Sci.* 2013;38(8):378–385.

2. Finkbeiner S. Huntington's disease. *Cold Spring Harb Perspect Biol.* 2011;3(6).

3. Ross CA, Aylward EH, Wild EJ, et al. Huntington disease: natural history, biomarkers and prospects for therapeutics. *Nat Rev Neurol.* 2014;10(4):204–216.

4. Walker FO. Huntington's disease. *Lancet.* 2007;369(9557):218–228.

Suggested Papers or Journal Club Assignments

Clinical Papers

1. Randomized controlled trial of ethyl-eicosapentaenoic acid in Huntington disease: the TREND-HD study. *Arch Neurol.* 2008;65(12):1582–1589.

2. Tetrabenazine as antichorea therapy in Huntington disease: a randomized controllí trial. *Neurology.* 2006;66(3):366–372.

3. Rosas HD, Goodman J, Chen YI, et al. Striatal volume loss in HD as measured by MRI and the influence of CAG repeat. *Neurology.* 2001;57(6):1025–1028.

Basic Papers

1. Yamamoto A, Lucas JJ, Hen R. Reversal of neuropathology and motor dysfunction in a conditional model of Huntington's disease. *Cell.* 2000;101(1):57–66.

2. Bradford J, Shin JY, Roberts M, Wang CE, Li XJ, Li S. Expression of mutant huntingtin in mouse brain astrocytes causes age-dependent neurological symptoms. *Proc Natl Acad Sci USA.* 2009;106(52):22480–22485.

3. Okamoto S, Pouladi MA, Talantova M, et al. Balance between synaptic versus extrasynaptic NMDA receptor activity influences inclusions and neurotoxicity of mutant huntingtin. *Nat Med.* 2009;15(12):1407–1413.

4. Wang N, Gray M, Lu XH, et al. Neuronal targets for reducing mutant huntingtin expression to ameliorate disease in a mouse model of Huntington's disease. *Nat Med.* 2014;20(5):536–541.

5. Tong X, Ao Y, Faas GC, et al. Astrocyte Kir4.1 ion channel deficits contribute to neuronal dysfunction in Huntington's disease model mice. *Nat Neurosci.* May 2014;17(5):694–703.

6. Paul BD, Sbodio JI, Xu R, et al. Cystathionine gamma-lyase deficiency mediates neurodegeneration in Huntington's disease. *Nature.* May 1, 2014;509(7498):96–100.

SECONDARY PROGRESSIVE NEURODEGENERATIVE DISEASES

8

Multiple Sclerosis

Harald Sontheimer

1. CASE STORY

Amy knew that something was not quite right when she almost dropped her 10 pound poodle a year ago. Her right arm, while following her direction, appeared simply not strong enough to lift the puppy. How was that possible? She had always been an athlete, an avid tennis player who worked out at least three times a week. Blaming it on a strained muscle, she brushed off the episode with little concern. However, for the next couple of weeks her arm remained weak with only rare episodes of strengthening. She noticed herself using the left arm to carry bags, and playing tennis was impossible. That Saturday she woke up with a numb left foot. She had been on a long walk with her dog the night before and all of a sudden she could barely feel the floor beneath her foot. In panic she called her mother who, not surprisingly, recommended she take it easy for the weekend and rest. It is true that she had been putting in long hours at work. At 36, she was driving herself hard, hoping to gain a partnership in the law firm where she had been working since law school. Luckily, it was Saturday, a rare day off. After a long, hot bath, her symptoms became even worse and she was barely able to get up on her feet. She was able to see her primary care doctor the following morning, who suggested that she might have a pinched nerve but was suspicious that this really didn't explain all her symptoms. He referred her to a neurologist for magnetic resonance imaging (MRI) to rule out a more serious condition. Like what? Lyme disease, a brain tumor, amyotrophic lateral sclerosis (ALS). What a shocking list of possibilities. Those were the worst three weeks of my life, Amy recalled. She Googled brain tumors and ALS; she Googled her symptoms and feared for the worst. She hadn't even started a family yet; this can't be happening to her. After a long wait, her MRI came out negative, leaving her neurologist at a loss. He ordered extra tests including blood work, a spinal tap, and electroencephalogram monitoring. After Amy had endured two weeks of missed work, numerous needle stings, and wearing a EEG cap with a hundred wires running from her head, her neurologist called her in for a consultation. "You may have multiple sclerosis, but we won't know for sure for several weeks or months." Amy didn't know how to react; relief it wasn't a brain tumor? The consequence of her diagnosis took weeks to sink in. She started injecting herself with Avonex and was able to work normal hours for the most part. She has been feeling tired in the afternoon, but largely managing her workload. However, she has spent every free hour researching her disease, and what she reads tells her that her time at work is limited. It was not a question of whether she would end up in a wheelchair, but when. She had not told her co-workers yet, but felt that she should, though there just hadn't been a good opportunity. Alone in bed, she has so many questions. Why did her body turn against her? Was it anything she did? Could it have been something she contracted during her college internship in Guatemala? Was it her volunteer work at the animal shelter? Was it related to her mononucleosis in high school? Could it have been the pesticides used on grandpa's farm? Some days, she falls asleep in despair. Damaged goods! Would she ever find a man to love her and, if so, how would she explain her disease to him?

2. HISTORY

Is it possible that multiple sclerosis (MS) was passed down to us from the Vikings through some infectious agent such as a virus? No one will ever know for sure, but our earliest accounts of MS trace back to the Vikings who populated medieval Iceland. A Norse saga reports that in the late twelfth century, Bishop Thorlak cured a woman named Halldora from debilitating transient paralysis reminiscent of MS. She experienced persistent pain and long periods of intermittent muscle weakness and eventually was bedridden for three years prior to her miraculous cure by Thorlak. Whether this was a spontaneous remission attributed to divine powers justifying Thorlak's elevation to sainthood will never be known, yet the saga shows that temporary paralysis was already known in medieval times. Another detailed case description is that of Lidwine, who lived in Schiedam, Holland from 1380 to 1433 and who developed weakness in her legs following a skating accident. After an initial recovery she suffered recurrent episodes of weakness, loss of balance, and visual problems, which were extensively documented as part of her later canonization.

Some of the most complete early descriptions of symptoms that more unequivocally resemble MS are contained in the personal diaries of Auguste d'Este (1794–1848), the illegitimate grandson of King George III of England. His disease began with visual problems, including spots floating before his eyes and double vision, followed by numbness on his lower abdomen. He describes imbalance and motor weakness: "when standing or walking I cannot keep my balance without a stick…for the first time in my life I was attacked by giddiness in the head (vertigo), sickness and total abruption of strength in my limbs." Later he developed tremors and spasticity, as well as bowel and bladder incontinence. Treatment included bloodletting, alcohol rubs, opium, and hot baths, none of which helped him.

Letters from the nineteenth-century German poet, Heinrich Heine, suggest that he too was a likely victim of MS. "For the last year and three quarters, I have been tortured day and night by the most horrible agonies, confined to bed, and paralyzed in all my members. Incessant cramps, most insufferable spasms, practically total blindness…" Heine passed away at age 59, 24 years after the onset of his symptoms, which included continued visual disturbances, numbness, blindness, paralysis, unsteady gait and muscle spasms, incontinence, and impotence.

As compelling as these cases are, neither was confirmed by autopsy and thus they cannot unequivocally be ascribed to MS. The first autopsy-confirmed cases of MS were described in 1838–1841 by Robert Carswell, who saw lesions in the white matter in autopsies of patients who had suffered from transient paralysis. The exact disease course and symptoms for these patients were, however, unknown to him. His colleague, Jean Cruveilhier, obtained autopsy specimens from four patients he had followed for many years,

allowing him to correlate specific symptoms with discrete pathological findings, an approach that would become known as the clinicopathological approach. Jean Martin Charcot (Figure 1), one of the most famous French physicians, who at the time practiced at Europe's largest hospital, the Hospital de la Salpetriere in Paris, became the master of this clinicopathological approach. Throughout the 1840s–1870s, he studied the case reports of many patients with periodic paralysis and also carefully examined their brain and spinal cord tissue on autopsies at the macroscopic and microscopic levels, allowing him to explain each symptom by the presence of specific lesions. He presented his findings in a series of published lectures that detail the clinicopathological features of MS and differentiate important symptoms from unrelated neurological disorders. In his lectures he distinguished the parkinsonian

FIGURE 1   Jean-Martin Charcot (1825–1893) was a French neurologist often regarded as the founder of modern neurology. He is generally credited with defining multiple sclerosis as an independent neurological disease. (public domain).

resting tremor from the tremors typical in MS patients, which only manifest with purposeful use of hands or legs. Most importantly, he was able to define stages of disease that were based on defined changes in pathology, an approach that survived essentially unaltered until today. Some of his observations, for example the transection of axons within plaques, have only recently been recognized to play a major role in disease. In 1868, Charcot suggested that MS should be considered a unique neurologic disease that he called "sclerosis en plaques."

Many of Charcot's pupils followed in his footsteps and contributed further to our current understanding of MS. These include, for example, Joseph Babinski, most well known for the toe reflex known as the Babinski sign, used to diagnose upper motor lesions. His medical thesis encompassed a meticulous study of lesions in the brain and spinal cord of MS patients.

Although early attempts were made to diagnose MS based on symptomatology alone, by and large an unequivocal diagnosis could not be reached until after a patient's death. This all changed with the introduction of noninvasive brain imaging using magnetic resonance imaging (MRI) in 1981. Since then, the progressive changes in the course of the disease can be visualized essentially in real time through repeated brain imaging. Not surprisingly, MRI has become truly a "game changer" in MS diagnosis.

Among the most fascinating findings in MS is its unique geographical distribution, with highest prevalence in the northern latitudes (Figure 2).

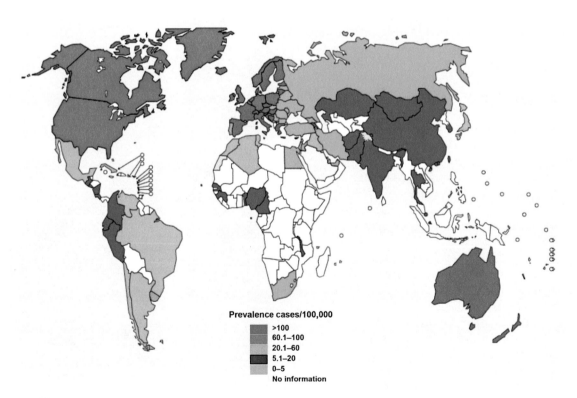

Prevalence cases/100,000

- >100
- 60.1–100
- 20.1–60
- 5.1–20
- 0–5
- No information

FIGURE 2   Geographic distribution of MS color coded for prevalence by country per 100,000 inhabitants. Disease prevalence is greatest toward the poles and lowest near the equator. *Reproduced from the World Health Organization (WHO), Atlas, Multiple Sclerosis Resources in the World, 2008.*

It may be surprising to the reader that the first evidence to suggest this geographical disease pattern dates back to 1882, when the British physician Sir Byron Bramwell published his empirical findings based on a relatively limited sample set. This has since been confirmed by large-scale epidemiological studies.

Similarly, as we are actively studying genetic causes of disease, a heritable genetic susceptibility of MS was already suggested by Ernst Leyden in the 1930s, then based on a relatively small number of families studied, but since solidified by a large collaborative Canadian study that enrolled almost 30,000 patients. In 1980, this study identified a 30% concordance rate for identical twins and 5% in nonidentical twins.

The notion that MS may be caused by an infection, too, has been around for a long time. An early proponent of this idea was Pierre Marie, one of Charcot's pupils, who was influenced by the germ theory that had just been developed by Pasteur and Koch. Marie initially believed that an infection akin to the spirochete bacterium that causes syphilis might also be involved in causing MS. Infectious agents proposed since include a multitude of bacteria and viruses. However, such pathogenic causes should allow the transfer of disease between humans and possibly between humans and animals. Such transmission has never been proven. However, this did not dissuade investigators from pursuing this question in sometimes reprehensible form. For example, in 1940, Georg Schaltenbrand injected cerebrospinal fluid (CSF) from MS patients into monkeys, some of which developed neurological symptoms. Then he transferred the monkey CSF back into unaffected humans, none of whom, fortunately, developed disease.

Louis Pasteur unwittingly demonstrated an immune response to myelin as the cause of MS in 1880 when his rabies vaccines contained central nervous system (CNS) antigens as contaminant. Sporadic cases of vaccine-induced paralysis resulted, which, upon examination of pathological specimens, showed clear signs of CNS demyelination.

In the 1930s, these findings were applied to monkeys and rabbits, which developed MS-like symptoms with repeated vaccination, suggesting that an immune-induced inflammatory process is at the center of disease. This approach led to the most commonly used experimental autoimmune encephalomyelitis (EAE) animal model of MS still used today. Quite surprisingly, it took until 1953 to recognize that myelin was not an axonal structure but was produced by oligodendrocytes. Advances since have been exhilarating, particularly the understanding of the complex interplay of the innate and adaptive immune system on the development of disease and how this understanding has afforded the development of disease-modifying drugs that interfere with various aspects of autoimmunity. Beginning with the introduction of interferon β in 1993, these drugs are changing the historical nihilistic outlook on MS and provide relief and improved quality of life for many MS patients. One of the most stunning (re)discoveries, however, is the recognition that progressive axonal loss is ultimately responsible for the symptoms of MS, clearly defining the disease as a neurodegenerative disease.

3. CLINICAL PRESENTATION/ DIAGNOSIS/EPIDEMIOLOGY

3.1 Epidemiology and Risk Factors

Multiple sclerosis is an inflammatory disease affecting myelinated nerve processes in the brain and spinal cord. In the United States approximately 350,000 people live with MS, and the worldwide prevalence is estimated at over 2.5 million. The disease can strike at any age, even in childhood, but the predominant time of diagnoses is between ages 20 and 40. No single symptom reliably characterizes MS, but telltale signs may include sudden-onset muscle weakness, loss of sensation, numbness, visual abnormalities, and pain. The disease typically progresses in stages separated by attacks lasting a few days and by periods of stable disease, a pattern called relapsing–remitting MS. Less commonly, MS may progress continuously, without any signs of remission. In most patients the disease ultimately becomes steadily progressive.

Women are three times more likely to develop MS than are men, and African Americans and Asians carry a significantly lower disease risk. Interestingly, the frequency of attacks during pregnancy is markedly reduced, suggesting a protective effect of female hormones released during pregnancy.

The disease is characterized by a unique pathology, consisting of multiple (>3) sclerotic lesions primarily in the white matter. These are several millimeters in diameter and are visible on MRI with or without contrast. Lesions affect myelinated nerve fibers most prominently and are believed to be associated with transient or permanent loss of myelination. This causes impaired signal conduction. Areas of sclerosis are dynamic and can disappear on repeated scans, only for new ones to emerge somewhere else. This may be due to the transient nature of inflammation or leakiness of the blood–brain barrier.

The cause of MS is unknown, but current research suggests a cooperation of environmental exposure and genetic susceptibility. A recurrence within families of 20%, and a 25% concordance rate for monozygotic twins, strongly supports genetic susceptibility. Recent large-scale gene association studies have identified a disease association with alleles of the major histocompatibility complex (MHC) involved in antigen presentation, suggestive of an immune/autoimmune cause. A possible viral etiology had been proposed for decades and is supported by rare cases of demyelination in response to infections with Epstein–Barr virus. The best support for a strong environmental influence, possibly an infectious agent, is illustrated by a spike in MS cases on the Faroe Islands immediately following their occupation by British soldiers during World War II.

Exhaustive epidemiological studies have thus far failed to identify any single disease-causing factor, yet they have revealed an interesting global geographic pattern of disease (Figure 2). MS prevalence is much greater in the Northern Hemisphere and increases progressively with distance from the equator. Overall highest prevalence is reported for the Orkney Islands north of Scotland. Compared to the rest of the world, the disease prevalence is 10– to 20-fold lower in the tropics, close to the equator. One way to explain this pattern relates to sun-derived (ultraviolet B [UVB]) vitamin D production in the skin, which obviously significantly decreases due to reduced sun exposure in northern latitudes. As discussed later, Vitamin D deficiency is indeed a risk factor for MS, and mutations in an enzyme involved in natural vitamin D synthesis have been linked to some cases of MS. Curiously, MS is virtually unknown among Alaskan Inuits, who may counteract their low sun exposure with a fish diet particularly rich in vitamin D.

Racial differences may be as important as or more important than diet. MS incidence differs 60-fold when comparing Britain (85/100,000) and Japan (1.4/100,000), two islands on the exact same latitude.

In addition to this fascinating geographic relationship to disease prevalence, there is a chronological (month of birth) relationship as well, whereby babies born in May are less likely to develop MS than babies born in November.

It therefore appears that several events must cooperate to trigger disease. The first risk is likely to arise either in utero or early in infancy, since the above-explained geographic disease risk maps to where an individual has spent his or her first 15 years of life. In other words, if you were born near the equator, yet move to Alaska at age 21, you are still 10- to 20-fold less likely to develop MS over your lifetime. The reverse is true as well. If you escape the cold northern climates as an adult, you still carry the disease risk of a Northerner for the rest of your life. The second risk must be a disease trigger, exposure or infection.

Overall there has been an increase in the incidence of MS globally, but particularly among

TABLE 1   Presenting Symptoms

Symptom	Percent of cases
Sensory loss	37
Optic neuritis	36
Weakness	35
Paresthesias (pins-and-needles pain)	24
Diplopia (double vision)	15
Ataxia (unsteadiness)	11
Vertigo	6
Paroxysmal attacks	4
Bladder dysfunction	4

Source: After WB Matthews et al., McAlpine's Multiple Sclerosis, New York, Churchill Livingstone, 1991.

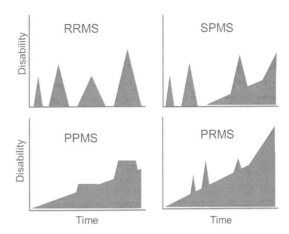

FIGURE 3   Graphical depiction of the four patterns in which MS progresses over time: RRMS, relapsing–remitting; PPMS, primary progressive; SPMS, secondary progressive; PRMS, primary relapsing.

white women. Many possible explanations have been suggested ranging from the enhanced use of UVB blocking sunscreen, which reduces natural vitamin D production in the skin, to increased sanitation, causing the immune system to experience less "priming" during childhood.

3.2 Clinical Presentation

The typical presenting features for MS are listed in Table 1, and frequently include a combination of sensory loss or numbness, motor weakness, and pins-and-needle pain. Vision problems in the form of a reduced visual field, cloudiness, or floating objects are common. Visual acuity is reduced and color perception may be impaired. These disease symptoms may come on suddenly or gradually, but all symptoms are asymmetric.

As the disease progresses, over 30% of patients have painful muscle spasms in their legs affecting their ability to walk. Ninety percent of MS patients complain of fatigue.

Similarly, pain is a chief complaint and can occur in any part of the body at any time. Some MS patients complain of a characteristic pain sensation triggered by forward extension of the neck. This is called Lhermitte's sign and feels like an electric current running down the spine. It is due to lesion in the spinal cord affecting the dorsal column at the level of cervical vertebrae, where nerve fibers make inappropriate connections with each other.

Over 90% of all MS patients have frequent bladder incontinence, and over one-third are constipated.

However, most patients with MS remain cognitively fully functioning with only minor signs of altered executive function and possibly slowness in overall problem-solving skills, although half of the patients develop depression.

The disease typically progresses in one of four distinct patterns, graphically illustrated in Figure 3. The most common disease course that accounts for 85% of all cases is called relapsing–remitting MS (RRMS). Here, flare-ups of symptoms that last from a day to a few weeks, called an attack or relapse, are separated by periods of remission during which most symptoms resolve spontaneously. Remission periods can last for many months. Although many patients stay in an RRMS pattern of disease for 15–25 years, over time the baseline disability worsens, with relapses occurring more frequently. Eventually many patients develop secondary progressive

MS (SPMS), where symptoms worsen gradually over time without remittance periods in between. Typically a patient with RRMS has a 2% annual chance to change disease course to SPMS.

About 15% of patients never have any attacks or remissions but instead have symptoms that gradually worsen over time, a course named primary progressive MS (PPMS). A final variant, progressive/relapsing MS, affects about 5% of patients where the progressive worsening is overlaid by occasional attacks.

3.3 Diagnosis

On first presentation a physician looks for recurrence of isolated neurological symptoms spaced several weeks or months apart. For a suspected MS diagnosis, at least two attacks must have occurred that correlate with suspected white-matter lesions. MRI has become a game changer with regard to identifying such lesions. Multiple sclerotic lesions are typically found throughout the white matter of the cortex and spinal cord (Figure 4). Since lesions must be at least 5 mm to be visible on MRI, early demyelination is likely always missed. The presence of three or more such lesions predicts development of MS with about 80% accuracy, whereas the absence of visible lesions similarly rules out the future development of MS with similar accuracy. Although somewhat counter-intuitive, the severity of symptoms is not necessarily correlated with the number and size of the lesions but is more related to the location of

FIGURE 4 Typical MRIs from four different patients with MS. Multiple lesions are found throughout (A) the white subcortical matter cortex, (B) the corpus callosum, and (C) the spinal cord. (D) Blood infusion of the contrast medium gadolinium reveals breaches of the blood–brain barrier primarily associated with lesions in the white matter. *Reproduced with permission from Ref. 1.*

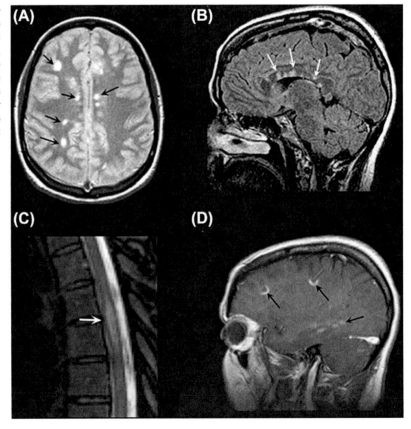

the lesion. That said, a steady increase in lesions is nevertheless a biological marker for disease progression. The use of blood contrast media such as gadolinium allows detection of focal breaches in the blood–brain barrier (Figure 4(D)), where blood-borne immune cells are actively infiltrating on MRI. A complementary strategy to help diagnose MS is the direct measurement of signal conduction velocity. Since demyelinated axons conduct much slower than myelinated ones, nerves can be stimulated and the signal conduction velocity measured over a certain distance along the nerve. These so-called "evoked potentials" can quantify the functional loss in affected limbs or the optic nerve and are useful in demonstrating clinically silent lesions, whose presence can substantiate a diagnosis of MS.

Biomarkers that can assist in an unequivocal diagnosis are, unfortunately, few and unreliable. CSF samples obtained through spinal tap show normal total protein content but, when probed on agarose gels, typically reveal so-called oligoclonal bands that consist of immunoglobulin-bound antibodies. These are suspected to be released by B cells and may be targeting myelin components, but the search for such antigens has been largely elusive. Antibodies to two glial proteins, the water channel aquaporin-4 and the potassium channels Kir4.1, are emerging as disease relevant in some forms of MS.[2]

Multiple sclerosis is sometimes misdiagnosed, since no two patients present the same. The seemingly random pattern of demyelinating lesions causes a variety of symptom constellations. In fact, some patients with visible lesions on MRI never develop any symptoms, and are diagnosed with benign MS. However, the presence of multiple sclerotic lesions does predict future MS in 30% of patients. The most frequent differential diagnoses for MS, listed in Table 2, include infections of the nervous system, such as Lyme disease, human immunodeficiency virus, or syphilis, but also brain tumors and vitamin B12 deficiency.

Patients with MS have an almost normal life expectancy, and death as a direct consequence

TABLE 2 Disorders with MS-Like Symptoms

Acute disseminated encephalomyelitis

Adrenoleukodystrophy

CADASIL (cerebral autosomal dominant arteriopathy with subcortical infarcts and leukoencephalopathy)

Human immunodeficiency virus infection

Ischemic optic neuropathy

Lyme disease

Mitochondrial encephalopathy (MELAS)

Tumors (e.g., lymphoma, glioma, meningioma)

Stroke and ischemic cerebrovascular disease

Syphilis

Systemic lupus disorders

Vitamin B_{12} deficiency

of MS is exceedingly rare. Much more common causes of death are opportunistic infections, and sadly, 15% of MS patients take their own lives.[3] An increasing arsenal of drugs, reviewed in detail later in this chapter, make the disease increasingly more manageable and particularly delay the time course of early progression. These disease-modifying agents have had a very positive impact on the quality of life for patients with MS. Nevertheless, most patients will eventually need ambulatory assistance and will be dependent on a caregiver.

4. DISEASE MECHANISM/CAUSE/ BASIC SCIENCE

There is no question that MS is a chronic inflammatory disorder that involves essentially all cells of the CNS over the course of the disease. As the name implies, multiple sclerotic lesions, often called plaques, are the hallmark of MS, and these are sizable enough to be visualized by MRI (Figure 4). Lesions are heterogeneous in cellular composition, size, and location, and are not predictive of disease severity. Generally speaking, white matter

FIGURE 5 Examples of the typical histopathology seen in MS lesions. (A) Example of an active MS lesion surrounded by macrophages at the rim. The lesion is stained for myelin oligodendrocyte glycoprotein (MOG), which, due to myelin loss, is absent in the lesion. (B) 700× magnification of an active lesion stained with an antigen for myelinated fibers, showing both myelinated fibers and ones that have lost their myelin. *Reproduced with permission from Ref. 4.*

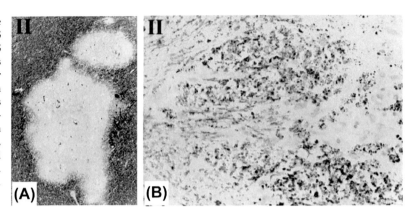

4.1 Histopathological Changes

Histopathologically, lesions show an abundance of immune cells, including T lymphocytes, B lymphocytes, plasma cells, activated microglia, and reactive astrocytes (Figure 5). This combination of cells encompasses the innate, CNS-specific immune cells (microglia/astrocytes) and peripheral, invading immune cells (T cells, B cells, macrophages). Pathologists distinguish four different lesion types, based on the distribution of myelin, protein loss, plaque geography, pattern of oligodendrocyte destruction, and evidence for IgG antibodies.[4] As illustrated in Figure 6, the disease is believed to start with the infiltration of autoreactive lymphocytes, which are lymphocytes that react with the body's own self-antigens and are usually eliminated in the thymus. Once in the brain, they release proinflammatory cytokines that attract macrophages and microglial cells. Under the influence of these cytokines, they kill oligodendrocytes via membrane-bound tumor necrosis factor alpha (TNFα). Macrophages are monocyte-derived debris-eating cells that assist cytotoxic T cells in the killing of the antigen-presenting cells. The inflammatory cytokines also induce a reactive transformation of the surrounding astrocytes,

lesions predominate in the early phase of disease, while later disease stages involve the gray matter as well.

causing astrogliosis (see Trauma, Box 2 in Chapter 2), where astrocytes lose their polarity and many molecules involved in water, ion, and transmitter homeostasis. The inflammatory milieu causes the release of oxygen and nitrogen radicals to form macrophages and microglia, damaging mitochondria in oligodendrocytes and the axons. The resulting energy loss causes glutamate (Glu) release from axons (discussed below) and Glu excitotoxicity. These factors are each damaging to oligodendrocyte health and possibly to the axon as well. As oligodendrocytes die, myelin debris is removed by microglia and macrophages, leaving the axon denuded and unable to propagate APs. Since each oligodendrocyte services 30–50 axons, the extent of the functional damage is magnified many-fold, with few dying oligodendrocytes leaving a large nerve trunk essentially unable to signal. One may argue, as I have in other places, that the evolutionary advances that produced the myelinated axon to allow rapid signaling with small-caliber fibers has put humans at risk of suffering catastrophic disease as exemplified here. On the other hand, there was probably little choice, since replacing our myelinated spinal cord with equally fast-conducting nonmyelinated axons would require a spinal cord the diameter of the trunk of a giant sequoia tree.

Although we do not know what initiates the immune response that forms the sclerotic plaque, the initial immune attack appears to be clearly

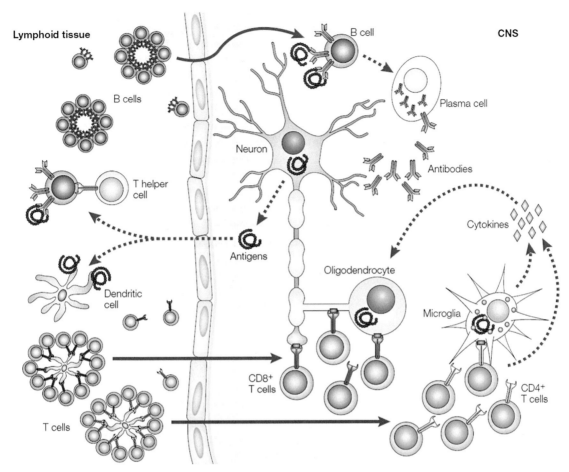

FIGURE 6   Hypothetical view of immune response in an acute MS lesion. Blood-derived CD4[+] T and B cells enter the brain by crossing the endothelial wall of the cerebral blood vessels. Once inside the brain, they recognize their respective myelin antigen(s) on microglia or astrocytes. T cells release proinflammatory cytokines that direct CD8[+] killer T cells or microglial cells toward myelin, oligodendrocytes, or axons, which they destroy. *Reproduced with permission from Ref. 5.*

mounted against the oligodendrocytes and their myelin sheath. It is therefore pertinent to first ask why and how oligodendrocytes are targeted.

4.2 Toxicity to Oligodendrocytes and Myelin

The search for a disease-initiating factor dates back at least a century. It began in earnest in 1925 when rabbits, inoculated with human spinal cord tissue, showed extensive spinal cord inflammation, muscle weakness, and paralysis remarkably similar to MS. Subsequent studies showed that myelin antigens were sufficient to induce such experimental demyelination in many species. In the 1960s, scientists began to look at the active process of demyelination in the dish using explant tissue cultures of brain or spinal cord from rats. When CSF samples taken from MS patients were added to explants containing normally myelinated nerves, their myelin sheath was attacked, resulting in nearly complete demyelination.[6] This suggested that a soluble component, later identified as antibodies against myelin oligodendrocyte glycoprotein (MOG), was the disease-initiating factor. Indeed, in rats and rabbits, antibodies against various myelin components are sufficient to mount an immune response similar to that seen in an MS attack. However, as of today, no similar antibody has been identified that initiates MS in human disease.

An important lesson learned from studies in animals was that the disease can be propagated from one animal to another by adaptive transfer, whereby CD4[+] T lymphocytes harvested from an affected animal are sufficient to induce disease in an otherwise naïve animal (see Section 4.3 below). This establishes T cells as drivers of the immunological attack, yet falls short of making them responsible for the initial attack.

CSF samples taken from MS patients show evidence of intrathecal antibody production by B lymphocytes. These appear on gels as oligoclonal bands that are similar to those seen in other nervous system infections such as encephalitis or neurosyphilis. However, the antigen(s) recognized by these oligoclonal IgG antibodies and their disease relevance are still unknown. Interestingly, B lymphocytes harvested from the blood of MS patients release a toxic substance that kills oligodendrocytes in the dish, yet this unknown molecule is neither an antibody nor one of the frequently studied cytotoxic molecules such as TNFα.[7]

Taken together, the initial attack on oligodendrocytes is mounted by immune cells, with T and B cells probably contributing to various degrees. No single initiating agent, antibody, or cytokine has yet been identified; yet as we learn below, suppression of various aspects of the immune cascade each provides some disease attenuation.

4.3 The Immune Attack in Multiple Sclerosis

The long-held notion that the nervous system is immune-privileged and lacks immune cells is clearly inaccurate and antiquated. Instead, the brain contains its own innate immune cells in the form of resident microglial cells, which, like the macrophages that they derive from, express MHC class II molecules and can therefore present antigens. Reactive astrocytes, too, can express MHC class II and assist in antigen presentation. Moreover, both microglia and astrocytes can release cytokines, both pro- and anti-inflammatory. Although leukocytes (white blood cells) are typically not present in normal brain, they can gain access under conditions of injury, disease, or inflammation. As pointed out above, the immune cells at work in MS are T and B lymphocytes.

B lymphocytes are antibody-producing cells that are born in the bone marrow (B = bone). They recognize antigens bound to antigen receptors via an immunoglobulin (Ig) link. Upon cross-linking such Ig molecules to their surface, B cells proliferate and form a clone of antibody-producing cells. Oligoclonal bands, single clones of B cells presumably producing the same antibody, are a reliable biomarker for MS, suggesting a contribution of B cells to the disease.

T lymphocytes are born in the thymus (T = thymus) and take part in the adaptive immune response. They are important in the identification and presentation of foreign antigens and, through the release of cytokines, activate or suppress other immune cells. There are three types of T cells. T helper cells (Th) recognize foreign antigens and stimulate the production of antibodies by B cells. T regulatory cells (Treg) oppose antigen presentation and dampen the immune response. Killer T cells attack and destroy cells marked as foreign.

The recognition of foreign antigens requires their digestion into small peptide fragments that are then bound to the cell surface to a protein complex called the major histocompatibility complex, specifically MHC class II. MHC class II is only expressed on antigen-presenting cells.

As already alluded to, antigens are presented by microglial cells and activated astrocytes.

The various T cells express cell surface glycoproteins that are involved in their specific function and these are often used as part of their name. For example, CD4 belongs to the immunoglobulin superfamily and is found on the surface of monocytes, macrophages, dendritic cells, and T cells. CD4 has four domains, of which one interacts with MHC class II molecules involved in antigen presentation. Therefore any cell expressing CD4, such as the CD4+ T helper cells, specifically recognizes antigen-presenting cells. Their main role is to recruit and activate CD8 killer T cells toward the antigen-presenting cells. CD8 is also a glycoprotein, but it binds MHC class I proteins. MHC class I proteins mark cells for destruction.

The immune response that causes MS lesions has been studied for decades and our understanding continues to evolve. The simplest depiction, presented in Figure 6, suggests that an MS attack starts with autoactive CD4+ Th cells that are reactive to myelin. How they become reactive to myelin antigens is unknown. It may involve molecular mimicry whereby an antigen similar to myelin, present for example on measles virus, is presented by an antigen-presenting cell. Alternatively myelin components may be shed from the CSF after injury and recognized

by dendritic cells that process antigens and activate T cells. Two types of activated Th cells (Th1 and Th17) with different cytokine profiles enter the brain. Upon encountering antigens, for example presented by microglial cells, the Th cells release proinflammatory cytokines that direct the eventual killing of oligodendrocytes or axons either via either CD8+ killer T cells or via activated microglial cells that kill cells via membrane-bound TNFα. As the inflammation proceeds, activated B cells contribute to cell death by releasing large amounts of antibodies directed toward myelin antigens, which mount a complement-dependent cell lysis. As we will see later, various aspects of this cascade are targeted by disease-modifying drugs, for example, the maturation of T and B cells, their entry into brain, and their release of proinflammatory versus anti-inflammatory cytokines.

4.4 Demyelination and Action Potential Block

Myelin provided by oligodendrocytes assures secure and rapid AP propagation in the CNS through saltatory conductance, whereby the electrical signal jumps along the segments of the myelinated axon (Figure 7; see also Box 1, Trauma Chapter 2). The advantages of this process are that small-caliber fibers can be used

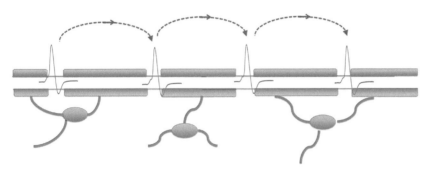

FIGURE 7   Saltatory conduction refers to action potentials jumping along the axon from one node of Ranvier to the next. This greatly enhances signal conduction even in small-caliber nerve fibers. Each myelin segment separating two nodes is provided by the processes of one oligodendrocyte, yet each oligodendrocyte can produce myelin for up to 50 axons myelin segments.

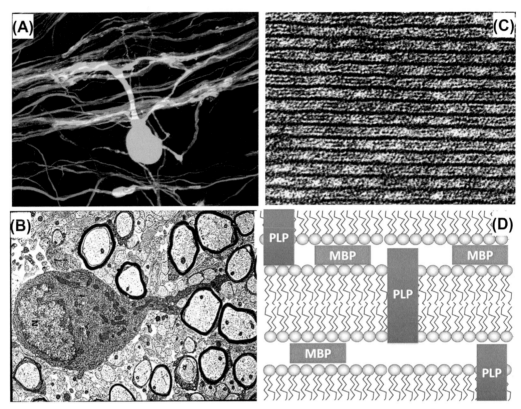

FIGURE 8   Myelin. (A) An individual oligodendrocyte (yellow) can form myelin segments for up to 50 axons (red). (B) Myelin wraps are continuous with the processes of an oligodendrocyte, as shown by electron microscopy (EM). (C) EM reveals the tight stacking of adjacent myelin wraps, which (D) are separated by myelin basic protein acting as extracellular spacers and proteolytic protein (PLP) anchoring adjacent membrane wraps. (A, kindly provided by Dr Partizia Casaccia, B & C by Dr Cedric S. Raine as used in Basic Neurochemistry 5th edition, reproduced with permission).

to transmit very fast signals, and energy is conserved as ionic gradients are only altered at each segment. Loss of myelin segments can be catastrophic as this slows, or even blocks, nerve conduction. Myelin is a lipid-rich substance formed by the compaction of many membrane layers synthesized by oligodendrocytes. Processes flatten out into sheaths that wrap around axons (Figure 8(A)). As they do, they squeeze essentially all cytoplasm out, making adjacent membrane wraps seem as if they are glued together. Wraps are spaced at 100A distance and held together by myelin proteins, namely proteolytic protein (PLP), a transmembrane protein that may anchor adjacent membrane wraps, and

myelin basic protein (MBP), a surface protein possibly acting as spacer (Figure 8(D)). Once compaction is completed, the structure is readily identified as being electron dense on electron microscopy (Figure 8(B) and (C)). Electrically, myelin enhances the axonal membrane resistance and reduces the capacitance. Not all axons are myelinated or need to be myelinated. Typically, only fast-conducting larger-caliber fibers >1 µm receive myelin sheathing during development. How oligodendrocytes recognize which axons to myelinate is unknown.

Each of the myelin segments spans approximately 100–200 µM (100 times axon diameter). Between adjacent segments, provided by

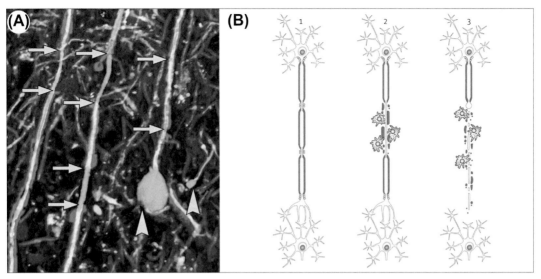

FIGURE 9   Degeneration of chronically demyelinated axons (green, axon; red, myelin). Most axons survive demyelination and redistribute Na$^+$ channels to recover signal conduction. Others, like the ones seen with the large green bulbous expansion in (A), are transected with accumulating organelles as their distal axon degenerates and is eaten by microglial cells, as schematically depicted in the cartoon in (B). Owing to loss of myelin trophic support, chronically demyelinated axons exhibit slowly progressive swelling and cytoskeletal disorganization. *Reproduced with permission from Ref. 8.*

different oligodendrocytes, remains a narrow myelin-free gap called the node of Ranvier. Here, Na$^+$ channels, primarily Nav1.6, cluster at high density (>1000/μm^2) and are the site of AP generation. The node is flanked by the paranode, which is rich in K$^+$ channels, specifically Kv3.1 and Kcnq2, that are responsible for the repolarization of the AP. Na$^+$ and K$^+$ channel clusters are maintained at the node of Ranvier by molecular anchors such as ankyrine, spectrin, and contactin that interact with the axon cytoskeleton. The clustering of channels precedes the compaction of the myelin, whereby the edges of the myelin that will eventually form the paranode push the channels along the axonal membrane into the node. The nodal versus internodal segment is delineated by one of several contactin-associated proteins (Caspr). Should the myelin sheath disintegrate, as is the case in MS, Na$^+$ and K$^+$ lose their anchors and diffuse into the bare axon. This makes propagation of APs more difficult and unreliable. If the

area of demyelination is too large, AP propagation becomes impossible and conduction is blocked.

Myelination is an essential maturation process in development and, in the human brain, typically begins after birth and proceeds into late adolescence. Some have argued that the late myelination of the frontal lobes explains the delayed development of executive functions. Whether some of the questionable decisions that teenagers make are attributable to incomplete myelination remains fodder for debate.

Myelin is part of living cells rather than a dead deposit like our fingernails or hair. It is connected to the oligodendrocyte and still metabolically supported by it. Therefore when oligodendrocytes die following injury or disease the vanishing myelin leaves the axon uninsulated (Figure 9). As a result, ion channels at the node are no longer anchored and diffuse into the bare axonal membrane. In addition to Nav1.6, new

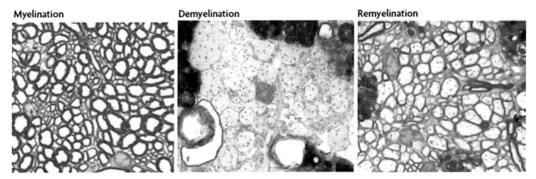

FIGURE 10   Remyelination produces thinner myelin sheaths with a reduced g-ratio. The g-ratio is a measure to compare myelin thickness by dividing the circumference of the axon by the circumference of the myelin. *Reproduced with permission from Ref. 9.*

Na^+ channels including Nav1.2, normally only found in nonmyelinated axons, are inserted throughout the bare membrane and the axon switches to continuous as opposed to salutatory conduction. For small lesions this is sufficient to sustain signaling at a reduced velocity. The downside of this compensatory insertion of Na^+ channels is a markedly increased influx of Na^+. This increases the requirement for Na^+/K^+ ATPase to extrude Na^+ which, in turn, increases the energetic demand. The increased Na^+ also increases the likelihood that the Na^+/Ca^{2+} exchanger operates in reverse, leading to pathophysiological increases of Ca^{2+} in the axon that can activate axonal proteases and destroy the axon (Figure 12).[9] Taken together, demyelination puts the axon at a significant risk of failing to signal and possibly dying.

4.5 Remyelination

In light of the negative consequences of myelin loss on axonal function, repair of the denuded axon through remyelination would be advantageous. For the longest time we had assumed that once lost, oligodendrocytes could not be replaced and the axon would stay unmyelinated or die. We since learned that the brain contains a large number of oligodendrocyte precursor cells (OPCs). These are highly mobile and proliferative cells that are present throughout life. Following injury or disease, they can move into a wound, differentiate into oligodendrocytes, and lay down myelin sheaths.[10] This indeed happens in MS during periods of remission where remyelination contributes to almost-complete functional recovery. The myelin that forms during regeneration is typically thinner, with fewer layers than originally present. A way in which this is often expressed is the g-ratio (circumference of the axon: circumference of the myelin sheath), a measure of the relative thickness of the myelin sheath. Examples of axons before and after remyelination are shown in Figure 10.

Intuitively, one would assume that the chief role for remyelination is to restore saltatory fast signaling of the axon. It appears that equally important is trophic and metabolic support provided by the myelin to the axon. Axons receive trophic support in the form of growth factors that include insulin-like growth factor 1 (IGF1) and glial cell-derived growth factor (GDNF). Furthermore, quite surprisingly, the oligodendrocytes also supply metabolic high-energy substrate to the axon.[11] Specifically, oligodendrocytes utilize glucose anaerobically and produce lactate, which is released under the myelin sheath and taken up by axons for their mitochondrial synthesis of ATP. It is possible that even incomplete remyelination that is insufficient

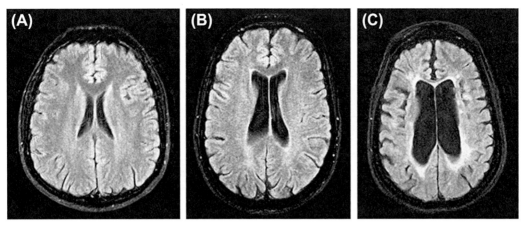

FIGURE 11   White matter atrophy visualized by MRI. (A) Normal brain, (B) the brain of a patient with relapsing–remitting MS, and (C) the brain of a patient with secondary progressive MS with end-stage disease. The progressive increase in ventricular volume indicates brain atrophy. *Reproduced with permission from Ref. 8.*

to restore axonal signaling provides sufficient support to prevent the axon downstream of the lesion from degenerating.

Successful remyelination entails recruitment of OPCs to the lesion, where they expand by proliferation, differentiate into oligodendrocytes, and ensheath axons. Genes and trophic factors, many of which are still unknown, regulate each step. Failure in each step results in failed remyelination. We know that as MS progresses, remissions are fewer and eventually absent. This is parallel to a progressive decline in remyelination. Why OPCs eventually fail in myelin repair is not entirely known, yet age may be one impediment, as the ability of OPCs to proliferate and differentiate gradually declines with age. As the repair process gradually dissipates, it leaves permanently demyelinated lesions behind. Therefore, understanding signals involved in remyelination has tremendous untapped therapeutic potential, a strategy that is thus far totally unexplored.

4.6 White Matter Atrophy and Axonal Injury

Whereas early MS lesions show relatively little evidence for axonal injury, axon loss is widespread in advanced MS. Eventually, up to 70% of the axons of the motor corticospinal tract are lost, causing a visible reduction in overall white matter, seen as brain atrophy by MRI (Figure 11). Transected axons can be seen in acute inflammatory lesions (Figure 10). It is likely that the transition from RRMS to SPMS represents the inflection point at which axonal loss becomes prominent and where remyelination no longer occurs and compensatory changes through functional cortical plasticity are exhausted.[9]

Axonal death appears to have two contributing components. The first is the already discussed loss of trophic and metabolic support when myelin disappears. The second component is a direct destruction of the axon by toxic signals such as NO, glutamate, and TNFα from microglia and macrophages. These potentially deadly signals are compounded by a compromised energy state akin to a stroke, a state that has been called virtual hypoxia[8] (Figure 12).

Activated immune cells release both glutamate and NO. Elevated Glu can be measured in the CSF of MS patients, and is particularly localized to lesions by magnetic resonance spectroscopy imaging. Glu is directly toxic to oligodendrocytes that express a variety of Glu

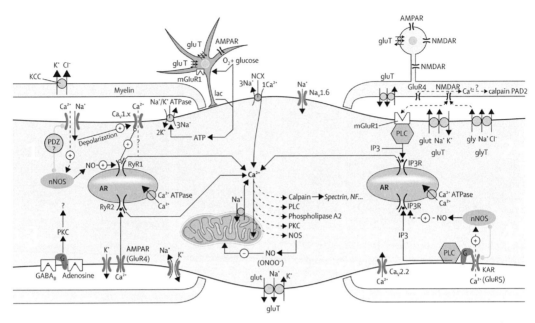

FIGURE 12   A multitude of signaling cascades contribute to axonal injury in MS. Central to all of them is depolarization of the axon causing influx of Ca^{2+}, which in turn causes destructive enzymes in the axolemma to be activated.[8]

receptors (GluRs). NO causes mitochondrial failure in both axons and oligodendrocytes, causing their membranes to depolarize and Na^+ to enter the cell, which, due to reduced ATP, is not sufficiently cleared. The resulting increase in Na^+ causes the axonal Na^+/Ca^{2+} exchanger to run in reverse and import Ca^{2+} rather than export Ca^{2+}. This in turn causes a pathological Ca^{2+} load in the axolemma, which leads to aberrant activation of destructive enzymes such as calpases and phosphatases, which damage the axon membrane and its cytoskeleton.

Important targets of uncontrolled axolemmal Ca^{2+} are neurofilaments and microtubules. Neurofilaments provide structural integrity of the axon and are normally heavily phosphorylated, yet upon demyelination these are dephosphorylated by Ca^{2+}-activated phosphatases. Microtubules are the substrate for axonal transport of proteins, lipids, and organelles. This transport is Ca^{2+} dependent and pathological increases in Ca^{2+} cause accumulation of transported molecules including organelles along the length of the axon (Figure 13).

It must be emphasized that the presence of GluRs in white matter was surprising at first. However, it has since become apparent that Glu is released from axons in an activity-dependent manner and serves as an axon–oligodendrocyte signal and is likely involved in myelination. Glu released activates NMDA and AMPA type GluRs. Expression of these ligand gated channels, which can permeate Ca^{2+}, makes oligodendrocytes also vulnerable to Glu toxicity under conditions of aberrant extracellular Glu.

Together these processes are now called white matter excitotoxicity, since they are strikingly similar to some of the toxicity observed in stroke and, indeed, white matter strokes present with similar axonal loss as proposed for MS. Ca^{2+} kills both oligodendrocytes and axons, although the entry pathway for Ca^{2+} appear to be different: GluRs in oligodendrocytes, reverse operation of the Na^+/Ca^{2+} exchanger on axons. Protecting axons from this toxicity will be among the most important challenges.

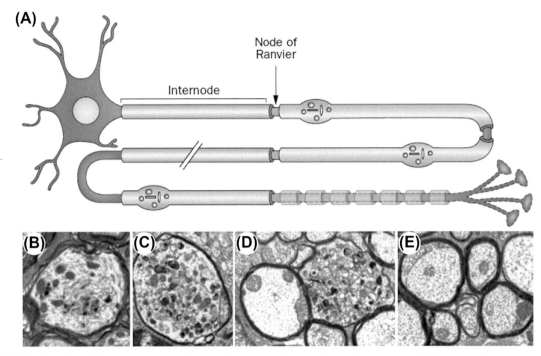

FIGURE 13 Axonal degeneration can also result from impairment of axonal transport, which causes the aggregation of organelles, depicted schematically in (A) as bulbous expansions. On electronmicrographic sections in B–E, intracellular accumulations of organelles can be visualized along the axon. *Reproduced with permission from Ref. 10.*

4.7 Genetics

As already alluded to, fascinating observations have been made regarding the geographic distribution of MS, which strongly suggests a confluence of genetic and environmental factors contributing to the disease. We no longer expect to be able to identify a single disease cause. However, epidemiological studies can help us establish risk factors, and, if these are concise and significant, allow us to take action so as to reduce an individual's risk.[12]

The best evidence for a genetic cause of MS stems from familial recurrence. As illustrated in Table 3, the chances of identical twins both having MS is about 1 in 3; that of fraternal twins about 1 in 20; compared to a 1:1000 chance in the general population. This inheritance pattern does not support a Mendelian trait but instead suggests that multiple genes are likely to contribute to the risk of developing MS.

TABLE 3 Familial MS Risk

>33%	Identical twins
5%	Fraternal twins
4%	Siblings
2%	Affected parent
1%	Cousin
0.1%	General population

A number of large-scale studies have searched for genes that increase MS risk. These led to only one gene that has a strong association. A component of the human histocompatibility complex, the HLA-DRB1 gene, particularly the HLA-DRB1*1501 allele, increases disease risk up to 3-fold.[13] A further 25 gene loci with modest changes in risk, on the order of 0.8–1.2, all map to the HLA region of the chromosome.

How changes in the HLA gene influence disease risk for MS is unclear, yet the involvement of HLA in antigen presentation makes a disease link via the T cell immune axis likely.

4.8 Environment

Epidemiological studies have suggested environmental triggers to explain various clusters of disease outbreak, the peculiar geographic distribution, and the critical time period of exposure prior to age 15. The range of factors includes, among others, diet, water, pollutants, radiation, and viruses. Of these, three environmental factors have strong enough epidemiological data so as to support the notion that they contribute causally to MS: smoking, vitamin D deficiency, and infection with the Epstein–Barr virus.

Smoking increases the lifetime risk for MS 2-fold in women and 3-fold in men. The decline in the percentage of men smoking in North America, with female smoking remaining unchanged, may be sufficient to explain the apparent increase in the female-to-male ratio in MS over the past decades. How smoking mediates the enhanced risk in MS is not known.

Vitamin D status, particularly insufficiency, has a strong correlation with MS incidence. Humans synthesize vitamin D in the skin from cholesterol in a process requiring UVB light. The exposure to UVB differs by latitude, resulting in much higher vitamin D production by people living near the equator than in the far northern or southern latitudes, possibly explaining the peculiar north–south gradient in MS incidence. The strongest evidence that low vitamin D levels correlate with increased MS risk comes from a large prospective study involving 182,000 female nurses that monitored their vitamin D intake as well as plasma levels of vitamin D over 20 years. One hundred seventy-three of these nurses developed MS. After taking all other differences into account, those nurses who took 400 IU of vitamin D as a supplement each day had a 40% lower risk of developing MS. Note that just

15 min direct sun exposure during midsummer can produce 20,000 IU of vitamin D naturally. How vitamin D protects is not known. However, increased vitamin D levels increase the number of regulatory T cells, which suppress immune responses, and enhance the level of anti-inflammatory cytokines such as TGF-β1 and IL-10. It also reduces activation of antigen presentation via IFN-γ and IL-17. Interestingly, a mutation in an enzyme responsible for converting 25(OH)D into the active vitamin enhances the risk for heterozygote carriers to develop MS.

The *Epstein–Barr virus* (EBV) is among the most common viruses in humans, present in an estimated 95% of the adult population. Most people develop immunity to it during childhood. If not, the virus can cause mononucleosis ("kissing disease") later in life. Rare individuals who have never been affected by EBV have a reduced risk of developing MS. How EBV infection is causally linked to MS is not clear. However, EBV may share sufficient homology with a myelin-associated antigen to trigger antibody production via molecular mimicry.

4.9 Animal Models of Disease

To fully understand the pathobiology of disease and to examine disease-altering therapies, animal models that accurately replicate the human disease are essential.[14,15] As we see throughout this book, animal models typically only replicate some aspect of disease, and multiple models must be consulted, each with inherent advantages and disadvantages.

The most important and most highly studied model system in MS is called *experimental autoimmune encephalomyelitis* (EAE). As already alluded to in the introduction, this model had its origin in some unintended vaccinations contaminated with CNS antigens causing sporadic paralysis. Follow-up studies clearly identified myelin as the culprit and showed that single immunizations, when given with Freud's adjuvant (an oil–water mixture containing inactivated bacteria and pertussis toxin), were sufficient to induce disease. It is now clear that immunization with any number of myelin antigens reliably causes inflammation in mice, rats, rabbits, pigs, and monkeys. The immune response is largely driven by CD4+ T cells, which are sufficient to transfer disease to naïve animals. Different genetic strains of mice respond preferentially to certain myelin antigens. For example, MBP works well to induce EAE in B10.PL mice; in SJL mice proteolipid protein is the best inducer; and in the widely used C57BL/6 mice a peptide representing amino acids 35–55 of the MOG is the best inducer. Given the wide use of C57BL/6 for transgenic studies, the latter model is by far the most widely used today. While these animals show a similar time course of disease progression and inflammation, they show the majority of pathology in the spinal cord rather than cerebrum.

Another popular model to induce widespread demyelination is the copper chelator, *cuprizone*. When cuprizone is added to food, mice develop almost-complete demyelination of the corpus callosum and other brain regions. Importantly, when switched to normal food, animals show extensive remyelination, allowing the study of myelin repair. A shortcoming of the model is that, unlike in MS, remyelination occurs without the background of continued inflammation. However, important insight into myelin repair can be gleaned through the use of such animals.

Yet another way to induce paralysis and CNS inflammation in mice is via a naturally occurring mouse virus named after its discoverer, the Theiler virus. Infected animals develop inflammation, brain and spinal cord atrophy, and overall loss of myelin. The mechanism(s) by which either cuprizone or Theiler virus cause demyelination are largely obscure.

Finally, a number of specific knockout mice have been developed. These include, for example, mice in which T cells were mutated to specifically recognize certain myelin antigens, or mice with complete loss of B cells.

As we discussed for other diseases in this book, many drugs or treatment approaches that worked well in animal models of MS failed in human trials. Clearly, more refined animal models are still needed, and a more careful approach to their use and interpretation of data from them is warranted. Not least important may be use of older animals that more closely mimic the age of the affected patient population.

5. TREATMENT/STANDARD OF CARE/CLINICAL MANAGEMENT

5.1 Supportive Treatments

There are currently no drugs that directly affect myelination, prevent demyelination, or encourage remyelination. Instead, the treatment for MS encompasses the amelioration of acute attack symptoms, followed by chronic administration of disease-modifying agents in conjunction with symptomatic treatments to make disease more bearable.

Acute treatment during an attack relies almost exclusively on glucocorticoids such as methylprednisolone, which are given for 3–5 days. As is the case in other forms of injury, this attenuates inflammation and reduces pain, but glucocorticoids cannot be given chronically.

Chronic treatments attempt to change the course of disease and these drugs are therefore called disease-modifying drugs. Most of these (Table 4) are relatively new, with the first one, interferon β, having been introduced in 1993. These drugs are often based on humanized monoclonal antibodies (and carry the ending – ab), on cytokines or growth factors. Since these are molecules identical or similar to those made by our body, these are often called "biological" drugs. In general they are very expensive and their long-term clinical benefit is still largely unknown, since most drugs were studied only in small cohorts of patients and typically only for 24 months. Nevertheless, given the progressive

TABLE 4   Disease-Modifying Drugs for MS

Drug	Trade name	Use	Decrease in attack rate	Change overall	Suspected mechanism
Intereferon-β-1a	Avonex Refib	RRMS	18%	37%	Attenuates inflammation by shifting balance of Th-Treg cells
Interferon-β-1b	Betaseron Extavia	RRMS	34%	29%	Attenuates inflammation by shifting balance of Th-Treg cells
Glatiramer acetate	Copaxone	RRMS	29%	12%	Reduces autoimmunity by competition with myelin antigens
Natalizumab	Tysabri Antegren	RRMS PRMS	68%	42%	Prevents entry of B cells
Fingolimod	Gilenya	RRMS	55%	27%	Retaining lymphocytes in the source organ
Mitoxantrone	Novantrone	SPMS PRMS	66%	75%	Anti-cancer drug, disrupts proliferation of T cells and B cells

Currently used disease modifying drugs that are approved for the treatment of MS target different disease mechanism.
Abbreviations: RRMS, relapsing–remitting MS; SPMS, secondary progressive MS; PRMS, primary relapsing MS.

nature of the disease, most MS patients who can afford them are eager to use any number of these drugs, and most insured patients receive coverage for the approved drugs. As of 2014, the average single drug cost per MS patient in the US was approximately $60,000.

5.2 Disease Modifying Drugs

Disease-modifying drugs by and large follow the same therapeutic strategy, namely, reducing our body's immune response and attenuating inflammation.[16] This can be done in various ways, through masking the antigens or cytokines that cause T cell activation, trapping these T cells at their site of birth, or preventing their entry into the CNS. The long-used interferon drugs shift the balance of cytotoxic to regulatory T cells to attenuate inflammation. These strategies are graphically illustrated in Figure 14.

Interferons: The biological drug with the longest history is interferon β. It now comes in a 1a and 1b form. The former is produced in cells, the latter in bacteria. They also differ by injection mode (subcutaneous [1b] versus intramuscular [1a]) and frequency of administration (daily 1b, weekly 1a). Overall, the clinical efficacy of this class of drugs is on the order of a 30% reduction in the rate of attacks and overall disease severity. Side effects are few and tolerable, but given the overall immune-suppressive nature of these drugs, it is not uncommon for patients to experience flu-like symptoms. Interferons are the first line of treatment given essentially to all MS patients.

Glatiramer acetate (GA), too, is frequently added to the treatment regimen early on, and patients may take both interferon and GA in combination. GA is a random tetrapeptide composed of glutamate-lysine-alanine-tyrosine and suppresses T cell activation. Its mechanism of action is believed to involve binding to MHC class II molecules on antigen-presenting cells, thereby competing with myelin antigens. As a result, T cell autoreactivity is attenuated.

Natalizumab is among the most efficacious new treatments. It is a humanized monoclonal antibody to α4β1 integrin, an adhesion receptor used by T cells to bind to blood vessel endothelial cells as they enter the brain. The antibody prevents T cells from entering the brain. Natalizumab is given intravenously once a month and, on average, reduces attack rates by two-thirds. Although the drug is well tolerated by most patients, long-term use (>2 years) carries

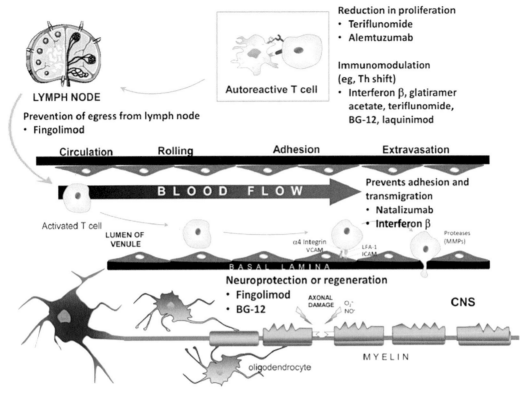

FIGURE 14 Schematic depiction of the sites of action where currently available drugs that treat MS are presumed to work. *Reproduced with permission from Ref. 16.*

a small risk (0.2%) of causing progressive multifocal leukoencephalopathy (PML), a severe and typically fatal virally induced demyelination. In light of this risk, natalizumab is currently given only as a last resort in patients who have failed all other forms of therapy, and is typically only given for less than 18 months to avoid PML.

Rituximab is a monoclonal antibody that recognizes CD20, which is primarily expressed on B lymphocytes but not on antibody-producing plasma cells. It destroys B cells in circulation and was developed to treat B cell non-Hodgkin lymphoma. It has since been demonstrated as effective in autoimmune diseases such as rheumatoid arthritis and lupus and, in light of the recent emergence of a role for B cells in MS, it is used in severe cases of MS, typically off label.

Rituximab has been shown to decrease lesion size and relapse frequency.

Fingolimod is a recently introduced orally administered sphingosine-1-phosphate inhibitor that reduces lymphocyte trafficking to the CNS. It does so by trapping them at their site of birth and maturation, that is, the spleen and lymph nodes. It significantly reduces the frequency and severity of attacks, but long-term safety is not yet established.

Mitoxantrone is a chemotherapeutic that inhibits DNA synthesis and is used in metastatic breast cancer and various leukemias. It is one of the few drugs that are used in MS patients who have advanced to the secondary progressive stage of disease.

As can be seen from this listing, most drugs target the early relapsing–remitting stages of disease and, unfortunately, no disease modifiers exist yet for primary progressive disease.

6. EXPERIMENTAL APPROACHES/ CLINICAL TRIALS

There are currently over 800 clinical trials registered with clinicaltrials.gov for MS. Many of these pursue the same strategies as the above-described approved biological drugs, yet using different targets or delivery strategies. These are typically phase III studies and seek as a primary endpoint a reduction in the rate and severity of attacks, often with MRI imaging of lesion size as a quantitative measure. We will review a couple of promising new drugs briefly before exploring very different disease-altering mechanisms, many of which are at much earlier stages of preclinical or clinical investigation.

6.1 Alemtuzumab

Alemtuzumab is a monoclonal antibody that was developed primarily to treat lymphocytic leukemia and binds to CD52 expressed on lymphocytes and monocytes. It is infused over a 5 day period with the objective of depleting the body of these immune cells. The advantage of alemtuzumab over other drugs is primarily the fact that it is given only once a year. It provides approximately 50% better relief than interferon β1a, to which it was compared in clinical trials. The immune suppressive effect, however, requires co-administration of acyclovir to reduce complications from opportunistic herpes infections. It is likely to be considered for patients who have failed first-line treatment with interferon β or GA.[16]

6.2 Na+ Channel Blockers

Na+ channel blockers[17] have been extensively studied in animal models and, more recently,

in people. Initially, Na+ channel blockers such as phenytoin and carbamazepine, which are commonly used antiepileptic drugs, were used in MS to treat comorbidities such as spasms and trigeminal neuralgia. However, following the discovery that Nav1.6 channels expressed along the demyelinated axon are somewhat leaky and do not close completely after their activation, it became clear that this Na+ entry might contribute to disease. It requires the Na+/K+ ATPase to work in overdrive, possibly exhausting the axonal ATP supply. In addition, Na+ channels appear to do more than just propagate the AP. They are expressed quite commonly on cells of the innate immune system and are particularly elevated in activated microglia, where they facilitate cell migration and cytokine release. Therefore, Na+ channel blockers could be a double whammy, blocking aberrant Na+ entry into axons while also attenuating immune cells (Figure 15). In preclinical studies a number of blockers including phenytoin, lamotrigine, and carbamazepine were quite effective in EAE models. A clinical trial tested lamotrigine in patients with SPMS where the objective was a reduction in brain atrophy by MRI. Unfortunately the trial failed this objective, but this may have been due to a flawed design,[10] and it would be expected that future trials may yield better results.

6.3 K+ Channel Blockers

K+ channels mediate the repolarization after an AP, and their blockage broadens the AP. This can be elegantly achieved using 4-aminopyridine (4-AP), and the resulting broader APs are more likely to jump over a demyelinated axon segment, as shown in Figure 16. 4-AP is widely used in neuroscience laboratories. It has now been formulated in a slow-release tablet (Fampridine-SR) that has shown clinically significant improvements in walking speed in a number of clinical trials,[18] although it carries

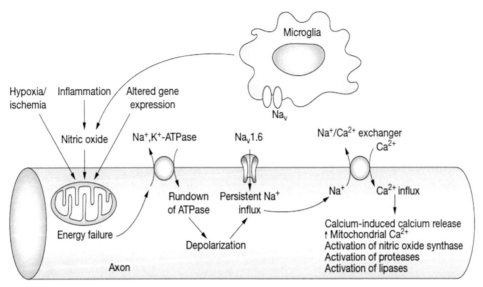

FIGURE 15   Role of Na$^+$ channels in MS. On the axon, energy failure, that is, loss of production of ATP, causes the Na$^+$/K$^+$ ATPase to decrease its function, thereby depolarizing the cell membrane. This in turn causes persistent Na$^+$ influx through Nav1.6 channels, leading to a rise in intracellular Na$^+$. As a consequence, the Na$^+$/Ca^{2+} exchanger now runs in reverse, thereby taking in Ca^{2+} rather than removing it. Na$^+$ channels on microglial cells are involved in microglial activation. *Reproduced with permission from Ref. 17.*

significant side effects due to the lack of specificity of the blocker, affecting many peripheral axons as well.

6.4 Hypothermia

Temperature: Just as K$^+$ channel blockers lengthen the duration of an AP, thereby moving more charge across the membrane, cooling an axon does the same. With this rationale, it stands to reason that temperature will affect the motor performance of an MS patient, being potentially improved at low temperature and worse at high temperature. Indeed, patients frequently report doing much worse in warm temperatures than in the cold, and physicians commonly recommend their patients stay in cooler climates or air-conditioned environments. Various active and passive cooling vests have since been devised and are marketed directly to consumers. Recently, a small clinical study compared several models and found

them somewhat effective, with active liquid cooling devices showing the greatest improvement. Typically patients reported better vision and being less fatigued and more mobile, at least for short periods of time, typically between 30 min and a couple of hours.[20]

As we have already seen in stroke and traumatic brain injury, a reduction in CNS temperature can also be achieved pharmacologically, for example using talipexole,[21] a D2 dopamine antagonist. This approach may similarly benefit MS patients with longer duration.

6.5 Medical Marijuana

Cannabis is widely used among MS sufferers and is believed to have a variety of physiological targets, including immune function and a reduction in CNS temperature. A placebo-controlled phase III clinical study in Great Britain showed a significant 30% reduction in muscle stiffness in patients taking cannabis orally compared to a

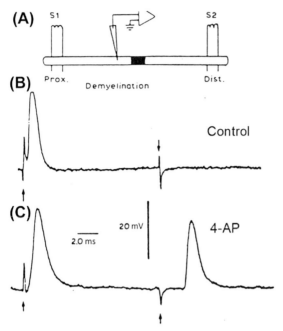

FIGURE 16   K+ channel block with 4-AP (now clinically used as fampridine-SR) broadens the action potential, allowing it to jump across a demyelinated axonal segment. The Control trace illustrates an experimentally demyelinated axon stimulated proximally at the electrode labeled S1 and recorded at S2 where the action potential fails to travel. In 4-AP the action potential bridges the demyelinated gap. *Reproduced with permission from Ref. 19.*

15% reduction in the placebo group.[22] With the broad legalization of medical marijuana and the permission to use it as a recreational drug in some states, its use among individuals with MS will likely increase.

6.6 Vitamin D

Vitamin D plays important roles in bone and immune health. Relevant to MS, antigen-presenting cells, monocytes, and T cells express vitamin D receptors and their biology is regulated by vitamin D levels. Vitamin D downregulates MHC class II, inhibits proliferation of B cells, enhances cell death of activated B cells, and reduces immunoglobulin levels, indicative of lower antibody titers.[23] These data suggest that vitamin D sufficiency is important for a normal immune response. For most people, dietary intake of vitamin D provides only 25% of the body's daily requirements. The majority of their daily requirement is synthesized in the skin from cholesterol in a UVB-dependent manner. The resulting vitamin D3 is hydroxylated to calcitriol (1α,25-dihydroxyvitamine D), a hormone critical for bone health and regulation of immune function. People who are not sufficiently exposed to sunlight, even when living in southern climates, will have insufficient vitamin D3 and should supplement their diet with vitamin pills or add vitamin D-rich food items such as fish. The daily intake of 400 IU of vitamin D has been shown to reduce the likelihood to develop MS in women by 40%. Moreover, of patients who have MS, those with low blood levels of vitamin D have more brain lesions and signs of active disease.[23] In children with MS, vitamin D levels are inversely proportional to disease severity. Therefore a compelling case can be made for rigorous supplementation in the general population and in persons with MS in particular. A 6-week dose-escalation study, in which patients took up to 14,000 IU/day showed it to be safe and well tolerated. Eight clinical trials are currently investigating the effect of vitamin D3 either alone or in combination with disease modifiers on the disease burden. These data should be available in the near future.

6.7 Diet

The Swank Diet: In the 1940s, Roy Swank observed that in Norway, MS incidence was much higher in people who consumed diets rich in dairy and animal fat as opposed to coastal communities consuming mainly fish. Assuming a dietary link to disease, he designed a diet based on very low overall fat intake and fish oils that has since been used by countless MS patients, many of whom report symptomatic improvements. Although Dr Swank has periodically published reports on his successes, there has not been a rigorous clinical trial to assess the effectiveness of his diet, so the available data are only anecdotal.

6.8 Estriol to Induce a Pregnancy-Like State

During the third trimester of pregnancy, women with MS experience an 80% reduction in the frequency of attacks and return to baseline levels post partum. However, when examined over a 3-year period that includes pregnancy, no significant change in disease frequency and severity since was apparent.[24] Nevertheless, a study that followed 200 women who had a child during their disease showed a delay in the time when a wheelchair was required for mobility from 12.5 to 18.6 years.[25] Obviously a number of factors change during pregnancy, yet the hormone estrogen continues to surface as the most likely candidate to afford a neuroprotective effect. Indeed, studies in mice with EAE show significant neuroprotection and attenuation of disease severity. Although estrogen replacement therapy was stopped in postmenopausal women due to unacceptable increases in the risk for heart attacks, stroke, and breast cancer, these risks are acceptable in a sick population of much younger individuals. Reports from one small clinical study support a favorable effect of estriol, the weakest of the three available estrogens that binds to the nuclear ERβ receptor. Women who have RRMS in particular showed fewer enhancing lesions, decreased circulating CD4 T cells, and increased cognitive function. Larger follow-up studies are ongoing.

7. CHALLENGES AND OPPORTUNITIES

MS is an excellent example that frank neurodegeneration can present at a very young age, far too early for the nervous system to have accumulated "toxic" misfolded proteins or have worn out its mitochondria, as is the case in Alzheimer or Parkinson disease. Instead, in persons with genetic risk factors, unknown environmental triggers turn our immune system toward destructing our own cells. It is likely that other autoimmune diseases such as lupus, psoriasis, or arthritis share some of the same triggers and possibly disease mechanisms. I expect that research on these diseases will synergize in coming years. With the recognition that axonal injury occurs early in disease, MS has unexpectedly morphed from an autoimmune condition into a true neurodegenerative disease. Unfortunately, this changed mindset has not influenced treatment strategies at all. Current disease-modifying drugs target the body's immune system exclusively. These drugs, celebrated as breakthroughs, are incredibly expensive and provide only modest and temporary relief to only a fraction of patients. Not a single current approach seeks to alter the underlying neurodegenerative process and, consequently, no drugs exist for patients who immediately develop progressive disease. Only when we protect axons from dying will the course of disease actually change. Current MS therapies are like treating a stroke patient with anti-inflammatory drugs as opposed to restoring blood flow to rescue dying neurons. This situation is clearly not acceptable.

Much effort is focused on identifying genetic risks for MS, and rightly so. However, some of the most promising ideas to reduce MS risk through simple lifestyle and nutritional changes are rarely heard in the media and appear to even have a select scientific and medical audience. Yet the epidemiological data are so compelling. Vitamin D supplementation with a single 400 IU pill daily reduced the MS risk by over 40% in a very large clinical study. This is a far greater benefit than any of the preventive measures taken in stroke, where the baby aspirin regimen yields only a 23% reduction in stroke risk. Why are we not more aggressive with vitamin D supplementation when there is essentially no risk? Should we not make vitamin D mandatory in pregnant mothers, as we have done by adding folic acid to bread? These findings, together with a 3-fold reduction in disease risk by smoking cessation, must be more widely publicized so that the public can embrace them. Importantly, we do not fully understand the beneficial effects of vitamin D, a gap that must be closed through further research, which could yield much more efficacious and specific drugs that are neuro- or glioprotective.

Any success in reversing the demyelination in MS and protecting axonal health carries a tremendous bounty, since such strategies are likely to be equally effective in trauma and stroke. A more neurocentric as opposed to immunological approach to MS appears warranted.

Acknowledgments

This chapter was kindly reviewed by Stephen Waxman, MD, PhD, Bridget Marie Flaherty Professor of Neurology, Neurobiology, and Pharmacology; Director, Center for Neuroscience & Regeneration/Neurorehabilitation Research, Yale University School of Medicine and VA Connecticut Healthcare System.

References

1. Hauser SL, Oksenberg JR. The neurobiology of multiple sclerosis: genes, inflammation, and neurodegeneration. *Neuron*. 2006;52(1):61–76.
2. Wingerchuk DM, Weinshenker BG. Acute disseminated encephalomyelitis, transverse myelitis, and neuromyelitis optica. *Continuum (Minneapolis, Minn)*. 2013;19(4 Multiple Sclerosis):944–967.
3. Marrie RA, Hanwell H. General health issues in multiple sclerosis: comorbidities, secondary conditions, and health behaviors. *Continuum (Minneapolis, Minn)*. 2013;19(4 Multiple Sclerosis):1046–1057.
4. Lucchinetti C, Bruck W, Parisi J, Scheithauer B, Rodriguez M, Lassmann H. Heterogeneity of multiple sclerosis lesions: implications for the pathogenesis of demyelination. *Ann Neurol*. 2000;47(6):707–717.
5. Hemmer B, Archelos JJ, Hartung HP. New concepts in the immunopathogenesis of multiple sclerosis. *Nat Rev Neurosci*. 2002;3(4):291–301.
6. Hughes D, Field EJ. Myelotoxicity of serum and spinal fluid in multiple sclerosis: a critical assessment. *Clin Exp Immunol*. 1967;2(3):295–309.
7. Lisak RP, Benjamins JA, Nedelkoska L, et al. Secretory products of multiple sclerosis B cells are cytotoxic to oligodendroglia in vitro. *J Neuroimmunol*. 2012;246(1–2):85–95.
8. Trapp BD, Stys PK. Virtual hypoxia and chronic necrosis of demyelinated axons in multiple sclerosis. *Lancet Neurol*. 2009;8(3):280–291.
9. Franklin RJ, Ffrench-Constant C. Remyelination in the CNS: from biology to therapy. *Nat Rev Neurosci*. 2008;9(11):839–855.
10. Franklin RJ, ffrench-Constant C, Edgar JM, Smith KJ. Neuroprotection and repair in multiple sclerosis. *Nat Rev Neurol*. 2012;8(11):624–634.
11. Lee Y, Morrison BM, Li Y, et al. Oligodendroglia metabolically support axons and contribute to neurodegeneration. *Nature*. 2012;487(7408):443–448.
12. Ascherio A, Munger KL, Lunemann JD. The initiation and prevention of multiple sclerosis. *Nat Rev Neurol*. 2012;8(11):602–612.
13. Bronson PG, Caillier S, Ramsay PP, et al. CIITA variation in the presence of HLA-DRB1*1501 increases risk for multiple sclerosis. *Hum Mol Genet*. 2010;19(11):2331–2340.
14. Simmons SB, Pierson ER, Lee SY, Goverman JM. Modeling the heterogeneity of multiple sclerosis in animals. *Trends Immunol*. 2013;34(8):410–422.
15. van der Star BJ, Vogel DY, Kipp M, Puentes F, Baker D, Amor S. In vitro and in vivo models of multiple sclerosis. *CNS Neurol Disord Drug Targets*. 2012;11(5):570–588.
16. Freedman MS. Present and emerging therapies for multiple sclerosis. *Continuum (Minneapolis, Minn)*. 2013;19(4 Multiple Sclerosis):968–991.
17. Waxman SG. Mechanisms of disease: sodium channels and neuroprotection in multiple sclerosis-current status. *Nat Clin Pract Neurol*. 2008;4(3):159–169.
18. Ruck T, Bittner S, Simon OJ, et al. Long-term effects of dalfampridine in patients with multiple sclerosis. *J Neurol Sci*. 2014;337(1–2):18–24.
19. Targ EF, Kocsis JD. 4-Aminopyridine leads to restoration of conduction in demyelinated rat sciatic nerve. *Brain Res*. 1985;328(2):358–361.
20. Ku YT, Montgomery LD, Lee HC, Luna B, Webbon BW. Physiologic and functional responses of MS patients to body cooling. *Am J Phys Med Rehabil/Association of Acad Physiatrists*. 2000;79(5):427–434.
21. Johansen FF, Hasseldam H, Rasmussen RS, et al. Drug-induced hypothermia as beneficial treatment before and after cerebral ischemia. Pathobiology. *J Immunopathol Mol Cell Biol*. 2014;81(1):42–52.
22. Zajicek JP, Hobart JC, Slade A, Barnes D, Mattison PG. Multiple sclerosis and extract of cannabis: results of the MUSEC trial. *J Neurol Neurosurg Psychiatry*. 2012;83(11):1125–1132.

23. von Geldern G, Mowry EM. The influence of nutritional factors on the prognosis of multiple sclerosis. *Nat Rev Neurol.* 2012;8(12):678–689.
24. Gold SM, Voskuhl RR. Estrogen treatment in multiple sclerosis. *J Neurol Sci.* 2009;286(1–2):99–103.
25. Vukusic S, Hutchinson M, Hours M, et al. Pregnancy and multiple sclerosis (the PRIMS study): clinical predictors of post-partum relapse. *Brain.* 2004;127 (Pt 6):1353–1360.

General Readings Used as Source

1. Harrison's Online, Chapter 380, Stephen L. Hauser & Douglas S. Goodin.
2. Compston A, Coles A. Multiple sclerosis. *Lancet.* 2008;372(9648):1502–1517.
3. Franklin RJ, et al. Neuroprotection and repair in multiple sclerosis. *Nat Rev Neurol.* 2012;8(11):624–634.
4. Ascherio A, et al. The initiation and prevention of multiple sclerosis. *Nat Rev Neurol.* 2012;8(11):602–612.
5. Lassmann H. Multiple sclerosis: lessons from molecular neuropathology. *Exp Neurol.* 2014;262(Pt A):2–7.
6. Nave KA. Myelination and support of axonal integrity by glia. *Nature.* 2010;468(7321):244–252.
7. Rasband MN. Composition, assembly, and maintenance of excitable membrane domains in myelinated axons. *Semin Cell Dev Biol.* 2011;22(2):178–184.
8. Poliak S, Peles E. The local differentiation of myelinated axons at nodes of Ranvier. *Nat Rev Neurosci.* 2003;4(12):968–980.
9. van der Star BJ, et al. In vitro and in vivo models of multiple sclerosis. *CNS Neurol Disord Drug Targets.* 2012;11(5):570–588.
10. Lucchinetti C, et al. Heterogeneity of multiple sclerosis lesions: implications for the pathogenesis of demyelination. *Ann Neurol.* 2000;47(6):707–717.
11. Trapp BD, Nave KA. Multiple sclerosis: an immune or neurodegenerative disorder? *Annu Rev Neurosci.* 2008;31:247–269.
12. Sherman DL, Brophy PJ. Mechanisms of axon ensheathment and myelin growth. *Nat Rev Neurosci.* 2005;6(9):683–690.
13. Hauser SL, Oksenberg JR. The neurobiology of multiple sclerosis: genes, inflammation, and neurodegeneration. *Neuron.* 2006;52(1):61–76.
14. Trapp BD, Stys PK. Virtual hypoxia and chronic necrosis of demyelinated axons in multiple sclerosis. *Lancet Neurol.* 2009;8(3):280–291.

Suggested Papers or Journal Club Assignments

Clinical Papers

1. Munger KL, et al. Vitamin D intake and incidence of multiple sclerosis. *Neurology.* 2004;62(1):60–65.
2. Ascherio A, et al. Vitamin D as an early predictor of multiple sclerosis activity and progression. *JAMA Neurol.* 2014. PMID: 24445558.
3. Vukusic S, et al. Pregnancy and multiple sclerosis (the PRIMS study): clinical predictors of post-partum relapse. *Brain.* 2004;127(Pt 6):1353–1360.
4. Calabresi PA, Radue EW, Goodin D, et al. Safety and efficacy of fingolimod in patients with relapsing-remitting multiple sclerosis (FREEDOMS II): a double-blind, randomised, placebo-controlled, phase 3 trial. *Lancet Neurol.* 2014;13(6):545–556.

Basic Papers

1. Axtell RC, et al. T helper type 1 and 17 cells determine efficacy of interferon-beta in multiple sclerosis and experimental encephalomyelitis. *Nat Med.* 2010;16(4):406–412.
2. Lee Y, et al. Oligodendroglia metabolically support axons and contribute to neurodegeneration. *Nature.* 2012;487(7408):443–448.
3. Lisak RP, et al. Secretory products of multiple sclerosis B cells are cytotoxic to oligodendroglia in vitro. *J Neuroimmunol.* 2012;246(1–2):85–95.
4. Deshmukh VA, Tardif V, Lyssiotis CA, et al. A regenerative approach to the treatment of multiple sclerosis. *Nature.* 2013;502(7471):327–332.
5. Kleinewietfeld M, Manzel A, Titze J, et al. Sodium chloride drives autoimmune disease by the induction of pathogenic TH17 cells. *Nature.* 2013;496(7446):518–522.

Brain Tumors

Harald Sontheimer

1. CASE STORY

The summer was coming to an end and Alex Baldwin was reviewing his reading assignments for the upcoming school year. He was teaching his favorite class, 12th grade AP History, where he would cover the modern Western civilization since World War II. Having recently gone through a rough divorce from his wife of 11 years, he was longing for some normalcy in his life. The classroom had always been his sanctuary and the structured school day, punctuated by regular hourly bells, was a welcome distraction from his misery. Six weeks into the school year, approaching Thanksgiving, Alex noticed that he was frequently unable to hold his train of thought during class and, on occasion, did not remember questions students had just asked. These episodes were becoming increasingly frequent, occurring almost daily. In fact they had become frequent enough that his students were making fun of the "absent-minded professor." Growing increasingly worried, Alex sought the council of his physician, who suggested that his condition was most likely stress-related and recommended that he try yoga, sleep more regularly, and cut back on coffee, as well. In spite of his best attempts, his condition worsened over the following months. From time to time, his fading spells occurred when he was doing simple chores around the house. On more than one occasion he wound up hitting his head against a wall or pillar. One day he was taking out the trash and when he came back to the house he found himself covered in mud, without any recollection of what had happened. His memory lapses now required him to keep a hidden list with the names of his students, even those he had taught for the past 3 years. He was getting easily annoyed and agitated, and from time to time yelled at students for rather minor infractions.

Upon the recommendation of his physician, Alex underwent a magnetic resonance imaging (MRI) exam during Easter break, followed by an 8-h electroencephalographic monitoring session. The clinical result suggested an abnormal electroencephalogram consistent with periodic absence seizures. A few days later his physician called him in for a visit to review the imaging findings. By that time Alex had already consulted every Internet site and blog describing similar seizure episodes and had prepared a long list of questions ranging from the side effects of antiepileptic drugs on memory function to the promise of the Atkins diet for his condition. Much to his shock, he learned that the epilepsy was the least of his problems. He had a brain tumor, located in the deep layers of the white matter of the left temporal lobe. Further assessment would require a needle biopsy, to be performed the following week. The histopathological report classified the tumor as a stage II oligodendroglioma. Alex was a wreck, particularly once he found out that the tumor would be inoperable since it was too close to vital brain regions. After several weeks, during which his mood fluctuated between anger, grief, and self-pity, he regained hope and the resolve to fight this cancer with the help of a local cancer support group. He started daily radiation treatments for 6 weeks, which resulted in baldness but caused the tumor to stabilize. As summer ended, he had gained enough strength to continue work and, in many ways, felt like a changed man. No longer was he focused on the failings in his personal life; he was finally able to concentrate on the job at hand, living day by day. His memory function had normalized and the seizures were largely controlled by daily doses of the antiepileptic drug Keppra. He was given a second lease on life.

After an eventless 3 years, punctuated by quarterly visits with his neuro-oncologist, an MRI scan showed evidence that the tumor was growing again. The mass had almost doubled since the last scan, which was taken just 6 months earlier, suggesting a progression toward a higher, more malignant grade tumor. Alex summoned his courage and spent the following 4 months in chemotherapy, receiving a drug combination called PCV that was administered once weekly over a 4-month period. He seemed to have dodged a bullet once again as the tumor growth seemed contained. Alex learned from other members of his support group that, in spite of harboring an incurable, inoperable brain tumor that will sooner or later claim his life, his is one of the few gliomas that are responsive to chemotherapy and radiation.

Some friends from his support group were not so fortunate and have long since passed away.

2. HISTORY

The first documented successful removal of a brain tumor dates to 1879, when William Macewen removed a "fungous tumor of the dura mater," most likely a meningioma, from the brain of a young woman. However, archeological records suggest that brain surgery was attempted long before the beginnings of neurosurgery and before the introduction of anesthetics. Skulls found in an archeological site in France dating back to 6500 BC revealed trepanation holes, round holes of various sizes that seem to have been purposefully drilled or scraped from the skull[1] (Figure 1). Similar holes have been found in skulls associated with different cultures and continents from prehistoric to modern times. Most patients probably did not survive these procedures, although skulls of Mayans in Mexico and Central America show evidence of near complete healing of the bone, suggesting some successful trepanations. These may have been conducted with local cocaine-induced anesthesia derived from coca leaves. To what extent trepanation holes served any true medical purpose, such as relieving pressure or removal of superficial tumors, or whether they were largely spiritual in nature, is not entirely clear.

It may be of some historic interest to frame brain tumors within the history of cancer in general, particularly in light of the fact that anyone affected by cancer almost reflexively asks the question of whether the disease may have been caused by exposure to any of the many man-made carcinogens or environmental hazards that came with the industrial revolution. For brain cancer these may include chemicals and radiation. It is interesting, therefore, to learn that the first documented cases of cancer predate the industrial revolution by several thousand years; early accounts

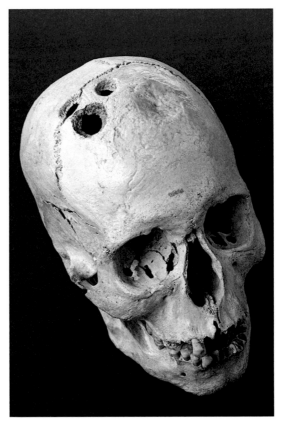

FIGURE 1 Skull excavated from a tomb in Jericho in January 1958, showing four separate holes made by the ancient surgical process of trephination. They had clearly begun to heal. This suggests that although highly dangerous, the procedure was by no means fatal. Also known as trepanation or trepanning, the process of making a hole through the skull to the surface of the brain might be carried out to treat a range of medical conditions or for more mystical reasons. *Science Museum, London, Wellcome Images.*

of breast cancers are documented on Egyptian papyri from 1500 BC. Even more astonishing is the earliest known brain tumor found in a dinosaur fossil, a 72-million-year-old gorgosaurus on display at the Children's Museum of Indianapolis. It harbors a golf ball-sized brain tumor within the skull cavity, and it is believed that the tumor's location would have compromised the animal's balance, explaining an abundance of wounds and fractures on her skeleton.

The early 20th century is typically regarded as the heyday of neurosurgery, and indeed, brain tumors helped to define neurosurgery as a surgical subspecialty that has become almost synonymous with tumors of the nervous system. The person most frequently credited with the birth of brain tumor surgery is Harvey Cushing, who advanced operative techniques and adopted numerous devices that enhanced surgical quality and improved outcomes. He developed a tumor classification system that became the foundation for modern pathology. The discovery of X-rays in 1895 allowed noninvasive tumor diagnosis and presurgical localization, which was advanced by Fedor Kruse, a German neurosurgeon who published the use of X-ray imaging to diagnose brain tumors in 1911. Although the careful study of autopsies by Pierre Paul Broca (1864) and Carl Wernicke (1876) had already identified areas in the posterior regions in the left frontal lobe of the brain that were required to articulate and understand language, a complete understanding of the "eloquent" brain came from the sensory and motor maps constructed during awake craniotomies using electrical stimulation. This approach was used extensively by Wilder Penfield (Figure 2) while working at the Montreal Neurological Institute. He mapped the primary motor cortex to the anterior frontal sulcus of the cortex and identified the posterior sulcus as the primary sensory cortex. This map, called the homunculus, is still used today, essentially unaltered.

The two greatest challenges to brain tumor surgeries at the beginning of the 20th century were infection and blood loss. Harvey Cushing made it his mission to reduce both conditions. Through meticulous surgical procedures he was able to almost eliminate infections, reporting only two cases among the several hundred surgeries he conducted. Achieving hemostasis (preventing bleeding) remained a challenge. Blood transfusions were rarely done, in part because blood types and compatibility were still unknown, and when transfusions were attempted they frequently resulted in death due to acute hemolytic reactions. Cushing introduced small clips made of silver wire to temporarily stop blood flow in larger cranial vessels. The wartime discovery that adrenaline acts as a

FIGURE 2   McGill neurosurgeon **Wilder Penfield** developed the groundbreaking Montreal procedure, a surgical technique that uses local anesthetic to keep the patient conscious and responding to questions while the surgeon stimulates parts of the brain. Using this method, Penfield created functional maps of the cortex (surface) of the brain, now called the homunculus (www.mcgill.ca/about/history/more-history/firsts/1950).

vasoconstrictor was rapidly deployed in brain surgery, as well. Surprisingly, removal of a meningioma, which is among the most curative brain tumors today, was the most problematic in those days because it was so highly vascularized and hemostasis such a challenge. This changed upon introduction of electrocauterization, which rapidly occludes vessels through application of electric current.

In parallel with the development of improved surgical techniques, tumor biology was becoming increasingly better understood as surgeons and pathologists carefully documented their findings. One of the most comprehensive descriptions was provided by the neurosurgeon Tooth (1912) (reviewed by Scherer[2]), who described gliomas as the most frequent brain cancer, accounting for close to 50% of tumors in his 258 patients. He correlated histopathological findings with clinical outcomes and first recognized that cerebral astrocytomas, which lack mitotic figures, carry a relatively favorable prognosis, whereas any evidence of tissue necrosis is characteristic of highly malignant gliomas with poor outcome. Based on his experience with repeated resections in the same patients, he also proposed that "benign," low-grade tumors often morph into more malignant ones upon recurrence, and that the propensity to do so may be enhanced by incomplete tumor resection. Extensive microscopic examination of autopsy materials provided additional insight into the different classes of gliomas, their invasive nature and routes taken as they invade, as well as early evidence for the induction of new blood vessels through angiogenesis. As will become clear throughout this chapter, many of the fundamental biological traits of tumors were already recognized by early pioneers in neurosurgery and pathologists such as Hans Joachim Scherer almost 100 years ago.[2] (A comprehensive historic account can be found in *Neurosurgical Focus* 2005,[3,4] from which highlights have been extracted here.)

3. CLINICAL PRESENTATION/DIAGNOSIS/EPIDEMIOLOGY

Brain tumors do not discriminate and can strike at any age. Among the roughly 150,000 patients diagnosed each year in the United States, 60% present with metastatic disease (tumors originating in the body), whereas 67,000 patients carry a primary brain tumor where the abnormal tissue started growing within the central nervous system (CNS).

3.1 Metastatic CNS tumors

Systemic cancers of the lung, breast, skin, or kidney frequently metastasize to the brain. Unlike primary brain tumors, metastases are typically well delineated and can be identified by a distinctly different cellular appearance during autopsy or based on biopsy tissue. These tumors represent a foreign tissue mass in a given organ; for example, dark pigmented kidney cells growing as a cancerous mass within pale brain tissue. Brain metastases typically occur late in the course of systemic disease progression and accordingly are highly malignant. Tumor metastases reach the brain via the circulation, and there is little evidence that these cancers subsequently invade surrounding brain tissue. If multiple metastases are present, they have formed independent of each other. Metastatic brain tumors are mostly found in the brain parenchyma, with approximately 80% growing in the cerebral hemispheres, 15% in the cerebellum, and 5% in the brain stem.[5] If possible, tumors are surgically removed and radiated by whole-brain radiation to relieve immediate CNS symptoms such as seizures and headaches. However, treatment is largely palliative; the main focus is on the primary systemic disease. These tumors are not discussed further because the neurological sequelae they present are comorbidities of the systemic cancer, and patients with metastatic brain tumors typically remain the domain of the oncologist rather than neurologist.

3.2 Primary CNS tumors

Primary CNS tumors are cancers originating and typically remaining in the brain. Primary tumors are divided into malignant or benign, and each group has different subtypes. This heterogeneous group of tumors includes tumors of the meningeal coverings, the ependymal cells lining the ventricles, myelinating cells of the nerve sheets, and specialized support cells in the pituitary (Table 1). Their relative contribution to the overall population is illustrated in the pie chart in Figure 3.

Meningiomas, tumors growing between the skull and brain, are the most common primary tumor. Since they typically do not invade the underlying brain, they are considered benign and can be surgically cured in most patients.

Ependymomas arise from the cells lining the ventricles. They typically grow into the ventricular space and displace cerebrospinal fluid. Ependymomas account for 5% of childhood tumors and are often curable through surgery and radiation.

Medulloblastoma is the most common primary brain tumor in children. They arise from neural progenitor cells in the cerebellum, and 5% of patients carry a germ-line mutation in patched-1, a gene involved in the hedgehog signaling pathway that is important in normal embryological development. Medulloblastomas are malignant tumors that require maximal resection and aggressive radiation and chemotherapy, often leaving children with significant neurocognitive impairments.

Neuromas are tumors of the nerve sheets that cover the cranial nerves; hence the affected cells are myelinating Schwann cells, and the term **schwannoma** would be more accurate. The most common neuromas are acoustic and/or vestibular neuromas affecting cranial nerve VIII. These are typically benign and can often be resected, albeit sometimes with lasting sensory impairments to the affected cranial nerves.

Neuroblastomas deserve mention primarily because, though implied by their name, these

TABLE 1　Common Primary Brain Tumors Grouped as Benign or Malignant[a]

Primary brain tumors	
Benign	**Malignant**
• Meningioma	• Astrocytoma
• Pituitary adrenoma	• Anaplastic astrocytoma
• Acoustic neuroma	• Glioblastoma multiforme
• Epidermoid tumor	• Oligodendroglioma
• Choroid plexus papilloma	• Oligodendro-astrocytoma
• Epenymoma	• Choroid plexus carcinoma
• Pilocytic astrocytoma	• Pineal tumors
	• Medulloblastoma

[a] *Often all of these are lumped under the term* glioma, *although only those in blue are clearly related to glial cells and form proper gliomas.*

typically do not present as CNS tumors but are the most common extracranial solid tumor in children. They are neuroendocrine tumors arising from neural crest cells that give rise to the sympathetic nervous system. These typically present in children younger than the age of 2 years and arise from the adrenal gland. They frequently differentiate into benign lesions, although if they occur later in life they can be malignant and incurable.

Gliomas: The largest group of primary brain tumors, and the one discussed from here on, encompasses **astrocytomas, oligodendrogliomas, and glioblastomas.** These are collectively called **gliomas** because they share a presumed glial origin. Gliomas make up one-third of all primary tumors but account for 80% of all malignant primary brain tumors. With an incidence of 5 per 100,000, approximately 40,000 new malignant gliomas are diagnosed each year in the United States, and about 16,000 patients die from the disease annually. Although gliomas can arise at any age, incidence is highest among the very young and the elderly. They are slightly more common in males versus females (~1.7 fold) and affect Caucasians two times more often than African Americans or Hispanics. Gliomas are the leading solid cancer in children and the second leading cause of cancer-related

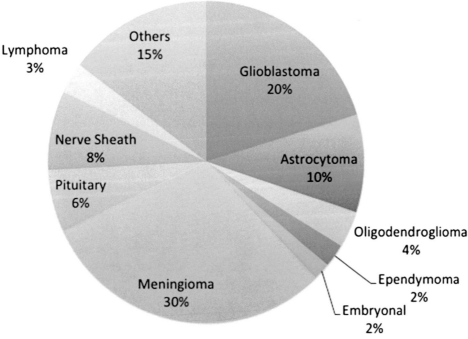

Distribution of Primary Brain Tumors

FIGURE 3   Distribution of primary brain tumors as a percentage of the total, according to the Central Brain Tumor Registry of the United States, 2005–2006.

deaths in young men 20–40 years of age. Incidence and prevalence have changed little since the 1970s, suggesting that risk factors for developing the disease have not changed, but unfortunately thus far treatments have not improved the outlook. Gliomas are always caused by multiple mutations in genes that control cell growth and differentiation. Such mutations occur sporadically, and there is no evidence of the general heritability of brain tumors, although a few rare genetic diseases exist that present with an increased risk of glioma; neurofibromatosis and Li-Fraumeni syndrome are two examples. A number of risk factors have been suspected to contribute to mutations, including exposure to petrochemicals, radiation, and electromagnetic fields. The only confirmed environmental risk factor to date is prior brain exposure to X-rays. In spite of much publicity, the use of cell phones does not increase the risk of developing a brain tumor.

3.3 Patient Presentation and Diagnosis

The initial symptoms displayed by patients with gliomas are highly varied and obviously very much related to the site of growth. Some common symptoms, however, such as sudden unexplained headaches, nausea, and vomiting, are the result of increased intracranial pressure. Other signs of general brain dysfunction include seizures, muscle weakness, visual field defects, or sudden changes in personality. Typical glioma headaches with pulsating (throbbing) pain localized to the tumor-bearing side of the head are worse in the morning and may be easily confused with a migraine. Visual field defects (scotomas) are common but may initially

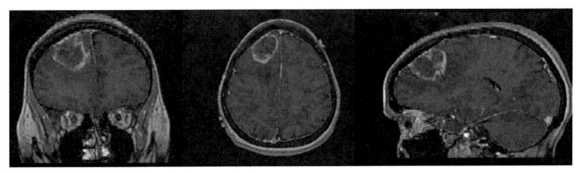

FIGURE 4 Contrast-enhanced magnetic resonance images reveal a frontal lobe intracranial glioma with a characteristic ring of enhancement seen on a T$_2$-weighted image. Coronal (left), sagittal (middle), and longitudinal axis (right) images are from the same patient. *Images courtesy of Dr James Markert, Department of Neurosurgery, University of Alabama Birmingham.*

go unrecognized. Personality changes include loss of interest, social withdrawal, and depression. Focal seizures are common and may be the presenting symptom in 50–90% of patients. Gliomas are best detected by noninvasive imaging techniques, typically via computed tomography (CT) or, preferably, MRI, which reveal abnormal masses or lesions in the brain. Gliomas present as hypodense, bright lesions on CT and MRI, with the tumor often surrounded by edema. If imaged with an electron-dense contrast medium, such as gadolinium injected into the blood stream, the tumor may be surrounded by a ring of contrast enhancement, indicative of increased vascularization or leakage of contrast medium from the blood, as clearly visible in the example in Figure 4. Although a trained radiologist can frequently make an informed and reliable diagnosis, a comprehensive assessment and unequivocal staging relies on a histopathological examination of a tissue biopsy, extracted with a needle that is stereotactically guided into the tumor through a small hole in the skull. Alternatively, resection tissue is examined at the bedside in cases where immediate surgery is warranted.

Once tissue is isolated, several pathological hallmarks are used for staging: atypical cellular appearance, nuclear atypia, mitotic activity, newly formed blood vessels (neovascularization),

and evidence of dead tissue (necrosis). Examples of the pathological features are illustrated in Figure 5. The World Health Organization (WHO) defined a common staging paradigm that ensures uniform grading of malignant gliomas on a scale of I–IV, with increasing number reflecting a higher degree of malignancy. The presence of at least three of these features characterizes the most malignant, rapidly growing WHO grade IV tumor, which always includes evidence of mitotic activity and necrosis. The presence of newly formed blood vessels reduces mean life expectancy from 5 to 3 years, and just a single mitotic cell in every 10 high-power microscopic fields further reduces this to just 1 year.[6] If only two features are present, most frequently mitotic activity and nuclear atypia, tumors are staged as WHO grade III. Lower-grade tumors do not show mitotic activity but are abnormal in cellular and nuclear appearance. This pathological classification is valuable for the patient and physician because it has predictive value regarding disease progression and life expectancy. As illustrated in Figure 6, malignancy grade correlates significantly with expected life span. It is common for lower-grade gliomas to morph into higher-grade tumors over time, and even lower-grade gliomas will in all likelihood cause premature death. Moreover, depending on the location within the brain, even low-grade tumors can severely impair

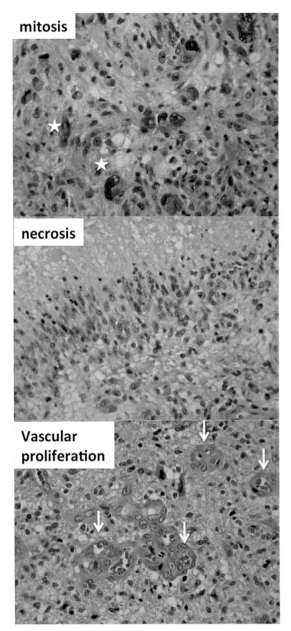

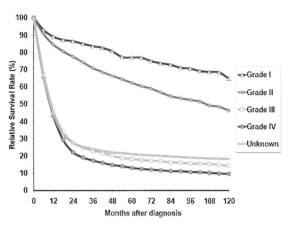

FIGURE 6   Patient survival as a function of tumor grade. *From Ref.* [7].

vital functions and hasten death. Harvested tissue also allows evaluation of genetic changes that can be useful for staging and carry some predictive value regarding tumor progression. Of most interest are the presence of mutations in the retinoblastoma gene *RB*, the tumor suppressor genes *p16* and *p53*, and oncogenes of the RTK family.

4. DISEASE MECHANISM/CAUSE/ BASIC SCIENCE

Over the past three decades, great strides in the understanding of brain tumor biology have been made. This includes improved knowledge concerning their cell(s) of origin, the various genetic mutations that are required to cause a glioma, routes of tumor invasion, mechanism of metastasis, and their ability to change the vasculature through angiogenesis. Many scientific findings are beginning to paint a comprehensive picture of the disease, and the insights gained are influencing new treatment approaches.

4.1 Cell(s) of Origin

Cancer forms only from growth-competent cells that have not permanently differentiated and are still capable of proliferating. In the brain,

FIGURE 5   Histopathological features used to diagnose and stage brain tumors include increased cellularity (top), with evidence of mitosis (stars); necrosis, often leaving palisading glioma cells around the necrotic area (middle); and vascular proliferation (bottom, arrows). *Images kindly provided by Dr Stephen Carroll, Department Pathology and Laboratory Medicine Faculty, Medical University of South Carolina.*

that excludes most neurons. Consequently, it had long been assumed that all primary brain tumors derive from brain support cells, including glial, ependymal, or meningeal cells. This requires multiple mutations in tumor suppressor and oncogenes that collectively cause a loss of growth control. The resulting tumors are historically named according to their presumed cell of origin. However, evidence that an astrocytoma formed as a result of a transformation of an astrocyte or that an oligodendrocyte gives rise to oligodendrogliomas is weak at best. Instead, the discovery of multipotent neural stem cells in the adult human brain has led to the hypothesis that gliomas derive from stem cells or lineage-restricted progenitor cells arising from stem cells.[8]

Cell proliferation along the ventricular wall normally stops after embryonic development. However, two proliferative niches persist along the wall of the fourth ventricle in the adult brain, the subventricular zone (SVZ) and the subgranular zone in the hippocampus (Figure 7). The SVZ contains four cell layers. The innermost are the ependymal cells lining the ventricles. The second layer contains astrocytes that are highly linked by gap junctions. The third layer is a ribbon of astrocyte cell bodies intertwined with some oligodendrocytes and ependymal cells. These astrocytes are highly proliferative and form multipotent neurospheres with self-renewing ability when isolated. Hence this layer is a prime candidate to give rise to tumors of astrocyte or oligodendrocyte lineage should a mutation occur. Importantly, these proliferative "niches" also contain a mixture of cells including ependymal cells, various glial cells, microglia, and vasculature. Among these, however, astrocytes seem to be the dominant cell type, at least in the rodent SVZ, where they function as a neural stem cell and as a support cell that releases growth-promoting factors.[9]

If either stem cells or neural progenitor cells in the SVZ or subgranular zone are mutated, cancers derived from them are likely to express proteins or antigens that are typically found in nonmalignant cells. For example, astrocytomas variably express the astrocytic intermediate filament protein GFAP, whereas NG2 and PDGF receptors characterize oligodendrocytes and are also expressed in oligodendrogliomas. Both cancers could have originated from a multipotent progenitor, which, if mutated, may divide uncontrollably. Indeed, the experimental introduction of mutations into the *p16* and *EGFR* genes into glial progenitor cells is sufficient to induce gliomas in mice.[10,11] While these genetic models favor the notion that progenitor cells are the most likely cell of origin for a glioma,[12] it is equally possible that some astrocytomas may develop from astrocytes, which can divide under conditions of injury and disease, often giving rise to glial scars. If mutated, these could develop into astrocytomas.

A variation of the above scenarios would be to consider that the cell of origin might not be the cell of mutation. The mutation may occur in a stem cell that is destined to become a glial progenitor cell. In turn, any offspring from this cell now have acquired an enhanced potential for malignancy, but only if placed in the appropriate oncogenic environment will they begin to form tumors, and indeed they may never do so.[14] These various possible scenarios are schematically illustrated in Figure 8 and are not mutually exclusive.

A majority of adult gliomas develop above the tentorium (a demarcating line that separates the cerebellum from the cerebrum) either in the subcortical white matter or in the laminar gray matter of the cortex. This region happens to be the major area normally populated by astrocytes and NG2 cells but also is the destination of progenitor cells from the ventricular zone (Figure 9). By contrast, most pediatric tumors develop below the tentorium, often in the midbrain or brain stem, and lack the characteristic EGFR amplification and mutations in the PTEN and p53 genes of adult tumors. It is therefore possible—even likely—that pediatric tumors originate from different progenitor cells, likely remnants of embryologic stem cells. Glial birth and differentiation is

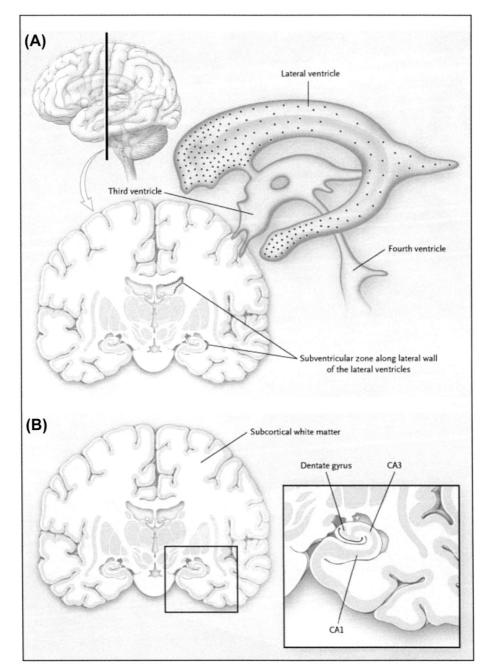

FIGURE 7   Two proliferative zones that persist in the adult brain—the subventricular zone that lines the wall of the first and second ventricle (A), and the subgranular zone that is located in the Dentate gyrus of the hippocampus (B), *From Ref.* [8].

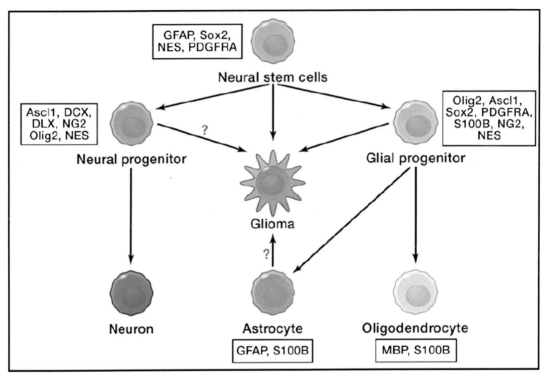

FIGURE 8   Cell of origin for glioma. Several possible cell types of origin have been discussed, and evidence supports each as a possible source. They include true multipotent adult stem cells, transit amplifying stem cells, lineage committed progenitor cells, and dedifferentiated astrocytes.[13]

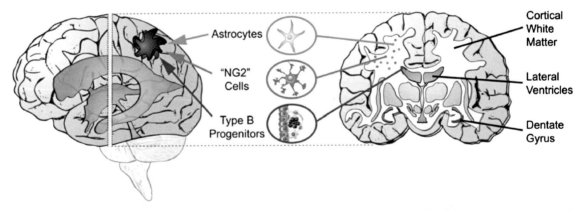

FIGURE 9   Glioma cell of origin. If gliomas are derived from multipotent type B cells of the subventricular zone in humans, then brain cancers might be expected to cluster within or near this germinal region, at least at the time of inception. Subsequently, they could migrate to the cerebrum. Alternatively, gliomas might arise from committed astrocytes or NG2 cells. *From Ref.* [12].

incomplete at birth and, in humans, oligodendrocytes are still generated until puberty. Mutations in these cell populations may give rise to pediatric astrocytomas and explain their different genotype and biology.

4.2 Cancer Stem Cells

Histologically, most gliomas show a remarkable cellular heterogeneity that suggests that they may contain morphologically and functionally different cells. After resection, an isolated tumor can be separated into two cell classes based on expression of the surface glycoprotein CD133, also known as prominin, resulting in CD133+ and CD133− cells. Implantation of just a few hundred of the pure CD133+ cells into a mouse brain is sufficient to give rise to a tumor. By contrast, even 100,000 CD133− cells failed to grow tumors after implantation. This suggests that within a tumor, a privileged population of CD133+ cells serves to propagate the tumor. These cells are often called tumor stem cells, although calling them tumor-propagating cells would be more accurate. Interestingly, tumor-propagating cells are highly resistant to radiation treatment, possibly explaining the poor response to radiation typical of patients with glioma.[15] A number of additional studies have examined the "stemness" of these glioma cells, that is, their ability to differentiate into other cell types. Surprisingly, some of the CD133+ cells are able to form blood vessel endothelial cells, essentially allowing the tumor to furnish its own blood vessels; other studies show that they develop into pericytes, another cell type that supports cerebral blood vessels.

4.3 Glioma: A Genetic Disorder

The uncontrolled proliferation of resident brain cells, whether they are derived from stem cells, progenitor cells, or glial cells, requires multiple mutations in genes that regulate fundamental aspects of cell growth. Most notably, in all cancers, mutations are present in two groups of genes that are generally called oncogenes and tumor suppressor genes. **Oncogenes** drive a cell toward enhanced proliferation. Examples are receptors that respond to growth factors such as epidermal growth factor (EGF) or platelet-derived growth factor (PDGF), often called receptor tyrosine kinases (RTKs) that serve important roles in stimulating cell division and growth of cells, ranging from macrophages to neuroblasts, throughout life. Two types of changes are generally observed in brain tumors. One type is gene multiplication, often called amplification, whereby too many receptors are being produced. The other type is a mutation in the receptor itself, making it constitutively (always) active, regardless of whether the ligand growth factor is present. Either type results in too much stimulation. In addition to oncogenes, there are numerous proto-oncogenes, which can stimulate growth by overexpression without mutations. These include, for example, the simian sarcoma virus, the transcription factor *myc*, the GTPase *Ras*, or the Raf kinase. Neither overexpression nor mutations affect the germ line, and hence these are not heritable.

The second class of common mutations occurs in **tumor suppressor genes**, which act as negative regulators of cell growth. The best examples are the retinoblastoma gene *RB* and the *p53* gene, frequently called the "guardian of the genome". These genes serve as monitors of orderly cell division and become active should any DNA replication errors occur, upon which they induce regulated cell death or apoptosis. Mutations in tumor suppressor genes are called loss-of-function mutations. Since the genome contains two copies of each gene, both alleles must be mutated to lose growth control.

It seems therefore that mutations in three signaling axes are a core requirement shared among all gliomas. These are loss of function of RB or, p53, and gain of function of receptor tyrosine kinases (RTKs) (Figure 10). Note that the mutations may occur in any gene along each signaling axis without directly involving EGFR, RB, or p53.

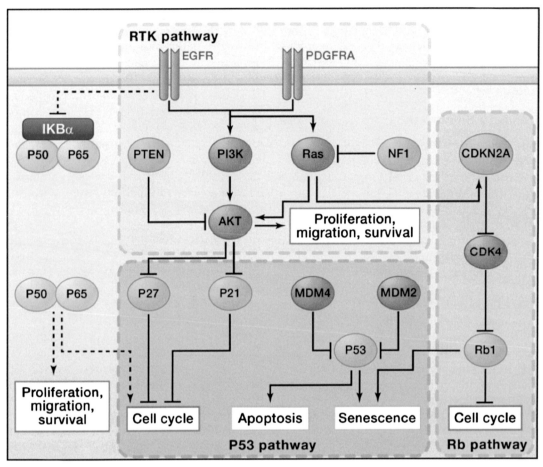

FIGURE 10 The core mutations present in gliomas affect three signaling axes: retinoblastoma (Rb), p53, or receptor tyrosine kinases (RTKs). *From Ref.* [13].

An extensive and growing list of mutations and genomic abnormalities has been cataloged through the diligent work of glioma researchers across the United States who deposited their data in the Cancer Genome Atlas. Through cluster analysis, patterns of gene expression emerged that were predominantly found in glioma subtypes: classical, mesenchymal, proneural, and neural.[16] As these names imply, there was a preponderance of genes typically associated with a certain cell type in each group. For example, the neural subtypes showed neural markers such as GABRA1, KCC2, and SYT1. The proneural group was enriched by neural progenitor or oligodendrocyte genes such as *PDGFRa* and *OLIG2*. Mesenchymal gliomas had deletion or mutations in *NF1*, expression of mesenchymal marker MET, and elevated expression of genes in the nuclear factor-κB pathway. Finally, classical gliomas showed an overall enrichment in astrocytic genes, with alterations in EGFR CDKN2A deletion but a lack of p53. These studies also confirmed the importance of mutations in the three signaling axes (RB, p53 m, and RTK), illustrated in Figure 10 as core requirements shared among all gliomas.

The above mutations are typically somatic and hence not heritable. However, there are a few rarer conditions where heritable germ-line mutations present a significant (>5%) risk factor for glioma. These include mutations in the neurofibromatosis type 1 and 2 genes (NF-1 & NF-2), which cause autosomal dominant disorders presenting with a wide variety of tumors and skin abnormalities, and tuberous sclerosis, caused by autosomal dominant mutations in the tuberous sclerosis genes *TSC1* and *TSC2*, which cause tubers, enlarged growths in the cortex that cause seizures and mental retardation.

4.4 Growth Control (Cell Proliferation and Evasion of Apoptosis)

The defining feature of brain tumors is their dysregulated growth; as we learned above, mutations in the core pathways that regulate normal growth are responsible for that. Cell proliferation is a normal and essential property. Of the ~50 trillion cells in our body (5×10^{18}), at least 10^{13} cells undergo a division daily. The blood alone produces 2×10^{12} red cells and 4×10^{12} platelets/day. Every time DNA is replicated, there is a possibility of a somatic mutation. The probability of a mutation in the coding region of a gene per cell division is estimated to be 10^{-2}. In other words, of every 100 dividing cells, one will have a somatic mutation.[17] If we multiply this by 10^{13} cells we have 10^{11} cells per day with a mutation. Given that each cell harbors 10^5 genes suggests that up to 1 million cells experience at least a single gene mutation each day. Not a rare event at all! The probability is lower in the brain than in the other organs because of the much-reduced number of proliferating cells and a relatively long cycle. For example, NG2 progenitor cells are thought to proliferate once every 2 days soon after birth but then declines to divide only once every 150 days in the aged mouse.[18] This may explain the relatively low incidence of brain cancer compared with, for example, breast cancer. However, a significant probability for a mutation in one of the growth-regulating genes for each organ exists, and this probability far exceeds the actual incidence. The many checks and balances that monitor our genome solve the puzzle. Most important are the cell cycle regulatory proteins that monitor DNA replication and induce apoptosis if necessary. These include RB and p53, which induce apoptosis should irreparable damage be detected. Moreover, DNA repair enzymes that interfere if damage occurs are constantly at work. As alluded to above, genetic mutations in any one of these groups of proteins has been identified in one or the other glioma, yet the core essential group of mutations are in the oncogenes, primarily RTKs, the pro-apoptotic genes in the RB cascade and tumor-suppressing p53 signaling axes. Since none of these are truly glioma-specific, but rather shared commonly among cancers, these are not discussed further here. Instead I focus on glioma-specific biology that gives unique insight into the disease.

4.5 Tumor Invasion/Intracranial Metastasis

Most malignant tumors form metastases at some stage of disease. This typically involves tumor cells entering the systemic blood stream or lymphatic system and passively propagating to a site distant from the primary tumor. By contrast, malignant gliomas are not believed to enter into circulation, and a lymph system is absent in the brain. Nevertheless, gliomas infiltrate the brain and seed secondary tumors throughout the brain, sometimes extending into the spinal cord. These secondary tumors form as a result of active cell migration through the extracellular brain spaces. In the 1930s German pathologist Hans Scherer systematically examined the brains of 100 patients with glioma by serially sectioning the entire brain. He described the brain structures that gliomas most commonly associate with, concluding that gliomas most likely invade along four major routes: (1)

the perivascular space surrounding blood vessels, (2) the perifasicular space along nerve bundles, (3) the subarachnoid space below the meningeal covering of the brain, and (4) the brain parenchyma. He also suggested that each cell's morphology is defined by the space that the cell occupies within a structure rather than being characteristic for a given tumor type. Consequently, when migrating along the perivascular space where they encase blood vessels, cells from the same tumor assume a thin sheetlike structure, whereas in a nerve bundle they may appear as an elongated solid wedge (Figure 11). It is not uncommon for malignant gliomas to cross from one hemisphere to the other, most likely using the white matter tracts of the corpus callosum as a pathway.

Invasiveness within the brain is a major impediment to surgery and typically precludes complete resection. In addition, the rapid spread throughout the brain can cause diverse neurological symptoms because tumors may occupy multiple cortical areas involved in higher brain function. The extent to which tumors have infiltrated the brain is typically not visible on MRI or CT but can be assessed only on histological sections. From such sections we know that tumors with a higher malignancy grade are typically more infiltrative than lower-grade tumors.

The process of glioma cell invasion is fascinating and has been extensively studied. Extracellular as opposed to intravascular, hematogenous spread poses significant challenges for invading cells. The extracellular space between brain cells is very small, on the order of a micrometer or less—far smaller than the typical diameter of even a cell nucleus (2–3 μm). Moreover, the extracellular space contains extracellular matrix, a heterogeneous mixture of substance rich in proteins and hyaluronic acid. This matrix is important for the development and function of multicellular organisms. It could best be described as a gel or soft putty-like substance that often attaches the basement membranes of adjacent cells. For gliomas to move along the extracellular space, they digest the extracellular matrix through the release of matrix-degrading ectopeptidases called matrix metalloproteinases. Twenty-three metalloproteinases with different substrate specificity have been described, of which at least seven are expressed by gliomas; the most commonly studied are MMP-2, -3, and -9. The composition of the extracellular matrix varies greatly and may be a major determinant of the route that a glioma cell chooses. For example, nerve fascicles are rich in laminin, which is readily degraded by the gelatinase MMP-2. Importantly, by interacting with integrin

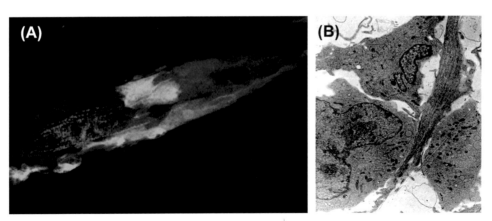

FIGURE 11   Examples of tumors invading along blood vessels (A) and through the white matter (B). *Courtesy, Dr Stefanie Robel, Department of Neurobiology, University of Alabama Birmingham.*

receptors on the cell surface, gliomas can recognize different extracellular matrix molecules. Hence matrix molecules are not just obstacles to movement but important extracellular cues.

Even after the digestion of extracellular matrix, the space for invasion is still constrained. However, gliomas have an unusual ability to dynamically adjust their volume and shape to squeeze through even the smallest spaces. This involves the concerted movement of chloride and potassium ions and water to move cytoplasm across the membrane (Figure 12). Inhibitors of these ion channels retard the ability of tumor cells to move in culture and contain tumor invasion in mouse models of the disease. Hence ion channel blockers are being explored as drugs to treat gliomas in clinical trials (see Section 6.5).

One of the fascinating questions is whether an invading tumor is directed in its movement toward an attractant or whether invasion is simply a random stochastic process. Studies of cultured gliomas or tumor-bearing brain slices suggest that glioma cells are capable of responding to concentration gradients to perform chemotaxis. For example, when placed in a two-compartment chamber separated by a filter containing small holes, gliomas cross these holes only if they find an extracellular matrix such as laminin or vitronectin on the other side. Moreover, when EGF is added, the speed with which cells migrate to the other side increases. Similarly, when placed in a fluid chamber that contains a gradient of bradykinin,

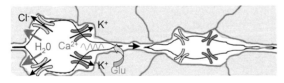

FIGURE 12 Cell volume changes facilitate glioma invasion through narrow brain spaces. Ion channels serve to secrete salt (potassium chloride), which moves cytoplasmic water across the membrane. The process is controlled by oscillatory changes in intracellular Ca^{2+} resulting from binding of bradykinin or glutamate (Glu).

a neuropeptide often associated with blood vessels, glioma cells migrate toward the bradykinin source. Bradykinin may be an important signal to seek out blood vessels in the brain since blockade of the receptor for bradykinin disables gliomas from finding blood vessels in experimental models. Other guidance molecules include the chemokines stromal-derived factor 1a and CXCL-12, the blood-borne phospholipid lysophosphatidic acid, and the neurotransmitter glutamate (Glu).

4.6 Glioma Vascular Interactions/ Angiogenesis

Gliomas have a complex relationship with blood vessels that changes and evolves as the disease progresses. Many tumors form on blood vessels, often occupying branch points. As they spread into the brain, gliomas often stay associated with the blood vessel and move along the vascular tree. This is a property shared with some of the stem cells from which they may derive. It is therefore possible that tumors found on blood vessels developed from vessel-associated stem cells and never leave the vasculature. On the other hand, cells forming in the parenchyma or in white matter sooner or later seek out blood vessels as they begin to grow satellite tumors. It is easy to envision why being close to a blood vessel would be advantageous. Gliomas, like other cancers, rely on a constant supply of nutrients, most importantly glucose, from the blood stream. Cell proliferation requires constant protein and lipid biosynthesis, which poses a significant energy demand on the cell and consequently enhanced ATP production. Like their nonmalignant astrocytic counterparts, and most other cancers, gliomas are wasteful when it comes to generating ATP from glucose. Rather than using oxidative metabolism, which generates up to 36 mol ATP/mol glucose, gliomas generate ATP through glycolysis, yielding only 2 mol ATP/mol glucose. They perform glycolysis even in the presence of oxygen, as first described in the 1950s by the chemist Otto Warburg.[19] The proximity to blood vessels ensures that the glucose requirements of a growing tumor are met. Not surprisingly, even early in disease many gliomas are found in close proximity, literally ensheeting blood vessels (Figure 13(A)). On histopathological sections these cells appear like cuffs (Figure 13(C)), a stage that is often called vascular co-option.[20] It is believed at this stage of disease, glioma cells associated with the blood vessel are degrading the basement membrane through

the release of MMPs. Subsequently, under the influence of angiopoietin, the tumor releases vascular endothelial growth factor (VEGF), which induces vascular cell proliferation and the spouting of additional vessel branches via a process called angiogenesis (Figure 13(B) and (D)). As the tumor continues to grow, this process also continues, thereby creating a complex and often bizarre vasculature frequently characterized by grossly enlarged vessels with abnormal perfusion. Under these conditions, VEGF release is stimulated by tissue hypoxia, which is sensed by hypoxia-inducible

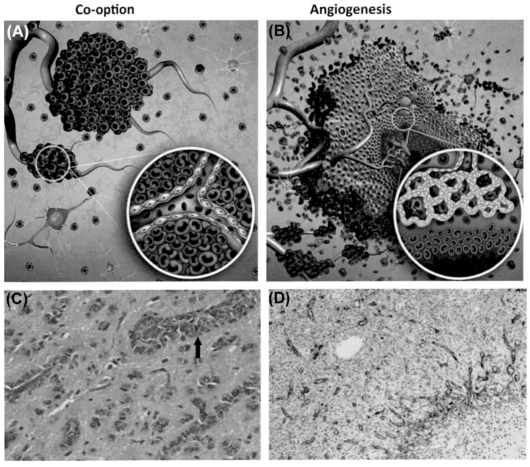

FIGURE 13   Vessel co-option (A) followed by angiogenesis (B,C). Cells cuff around an existing vessel. Sprouting of new vessels is visualized by tenascin staining (D). *From Ref.* [20].

factor-1a, an oxygen-sensitive transcription factor that induces transcription of the *VEGF* gene under low oxygen conditions. Newly formed blood vessels typically lack tight junctions that seal endothelial cells, thereby creating the blood–brain barrier. As a result, tumor-induced vessels are often leaky, possibly a desirable property for the tumor because it provides unimpeded access to blood glucose and nutrients. To what extent breaches in the blood–brain barrier occur and whether they are restricted to the tumor mass is currently unknown. Dogma holds that many chemotherapeutic drugs fail in part because they only reach the main tumor mass, where leaky blood vessels are common, but fail to reach distant cells that are actively invading in association with intact host blood vessels. However, this notion is being challenged by recent studies showing that the invading gliomas insert themselves between the basement membrane of blood vessels and the astrocytic endfeet attached to them. These endfeet are believed to be important stabilizers of the endothelial tight junctions that form the blood–brain barrier (see Chapter 2, Box 1). As a result, focal breaches in the blood–brain barrier likely provide the opportunity for targeted therapy.

In 1971 Judah Folkman hypothesized that angiogenesis is essential to tumor growth, a fact we take for granted today, and suggested that cancerous growth may be halted by depriving tumors of new vessels. As a result, a number of antiangiogenic therapies that attempt to bind tissue-released VEGF, block its receptor on endothelial cells, or the inhibit the downstream signaling cascade that leads to endothelial proliferation have been developed. The most well known drug is bevacizumab (Avastin), a humanized monoclonal antibody that is designed to bind VEGF released from the tumor. It was approved for experimental use in gliomas in 2009. Unfortunately, it has not been as effective as initially expected, as will be further discussed in Section 5.3.

4.7 The Many Roles of Glu in Glioma Biology

Glu excitotoxicity: Growing in the cranium, gliomas are the only cancer that is physically constrained in their growth by the bony cavity provided by the skull.[21] This space is densely packed with brain cells, leaving only the small, fluid-filled extracellular space between cells and the fluid-filled ventricles. Together, they account for only one-tenth of the total brain volume. As tumors form, they must have mechanisms to create room. We intuitively assume that the growth simply pushes adjacent brain aside. This does occur to some extent; the displacement of brain tissue into the ventricular space can be seen by a midline shift in MRI images (see, for example, Figure 3, left). However, some gliomas grow bigger than the available space, and some do not compromise the ventricles at all. Instead, the growing tumor gradually kills surrounding brain cells as it grows. One molecule that has been suggested to aid in this is the excitatory neurotransmitter Glu. As discussed more extensively in Chapter 1, Glu can inflict excitotoxic death on neurons and oligodendrocytes. The brain is organized to prevent Glu from diffusing away from its primary site of action, the synapses. Therefore astrocytes closely encase synapses and express transporters that remove Glu from the extracellular space and transport it into their cytoplasm. Here Glu is deaminated by glutamine synthetase to form glutamine, which then serves as a substrate for the neuronal synthesis of Glu or γ-aminobutyric acid. Measurement of Glu in patients with gliomas suggests that this process is disrupted near the tumor, where Glu can reach concentrations over 100-fold higher than normal.[22] Such concentrations are toxic and readily explain why neurons in the vicinity would be subject to excitotoxicity. Studies of glioma in animal models suggest that most of the Glu is actively synthesized by gliomas and released through an abundantly expressed transport system (system X^c). This transporter imports cysteine into the glioma cell for synthesis of the cellular antioxidant glutathione and couples it to the Glu being released. Metabolically active cells, particularly tumor cells, express elevated concentrations of glutathione to reduce the many metabolites formed as a byproduct of cellular growth. The Glu consequently released from the tumor is a byproduct of the redox biology of the glioma, yet has catastrophic consequences for the surrounding brain (Figure 14). Fortunately, a number of drugs that interfere with the activity of the system X^c transporter have been discovered. One of these, sulfasalazine, is already approved for different disease indications and, after promising preclinical results in tumor-bearing mice, is now being explored in clinical trials for patients with glioma.

Another strategy is to limit Glu accumulation by enhancing its removal. The observed accumulation of Glu near the tumor suggests that the astrocytes surrounding the tumor either fail to do their job of mopping up excess Glu or are simply overwhelmed by the amount of Glu present. The expression of the astrocytic Glu transporter EAAT2 on the membrane can be increased by the administration of β-lactam antibiotics such as ceftriaxone, which act as

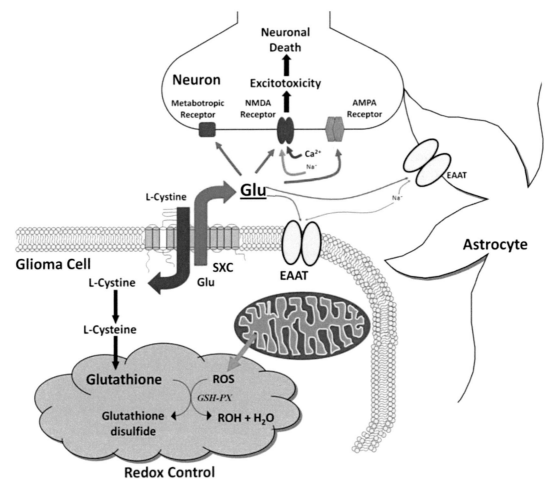

FIGURE 14   Schematized role of glutamate (Glu) release from gliomas via the system X^c (SXC) transporter. *From Ref.* [23].

transcriptional activators of EAAT2, thereby enhancing Glu uptake. This approach was pursued with a similar rationale in a phase I/II clinical trial for amyotrophic lateral sclerosis (ALS; discussed in Chapter 6) but since it has shown promise in animal models of glioma[24] should be considered for patients with glioma as well.

Glu and peritumoral seizures: In Chapter 3 we learned that seizures result from an imbalance in GABAergic inhibition and glutamatergic excitation. In light of the assiduous release of Glu from gliomas, it is not surprising that patients with gliomas frequently suffer an excitation–inhibition balance that may explain their seizures. Indeed, about 80% of patients with glioma experience at least one seizure during their illness, and up to 50% develop recurrent seizures or tumor-associated epilepsy. Gliomas belong to the group of acquired epilepsies and account for about 4% of all cases of epilepsy. In many patients surgical resection, which reduces the main source of Glu release, leads to near complete seizure control. However, one-third of patients continue to have seizures that in many instances do not respond to traditional antiepileptic drugs. Although the cause of seizures in patients with glioma is multifactorial, Glu is a likely contributor. In light of our understanding of Glu release from gliomas, and the role that it plays in redox regulation and excitotoxicity, one can consider seizures a biological signature of a growing tumor and possible evidence of Glu release. Indeed, in 2012 the American Epilepsy Association hosted an annual course on tumor-associated epilepsy that concluded with the suggestion to treat this condition as a separate disease entity as opposed to a mere comorbidity of the cancerous growth. This will hopefully stir further research to explain the failure of seizure control in many patients. Once a conclusive link between tumor-associated seizures and Glu release is established, novel treatment approaches can be developed. One open clinical pilot study that specifically aims at reducing glutamate release to treat patients with tumor-associated epilepsy is mentioned later in Section 6.1.

Glu signaling promotes glioma invasion: During cerebellar development granule neurons proliferate in the external granule layer where, upon their final division, they migrate as bipolar cells across the molecular layer to the inner granule layer. Their migration is associated with cyclical, oscillatory changes in intracellular Ca^{2+}, which are due to Ca^{2+} influx via NMDA-type Glu receptors and voltage-gated Ca^{2+} channels. The amplitude and rate of Ca^{2+} fluctuations control the velocity of cell movement.[25] Similar Ca^{2+} oscillations occur in migratory glioma cells, where they are also mediated by Glu receptors. Here it is the Ca^{2+}-permeant AMPA Glu receptor that mediates Ca^{2+} entry. If pharmacologically inhibited or if the Ca^{2+} permeability is genetically removed, the resulting tumors are unable to invade mouse brain.[26] Glu released by one migratory glioma cell can act as a paracrine signal to encourage movement of adjacent cells, causing chain migration. Such paracrine Glu signaling has been shown for isolated glioma cells, and disruption of Glu release has reduced glioma invasion in mice. These studies all converge on changes in intracellular Ca^{2+} as the major signal. Ca^{2+}, of course, mediates the dynamic polymerization of tubulin as well as actin–myosin interactions, which constitute the molecular motors of a cell. In addition, Ca^{2+} entering gliomas through an AMPA receptor activates focal adhesion kinase and promotes cellular detachment from its substrate. Cyclical changes in Ca^{2+} therefore cause cyclical changes in cell attachment and detachment, which work in tandem with the contractile forces of the cell to cause directed movement.

Glu as growth factor? The signaling role of Glu extends further to the regulation of cell proliferation. Specifically, removing Glu from the culture medium or blocking Ca^{2+} entry via Glu receptors arrests glioma cell proliferation. Ca^{2+} entry via these receptors activates the AKT/phosphoinositide 3-kinase signaling pathway, which is normally activated by EGF binding to the EGF or PDGF receptors. A second pathway activated is the extracellular signal-related kinase/mitogen activated protein kinase pathway, which enhances proliferation via c-myc, mnk, and creb, known transcription factors that regulate cell cycle-associated proteins such as p53 and cyclin-dependent kinases. Hence Glu may substitute as a growth factor by stimulating the same signaling pathway that is normally engaged in growth regulation by growth factors, essentially bypassing a requirement for these factors to be present. Since Glu may stem from the same or neighboring glioma cells, it establishes an autocrine or paracrine feedback loop that stimulates tumor growth in the complete absence of traditional growth factors.

4.8 Tumor Stromal Interactions

Much of the research to date has focused on the intrinsic biology of gliomas, largely ignoring other brain cells (stromal cells) that provide the immediate environment in which they develop. Yet the above examples allude to glioma cells interacting with extracellular matrix, which they remodel with co-opted blood vessels. One often neglected interaction is with the resident immune cells of the brain. Microglial cells are constantly surveying the brain, seeking out areas of injury and inflammation, and they play a major role in the resolution of an acute insult.[27] As recognized by Penfield in 1925, gliomas contain a large number of microglial cells, and it is now estimated that up to 30% of the tumor mass may be microglial.[28] These are attracted by molecules such as monocyte-chemotactic protein-1 released by the tumor. Microglial cells typically become activated during disease, yet gliomas suppress microglial activation by releasing cytokines such as transforming growth factor-β. They further coerce microglia into releasing matrix degrading enzymes, thereby helping gliomas

degrade the matrix as they invade.[28] Gliomas also recruit neural progenitor cells (NPCs) from the SVZ. These recognize stromal-derived factor 1 via their CXCR4 receptors. Interestingly, NPCs suppress glioma proliferation through the release of bone morphogenic protein 7, and it has been suggested that the developmental decrease in NPCs may explain the typical late onset of gliomas, a time when NPCs have almost disappeared from brain. The finding that both microglia and NPCs are attracted to gliomas may harbor some untapped therapeutic potential. The latter may be used to directly suppress growth, whereas the former may be used to shuttle cytotoxic molecules to the tumor.

4.9 Glioma: A Primary Neurodegenerative Disease?

Gliomas expand in the brain by progressively killing neurons that normally occupy the brain. Progressive neuronal cell loss is the defining hallmark of neurodegenerative diseases. The poster child for this group of diseases is Alzheimer disease, yet ALS, Parkinson disease, and Huntington disease are other examples. What neurodegenerative diseases share in common is that they originate in the brain, typically due to genetic mutations. These have been identified in familial forms of Alzheimer disease and ALS, and, even if currently unknown, all neurodegenerative diseases are expected to have a genetic basis. As discussed above, gliomas also form only as a result of mutations in various genes involved in growth control and differentiation. The final death pathway for most neurons undergoing a neurodegenerative process varies, but it frequently involves Glu, which often causes seizures and becomes a toxic agent. Gliomas avidly release Glu to cause seizures and inflict Glu toxicity on adjacent neurons. Neurodegenerative diseases are characterized by a reactive glial response, microglial recruitment, and changes in the brain vasculature, all of which are prominently displayed in gliomas. Hence the similarities with other neurodegenerative diseases are overwhelming and may justify categorizing glioma among this group of diseases as opposed to merely a cancer growing in the brain. This distinction may be more than semantics; it suggests that future understanding of disease pathology and treatment should be inspired by neuroscience rather than oncology, as is currently the case.

5. TREATMENT/STANDARD OF CARE/CLINICAL MANAGEMENT

Currently available treatments can be divided into definitive and symptomatic. Definitive treatments are those targeting the tumor directly and include radiotherapy, chemotherapy, and surgery. Symptomatic treatments are aimed at reducing a tumor's effect on the surrounding brain and include strategies such as administration of steroids to reduce swelling or edema.

5.1 Surgery

Surgery is curative for the vast majority of meningiomas. These tumors typically do not invade and grow superficially, allowing the surgeon to cleanly resect them. By contrast, most astrocytomas, oligodendrogliomas, and glioblastomas grow deep within the brain parenchyma, and thus complete surgical resection is rarely feasible. However, aggressive maximal resection of tumor tissue is significantly improving outcomes. High-resolution imaging, even intraoperative imaging, now allows the surgeon to resect a larger portion of the tumor without accidentally removing normal tissue. Intraoperative stimulation is used extensively to protect eloquent cortical structures (Figure 15). Retrospective analyses of the surgical outcome suggest that it almost doubles life expectancy.[30] In addition, removal of the tumor typically relieves the patient of some tumor-associated symptoms such as seizures, headaches, or change in personality. Because of their invasiveness, most tumors eventually recur at or near the site of resection. A second debulking surgery is performed primarily to relieve acute symptoms or if the tumor impinges on vital brain structures.

5.2 Radiation

Ionizing radiation continues to be an effective weapon against most cancers, and this holds true for gliomas as well. The typical treatment involves 60 Gy of radiation, administered as daily fractions of 1.8–2.0 Gy. Responsiveness varies from patient to patient, although an initial phase of remission is typical. Treatment is often followed by a variable period of stable disease followed by regrowth. Given that gliomas invade beyond the visible tumor margin, whole-brain radiation is more effective than focal radiation. Nevertheless, in up to 90% of patients, the tumor recurs within 2 cm of the original tumor margin. The benefit of radiation treatment varies considerably from patient to patient and by tumor type. For reasons that are not well understood, oligodendrogliomas and mixed gliomas are much more responsive to radiation than most astrocytomas and glioblastomas. Radiation treatment of gliomas increases average life expectancy by several weeks and radiation remains the most effective nonsurgical treatment for glioma. In many patients radiation treatment is administered in conjunction with chemotherapy, with the hope that the latter weakens the tumor and sensitizes it to radiation. This is certainly supported by clinical trials that suggest the success of radiation plus temozolamide, for example, is significantly, albeit not dramatically, better (14.6 vs 12.1 months) than radiation treatment alone.[31] Radiation carries some risk of acute toxicity, most notably the potential to disrupt the blood–brain barrier, resulting in edema and elevated intracranial pressure. As more delayed responses, demyelination and radiation-induced tissue death are not uncommon. Radiation exposure can decrease mental function, although separating side effects of treatment from the deleterious effects of the tumor is difficult.

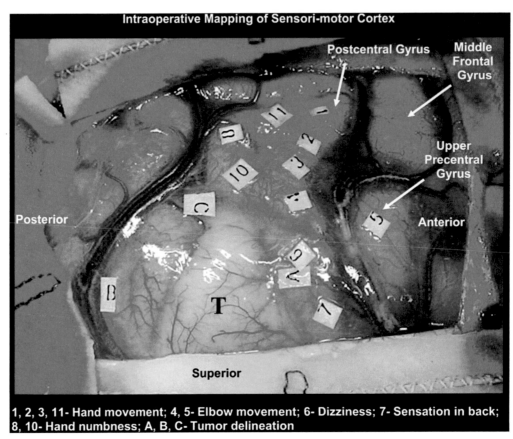

1, 2, 3, 11- Hand movement; 4, 5- Elbow movement; 6- Dizziness; 7- Sensation in back;
8, 10- Hand numbness; A, B, C- Tumor delineation

FIGURE 15   Intraoperative brain mapping identifies the eloquent cortex in a patient with a glioma. Stimulation at these electrodes resulted in the following responses: 1, 2, 3, 11: Hand movement; 4, 5: elbow movement; 6: dizziness; 7: sensation in back; 8, 10: hand numbness; A, B, C: tumor delineation. *From Ref.* [29].

5.3 Chemotherapy

For decades, glioma therapy has involved the same classes of drugs that are used to treat systemic cancers and target DNA. These are the alkylating agents cisplatin, carmustin, lomustin, and procarbazine or the antimicrotubule agents vincristine and taxol. There are additional drugs that target proteins involved in DNA repair, such as the telomerase inhibitor etopiside. To enhance chemotherapeutic success, drugs are frequently administered as a cocktail of procarbazine, lomustine (CCNU), and vincristine sulfate, a regimen designated PCV. By and large these are nonspecific for dividing cells and explain the side effects experienced by most patients, such as acute hair loss, nausea, and peripheral neuropathy. Although these chemotherapeutics stifle tumor growth in up to 75% of oligodendrogliomas, they are not very effective in treating astrocytic gliomas. Effective delivery across the blood–brain barrier and the inability to reach small, invasive cell populations remains a major challenge.

Currently, the best available chemotherapeutic for glioblastomas and astrocytomas is temazolamide (Temodar), an alkylating agent also

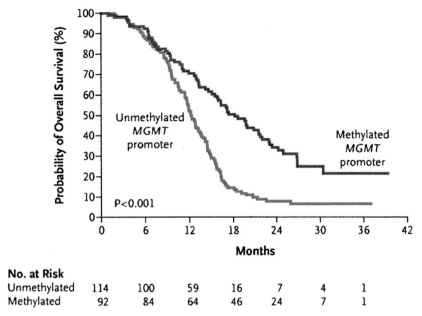

FIGURE 16 Combined chemotherapy for glioma using radiation and temozolamide. Patients with a methylated methylguanine methyl transferase (MGMT) promoter and therefore silenced *MGMT* gene have a significantly enhanced survival benefit. *From Ref. [32].*

used for melanoma. After initial studies showed an average increase in survival by just 8 weeks when added to conventional radiation therapy, further analysis suggested that a subgroup of patients benefitted significantly from this treatment. These are patients who have an inactive DNA repair enzyme, which is silenced by methylation of its promoter. Since DNA replication is such an important process in cell division that requires absolutely precise copies to be made, cells contain enzymes that can repair base pair changes. One of these enzymes is a methylguanine methyl transferase (MGMT). In some patients the promoter for the gene encoding the *MGMT* gene is turned off by methylation; as a result the cells are deprived of this important repair mechanism. Hence in patients who have a methylated or inactive *MGMT* gene, temozolamide-induced DNA damage is much more efficacious than in patients in whom the activity of this chemotherapeutic is essentially neutralized by the MGMT enzyme.[32] On average, patients achieved a 6-month survival benefit; 20% were short-term survivors (Figure 16). This finding is quite exciting and is an example of the power of personalized medicine, which allows the most efficacious treatment for a patient to be selected based on the genetic changes present.

Another promising class of novel chemotherapeutic drugs targets the tumor vasculature. As discussed above, tumors tend to induce new blood vessels to satisfy their energy needs. This involves the release of VEGF. Neutralizing humanized monoclonal antibodies such as bevacizumab (Avastin) can bind and thereby neutralize the released VEGF, and this approach has shown promise in many cancers. Unfortunately, following initial successful early stage clinical trials where tumors showed a radiologic response, the drug seems to have negligible effects on overall patient survival. While these drugs seem to hit their target and reduce angiogenesis, patient survival is not improving as expected, in large part because tumor cells that were prevented from forming solid masses now disperse into the brain.

5.4 Palliative Therapy

Depending on the location of the tumor, gliomas can cause a number of sequelae and comorbidities including seizures, depression, and cognitive decline. Some of the neurological symptoms are caused by edema. Corticosteroids such as dexamethasone are highly effective in reducing peritumoral edema and can rapidly improve neurological function. However, long-term use of steroids may cause insomnia, weight gain, and personality changes. Cognitive decline and fatigue may be treated with psychostimulants such as methylphenidate. Among the most important palliative treatment is the suppression of seizures. These are common among patients with glioma, particularly those with low-grade gliomas and oligodendrogliomas. In about two-thirds of patients, seizures can be well controlled with valproate (Depakote), topiramate, or levetiracetam (Keppra), but many patients are unresponsive to these treatments. For these, tumor-associated epilepsy becomes a disease of its own.

6. EXPERIMENTAL APPROACHES/ CLINICAL TRIALS

In light of the tremendous unmet needs of these patients there are countless clinical trials being carried out. Unfortunately, most of them focus on combination therapies using drugs with limited individual efficacy. Only a few active trials are truly inspired by new knowledge gained from basic research. These are mostly in early stages, where they are primarily studying feasibility and safety as opposed to efficacy. Hopefully this will change in the coming years. A few examples of trials that are inspired by basic neuroscience are briefly mentioned below.

6.1 Glutamate Release

Both tumor expansion and tumor-associated epilepsy are believed to result from excessive Glu release from tumors. Blocking one of the release pathways, namely, the system X^c cysteine/Glu transporter or vesicular release from neurons, may reduce release. Sulfasalazine (Azulfidine), a drug commonly used to treat Crohn's disease, is a nontransportable substrate analog for the transporter and blocks Glu release. In preclinical studies it reduced the frequency and severity of tumor-associated seizures and significantly reduced tumor growth. Given the availability and known safety of this drug, sulfasalazine is being used in early clinical trials for patients with glioma who present with seizures.

6.2 Ceftriaxone

Another way to contain Glu is to increase the rate of removal. This is the principal job of astrocytes, which import Glu through the EAAT2 transporter. EAAT2 is downregulated in glioma tissue, and the peritumoral buildup of Glu is in part due to ineffective Glu removal. The transcription of the *EAAT2* gene can be enhanced by a group of β-lactam antibiotics such as ceftriaxone, commonly used to treat pneumonia and bacterial meningitis. Ceftriaxone has been examined with a similar rational in patients with ALS in a phase III study, and it is now being explored for the treatment of glioma.

6.3 Levetiracetam (Keppra)

Levetiracetam (Keppra) is a relatively new antiepileptic drug that binds to SV2A, a glycoprotein associated with synaptic vesicles that contain Glu. Levetiracetam reduces neuronal excitability by reducing synaptic Glu release from peritumoral neurons. A current phase IV study is specifically seeking to show that patients who have elevated expression of SV2A near the tumor receive a selective benefit from this drug.

6.4 Cilengitide

Integrins are important membrane proteins that mediate cell–cell and cell–matrix interactions and are involved in multiple aspects of glioma biology, ranging from cell invasion to angiogenesis. Integrins are a superfamily of receptors that generally bind to RGD-containing (Arg-Gly-Asp) motives on extracellular matrix, such as fibronectin, laminin, and others. Clinegitide is a synthetic RGD pentapeptide that has a 100-fold higher affinity to some integrins than their natural substrates. As a result, the drug masks and thereby interferes with pathophysiological integrin signaling, particularly in the context of angiogenesis. It showed some success in phase II clinical trials as a single agent and is currently being evaluated as combinatorial treatment.[33]

6.5 Chlorotoxin

Ion channels play a supportive role in glioma invasion. K^+ and Cl^- ions permeate the cell membrane in a highly coordinated way and serve as osmolytes to regulate the shape and volume of invading cells. Preclinical studies identified a putative Cl^- channel-blocking peptide that specifically binds to gliomas of all malignancy grades and promises to be an effective anti-invasive drug. A synthetic version of this peptide, called chlorotoxin, has been explored in phase I and II clinical studies, where it showed target specificity and therapeutic efficacy. Figure 17 shows tumor-specific binding of the peptide in a patient with glioma and a second patient with transient tumor remission. Given that this drug was administered to patients with recurrent glioma who had failed all other options, the antiinvasive properties of the drug could not be tested. This will, however, be tested

in an upcoming phase III clinical trial in which the drug will be given to newly diagnosed patients as a first-line chemotherapeutic.

7. CHALLENGES AND OPPORTUNITIES

The diagnosis of a brain tumor is scary. Next to pancreatic cancer, gliomas present the bleakest outlook for patients, with an average survival of roughly 1–2 years, depending on tumor grade. The most frightening aspect is how the tumor gradually robs the patient of his or her brain function and has the potential to profoundly alter a person's behavior and personality. No other cancer does that. In spite of significant advances in both the understanding of the disease and surgical techniques, the disease's incidence, prevalence, and prognosis have all remained significantly unchanged over the past 50 years. Is it time to throw in the towel and give up? It may surprise the reader that an anonymous survey that I conducted several years ago among leading neurosurgeons and neurologists pointed in that direction. When asked what course of action they would choose if they were themselves diagnosed with a high-grade brain tumor, 9 of 10 answered that they would choose no treatment whatsoever.

This attitude of helplessness leads to complacency. This is comparable to what we experienced for decades with spinal cord injury among the rehabilitation medicine community. Not until Christopher Reeve challenged the clinical and scientific community to believe that regeneration might be possible did the overall attitude become much more positive. I, too, would take a much more optimistic stand regarding gliomas. Progress cannot always be judged simply by how many years of life have been given to patients suffering from a disease. Surgical techniques have improved substantially, particularly regarding the use of high-resolution imaging to aid surgery

Anterior Posterior

3 month post surgery

12 months, 6 doses of Cltx

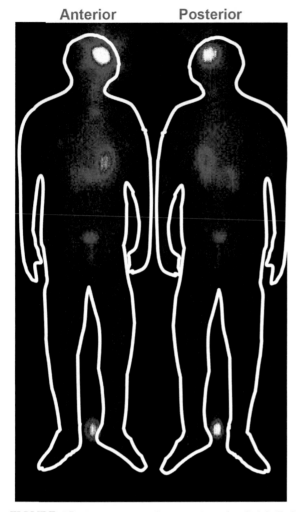
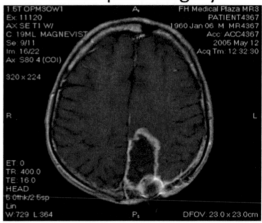
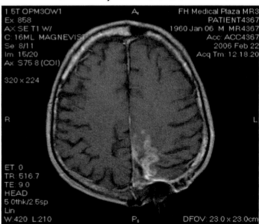

FIGURE 17 Intracavitary administration of radiolabelled chlorotoxin (Cltx) localizes to the tumor, as visualized in a whole-body radiation scan (left). Contrast-enhanced magnetic resonance imaging (right) before and after six doses of Cltx shows significant radiological response with evidence of tumor shrinkage. *Images courtesy of Dr John Fiveash, University of Alabama Birmingham.*

for maximal resection with minimal destruction. From a scientific point of view, we are gradually inching closer to a better understanding of where these tumors come from and how they grow and invade, and we have elucidated many of the signaling molecules that they rely on. It is just a matter of time until all these advances will bear fruit in the form of novel treatments. I suggest that the outlook for glioma is not much different than for other rapidly progressing neurodegenerative diseases, for example, ALS. Indeed, as repeatedly suggested in this chapter, it is time to consider brain tumors a true neurodegenerative disorder as opposed to simply a cancer growing in the brain. This is not just semantics but rather an alteration in the scientific and clinical approach to this disease.

Acknowledgments

This chapter was kindly reviewed by Dr Louis Burt Nabors, Chief of Neurooncology, The University of Alabama Birmingham.

References

1. Restak RM. *Mysteries of the Mind*. Washington, DC: National Geographic Society; 2000.
2. Scherer HJ. A critical review: the pathology of cerebral gliomas. *J Neurol Psychiatry*. April 1940;3(2):147–177.
3. Bulsara KR, Johnson J, Villavicencio AT. Improvements in brain tumor surgery: the modern history of awake craniotomies. *Neurosurg Focus*. April 1, 2005;18(4):1–3.
4. Voorhees JR, Cohen-Gadol AA, Spencer DD. Early evolution of neurological surgery: conquering increased intracranial pressure, infection, and blood loss. *Neurosurg Focus*. April 1, 2005;18(4):1–5.
5. Pekmezci M, Perry A. Neuropathology of brain metastases. *Surg Neurol Int*. 2013;4(Suppl. 4):S245–S255.
6. Fulling KH, Garcia DM. Anaplastic astrocytoma of the adult cerebrum. prognostic value of histologic features. *Cancer*. 1985;55(5):928–931.
7. Barnholtz-Sloan AS JS, Schwartz AG. Cancer of the brain. In: Ries LAC, Young JL, Keel CE, Eisner MP, Lin YD, Horner M-J, eds. *SEER Survival Monograph: Cancer Survival Among Adults: U.S. SEER Program, 1988-2001, Patient and Tumor Characteristics, J Natl Cancer Inst*. NIH Pub. No. 07-6215, Bethesda, MD, 2007:203–215.
8. Sanai N, Alvarez-Buylla A, Berger MS. Neural stem cells and the origin of gliomas. *N Engl J Med*. August 25, 2005;353(8):811–822.
9. Doetsch F, Garcia-Verdugo JM, Alvarez-Buylla A. Cellular composition and three-dimensional organization of the subventricular germinal zone in the adult mammalian brain. *J Neurosci*. July 1, 1997;17(13):5046–5061.
10. Dai C, Celestino JC, Okada Y, Louis DN, Fuller GN, Holland EC. PDGF autocrine stimulation dedifferentiates cultured astrocytes and induces oligodendrogliomas and oligoastrocytomas from neural progenitors and astrocytes in vivo. *Genes Dev*. 2001;15(15):1913–1925.
11. Dai C, Holland EC. Glioma models. *Biochim Biophys Acta*. 2001;1551(1):M19–M27
12. Stiles CD, Rowitch DH. Glioma stem cells: a midterm exam. *Neuron*. 2008;58(6):832–846.
13. Chen J, McKay RM, Parada LF. Malignant glioma: lessons from genomics, mouse models, and stem cells. *Cell*. March 30, 2012;149(1):36–47.
14. Zong H, Verhaak RG, Canoll P. The cellular origin for malignant glioma and prospects for clinical advancements. *Expert Rev Mol Diagn*. 2012;12(4):383–394.
15. Bao S, Wu Q, McLendon RE, et al. Glioma stem cells promote radioresistance by preferential activation of the DNA damage response. *Nature*. 2006;444(7120):756–760.
16. Verhaak RG, Hoadley KA, Purdom E, et al. Integrated genomic analysis identifies clinically relevant subtypes of glioblastoma characterized by abnormalities in PDGFRA, IDH1, EGFR, and NF1. *Cancer Cell*. 2010;17(1):98–110.
17. Frank SA, Nowak MA. Problems of somatic mutation and cancer. *Bioessays*. March 2004;26(3):291–299.
18. Psachoulia K, Jamen F, Young KM, Richardson WD. Cell cycle dynamics of NG2 cells in the postnatal and ageing brain. *Neuron Glia Biol*. November 2009;5(3–4):57–67.
19. Warburg O. On respiratory impairment in cancer cells. *Science*. 1956;124(3215):269–270.
20. Hardee ME, Zagzag D. Mechanisms of glioma-associated neovascularization. *Am J Pathol*. 2012;181(4):1126–1141.
21. de Groot J, Sontheimer H. Glutamate and the biology of gliomas. *Glia*. 2010;59:1181–1189.
22. Marcus HJ, Carpenter KL, Price SJ, Hutchinson PJ. In vivo assessment of high-grade glioma biochemistry using microdialysis: a study of energy-related molecules, growth factors and cytokines. *J Neurooncol*. 2010;97(1):11–23.
23. Watkins S, Sontheimer H. Unique biology of gliomas: challenges and opportunities. *Trends Neurosci*. 2012;35(9):546–556.
24. Sattler R, Tyler B, Hoover B, et al. Increased expression of glutamate transporter GLT-1 in peritumoral tissue associated with prolonged survival and decreases in tumor growth in a rat model of experimental malignant glioma. *J Neurosurg*. 2013;119(4):878–886.
25. Komuro H, Rakic P. Orchestration of neuronal migration by activity of ion channels, neurotransmitter receptors, and intracellular Ca^{2+} fluctuations. *J Neurobiol*. 1998;37(1):110–130.
26. Ishiuchi S, Tsuzuki K, Yoshida Y, et al. Blockage of Ca(2+)-permeable AMPA receptors suppresses migration and induces apoptosis in human glioblastoma cells. *Nat Med*. 2002;8(9):971–978.
27. Streit WJ, Walter SA, Pennell NA. Reactive microgliosis. *Prog Neurobiol*. 1999;57(6):563–581.
28. Charles NA, Holland EC, Gilbertson R, Glass R, Kettenmann H. The brain tumor microenvironment. *Glia*. March 2012;60(3):502–514.
29. Salvan CV, Ulmer JL, Mueller WM, Krouwer HG, Prost RW, Stroe GO. Presurgical and intraoperative mapping of the motor system in congenital truncation of the precentral gyrus. *Am J Neuroradiol*. March 2006;27(3):493–497.

30. Lacroix M, Abi-Said D, Fourney DR, et al. A multivariate analysis of 416 patients with glioblastoma multiforme: prognosis, extent of resection, and survival. *J Neurosurg.* 2001;95(2):190–198.
31. Stupp R, Hegi ME, Mason WP, et al. Effects of radiotherapy with concomitant and adjuvant temozolomide versus radiotherapy alone on survival in glioblastoma in a randomised phase III study: 5-year analysis of the EORTC-NCIC trial. *Lancet Oncol.* May 2009;10(5):459–466.
32. Hegi ME, Diserens AC, Gorlia T, et al. MGMT gene silencing and benefit from temozolomide in glioblastoma. *N Engl J Med.* 2005;352(10):997–1003.
33. Mas-Moruno C, Rechenmacher F, Kessler H. Cilengitide: the first anti-angiogenic small molecule drug candidate design, synthesis and clinical evaluation. *Anticancer Agents Med Chem.* December 2010;10(10):753–768.

General Readings Used as Source

1. Harrison's Online, Chapter 379 by Lisa M. DeAngelis and Patrick Y. Wen.
2. Pathology & Genetics. *Tumor of the Nervous System.* Lyon: Kleihues & Cavenee, International Agency for Research on Cancer; 1997.

Suggested Papers or Journal Club Assignments

Clinical Paper

1. Hegi, et al. MGMT gene silencing and benefit from temozolomide in glioblastoma. *New Engl J Med.* 2005;352:997.
2. Lai, et al. Phase II study of bevacizumab plus temozolomide during and after radiation therapy for patients with newly diagnosed glioblastoma multiforme. *J Clin Oncol.* 2011;29:142–148.

Basic

1. Bao, et al. Glioma stem cells promote radioresistance by preferential activation of the DNA damage response. *Nature.* 2006;444:756–760.
2. Lahtia, et al. Direct in vivo evidence for tumor propagation by glioblastoma cancer stem cells. *PLOS One.* 2011;6.
3. Stock, et al. Neural precursor cells induce cell death of high-grade astrocytomas through stimulation of TRPV1. *Nature Med.* 2012;18:1232–1238.

Infectious Diseases of the Nervous System

Harald Sontheimer

1. CASE STORY

Steve doesn't remember when and where it all started, but it must have been at least 5 years ago. While on a skiing vacation in British Columbia, Steve suffered a minor seizure. It was not terrifying, but being barely 30 years old, he was caught by surprise nevertheless. His skiing friends blamed it on exhaustion and dehydration and, since he did not have health insurance at the time, he did not see a doctor right away. A few months later, back in Chicago, Steve started to have abdominal pains. At first they were just a nuisance and Pepto-Bismol usually did the trick. However, over the following 2 weeks the pains worsened. Then Steve started to wake up in the middle of the night with nightmares, soaking wet, with a racing heart. He paced the bedroom but could not settle down and go back to sleep. He canceled a presentation he was supposed to give one morning, feeling too tired and irritated to leave the house. Feeling better the following day, Steve rescheduled the meeting, but the racing heart returned and kept him up another night. Steve had no idea what time it was but he called his friend Austin, who was a medical resident. Steve explained everything that was happening to him. Austin frightened him when he suggested that Steve should see a psychiatrist and actually scheduled a consultation for him the following day. Steve was irritated and angry, to say the least, and probably did not cooperate well with the doctor, giving rude answers throughout the exam that lasted several hours. In the end he was given the antidepressant Prozac, which would help him with what the doctor labelled a "professional crisis." While the medicine allowed Steve to sleep through the night, he always felt groggy in the mornings and had a difficult time remembering things or staying on task. On several occasions he made a fool of himself in business meetings by not even recognizing the very slides he had prepared just a few days earlier. Surprisingly, while most people gain weight while taking antidepressants, Steve lost almost 20 lb in a 2-month period and had to buy new clothes. Finally, almost 4 years after his first seizure episode, he lost complete control over his body at a business lunch in downtown Chicago. Steve has no recollection of what happened other than shaking uncontrollably and falling. When I woke up he was in an ambulance. Following a number of procedures, magnetic resonance imaging (MRI) scans and tests, Steve learned that he had a worm living in his body and that its larvae had invaded his brain. It still

freaks him out to even think about it. How can something like this happen in Chicago? Steve had no idea how he may have contracted this parasite. After 6 months of taking praziquantel, which kills these parasites, and daily Dilantin to control his seizures, Steve finally began to function again. He regained weight, and was able to sleep and concentrate. However, Steve's brain scans look like Swiss cheese, littered with holes where he once had brain cells.

2. HISTORY

We must assume that infections of the nervous system have occurred since humans evolved. Therefore our historical understanding is limited to the few cases reported in the literature that allow us to deduce the infectious organism from the symptoms it presents. In some instances this takes us back to ancient Egypt, where steles depict polio-crippled people (Figure 1), suggesting that the poliovirus has most likely been around since the origins of mankind. In other instances, the source of infection may have only recently crossed species, as is the case for human immunodeficiency virus (HIV), which was entirely unknown in humans 50 years ago. The earliest credible reports of **syphilis** point to a massive outbreak during the French invasion of Naples in 1494, before which the disease was seemingly unknown. It is hypothesized that the disease-causing spirochete bacteria were brought back from the Americas with Christopher Columbus's voyages and remained rampant throughout the Middle Ages, being treated with mercury, which we now know to be a potent neurotoxin that was almost more debilitating than the disease itself. Following the introduction of the theory of germs by Louis Pasteur and Robert Koch in the 1860s and the development of methods to contain sepsis by Lister, much of the early twentieth century was devoted to identifying disease-causing pathogens.

Among early successes was the isolation of the spirochete bacterium that causes **syphilis** by Schaudinn and Hoffmann in 1905. Just 5 years

FIGURE 1   An Egyptian stele thought to represent a polio victim, 18th dynasty (1403–1365 BC). Carving shows an Egyptian with a crippled leg, likely the result of poliomyelitis infection.

by muscle paralysis and often death. Kerner was able to isolate the toxin from contaminated food, administer it to animals, and replicate the paralysis seen in humans. The German physician Mueller named the disease after the Latin word for sausage, *botulus*, in 1870. Emile van Ermengem identified the disease-causing bacterium in 1897. To date, sausages are rarely the cause of botulism; instead, the toxin more likely spreads via contaminated salad or herbs.

Infantile paralysis, or **poliomyelitis**, may be among the most defining neurological infections of all time. Not only did it demonstrate how severely and indiscriminately an infection can cripple a healthy child, but the race toward a cure came to define biomedical research in the United States to date. The illness was perceived as such a threat to society that the US government, for the first time, awarded research contracts to universities and hospitals to support targeted research. This practice ultimately led to the birth of the National Institutes of Health, now the largest sponsor of biomedical research in the world. Franklin D. Roosevelt, the 32nd president of the United States, attempted to hide his own illness, being ashamed of any public portrayal of his polio-inflicted weakness. He founded the March of Dimes, a national fundraising campaign through which the entire nation supported the race to find a cure for infantile paralysis in 1938. In the US and much of the western world, polio was ultimately eradicated through an aggressive vaccination campaign in the 1950s and 1960s, with two different vaccines independently developed by Jonas Salk and Hilary Koprovski. Salk used an injectable, dead virus vaccine, whereas Koprovski used a live attenuated virus administered orally. Although Jonas Salk is better known with regard to the development of the polio vaccine in the United States, the oral attenuated virus vaccine was much more widely used in the worldwide eradication of polio. Unfortunately, polio still occurs at an alarming and increasing rate in some developing countries. Although depictions of polio-crippled people date the disease back thousands of years (Figure 1), polio was probably not very common until the

later, Paul Ehrlich, who won the Nobel Prize in Medicine for discovering the blood–brain barrier, succeeded in synthesizing the first organic antisyphilis treatment in 1910. This was an arsenic derivative called Salvarsan, which had significant side effects. It was, however, much more effective than mercury, which had been used for centuries to treat the disease. The mass production of the antibiotic penicillin in the 1940s provided a curative agent for syphilis and many other bacterial infections, yet in spite of this effective and cheap cure, syphilis remains a major health concern in the developing world, with over 12 million people infected to date.

In 1822, Justinus Kerner described 155 cases of "**sausage poisoning**" in Germany,[1] with people suffering severe abdominal cramps followed early twentieth century, since most children who survived childbirth acquired immunity in early life. Increasingly, hygienic living conditions are believed to have contributed to a loss of immunity through early exposure and the eventual epidemic that characterized the Great Depression. The fight of Polio taught society about the power of mass vaccination, a lesson seemingly forgotten by many parents who, today, tragically elect to forgo vaccinating their children for common communicable disease such as measles.

3. CLINICAL PRESENTATION/ DIAGNOSIS/EPIDEMIOLOGY/ DISEASE MECHANISM

The list of pathogens that can invade the nervous system and elicit an adverse response is almost endless. Some, such as the brain-eating amoeba, are infrequent causes of disease, yet others, such as the *Streptococcus pneumoniae* bacterium or the Cocksackie virus, affect thousands of patients each year. Some infections produce only a short, acute illness; others result in chronic, lifelong maladies that may not become fully apparent until long after the infection has occurred. Some pathogens stay confined to the meningeal covering of the brain (meningitis); others inflame the brain tissue itself (encephalitis). Instead of providing a comprehensive review of nervous system infections, which would be impossible, I have elected to highlight some representative examples of nervous system infections to provide a sample of the diversity of the infectious agents and underlying disease processes. These examples, listed in Table 1, include infectious proteins (prions); bacterially produced toxins (botulism); bacterial, viral, and fungal infections (meningitis, tetanus, syphilis, polio, HIV); and invasion by either single (amoeba) or multicellular organisms (tapeworm). Varying from the format of all other chapters in this book, each infection is discussed by including the relevant disease mechanisms, epidemiology, basic science, and treatment together.

3.1 Meningitis

Meningitis is the most common **bacterial** infection of the nervous system, occurring with a relative incidence of >2.5 cases per 100,000 (8000 cases each year in the United States). After the introduction of an efficacious vaccine against *Haemophilus influenzae*, the majority of cases in the United States are now caused by *Streptococcus pneumoniae*, even though the epidemiology of this infection has been altered by an effective vaccine. The second most common bacterial meningitis is caused by *Neisseria meningitides*, also called meningococcus. Its incidence has declined significantly because of routine childhood vaccination and a vaccine requirement for college attendance. The telltale signs of bacterial meningitis include high fever (101–103 °F), sudden-onset headache, and a stiff neck, possibly associated with nausea and followed by seizures and decreased consciousness, even coma. These infections are life-threatening, can cause long-term neurological deficits, and require immediate recognition and treatment.

Bacteria typically enter the brain via the circulating blood, where they are protected from lysis by neutrophils through a polysaccharide capsule. They enter into the cerebrospinal fluid (CSF) via the choroid plexus epithelium, a structure at the top of the third and fourth ventricle involved in the production of the CSF (Box 1, in Chapter 2) that bathes the brain. Once in the CSF, bacteria rapidly colonize in a compartment that is typically free of *bona fide* immune cells. Once associated with the meningeal covering of the brain that delineates the arachnoid space, the lysis of the bacterial wall and release of cell wall components such as lipopolysaccharide triggers a relatively nonspecific inflammatory response from the innate immune system that primarily involves microglial cells and the release of proinflammatory cytokines and chemokines, including tumor necrosis factor (TNF)-α and interleukin (IL)-1β. These factors increase the permeability of vascular vessels, causing vascular edema

TABLE 1   Examples of CNS Infections by Infectious Agent

Infectious Agent	Disease	Pathogen	Treatment	Prevalence (US/Global)	Outlook
Bacterium	Meningitis	*Streptococcus*	Antibiotics	Low/low	Good to excellent with treatment
	Neurosyphilis	*Treponema pallidum*	Antibiotics	Low/medium	
	Lyme disease	*Borellia*	Antibiotics	Low	
	Leprosy	*Mycobacterium*	Antibiotics	0/Medium	
Virus	Meningitis	Coxsackie	Antiviral	Medium/medium	Good
	Poliomyelitis	Poliovirus	None	0/Low	Stable
	Measles	Paramyxovirus	Supportive	Medium	Good
	Herpes	Herpes simplex	Aciclovir	High	Good
	Rabies	Lyssavirus	Vaccine	Low/medium	Fatal
	HIV/AIDS	HIV-1/2	HEART	Medium/high	Good to poor
Fungus	Cryptococcal meningitis	*Cryptococcus*	Amphoterecin B	Low	Poor to good
Toxin	Botulism	*Clostridium* B	Neutralizing	Low	Good if treated in time
	Tetanus	*Clostridium tetani*	Antibodies	Low/high	
Prion	Mad cow disease Creutzfeld-Jacobs Kuru	PsP^{sc} prion	None	Low Low Low	Universally fatal
Protozoa	Malaria	*Plasmodium*	Artemisinins	Low/high	Good
	Primary amoebic meningoencephalitis	*Naegleria fowleri*	Amphoterecin B	Low	Fatal
Metazoa	Neurocysticosis	*Taenia solium*	Praziquantel	Low/medium	Good

Prevalence: low = <10,000; medium = 10,000–100,000; high = >100,000.

and entry of blood-borne immune cells into the subarachnoid space. This amplifies the immune response, starting a vicious cycle. Worsening edema causes intracranial pressure to increase, inducing severe headaches, alterations in the level of consciousness, and, eventually, coma.

Bacterial meningitis can be unequivocally confirmed only by bacterial cultures from CSF obtained by spinal tap, which, in addition to ruling out other causes such as fungal or viral infections, aids in the selection of the most appropriate antibiotic. However, treatment cannot wait this long and must begin immediately using empirical antibiotic treatment before CSF findings are known because mortality ranges from 3 to 20%, even with aggressive intravenous antibiotic treatment.

Viral infections of the **meninges** are 10 times more common than bacterial infections, with approximately 75,000 cases/year; they show identical symptoms, but CSF samples contain an abundance of white blood cells and, obviously, an absence of bacteria. The vast majority of cases (85%) are caused by enteroviruses such as

Coxsackie. Less common are West Nile virus, arboviruses transmitted through mosquitos, HIV, and, rarely, neurotropic herpes viruses such as herpes simplex virus (HSV). Because antibiotics are ineffective for viruses, treatment is palliative, and the inflammation typically resolves spontaneously. Prognosis is usually good for most cases of enteroviral meningoencephalitis, and recovery is almost always complete. In cases of more severe involvement of the brain parenchyma, the clinical outcome can be complicated by long-term neurological deficits. In some 20,000 patients annually, the viral infection is not restricted to the meningeal coverings but affects the underlying brain as well, causing inflammation that results in diffuse neurological symptoms ranging from mild cognitive impairment to severe neurological conditions such as seizures and ataxia. The resulting encephalitis can be caused by hundreds of different viruses whose identity often remains obscure. Some cases caused by HSV can be effectively treated with antiviral drugs such as acyclovir, whereas others such as poliomyelitis cannot be treated at all.

In addition to bacteria and viruses, **fungi can also be the cause of meningitis**. Although rare among the general population, fungal infections represent the second most common opportunistic infection in HIV-infected individuals in the developing world and represent important infections in transplant patients and other individuals with weakened immune systems. Unlike bacterial and viral meningitis, which present with acute severe headaches, fungal infection of the meninges presents with slowly developing subacute symptoms that worsen over the course of several weeks. Mortality is high (30–70%), and treatment with an antifungal agent, such as amphotericin B, is often ineffective if the infection is of long duration.

3.2 Botulism

Botulism is an illness caused by a bacterial neurotoxin rather than a bacterium itself. It typically results from eating spoiled food that contains the bacterial toxin. After the gastrointestinal symptoms of food poisoning subside, a characteristic flaccid paralysis (loss of muscle tone) develops. Botulism toxin is produced by gram-negative bacteria of the genus *Clostridium*, which are abundant in the soil or in silt of streams and lakes. Bacterial spores are inert and survive for many years. When conditions are favorable, spores replicate and grow. Warm, moist temperatures and anaerobic, oxygen-free conditions activate the bacteria to produce toxin. Such conditions occur, for example, in home-canned food items, sauces, ketchup, and the like.

There are seven different botulism toxins, designated A–G, each produced by a different strain of *Clostridium*. These toxins are the most toxic substances known to man; an amount fitting on the tip of a needle is sufficient to kill an adult, and half a pound is sufficient to kill every human on earth. Only A, B, E, and F toxins cause disease in humans. Different geographic locations correlate with different toxin outbreaks; type A is the most prevalent in the western United States (California, Washington, Colorado, and Oregon), type B is most common in the eastern United States, and type E is most common in Alaska. This distribution correlates with the most prevalent *Clostridium* bacteria or their spores found in soil samples from these geographic regions. Spores are inert and can survive for years.

In general, we assume that botulism is the result of dietary intake of the toxin. However, in addition to food-borne botulism, there are two additional distinct clinical presentations of botulism. Wound botulism is caused by *Clostridium* colonizing a contaminated wound, where the toxin enters the bloodstream at the site of injury. Furthermore, children and adults can develop botulism as a result of *Clostridium* colonization of the gut, where the assiduous release of toxin causes paralytic symptoms in the complete absence of gastrointestinal signs. Indeed, in the United States, childhood botulism is the most common form of the disease. The source of contamination is not entirely clear, but in 20% of affected children exposure to the spores could be traced to honey that was applied to pacifiers.

However, soil, dust, or other contaminants are likely the predominant source of the bacterial spores. Infantile botulism may contribute to some of the unexplained causes of sudden infant death syndrome. The botulism toxin originates as a large, 150-kDa precursor polypeptide, which, upon bacterial lysis, is activated by proteases. The active toxin consists of a 100-kDa heavy chain and a 50-kDa light chain linked by a disulfide bond (Figure 2). The toxin enters cholinergic nerve terminals via binding to two receptors: a ganglioside and, depending on the toxin type, either synaptotagmin or the synaptic vesicle protein SV2.[2,3] The latter are typically inside synaptic vesicles. Upon fusion with the membrane to release transmitter, the light chain of the toxin binds to either synaptotagmin or SV2, which now transiently appears on the cell surface (see Figure 3). Upon vesicle recycling, the entire toxin becomes trapped in the vesicle. Because of the acidic environment, a conformational change of the toxin allows its heavy chain to integrate into the vesicle membrane, where it forms a pore. This pore is used for the light chain to leave the vesicle and enter the cytoplasm. Here it binds to and cleaves the vesicle-associated membrane protein (VAMP) and synaptosomal-associated protein (SNAP25), two molecules of a protein complex called SNARE that mediates Ca^{2+}-dependent vesicle docking and fusion. The B, D, F, and G toxins exclusively cleave VAMP, whereas the A, C, and E toxins cleave SNAP25 (Figure 4). The end result is the same, namely, insufficient acetylcholine release, causing muscle paralysis. Note that the rate of toxin uptake is a direct function of the synaptic activity since vesicle fusion is an essential step in toxin uptake. As a consequence, more active muscle groups

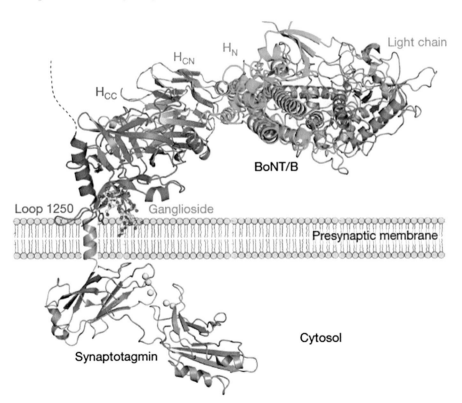

FIGURE 2   Botox/B binding to both synaptotagmin and the ganglioside receptor on the membrane of cholinergic neuron. *From Ref. 1.*

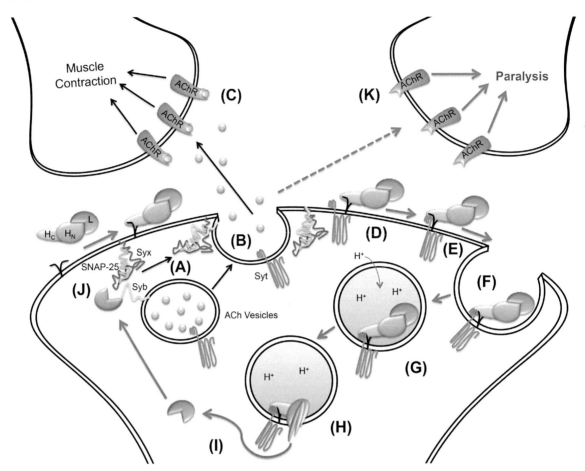

FIGURE 3   Model for the entry of Botulinum Neurotoxin (BoNT) into nerve cells. SNARE proteins mediate vesicular docking. Vesicular Synaptobrevin (Syb) and membrane Syntaxin (Syn) and SNAP-25 must interact (A) for vesicular docking on the cell membrane, which in turn allows vesicular release of acetylcholine (ACh) (A). Once released, ACh binds acetylcholine receptors (AChR), which mediates muscle contraction (C). The intra-vesicular parts of synaptic vesicle proteins become exposed on the plasma membrane as the vesicle integrates into the membrane (D). When BoNT is present it binds to a ganglioside molecule on the cell surface, which allows BoNT to access their protein receptor (exemplified here for synaptotagmin (Syt)) (E). After binding the protein receptor, the neurotoxins are endocytosed via retrieval of synaptic vesicles (F). The lumen of the recycling vesicles becomes acidified via the action of the vesicular proton pump (G). Acidification provokes a structural rearrangement in the neurotoxins, whereby the heavy-domain (HC and HN) forms a channel through the vesicular membrane (H). The light chain (L) passes through the channel due to partial unfolding and is released to the cytosol following reduction of the disulfide bond (I). Ultimately, the light chain cleaves its target SNARE(s), Syb, Syx, or SNAP-25, (J) and thus blocks the synaptic vesicle cycle, which it exploits for cellular entry. Preventing the further docking of ACh vesicles prevents ACh release and subsequent binding to AChR, leading to paralysis (K). The entry of BoNT into nerve cells requires the vesicular docking cycle to expose the appropriate protein receptor (here shown as Syt), hence the rate of toxin uptake is directly related to vesicular fusion and therefore directly related to synaptic activity. This results in more active muscle groups becoming paralyzed more quickly. *Figure courtesy of Stephanie Robert and adopted from Ref. 4.*

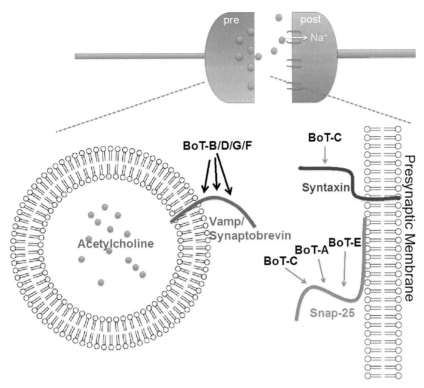

FIGURE 4   The various botulinum toxins have specific presynaptic targets, which are all involved in synaptic vesicle release.

paralyze faster. Tetanus toxin (discussed below) is produced by a related strain of *Clostridium* and is taken up in a similar fashion but transported into the spinal cord, where it disrupts the release of γ-aminobutyric acid (GABA) from interneurons, causing spastic as opposed to flaccid paralysis.

The clinical presentation of botulism, regardless of the source, is rather similar. In food-borne botulism, initial symptoms occur within 18–36 h after exposure to the toxin, yet incubation can range between 6 h and 10 days. Since the toxins cause sustained blockage of acetylcholine release, they affect both muscarinic and nicotinic cholinergic neurons. Hence, in addition to the nicotinic skeletal, facial, and respiratory muscles, the muscarinic autonomic and smooth muscles are also paralyzed. Facial and throat muscles are the first skeletal muscles affected, and paralysis progressively descends to include trunk and the respiratory and visceral muscles. Dry mouth, blurred vision, and diplopia (double vision) are also early symptoms, and dilated fixed pupils are typical. This is followed by drooping eyelids (ptosis), a hoarse voice, overall difficulty speaking (dysarthria), and difficulty understanding language (dysphasia). Severe cases affect the respiratory muscles and diaphragm, causing respiratory failure.

Food-borne botulism is rare, with about 10 cases each year in the United States. Infant botulism occurs in 80–100 cases annually, typically in children younger than 6 months of age. It is believed that immature intestinal microflora and low acidic bile content, which normally suppress *Clostridium* growth, increase susceptibility. As toxin enters the blood stream, these infants display symptoms including apathy, weakened cry, loss of appetite, constipation, and overall weakness (floppy baby syndrome), which develop over days to weeks.

Less than 1% of infants die from botulism, and survivors have no long-term sequela.

Treatment typically involves intravenous injection of an equine antitoxin containing either antibodies to A, B, E, or A–G toxins and that neutralizes molecules not yet bound to receptors. A single injection of 10 ml of antitoxin is sufficient to stabilize the disease. Recovery occurs over time through the regeneration of nerve terminals; it typically takes 2–8 weeks. The cure rate is high, and most individuals are left without any lasting deficits. It is important that suspected cases be reported to the US Centers for Disease Control and Prevention to ensure that potential sources of the toxin can be identified and destroyed.

Poisoning by food-borne toxin can be prevented by boiling; however, this requires a temperature of 121 °C, for example, in a pressure cooker for >3 min. It is important to stress that botulism is a rare neurological condition. Not counting infant botulism, fewer than 2500 cases have been reported in the United States since 1899, and the average number of disease outbreaks is around 10 per year, each affecting (on average) 2.6 people. Although historically botulism has been a deadly disease, with 60% of infected individuals dying, the introduction of antitoxins and improved supportive care, including mechanical ventilation, has reduced case fatality ratios to less than 1%.

3.3 Tetanus

Tetanus toxin[5] is similar to wound botulism in that it is a bacterial toxin produced by bacteria that germinate from bacterial spores found abundantly in the environment and that enter the body through open wounds or scrapes. *Clostridium tetani* produces two toxins, tetanolysin and tetanospasmin. Because only the latter causes the typical symptoms, it is the one simply referred to as tetanus toxin. As with botulinum toxin, tetanus toxin is produced as a 150-kD precursor protein with a light (50 kD) and heavy chain (100 kD) linked via disulfide bonds. The heavy chain binds to

gangliosides on presynaptic motor terminals and mediates the entry of the complex into the cell. From there it is transported (retrogradely) back to the motor neuron cell body via axonal transport, reaching the motor nuclei in the ventral horn of the spinal cord or motor nuclei of the cranial nerves in the brainstem. The major difference between botulinum toxin and tetanus toxin is that the former stays at the synapse and inhibits transmitter release, whereas the latter is transported back to the neuronal cell body and ultimately enters "one cell up" in the presynaptic terminal. How the toxin escapes lysosomal degradation and instead leaves the cell to enter its presynaptic GABAergic terminal is not well known, but transcytosis is a suspected mechanism. Once in the GABAergic terminal the light chain of the toxin, which is an endopeptidase, cleaves VAMP2 (synaptobrevin), thereby preventing the release of the inhibitory neurotransmitter GABA. This results in unregulated excitatory activity of the motor neurons, causing the tetanic muscle contractions that are the hallmark of the disease. Hence the difference of the site of action between these toxins makes one induce flaccid paralysis and the other spastic paralysis.

Depending on the motor nuclei affected, tetanus toxin infection can present with a wide variety of symptoms. It can cause spastic contractions of just about any muscle group. If motor nuclei innervating the laryngeal or respiratory muscles are affected, the infection is life-threatening because respiratory failure is likely; indeed, this is the most common form of death resulting from tetanus poisoning.

Although deaths related to tetanus are exceedingly rare in the Western world, where vaccination is common, it is estimated that tetanus kills over 1 million people worldwide every year. Once symptomatic, wound cleaning and administration of antitoxin, where available, along with benzodiazepines to control spasms, can stabilize the disease. A ventilator may become necessary if breathing problems develop. If caught in time, the prognosis is typically good, with a complete reversal of symptoms.

3.4 Neurosyphilis

Neurosyphilis (Tabes dorsalis)[6] is a late presentation of syphilis that occurs in 10% of infected and untreated individuals. Syphilis derives from the Greek word *syphlos*, meaning crippled or maimed, since untreated patients can develop bizarre-looking cutaneous, noncancerous growths on any parts of their bodies, called gummas (Figure 5).

Syphilis is a bacterial infection that is typically acquired through sexual contact and hence belongs to the sexually transmitted disease category. The infectious reagent is *Treponema pallidum*, a gram-negative bacterium called a spirochete because it looks like a corkscrew. The

FIGURE 5 Gummas disfiguring the head and face of an individual with syphilis. Bust in Musée de l'Homme Paris.

infection typically occurs through contact with open wounds or fluids during unprotected sex. Since the disease can be effectively controlled using the antibiotic penicillin G, incidence rates have decreased significantly since the 1940s in the developed world. Rates are currently increasing, however, presumably because of unprotected sex among men infected with HIV.

The disease usually involves three phases. The primary infection follows the inoculation of an individual with about 500–1000 bacteria. Within 36 h these replicate and result in a painless ulceration called a chancre. These typically occur in the genital areas. After 2–6 weeks the second stage of the disease continues, with wide infiltration throughout the body and nervous system. Afterward, during the latent stage of disease, patients are frequently asymptomatic for many years. About 10% of patients with untreated syphilis develop neurological symptoms called neurosphyilis, or tabes dorsalis, 10–15 years later. These include progressive muscle weakness, unsteady gait, and cognitive impairments typically including confusion, delusions, disorientation, and depression. Pupillary abnormalities (Argyll-Robinson pupils), dysarthria, and tremors in extremities can also be present.

The earliest stages of neurosyphilis involve inflammation of the meninges presenting with headache, nausea, vomiting, and, occasionally, seizures. This acute syphilitic meningitis responds well to aggressive penicillin treatment, whereupon the symptoms typically resolve completely. The classical neurosyphilis presentation was described by Romberg and includes the following "3 Ps": paresthesia, pain, and polyuria. Patients experience sharp, lightning pains (paresthesia) and have an unsteady ataxic gait. With eyes closed, patients typically cannot walk or even stand upright, but fall. This is called a Romberg sign and is caused by a loss of proprioception. Unlike the senses with which we perceive the environment, feedback information from stretch receptors in skeletal muscles tells us the relative position of each extremity

FIGURE 6 Cross-section of spinal cord showing loss of dorsal spinal neurons in the dorsal spinothalamic tracts (white areas encircled by dashed line) in a patient with neurosyphilis. *Centers for Disease Control and Prevention/Susan Lindsley.*

in space. Such receptors also report the state of contraction of visceral muscles such as the bladder, explaining polyuria (enhanced frequency of voiding).

The progressive degeneration of the spinal sensory nerve roots along the dorsal column of the spinal cord essentially causes a functional deafferentation, as seen in Figure 6. This includes the sensory nerve fibers innervating the skin and the viscera, causing pain comparable to amputation or diabetic neuropathies. Such deafferentation typically causes hyperexcitability in the ascending pain pathways and is irreversible.

A number of studies suggest that approximately one-third of individuals infected with syphilis show bacteria in their central nervous system (CNS) and are at risk of developing neurosyphilis. In contrast to the previously held assumption that bacterial spread to the CNS occurs late during the disease, this is not supported by research, which instead suggests that CNS infection occurs early in disease, possibly within just a few days after infection.

The natural history of neurosyphilis was comprehensively evaluated in the infamous Tuskegee studies that enrolled a large number of infected individuals in rural Alabama. This clinical study, conducted from 1932 to 1972 by the US Public Health Service, enrolled 600 sharecroppers, of whom 400 were infected and 200 were not. In spite of the introduction of a curative agent midway through the study in 1945, disease-carrying patients were never offered treatment, and the disease was allowed to run its course. This egregious lapse in ethical conduct led to the Belmont report, which in 1979 established the procedural and ethical framework for human experimentation, including the requirement for informed consent and oversight by an independent review board.

Since 1945, syphilis has been effectively treated with penicillin, and the bacterium has not developed any resistance over the past 70 years. High doses are necessary to reach the CNS, including large doses of intravenous penicillin G daily over 10–14 days. Two serological tests are available (Rapid Plasma Reagin and Venereal Disease Research Laboratory) that can detect established infection with high accuracy. CSF is analyzed only if neurosyphilis is suspected. Such infection presents with an increased number of white blood cells (>5 per ml) and CSF protein, which is normally very low and increases to >45 mg/dl. Additional serologic tests for syphilis should show reactivity.

3.5 Poliomyelitis

Poliomyelitis, also known as infantile paralysis, is caused by infection with the poliovirus, a nonenveloped RNA enterovirus that colonizes the gut. The virus is contracted through the mouth and, as an RNA virus, highjacks the body's own cells as it colonizes, coaxing cells into producing more virus. The virus is highly resistant to acidic environments and can survive for long periods of time in sewage or water. Hand-to-mouth contact is the typical mode of spread. Once ingested the virus enters the bloodstream, and 99% of infected people show either no response or only a short, mild fever. In 1% of patients, however, the virus enters the brain and causes an inflammatory response. The virus infects and kills motor neurons in the spinal cord, motor cortex, and brainstem. This typically occurs long after the virus has colonized the body. The virus shows an intrinsic neurotropism and specifically selects motor neurons over sensory neurons and neurons over glia cells. The entry of virus into the nervous system can occur via two routes: by crossing the blood–brain barrier or infecting the peripheral nerves, which transport the virus back to the spinal cord and brainstem via axonal transport. CD155, also known as the poliovirus receptor, can bind to a protein that is associated with the dynein complex, which moves cargo along the axonal microtubules. Once the virus has entered the brain though this Trojan horse strategy, it is free to replicate and infect additional neurons. Ultimately, the virus kills its host neurons by shutting down the cell's protein biosynthesis and activating the apoptotic caspase cascade. Remarkably, the virus primarily affects the motor pathways responsible for the lower extremities and completely spares sensory neurons.

Most polio cases occur in infants, rarely babies or adults. Until the mid-1950s, polio crippled on average 40,000–50,000 children each year. Through aggressive vaccination, polio has been essentially eradicated in the Western world, but it is still found in pockets of the developing word, particularly in Afghanistan, Pakistan, and West Africa. There are three poliovirus serotypes, of which type 1 is the one most often associated with paralytic disease. All three serotypes are contained in modern vaccines. There is no cure, and treatment is entirely supportive. Death is rare, but permanent disability affecting primarily the lower extremities is the norm (Figure 7). In some patients who have had stable disease for 30–40 years, muscle weakness can suddenly worsen, causing a poorly understood postpolio syndrome. There is no evidence of residual virus at this stage, and in all likelihood the remaining motor neurons are simply exhausted from overuse.

FIGURE 7 Children with polio at the Amar Jyoti Research Center, Delhi, India. *Photo: WHO/P. Virot (UN News Center).*

3.6 Rabies

A much more common RNA virus is the rabies virus,[7] which typically infects animals but can spread to humans through animal bites. Frequent carriers include bats, raccoons, skunks, and foxes, and disease is typically transmitted to people via stray dogs. Rabies is found throughout the world except Antarctica. After entering the body via a bite, the virus replicates in the surrounding muscle tissue. From there, the virus spreads centrally along peripheral nerve fibers, presumably via fast axonal transport. After the initial peripheral to central spread, the virus eventually replicates in the acinar cells of the salivary gland, which become the vector for disease transmission. Rabid animals have excessive saliva containing the virus, often seen as froth. As a result of this excessive salivation, infected animals or people have difficulty swallowing and show hygrophobia (fear of water). The CNS inflammation is mild and there is surprisingly little evidence of neuronal cell death given the severity of the disease. On pathological specimens, inclusion bodies, called Negri bodies, can be found in cerebellar Purkinje cells and pyramidal neurons. It must therefore be assumed that functional rather than anatomic changes explain the neurological symptoms. Rabies-infected animals and humans become aggressive and fearless and are quite dangerous to uninfected animals or humans, yet the biological underpinning of the "rabid" behavior is not understood.

The disease is typically first noticed by an infection around bite marks, although bat bites, common for transmission in the United States, may be too small to be identified, making a suspected diagnosis difficult. In 80% of patients the disease is encephalitic and involves the forebrain and cerebellum, causing furious behavior with combativeness and hallucinations. In 20% of cases it is paralytic, causing flaccid paralysis that is easily misdiagnosed as Guillain-Barré syndrome.

Rabies is almost always fatal; patients die within days of becoming symptomatic. However, if recognized early after infection during the incubation period, which lasts from 1 to 3 months, proper wound care and postinfection vaccination may be able to contain the disease.

Given the extensive vaccination of animals and particularly pets in the United States, rabies rarely infects humans. However, canine-transmitted rabies is common in Asia and Africa, where an estimated 55,000 people die from it each year.

3.7 HIV and Neuroaids

HIV is one of the most common viral infection worldwide and the single most important virus-induced neurological disease thus far described. HIV is a member of the retroviral family, which is characterized by a unique replication cycle whereby genes are encoded by RNA, rather than DNA, and must undergo reverse transcription and integration into the host DNA. Retroviruses almost exclusively infect vertebrates. Because they insert their genome into the host germ line, they are adept at regulating a host cell's behavior. The viral core contains two copies of single-stranded RNA and is surrounded by a capsid and enveloped by a lipid-rich membrane (Figure 8). The viral RNA is reverse transcribed into double-stranded DNA in the cytoplasm and trafficked to the nucleus, where it is irreversibly inserted into the host cell's genome. Some of these genes initiate the production of virions by the host cell, which then are released from the plasma membrane by budding (Figure 8). The host cell then has been reprogrammed to produce viral genes and is

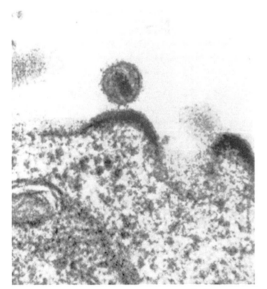
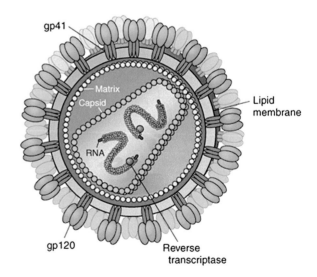

FIGURE 8   Human immunodeficiency virus I budding from CD4 lymphocytes (left). Schematic of the structure of the viral particle (right). *Reproduced with permission from Ref. 8.*

entirely under the regulation of the virus. HIV has been unequivocally determined to be the cause of acquired immune deficiency syndrome (AIDS). Although there are two distinct strains of HIV, HIV-1 and HIV-2, which differ genetically, the vast majority of AIDS cases are caused by HIV-1.

AIDS is a relatively new disease, dating back to late 1970s, when the virus must have jumped species from chimpanzees or gorillas to humans. The virus was first isolated in 1985 and is probably the most well-studied pathogen. HIV typically infects T-lymphocytes by attaching via its gp120 coat protein to the CD4 receptor. This results in a conformational change, allowing the complex to bind two coreceptors, CXCR4 and CCR5, leading to internalization through fusion with the cell membrane. Once inside the T cell the virus releases its RNA, which is transcribed into DNA and integrates into the host genome. The reprogramming of the T cell causes production of virions that are released from the lymphocytes by budding.

HIV is primarily transmitted through sexual contact, and the likelihood of infection is directly proportional to the viral load of the infected individual. Transmission through infected blood products is also possible but is, in light of recent extensive testing, exceedingly rare (one in 1 million). However, the risk of health care workers contracting HIV from job-related injuries is significant. Close to one million accidental needle sticks occur each year in the United States, each carrying a 0.3% chance of infection if the needle was used by an HIV-infected individual. There is a 20–30% likelihood that a pregnant HIV-infected women will infect her baby. The global scope of HIV is immense, approaching a current incidence of 35 million people. Almost two-thirds of infected individuals are in sub-Saharan Africa, where the infected are not predominantly homosexual men, as is the case in the United States. Instead, 50% of all HIV cases worldwide are in women, and 2.5 million children are HIV positive. Of the 1.1 million people confirmed to be HIV positive in the United States, 50% are homosexual men and 10,000 are children.

The systemic disease presents with profound immunodeficiency secondary to uncontrolled production of the virus and loss of T-lymphocytes. Over a period of a few weeks, fever, headache, swollen throat, and lymph nodes, as well as severe muscle pain and fatigue, develop and last for several weeks. This acute stage of disease is followed by a latent stage lasting from a few months to well over 20 years. In the final stage of the disease, most of the body's T cells have been lost, resulting in a drastic reduction in the blood concentration of T-lymphocytes (<200/ul). At this point the infected patient is at risk of dying from even trivial opportunistic infections. This is the stage that is defined as clinical AIDS. It can be prevented through the early use of a multiple drug cocktail called highly active antiretroviral therapy (HAART), which provides almost normal life-expectancy in treated patients.

Neuro-Aids[9,10]: While initially the focus related to HIV was entirely on the peripheral immunosuppressive aspects of the disease, patients with HIV frequently showed cognitive problems; some presented with frank dementia, suggesting that the virus infects the brain at some stage of disease and may escape antiviral therapy. It is now clear that HIV infection of the nervous system occurs early in disease, and in up to 10% of patients neurological symptoms are the initial presenting symptoms. Brain infection occurs most likely through virally infected lymphocytes that enter the brain and, to a lesser extent, through transcytosis of blood-borne virus across the blood vessel endothelial cells. Once in the brain, microglia, macrophages, and monocytes all become infected and activated. The virus is not found in neurons, however, and direct neuronal killing through lysis by the virus does not occur. Instead, neuronal death is likely due to extracellular virus presenting its toxic coat proteins, such as gp120, which is known to cause N-methyl-D-aspartate–mediated glutamate excitotoxicity, as well as trigger the release of toxic molecules such as TNF-α from resident

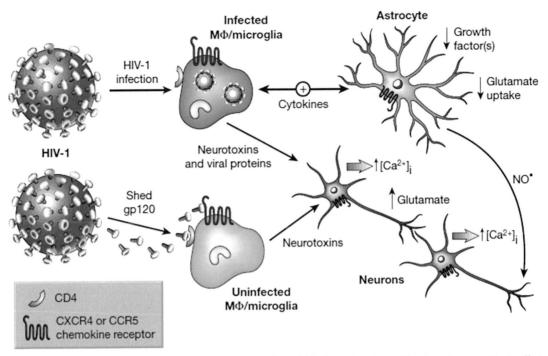

FIGURE 9   Human immunodeficiency virus infects microglia, which, in conjunction with astrocytes, exert toxic effects on neurons, causing indirect killing by the virus. *From Ref. 11.*

microglial cells (Figure 9). Both gp120 and the HIV-encoded protein transactivator (tat) also harm endothelial cells, leading to a breach of the blood–brain barrier. Entering monocytes then release toxic cytokines such as TNF-α, IL-1, and IL-6. Hence, a combination of direct toxicity from virus proteins and indirect toxicity from the activated immune cells contribute to disease.

The extent of the neurological deficit correlates with the initial viral load and can range from mild cognitive decline to full-blown dementia. The early use of HAART reduces the systemic viral load and also attenuates intracerebral viral activity, although in later stages of disease the cerebrum probably harbors its own viral ecosystem. Regardless of initial infection, 50% of HIV-infected persons show neurocognitive decline even when treated aggressively, a condition referred to as neuro-AIDS. In severe cases, MRI shows atrophy of subcortical structures including white matter (Figure 10) and basal ganglia presenting as leukoencephalopathy (white matter disease of unknown cause); neurocognitive deficits, namely, motor and cognitive slowing, and changes in mood and anxiety are consistent with a subcortical dementia.

In patients treated early and aggressively, systemic viral load is low, and so is the burden of virus hiding out in the CSF. Neurocognitive decline still develops but involves primarily cortical structures with memory loss imitating Alzheimer disease (AD). Although with HAART the percentage of patients developing full-blown HIV-related dementia has been reduced from 16% to fewer than 5%, millions of HIV-affected individuals in the developing world still have inadequate access to treatment and are at risk of suffering from severe forms of neuro-AIDS.

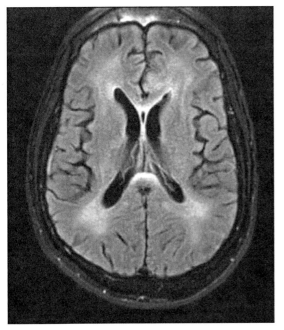

FIGURE 10   A fluid level-attenuated inversion recovery magnetic resonance image of a patient with human immunodeficiency virus-induced dementia presenting with subcortical white matter abnormalities throughout the brain. These changes, visible as bright brain tissue within the otherwise gray-appearing brain tissue, are typically seen in the late stages of the disease and have become rare with aggressive antiviral treatment. *From Ref. 9.*

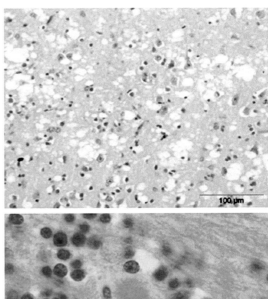

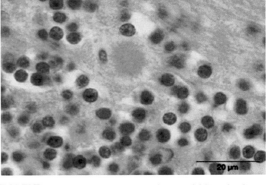

FIGURE 11   Cellular pathology of Creutzfeld Jacobs disease (top), and Kuru (bottom). Each shows an abundance of vacuoles, giving the brain a sponge-like consistency. *From Ref. 12.*

3.8 Prion Diseases

For several centuries a peculiar disease called scrapie occurred sporadically in sheep and goats, causing them to compulsively scrape their skin and fur against rocks and trees. Animals show ataxia and progressive muscle weakness and succumb to wasting. In 1986 a similar disease developed in cows and was called mad cow disease because the cows were highly uncoordinated, visibly agitated, and aggressive. Soon after the first cases were discovered, several young human adults developed paranoia and psychosis, along with motor problems, such as ataxia, similar to those seen in a rare neurodegenerative disorder called Creutzfeld Jacob disease (CJD), which typically affects much older individuals. Ultimately,

165 people became ill and died as a consequence of eating meat from mad cows. Some 30 years earlier, a group of scientists described similar neurological findings in a disease called Kuru that developed in members of the Fore tribe in Papua New Guinea who had been living completely isolated from modern society. It turned out that all of these diseases shared a remarkable brain pathology characterized by extensive neuronal loss, inflammation, and the appearance of vacuoles, giving the brain and spinal cord a spongy appearance; hence it is often called spongiform encephalopathy (Figure 11).

It was subsequently discovered that these diseases also share a common pathogenic mechanism, namely, an infectious protein called a prion. Prion proteins exist in every cell of the body, and their function is not well understood. They are large proteins encoded by the *PRNP* gene on chromosome 20 and form a tertiary structure with multiple α-helixes (Figure 12). Mutations of the wild-type PrP give rise to a mutated PrP^{SC} (for scrapie) prion protein that contains an unusually large β-sheet structure. These can easily aggregate into plaques. Indeed, we already discussed the appearance of protein aggregates in a number of degenerative diseases, including AD, Parkinson disease, amyotrophic lateral sclerosis (ALS), and Huntington disease, where misfolded protein aggregates are central to disease pathology. It turns out that the tertiary shape of the PrP^{SC} prions, which act like a mold to coax normally folded PrP proteins into a shape that leads to aggregation, can guide misfolding. Propagation of this aggregating, disease-causing phenotype therefore involves the PrP^{SC} protein entering cells and gradually changing them from wild type into a misfolded state. It is believed that such a mode of transmission, akin to the nucleation of a crystal, may explain the transmission of disease within an organism, but it may also explain transmission between affected sheep, cattle, and humans. This is particularly plausible because PrP^{SC} is incredibly stable and resistant to denaturing. Neither radioactivity nor heating, acids, or solvents can eliminate the infectious particle, which therefore readily survives bile and stomach acid when ingested. It is likely that prion disease is spread by either inoculation with the prion or by dietary exposure to prions in excessive quantities. Prion disease also has been shown to be transmitted by improperly sterilized surgical instruments that were used during surgery on a patient with CJD as well as to young children treated with growth hormone isolated from pooled collections of pituitary glands obtained from deceased donors. Mad cow disease, or spongiform encephalopathy, is caused by feeding cows the brains of scrapie-infected sheep. The disease spreads to humans through the consumption of infected cow meat, particularly brain and spinal cord. In Kuru, the Fore tribe honored their dead through the cannibalistic ritual of eating their corpses. While the men primarily eat the

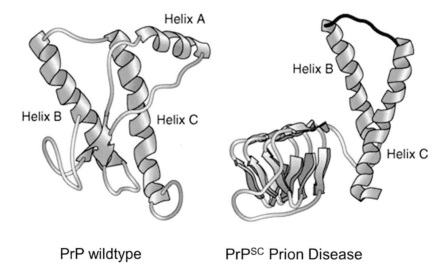

FIGURE 12 Three-dimensional shape of PrP and PrP^{sc} illustrating the conversion from primarily a-helical to b-sheet structure. *Reproduced with permission from Ref. 13.*

muscle, women and children not only prepared the human material, thus exposing them to large inoculums through broken skin secondary to insect bites, but also consumed the brain and were more likely to become infected. Transmission to other species or between species, particularly sheep and cows, was artificially introduced by "industrial cannibalism" as these animals are herbivores with no interest in consuming meat, let alone brain. CJD instead is a sporadically occurring neurodegenerative disease that presents with the de novo mutation of the *PRNP* gene to generate the PrP^SC protein.

Since these diseases share a common pathogenic mechanism and similar CNS pathology, they are now collectively called prion diseases. They are universally fatal. The exact mechanism of neuronal death inflicted by the presence of prions is still unclear, although an important role for microglial clearance of dying neurons leaving vacant vacuolar spaces in the brain is well established.[14] Interestingly, as we briefly discussed in Chapter 4 (Aging & AD), a prion-like mechanism of spreading misfolded amyloid-β has been suggested to underlie the gradual spread of Aβ plaques throughout the brain with aging.[15] Notably, however, AD is not a communicable disease, and unless you eat the brain of your demented professor, there is little risk of acquiring AD via an infectious prion. From an epidemiological point of view, prion diseases are something of a curiosity and not a significant health challenge, although they have hopefully taught us a lesson about proper agricultural practices. Incidence of CJD is one in 1 million, with about 300 annual cases in the United States, mostly in older individuals. Of note, however, CJD can be transmitted via organ or blood donations even when the source was a presymptomatic individual.[14] No further human infections with mad cow disease have been reported since the 1980s, in part because of aggressive testing of domestic and imported beef. Infections of humans by scrapie have not been reported, nor has transmission of prions via animal milk.

3.9 Brain-eating Amoeba (*Naegleria fowleri*)

Found in almost every freshwater lake in the southern United States during the summer months, the amoeba *N. fowleri* enters the body through the nasal passages and then actively moves along the olfactory nerve, through the cribriform plate, a sieve-like porous structure of bone, to the brain, where it takes up residence.[16] The point of entry explains why the major pathology is found in the frontal parts of the brain (Figure 13), adjacent to the olfactory nerve entry points. Although very rare at 5–8 cases a year, it is almost 100% lethal to those infected. The average age of affected people is 12 years, and males are more likely to be victims. The amoebae are 15–30 μm in size and normally feed on bacteria. They survive in water of almost any osmolality and prefer warm temperatures of 37–45 °C. In the absence of nutrients they form dormant cysts. A single amoeba that invades the brain is sufficient to set off disease, which, after a 1- to 7-day incubation period, presents with typical symptoms of a viral or bacterial meningitis, including bilateral frontal headaches, nausea, vomiting, and confusion. Given the delay, making the connection between recreational exposure and the disease is sometimes difficult. The disease is called acute meningoencephalitis or primary amoebic meningoencephalitis , also known as PAM, because it inflames both the meninges and the underlying brain parenchyma. On autopsy, the frontal parts of the cortex close to the olfactory show edema and inflammation and microscopic clusters of amoeba surrounded by edematous necrotic tissue (Figure 13). While typically contracted by swimming in contaminated water, two recent cases of primary amoebic meningoencephalitis resulted from nasal irrigation with tap water.[17] The presence of disease-causing amoeba can be confirmed by microscopic inspection of CSF in a wet mount. Highly motile amoeba should be visible. They kill neurons in the brain by puncturing their

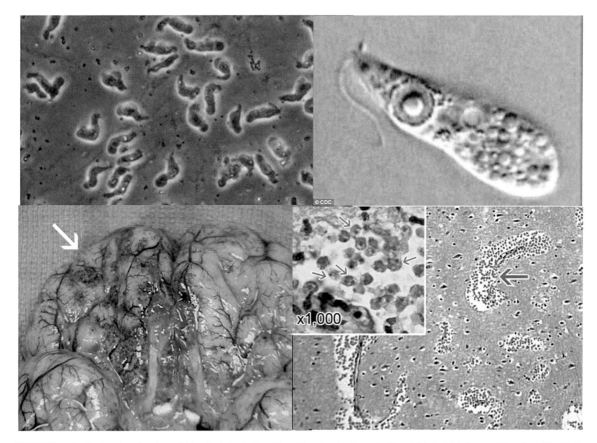

FIGURE 13   Isolated and cultured *Naegleria fowleri* trophozoites (top) at low (left) and high (right) magnification. The high-magnification image (top, right) shows the flagella with which the pathogens propel. Brain of a patient who died from primary amoebic meningoencephalitis (PAM), showing extensive hemorrhage and necrosis, mainly in the frontal cortex (bottom, right, arrow). Cerebral portion of the brain from a patient with PAM, stained with hematoxylin and eosin, showing large clusters of *Naegleria fowleri* trophozoites and destruction of the normal brain tissue architecture (bottom, right). Cysts are not seen (original magnification ×100). Inset: Higher magnification (×1000) of *N. fowleri* trophozoites (arrows) with the characteristic nuclear morphology. *Centers for Disease Control and Prevention,* http://www.cdc.gov/parasites/naegleria/naegleria-fowleri-images.html.

cell membrane, feeding on their cytoplasm, and gradually digesting them.

Unfortunately, the time course of this disease is fast, and no effective treatment has been identified. Although isolated amoeba are sensitive to amphotericin B, the drug is rarely curative in humans. The only known survivor, a 9-year-old girl, was rapidly treated with intravenous amphotericin, but she also had unusually high antibody titers against the pathogen. In general, the disease carries 100% mortality.

3.10 Neurocysticercosis

Neurocysticercosis[18] (tapeworm infection of the brain) is the most common parasitic infection of the CNS worldwide. It is caused by infection of the nervous system with the larvae of the pork tapeworm *Taenia solium*. This parasite's lives in both humans and pigs, but humans usually carry only the tapeworm, not the larvae. Tapeworms attach to the human intestine to continuously shed body segments containing thousands of eggs that are liberated with human feces. Roaming pigs that come into contact with human feces ingest the eggs. These form larvae, called cysticerci, which leave the gut and can infest almost any part of the pig's body but predominantly populate muscle tissue. Humans who eat undercooked pork digest the larvae, which form new tapeworms in the intestine, completing the normal life cycle. In the case of disease, humans erroneously become carriers of cysticerci after contracting the larvae through human feces. How the cysticerci enter the brain is not known but is likely hematogenous. Once there, they form collagen-rich capsules around them, which ultimately calcify into hard cysts surrounded by areas of edema and marked inflammation with reactive microglia and astrocytes. The cysticerci can be present in the brain for years without causing symptoms. The most typical symptoms are headaches and focal seizure. On MRI, or more commonly on computed tomography, cysts are easily identified as diffuse lesions that are often scattered throughout the brain, as shown for an extreme case in Figure 14. Surprisingly, cysts found throughout the body exist without causing symptoms.

The exact prevalence of neurocysticercosis is unknown, but it is estimated that millions of people in the developing world are currently infected with the tapeworm and may become symptomatic at some point in the future. The highest rates of infection are found throughout South America, Mexico, Africa, and certain parts of Asia. In these countries, neurocysticercosis is the leading cause of acquired epilepsy and frequently is misdiagnosed as a brain tumor. In the southern United States, incidence is on the rise. Ninety percent of all diagnosed cases are in migrants from Mexico. Treatment aims to contain the disease through the use of cysticidal drugs, such as praziquantel, together with corticosteroids. Antiepileptic drugs are used to contain seizures. If caught in time, and if neurological symptoms can be arrested, the prognosis is good, although damage done to the brain cannot be reversed.

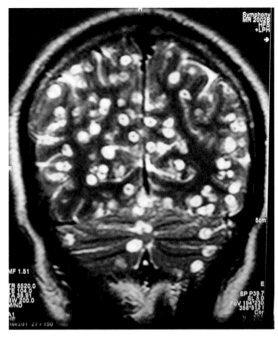

FIGURE 14 Coronal T_2-weighted image of the brain reveal multiple cystic lesions throughout the entire brain. That are markedly hyperintense with a hypointense mural nodule. *Image provided by Dr Rajesh Sharma DMRD, MD.*

4. BEYOND THE INFECTION: BONA FIDE BRAIN DISORDERS INVOLVING PATHOGENS

It is fascinating how microorganisms that may encode a single gene that leads to neurotropism (polio) can have such a profound reprogramming effect on the nervous system. Similarly, seemingly simple toxins can have such remarkable cell type and target specificity so as to selectively shut down, for example, GABAergic inhibition in tetanus. Equally interesting, and often much less understood, is the contribution that infectious agents may have to the etiology of nervous system disorders. This list is almost endless, yet the supporting evidence for a causal relationship between infection and disease is often weak. A few examples of infections strongly suspected in disease are described here.

4.1 Viral Etiology of Multiple Sclerosis: Epstein–Barr Virus

Being an immune or autoimmune disease at its core, a link to an infectious agent that initiates an immune attack on the body's own myelin being causal in MS is not far-fetched.[19] Unusual disease outbreaks, for example, that following the occupation of the Faroe Islands by British soldiers during World War II, are frequently attributed to an unknown infectious agent being imported into an otherwise unaffected population. Indeed, viral DNA can be detected in the CSF of patients who have MS. These include two common strains of herpes, namely the *Varicella zoster* virus that causes chickenpox and shingles and the Epstein–Barr virus that causes mononucleosis. However, given the prevalence of these viral infections in the general population (approaching 90%), a causal link to disease is difficult to fathom. That said, the 10% of the population not infected by the Epstein–Barr virus have a significantly lower risk of contracting MS. The most compelling argument for a virus causing MS comes from animal studies. Infection of SJL/J mice with the Theiler murine encephalomyelitis virus is sufficient to cause extensive demyelination and polio-like symptoms in these mice, yet other mouse strains, for example, C57/Bl6, are resistant to the virus. Similar differences in humans may explain differences in susceptibility to virally induced MS.

4.2 Viral Cause of Schizophrenia

The influenza epidemic of 1918–1919 took the lives of almost 100 million people worldwide. Among those who survived, there was a remarkable increase in the number of patients admitted to psychiatric hospitals suspected of suffering from what we now call schizophrenia. Since then, a viral etiology of schizophrenia has been debated, although a direct link has remained elusive. A link to viral infections has also been suggested to explain a peculiar correlation to month of birth, now well established, whereby babies born during the winter months, when rates of viral infection are increased, are at significantly higher risk of developing schizophrenia than those born during summer months. If indeed due to viral exposure, this of course would suggest the exposure of the unborn fetus through an infected mother. Numerous studies have examined the possibility that an infected mother exposes the unborn fetus to the influenza virus, thereby causing changes in the developing brain that result in an enhanced risk of developing schizophrenia. A sevenfold increase in schizophrenia risk was reported among children from mothers who were exposed to influenza in the first trimester of pregnancy and a threefold increase if exposure occurred later.[20] A recent genome-wide association study showed a fivefold elevated risk in mothers infected with cytomegalovirus whose child also harbors a newly identified genetic susceptibility locus.[21] Therefore it is possible that genetic predispositions to disease must coincide with an infection for the disease to develop.

4.3 Infection, Immunization, and Autism

As is the case in schizophrenia, a link to maternal exposure has been proposed for autism. This, too, is entirely based on epidemiological data, with a causal link being even weaker. It rests largely on the hypothesis that infection causes immune responses, which in turn cause the release of cytokines and activation of the brain's immune response. Until this can be demonstrated in animal models yielding neurocognitive symptoms consistent with autism, no compelling link can be established. Importantly, of course, the link between infection and autism gained unfortunate attention in 1998 from the now debunked claims by Dr Andrew Wakefield that falsely suggested that the childhood vaccine for measles, mumps, and rubella causes autism. These studies, published in the medical journal *The Lancet*, have since been retracted, and the author has been convicted of misconduct and stripped of his medical license, yet thousands of parents are still convinced that their child contracted autism via a childhood vaccine.

4.4 Prions in AD

As more extensively discussed in Chapter 4 (Aging, Dementia, and Alzheimer), a gradual spread of amyloid deposits containing misfolded amyloid β42 correlates with the gradual decline in memory function. More specifically, the disease typically originates in the hippocampus and surrounding entorhinal cortex and then gradually spreads into the frontal and parietal lobe, ultimately showing deposits throughout the cortical gray matter. Is it possible that cell-to-cell spread involves a prion-like mechanism whereby the misfolded amyloid β imprints its shape on amyloid β in adjacent cells, coaxing them to form amyloid deposits. In mouse models of AD containing a mutated amyloid precursor protein that encourages the production of amyloid β42, the unilateral inoculation of brain homogenate from symptomatic animals was sufficient to produce amyloid deposits throughout the cortex, even spreading to the contralateral brain.[15] A synthetically generated misfolded amyloid β was sufficient to infect both brain hemispheres. Consequently, it is conceivable that human disease begins in one part of the brain but then propagates from cell to cell in a prion-like fashion. That said, unlike in other prion diseases where the prion infects an unaffected individual, there is no evidence that AD can spread between people via prions.

5. EXPERIMENTAL APPROACHES/ CLINICAL TRIALS

Rather than focusing on experimental treatments for infections, all of which are immunological and microbiological in nature and do not teach us much about the nervous system, I focus instead on exciting advances where pathogens are used to treat neurological disease.

5.1 Medical Use of Neurotoxins (Botox)

In spite of being highly poisonous, Botox is now widely used clinically. Botox is particularly useful for hyperactive cholinergic nerves because it is selectively taken up by cholinergic neurons, where the toxin's uptake occurs in an activity-dependent manner. Both A and B toxins are commercially available under the trade names Botox and Neurobloc and are approved for a number of disease indications. The first medical uses of BTX-A in 1980 included treatment of crossed eyes, strabismus, and uncontrolled blinking (blepharospasm), with treatment effects lasting between 4 and 6 months. Medical use expanded and now includes spasticity associated with cerebral palsy, excessive sweating, cervical dystonia, anal fissure, diabetic neuropathy, chronic migraines, and many other conditions.[22] The benefit is the reversible, albeit long-lasting nature of the treatment without evidence of habituation.

Cosmetic use began in 1989 with the treatment of frown lines and has since expanded to include essentially all facial muscle groups. This use followed the accidental discovery that patients treated for blepharospasm (forced closure of the eyelid due to hyperactive muscles) showed decreased wrinkling around their eyes. A second serendipitous discovery followed when, with more extensive facial cosmetic use, patients with migraines were treated for wrinkles and showed a large decrease in the frequency and severity of their headaches. Note that for medical and cosmetic use the dosage is typically expressed as units, with 1 unit being the median lethal dose for a typical 20-g mouse. The median lethal dose for an average human is ~3000 units.

5.2 Viruses to Deliver Genes for Gene Therapy or Reprogramming of Endogenous Cells

Several viruses can deliver genes to the nervous system for insertion into the host genome. In the laboratory, lentivirus, adenovirus, adeno-associated virus (AAV), and HSV are widely used to transfect neurons.[23,24] Viruses are ideal vehicles that can carry foreign genetic information in the form of RNA or DNA to cells throughout the body and brain; the viral capsid protects the genetic information and specific surface receptors facilitating cell-specific entry. Moreover, depending on the virus used, replication at the site of treatment is possible, expanding the number of infected cells. Figure 15 summarizes the specific advantages of the most commonly used viruses for CNS delivery of genes and their applications for disease.

The most widely used virus is the **AAV**. It relies on coinfection with a helper virus to replicate and is therefore nonpathogenic and safe. AAV binds to heparin sulfate expressed on the surface of many nervous system cells and can integrate its DNA into a specific site on chromosome 19, thus permitting insertion of a foreign gene into the host DNA. AAV is used in ongoing clinical trials treating Parkinson disease, ALS, and epilepsy. Because they lack natural tropism, the virus is typically surgically injected and therefore gene delivery is regionally restricted. The virus is found in neurons, astrocytes, microglia, and endothelial cells.

Unlike AAV, the **HSV type 1 (HSV-1)** natively shows neurotropism, that is, it specifically infects neurons. HSV-1 is a large, double-stranded DNA virus that infects epithelial or mucosal tissue, where the virus replicates, assembles virions, and eventually lyses the cells, causing the typical superficial herpetic sores. Sensory nerve endings take up the virions and transport the viral genome by a retrograde mechanism back to the cell bodies contained within the dorsal root ganglia or the trigeminal ganglia. The virus often enters a latent stage during which essentially all viral genes are turned off. The infection may never reoccur, but certain factors can trigger reactivation, whereupon the virus generates virions that are transported back to the nerve terminal where they originally entered the cell, once again causing lytic infection resulting in the herpetic lesion, sometimes called fever blisters, on the lips. The virus enters neurons preferentially by binding to a nerve growth factor–like HSV receptor, conferring natural neurotropism. Genes can be packaged into the viral genome for delivery to the neuronal cell nucleus. Moreover, the virus can be constructed such that it is replication deficient and simply becomes a vector for the cell type–specific delivery of genes to be inserted into the neuronal nucleus. One immediate clinical application is treatment of chronic pain, originating from the very sensory neurons that normally harbor HSV-1 through delivery of an opioid peptide precursor such as pre-proenkephalin. Inoculation of the CNS allows retrograde transport to CNS neurons as well. This strategy is being now exploited in many preclinical studies,[25] including, for example, in Parkinson disease, where inoculation of the caudate nucleus allows the retrograde delivery of neuroprotective genes. In ongoing clinical studies HSV-1 is being used to deliver suicide genes to kill brain tumor cells.

5.3 Weapons and Bioterrorism

Although beyond the scope of this book, it is important to recognize that neurotropic pathogens such as polio or neurotoxins such as Botox could be used as weapons for bioterrorist attacks. It is therefore important to safeguard the reagent stocks, preventing them from falling into the wrong hands. Fortunately, some of the most toxic substances, such as Botox, are difficult to deliver. Simply adding it to drinking water, for example, would be insufficient because the toxin rapidly degrades in oxygenated and chlorinated water. However, if it enters the food chain, for example, in milk, hundreds of thousands of individuals could be exposed in a short period of time.[26]

Vector	Specifications	Application in the CNS
Adeno-Associated Virus (AAV)  ~20 nm	Genome: ssDNA Capacity: ~4.7 kb (~2.2 kb with scAAV, ~8 kb with dual vectors) Forms circular and linear episomes; integrates with very low frequency Shown to infect neurons, astrocytes, glial, and ependymal cells	Used extensively in clinical trials, including Parkinson, Alzheimer, Batten, and Canavan diseases. Preliminary studies suggest AAV vectors could also be used to treat mucopoly-saccharidoses (MPS), spinocerebellar ataxia, amyotrophic lateral sclerosis (ALS), epilepsy, and huntington disease.
Retrovirus: Human Immunodeficiency Virus (HIV)  ~100 nm	Genome: ssRNA Capacity: ~8 kb NIL vectors form linear and circular episomes; integration is low. Other HIV vectors integrate with high efficiency Shown to infect neurons and astroglial cells	Used in clinical trials for treatment of Parkinson and Alzheimer diseases. Vectors are being developed for use with Huntington and lysosomal storage diseases.
Adenovirus 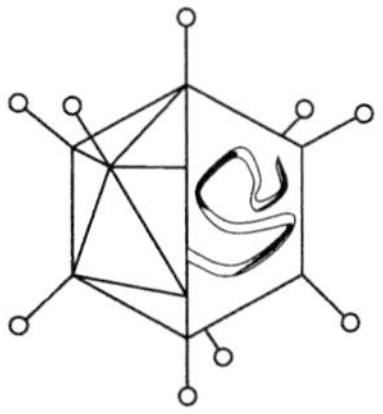 ~70–100 nm	Genome: dsDNA Capacity: ~36 kb Maintained as linear episomes; integration is minimal even with extensive homology to genome Shown to infect neural, astroglial, and human glioma cells	Not in clinical use for gene therapy in the CNS due primarily to vector toxicity. Has been used for oncolytic potential as an anti-cancer agent.
Herpesvirus: Herpes Simplex Virus-1 (HSV-1)  ~186 nm	Genome: dsDNA Capacity: ~150 kb Genome circularizes upon entering nucleus and is maintained episomally; integration is minimal Shown to infect neurons	Not in clinical use for gene therapy in the CNS due to problems with vector toxicity and production. Vectors are being developed for use with Parkinson disease. Has also been developed for anti-cancer therapy.

FIGURE 15 Viruses suitable for gene delivery to the central nervous system. dsDNA, double-stranded DNA; ssDNA, single-stranded DNA. *From Ref. 24.*

6. CHALLENGES AND OPPORTUNITIES

By their nature, infections of the nervous systems exist at the intersection of immunology and neurology. As evident in polio, measles, and meningitis, effective treatments and disease prevention rest entirely in the hands of immunologists and microbiologists, with little help from the neuroscience community. The success in eradicating polio and reducing the incidence of meningitis and shingles through vaccination in the Western world should be stark reminders of the power of disease prevention through vaccines. However, most of these efforts are successful only if the vast majority of the population participates so as to achieve "herd immunity." Unfortunately, this is not ensured in certain parts of the world where some of these diseases are making a comeback. Equally troubling, an increasing number of families in industrialized countries elect to forgo immunization of their children on dubious philosophic grounds, putting their children and others at an unnecessary risk of potentially deadly diseases. Egregious health disparities exist around the globe, with much higher incidence of preventable disease in developing countries, where even easily treatable diseases such as syphilis and tetanus still claim millions of lives.

From a neuroscience perspective, the study of infectious diseases has revealed fascinating insight regarding the action of neurotoxins and their selective transport along axons. It has provided valuable experimental tools, for example, neurotropic viruses, with which to deliver genes into specific cells, or therapies such as Botox, with which to treat muscle spasms and even migraines. It is likely that viral gene delivery will play a major role in the treatment of neurological diseases in the future, and such trials are already ongoing for ALS and Parkinson disease. While the neurology and neuroscience community is least likely to influence the course and outcome of infectious disease per se, it stands to benefit tremendously from a better understanding of host–pathogen interactions.

Most parasitic infections of the brain, be it through single or multicellular organisms, are medical curiosities that affect just a few individuals each year and for which treatment and prevention will likely remain elusive, in part because of the low incidence and urgency. By contrast, the single greatest challenge regarding CNS infections is the still expanding HIV crisis. Unknown just 40 years ago, and probably first transferred from monkeys to humans just 50 years ago, this virus has taken the world by storm, creating an epidemic of unimaginable proportion. CNS infection was initially not considered to be a major problem but is now recognized to be a major comorbidity. Particularly with the introduction of highly effective antiviral therapies, HIV-infected people now live long enough to develop significant cognitive decline, even dementia, and in some patients the neurological symptoms are the major impediment. Therefore, management of patients with HIV now increasingly involves neurologists as well. Ultimately, only the development of a vaccine with worldwide immunization campaigns, similar to those that eradicated polio, will contain his health epidemic.

Acknowledgments

This chapter was kindly reviewed by Dr William Britt, the Charles A. Alford Professor of Pediatric Infectious Diseases at the University of Alabama at Birmingham.

References

1. Erbguth FJ, Naumann M. Historical aspects of botulinum toxin: Justinus Kerner (1786–1862) and the "sausage poison". *Neurology*. November 10, 1999;53(8):1850–1853.
2. Chai Q, Arndt JW, Dong M, et al. Structural basis of cell surface receptor recognition by botulinum neurotoxin B. *Nature*. December 21, 2006;444(7122):1096–1100.
3. Colasante C, Rossetto O, Morbiato L, Pirazzini M, Molgo J, Montecucco C. Botulinum neurotoxin type A is internalized and translocated from small synaptic vesicles at the neuromuscular junction. *Mol Neurobiol*. August 2013;48(1):120–127.
4. Binz T, Rummel A. Cell entry strategy of clostridial neurotoxins. *J Neurochem*. June 2009;109(6):1584–1595.
5. Thwaites C, Yen L. Chapter 140. Tetanus. In: Longo DL, Fauci AS, Kasper DL, Hauser SL, Jameson J, Loscalzo J. eds. Harrison's Principles of Internal Medicine, 18e. New York, NY: McGraw-Hill; 2012.

6. Ghanem KG. Review: neurosyphilis: a historical perspective and review. *CNS Neurosci Therapeutics*. October 2010;16(5):e157–e168.

7. Jackson AC. Chapter 195. Rabies and Other Rhabdovirus Infections. In: Longo DL, Fauci AS, Kasper DL, Hauser SL, Jameson J, Loscalzo J. eds. Harrison's Principles of Internal Medicine, 18e. New York, NY: McGraw-Hill; 2012.

8. Fauci AS, Lane H. Human immunodeficiency virus disease: AIDS and related disorders. In: Longo DL, Fauci AS, Kasper DL, Hauser SL, Jameson J, Loscalzo J. eds. *Harrison's Principles of Internal Medicine*, 18e. New York, McGraw-Hill; 2012.

9. Clifford DB, Ances BM. HIV-associated neurocognitive disorder. *Lancet Infect Dis*. November 2013;13(11):976–986.

10. Mirza A, Rathore MH. Human immunodeficiency virus and the central nervous system. *Semin Pediatr Neurol*. September 2012;19(3):119–123.

11. Kaul M, Garden GA, Lipton SA. Pathways to neuronal injury and apoptosis in HIV-associated dementia. *Nature*. April 19, 2001;410(6831):988–994.

12. Kretzschmar H, Tatzelt J. Prion disease: a tale of folds and strains. *Brain Pathol*. May 2013;23(3):321–332.

13. Prusiner SB, Miller BL. Chapter 383. Prion diseases. In: Longo DL, Fauci AS, Kasper DL, Hauser SL, Jameson J, Loscalzo J. eds. *Harrison's Principles of Internal Medicine*, 18e. New York, McGraw-Hill; 2012.

14. Aguzzi A, Nuvolone M, Zhu C. The immunobiology of prion diseases. *Nat Rev Immunol*. December 2013;13(12):888–902.

15. Stohr J, Watts JC, Mensinger ZL, et al. Purified and synthetic Alzheimer's amyloid beta (Abeta) prions. *Proc Natl Acad Sci USA*. July 3, 2012;109(27):11025–11030.

16. Visvesvara GS. Free-living amebae as opportunistic agents of human disease. *J Neuroparasitol*. 2010;1:1–13.

17. Yoder JS, Straif-Bourgeois S, Roy SL, et al. Primary amebic meningoencephalitis deaths associated with sinus irrigation using contaminated tap water. *Clin Infect Dis*. November 2012;55(9):e79–e85.

18. Del Brutto OH. Neurocysticercosis. *Continuum (Minneapolis, Minn)*. December 2012;18(6 Infectious Disease):1392–1416.

19. Owens GP, Gilden D, Burgoon MP, Yu X, Bennett JL. Viruses and multiple sclerosis. *Neuroscientist*. December 2011;17(6):659–676.

20. Yudofsky SC. Contracting schizophrenia: lessons from the influenza epidemic of 1918–1919. *JAMA*. January 21, 2009;301(3):324–326.

21. Børglum AD, Demontis D, Grove J, et al. Genome-wide study of association and interaction with maternal cytomegalovirus infection suggests new schizophrenia loci. *Mol Psychiatry*. 2014;19(3):325–333.

22. Montecucco C, Molgo J. Botulinal neurotoxins: revival of an old killer. *Curr Opin Pharmacol*. June 2005;5(3):274–279.

23. Frampton Jr AR, Goins WF, Nakano K, Burton EA, Glorioso JC. HSV trafficking and development of gene therapy vectors with applications in the nervous system. *Gene therapy*. June 2005;12(11):891–901.

24. Lentz TB, Gray SJ, Samulski RJ. Viral vectors for gene delivery to the central nervous system. *Neurobiol Dis*. November 2012;48(2):179–188.

25. Berges BK, Wolfe JH, Fraser NW. Transduction of brain by herpes simplex virus vectors. *Mol Ther*. January 2007;15(1):20–29.

26. Wein LM, Liu Y. Analyzing a bioterror attack on the food supply: the case of botulinum toxin in milk. *Proc Natl Acad Sci USA*. July 12, 2005;102(28):9984–9989.

General Readings Used as Source

1. Harrison's Chapter 381 Meningitis, Encephalitis, Brain Abscess, and Empyema.

2. Karen L, Roos Kenneth L, Tyler.

3. Merritt's Neurology, 10th edition.

4. Botulism in the United States, Center for Disease Control (CDC) www.cdc.gov.

5. van den Pol AN. Viral infections in the developing and mature brain. *Trends Neurosci*. 2006;29(7):398–406.

6. Nathanson N. The pathogenesis of poliomyelitis: what we don't know. *Adv Virus Res*. 2008;71:1–50.

7. Colby DW, Prusiner SB. Prions. *Cold Spring Harb Perspect Biol*. 2011;3(1):a006833.

Suggested Papers or Journal Club Assignments

Clinical Paper

1. Yoder JS, et al. Primary amebic meningoencephalitis deaths associated with sinus irrigation using contaminated tap water. *Clin Infect Dis*. 2012;55(9):e79–e85.

2. Zink MC, et al. Neuroprotective and anti-human immunodeficiency virus activity of minocycline. *JAMA*. 2005;293(16):2003–2011. ......Together with.

3. Nakasujja N, et al. Randomized trial of minocycline in the treatment of HIV-associated cognitive impairment. *Neurology*. 2013;80(2):196–202.

Basic

1. Zhang K, et al. HIV-induced metalloproteinase processing of the chemokine stromal cell derived factor-1 causes neurodegeneration. *Nat Neurosci*. 2003;6(10):1064–1071.

2. Wein LM, Liu Y. Analyzing a bioterror attack on the food supply: the case of botulinum toxin in milk. *Proc Natl Acad Sci USA*. 2005;102(28):9984–9989.

3. Chai Q, et al. Structural basis of cell surface receptor recognition by botulinum neurotoxin B. *Nature*. 2006;444(7122):1096–1100.

PART IV

DEVELOPMENTAL NEUROLOGICAL CONDITIONS

Neurodevelopmental Disorders

Harald Sontheimer

1. CASE STUDY

We just celebrated Haley's third birthday. The gathering included grandparents and family. What was missing, however, were Haley's friends or, more accurately, the ones she once had. As a programmer at Google, I was fortunate to have access to free day care provided conveniently on the company campus. This allowed me to return to work barely 6 weeks after delivering Haley, leaving her with the talented staff at the day care. Whenever I could, I spent my lunch hours visiting her, primarily to put my mind at ease for leaving her with strangers so early in life. Haley developed normally, it seemed. She mumbled her first words by 7 months and started walking by 9 months. We celebrated her first birthday with 20 other children at the day care, singing songs and eating cupcakes. What a joy it was! The following year, however, was different. Whenever I visited for lunch, Haley became increasingly disinterested in my presence and kept moving toys around the table, hardly noticing me. She didn't like to be held or cuddled either. As a first-time mother I assumed that spending 8–10 h daily in day care was taking its toll on our relationship. I seriously contemplated quitting my job and becoming a full-time mother. The more I visited with her, the more I noticed how withdrawn she became. As we approached her second birthday, the teaching staff at the day care center asked for a meeting. Maika, the instructional director, explained that Haley was having trouble. She was not participating in group play and was either withdrawn or outright disruptive, throwing toys all over the floor. She did not pay attention when books were read to her and often ignored her teachers when called by name. I, too, had noticed that she typically did not respond if I called her by name. Again, guilty for pursuing my career, I assumed that it was Haley's way of complaining. Maika recommended a consultation with Haley's pediatrician or a child psychologist. She suggested that some of the features she noticed were common in children with autism. At her 2-year wellness visit her pediatrician noted that her growth rate had declined, and her head circumference dropped from the 95th to the 5th percentile. Her hands, too, seemed to be growing more slowly. He suggested an evaluation by a child psychologist and a genetic test. Shortly thereafter we learned that Haley carries a mutation in a gene called *MECP2* that causes Rett syndrome (RTT).

I had never heard of this disease and was eager to learn all about the condition. The geneticist who counseled us during the visit instructed us to refrain from Googling the condition because we would probably only find worst-case scenarios. Without question, the past months have been the most challenging of my life. Haley's mobility has worsened. I can't understand how she had been running around the house last Christmas and now can barely stand on her own feet. Just as troubling, what little language she had developed seems to be gone. She spends her days making cooing noises at best. She makes tick-like, hand-wringing motions and generally does not take part in our family life. I go between anger and grief, feeling like I have lost my child and blaming myself for not being with her more during the past 3 years. I often have nightmares just thinking about what lies ahead.

2. HISTORY

Genetic defects that impair the orderly development of the brain have coevolved with the brain's increasing complexity. This chapter explores several neurodevelopmental disorders that commonly present with intellectual disability ranging from mild to severe. Some of these conditions present with degrees of impairment requiring significant medical intervention and societal support. For example, many babies born with Down syndrome (DS) need early corrective surgery for a life-threatening heart defect, and girls born with RTT are at risk of dying in the second or third year of life due to respiratory syndromes, seizures, and the inability to feed themselves. Historically, individuals suffering such impairments may not have survived long, explaining a near absence of accounts of these illnesses in the historical literature. After all, infant mortality was generally very high in those days. At the same time, mild forms of autism may have gone largely unrecognized because their symptoms fall within the spectrum of behavioral

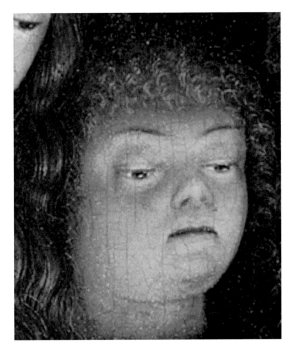

FIGURE 1   A person with Down syndrome depicted in a painting by an unknown painter entitled "The Adoration of the Christ Child," c. 1515. This image provides visual evidence that Down syndrome is a very old disease and occurred even during times when childbirth later in life was uncommon. Image has been cropped and magnified. ©*The Metropolitan Museum of Art.*

abnormalities that occur in the general population. A person may have simply been labeled odd or asocial. A sixteenth-century painting (Figure 1) shows a child with facial features consistent with DS, suggesting that this condition was known at least as far back as the Middle Ages. However, our first reliable accounts go back barely 150 years when DS, then called mongolism, was first described by John Langdon Down, a British obstetrician who served as medical superintendent at the Royal Eastwood Asylum for Idiots. In a paper published in 1866 Down described a group of people under his care with characteristic facial features that gave them the name mongolians. He established a private housing facility, Normansfield, to care for these patients. He advocated kindness, loving care, and patience

toward these institutionalized individuals. Normansfield remained in operation through several generations of John Landon Down's children and grandchildren, who followed in their father's footsteps, devoting themselves to the mentally disabled. In sensitivity to the Mongolian race and appreciation of Down's contributions to its early etiology, this condition was renamed DS in 1961, just 2 years after its genetic cause was revealed.

The discovery that a triplication of chromosome 21 causes DS led to the development of prenatal chromosomal screening through medical cytogenetics. While this technique was still in its infancy, two brothers who presented delayed development with intellectual impairments were referred to Yale Medical School in 1966 for chromosomal testing. Both brothers showed identical constrictions in the distal long arm of their X chromosome. When the boys' extended family was examined, an uncle and great-uncle were found to harbor the same chromosomal abnormality. This constriction was soon identified as a cytogenic marker for a disease named fragile X. Cloning of the disease-causing gene encoding the fragile X mental retardation protein had to wait until 1991.

In 1943 the American child psychiatrist Leo Kanner described eight boys and three girls in his care who showed common but strange behavioral abnormalities. These included a profound disinterest in people, a focus on objects, repetitive behaviors, and a delay or even absence of language development. The following year the Austrian pediatrician Hans Asperger described four boys with a similar focus on objects and a general disinterest in playing with other children. By contrast, however, they talked very early, although they lacked a deeper understanding of their own language. Both groups of children have features that are now commonly associated with autism spectrum disorder (ASD), a condition that only recently gained attention because of a false link to a common childhood vaccine and epidemiological studies suggesting an increase in disease incidence of epidemic proportion.

In 1966 the Austrian pediatrician Andreas Rett reported an unusual illness affecting only girls who, after a short period of normal development, lost language ability and purposeful movements of their hands, which were gradually replaced with stereotypic hand-wringing motions. This disease, too, received little attention until the late 1980s, when the fortuitous convergence of neurology and molecular biology showed that mutations in a single gene, the MECP2 transcriptional regulator, are responsible for this disease, named RTT in recognition of Andreas Rett.

Common to these four diseases, and indeed to a much larger group of related neurological conditions, is variably impaired intellectual function. Historically, this was labeled with derogatory terms that include idiocy, feeblemindedness, mental retardation, and mental subnormality. Many of these affected individuals were excluded from society and locked away in institutions for life or worse. Following the passage of the Americans with Disabilities Act in 1979, responsiveness and sensitivity to disability has gradually increased. This is reflected in the adoption of *developmental intellectual disabilities* as the umbrella term that characterizes inborn or developmentally acquired impairments of cognitive function. Integration of affected children in mainstream educational programs is now standard, but their long-term assimilation into society remains a challenge. In spite of their relatively short histories, the four diseases discussed in this chapter are among the most insightful and encouraging medical studies in neuroscience and neurology.

3. DEVELOPMENT OF SYNAPSES IN THE HUMAN CORTEX AND DISEASES THEREOF

The list of neurodevelopmental disorders is long and heterogeneous. It includes diseases with genetic causes as well as conditions resulting from pre- or postnatal insults such as hypoxia, alcohol or toxin exposure, nutritional deficiencies, infection, trauma, or metabolic changes. Many of these are poorly understood, and gene defects are often unknown. However, several diseases provide excellent examples of where careful neuroscience research has informed our understanding to the point where novel and unexpected treatments are emerging. Four diseases are singled out here: DS, fragile X syndrome (FXS), RTT, and autism; these share a common disease phenotype, namely, a change in synaptic structure and function. In spite of this shared phenotype, the affected synapses and the underlying genetic causes are different and illustrate a spectrum of disease-causing changes. In DS, an extra copy of chromosome 21 provides aberrant expression of numerous proteins involved in synaptic structure and function. In FXS, the loss of just one protein (fragile X mental retardation protein (FMRP)) causes dysregulation of the activity-dependent translation of many proteins underlying spine development and morphology. In RTT, a single mutated enzyme regulates the expression of numerous genes through DNA methylation, thereby indirectly affecting many proteins involved in synapse development and cognitive function. Finally, in autism, aberrant synaptic connectivity within the cortex leads to the suspected presence of mutations in synaptic genes, as shown in rare cases, affecting pre- and postsynaptic development and function. Following a brief introduction to normal cortical synapse development, each of these conditions is discussed in turn within self-contained sections, each encompassing clinical presentation, epidemiology, and neurobiology of the disease and current and future treatments.

Phases of Normal Synaptic Development

Three largely sequential developmental processes have lasting influence over the proper functioning of the cortex. These three phases are graphically illustrated in Figure 2 and proceed at different times and rates in different parts of the brain. First, during neurogenesis, birth,

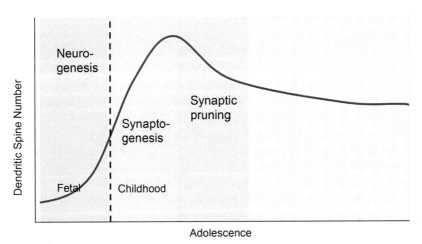

FIGURE 2   Changes in dendritic spine numbers during defined stages of brain development. A significant increase in the number of synaptic spines occurs during late fetal and early childhood development. Eventually, up to 40% of all spines are eliminated by a process called pruning. A correct number of synaptic spines is essential for normal cognition; both too few and too many spines cause cognitive abnormalities.

migration, and differentiation of neurons from neural progenitor cells ensure that an appropriate number of cells populate defined layers of the cerebral cortex. About twice as many neurons are born than are actually needed, and half of them eventually undergo programmed cell death, or apoptosis. It is believed that competition for neurotropic factors such as brain-derived neurotrophic factor (BDNF), generated by target neurons or surrounding glia cells, plays an important role in selective neuronal survival. Neurogenesis is followed by synaptogenesis, when neurons extend dendritic processes and make synaptic contacts, forming functional networks. Histological studies suggest that in the human cortex synaptogenesis begins in the fifth month of gestation and continues well into childhood. Initially, there are few, thick dendrites with very few spines. Over time the complexity of the dendrites increases and they become increasingly studded with spines (Figure 3). Once synaptogenesis is completed, the cortex undergoes an extended period of activity-dependent synapse elimination, or pruning, of unnecessary inactive synapses and selective strengthening of active sites of communication.

Throughout the brain, synaptogenesis and synaptic pruning commence at different times and proceed at different rates. For example, visual cortex synaptogenesis is most active 2–4 months after birth, with maximal synaptic density attained at 8 months. Then synapse elimination begins, and by 11 months, 40% of synapses are already lost. By contrast, in the frontal cortex, the maximal number of synapses is reached much later (24 months) and, while the extent of pruning is similar (40%), it occurs much more slowly, extending into late childhood.

These stages of development are tightly regulated by genes and environmental factors including neurotransmitters and tropic factors. Among the better known are metabotropic glutamate (Glu) receptors (mGluRs), N-methyl-D-aspartate receptor, and growth factors that signal via tropomyosin-related tyrosine kinase (Trk) receptors. These mobilize intracellular Ca^{2+}, causing the structural and functional changes required for development. Both genetic and environmental factors have the potential to disturb these signaling axes, causing aberrant cortical development and resulting in cognitive

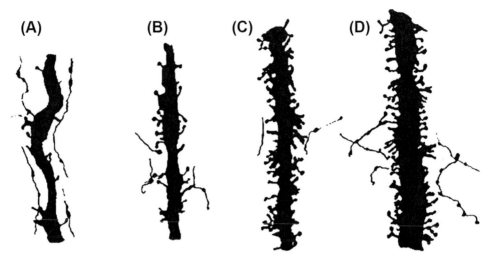

FIGURE 3   A developmental increase in the number and complexity of dendritic spines in layer 5 pyramidal cells. Microscopic drawings depict the appearance of dendrites in the human cortex at different stages of development: fifth gestational month (A), seventh gestational month (B), neonatal period (C), and second postnatal month (D). The complexity and number of spines change markedly. *Reproduced with permission from Ref. 1.*

impairments later in life. Several examples where this is the case are described below.

The development of these neural circuits goes hand in hand with the development of glia, and many, if not all, developmental processes require glial–neuronal interactions. Glial cells originate primarily from progenitor cells of the subventricular zone, and these continue to divide after they have migrated into the cortex. The first glial populations that form are astrocytes and radial glial cells, with processes that extend from the ventricular zone to the pial surface. These frequently serve as guides for neuronal migration. Oligodendrocytes are born after astrocytes, and some oligodendrocyte progenitor cells, called NG2 cells, are retained throughout life. The role of oligodendrocytes in development extends even further into the life span because functional nerve pathways require myelination and tropic support of the axon by oligodendrocytes. Myelination persists into early adulthood. Changes in glial development can directly or indirectly influence normal cortical development, and any aberration can contribute to disease.

Based on the phases of cortical development described above, there are defined stages of vulnerability during which developmental diseases can develop or first manifest. These include the prenatal (up to birth), perinatal (early after term), childhood, and adolescent phases. During these stages, there can be a primary insult or disturbance, such as a gene defect manifesting with altered synaptic pruning, or a secondary insult such as viral infection, trauma, hypoxia, neglect, or a combination thereof.

4. DOWN SYNDROME

DS is the most well-known chromosomal disorder in newborns. It affects approximately 1 in 800 children and accounts for 10% of all severe mental illnesses. Children have a characteristic appearance: a round head, poorly developed nasal ridge, and overall flattened face. Patients have an enlarged tongue, small but broad hands, and short fingers. Their growth is stunted and, as adults, they rarely grow taller than an average 10-year-old. 30% of patients with DS have congenital heart defects, with the most common being incomplete separation of the heart chambers (atrioventricular septal defect), which requires surgical correction. Seizures, too, are a common comorbidity, affecting ~5–10% of all patients. Developmental milestones are typically delayed, and children with DS begin to walk at age 3–4. They have delayed development of speech, yet 90% talk by age 5. Mental capacity is reduced, with an intelligence quotient ranging from 20 to 70; the average intelligence quotient for an individual is 50 and for a normal person the average is 100. Individuals with DS have specific memory defects involving explicit long-term memory, such as remembering names, facts, or formulas, as well as verbal short-term memory. However, procedural implicit long-term memory, such as using a keyboard or riding a bicycle, is unaffected. These facts point toward an impairment in learning circuitries involving the hippocampus and frontal lobe.

Non-nervous system comorbidities include gastrointestinal defects, compromised immune function, and obesity. DS significantly reduces the average life expectancy, with signs of Alzheimer disease (AD) developing by age 35–40 and death by age 50. The triplicated chromosome 21 harbors the amyloid precursor protein (APP), which gives rise to disease-causing amyloid plaques, as in AD (see Chapter 5); in DS, amyloid plaques begin to form in the second decade of life. The risk for premature onset of AD in individuals with DS is 100%. Since the duplication of chromosome 21 occurs more commonly in pregnancies of older mothers, amniotic chromosomal testing and genetic counseling are now routinely offered to expecting mothers older than 35. Individuals with DS are typically very affectionate and loving and, in spite of the challenges of raising children with a disability, their unique personality makes caring for people with DS more rewarding than for many other neurological diseases.

4.1 Neurobiology of DS

Individuals with DS have striking abnormalities in cortical development, including reduced dendritic branching and diminished synaptic density. The abnormalities also include changes in the lamination of the cortex that only become apparent after the 22nd week of gestation. Surprisingly, dendritic number is increased at birth, yet by 6 months of age and later the cortex has a reduced number of dendrites that are less elaborate. This applies particularly to the GABAergic neurons in layers II and IV of the cortex. Reduced GABAergic inhibition in the cortex may explain why 5–10% of individuals with DS present with epileptic seizures. Glutamatergic synapses also seem to be altered; the synaptic spines are larger and fewer in number.

Although the cause of DS being a triplication of chromosome 21 was identified in 1959, it was not until after the sequencing of chromosome 21 in 2000, followed by the entire human genome in 2003, that we gained a more comprehensive understanding of all the affected genes. Not surprisingly, a flurry of research has since provided a more comprehensive understanding of the cellular and molecular changes in DS, particularly relating gene amplifications to specific disease symptoms. Overall, the triplication affects up to 400 genes and several regulatory microRNAs. These genes are present at 1.5-fold normal levels (from three as opposed to two chromosomal copies), and because of gene dosage effects one would expect a 50% increase in transcribed proteins. However, the resulting changes in gene and protein dosage are much more complex.[2] Some of the triplicated proteins are transcriptional regulators, whereas others are involved in post-translational modifications. Consequently, changes in gene dosage can either suppress or enhance transcription of other genes or can change the nature of the resulting protein through post-translational changes. If one considers additional epigenetic regulations, the outcome at the protein level would be impossible to predict. Importantly, genes on chromosome 21 can influence the transcription and translation of genes on other chromosomes, causing genome-wide changes. This complexity led to two competing but not mutually exclusive hypotheses. The gene dosage effect hypothesis postulates that the disease phenotype is a direct consequence of dosage imbalances on chromosome 21, whereas the amplified developmental instability hypothesis argues that dosage imbalances in the hundreds of affected genes cause nonspecific changes in genome-wide expression. It is important to emphasize that all the affected genes are normal and the disease symptoms are entirely due to a change in the abundance of these proteins. The disease is therefore a result of quantitative rather than qualitative changes.

Some important genes are found on chromosome 21 and play a role in nervous system function and are listed in Table 1. Their potential roles in disease are discussed in more detail below. Most of these trisomic genes show transcriptional

increases of 1.5-fold, and only a small number of genes deviate from a gene dosage effect.

Not all patients have a full triplication of chromosome 21; some patients instead have a partial triplication. This gave rise to the idea that a critical chromosomal region called the Down syndrome critical region accounts for the principal hallmarks of DS. This notion is somewhat controversial. However, currently available mouse models of DS also triplicate only subsets of genes. The most common mouse model of DS, the Ts65Dn mouse, has triplicate copies of 150 of the affected genes, which, in the mouse, are largely found on chromosome 16. These mice are smaller in size and have similar craniofacial abnormalities as humans with DS and show impairments in hippocampal-dependent long-term memory tasks. They show a reduced number of neurons with reduced complexity in dendritic branching. As is the case in patients with DS, Ts65Dn mice show a reduced density

TABLE 1   Nervous System Genes/Proteins with Altered Expression due to Triplication of Chromosome 21 in Down Syndrome (DS)

Gene	Protein	Dosage in DS
APP	Amyloid βA4	1.5×
SOD1	Superoxide dismutase 1	Increased
SYNJ1	Synaptojanin 1	Increased
OLIG1/2	Oligodendrocyte transcription factor	1.5×
DSCR1	Down syndrome critical region gene 1	1.5×
DOPEY2	DOPEY2	1.5–2×
CLDN14	Claudin 14	Increased
KCNJ6	GIRK2 K$^+$ channel	1.5×
PCP4	Purkinje cell protein 4	Increased
S100b	S100 calcium-binding protein	Increased
SIM2	Transcription factor	1.5×
TRPD	TRPD/TTC3	>2×
DYRK1A	Drosophila minibrain gene homolog	1.5×

Table was established using data in Ref. 2.

of synaptic spines, indicative of reduced excitatory input, and an enlargement of spine heads (Figure 4). Moreover, the placement of GABAergic inhibitory synapses is altered. These now innervate at the neck as opposed to the dendritic shaft, which in all likelihood enhances their inhibitory effect in the hippocampus (Figure 4, Schema). As a consequence of the synaptic changes, the hippocampus of these mice exhibits long-term potentiation (LTP) and greatly increased long-term depression (LTD), the two cellular substrates for synaptic plasticity believed to underlie learning and memory (Figure 5; see also Box in Chapter 4). This seems to be due largely to enhanced GABAergic inhibition since the GABAa receptor antagonist picrotoxin restores LTP. γ-Aminobutyric acid (GABA) also activates GABAb receptors, which reduce neuronal excitability via G protein–coupled K+ channels (GIRK2). Since GIRK2 is expressed at 1.5-fold

normal levels in DS, this further enhances the inhibitory tone in the hippocampus.

4.2 Genes Affected by the Triplication of Chromosome 21 and Their Potential Link to DS Symptoms

The triplication of chromosome 21 affects a number of nervous system–specific genes. Most of these follow a gene dosage effect and are 1.5-fold more abundant, whereas others show even greater enhancement or near complete loss. Interestingly, several of these genes link their function to the observed structural and functional defects in DS.

Two proteins may directly explain the enhanced GABAergic inhibition in DS. These are the G protein–coupled K+ channel GIRK2, encoded by the *KCNJ6* gene that is upregulated

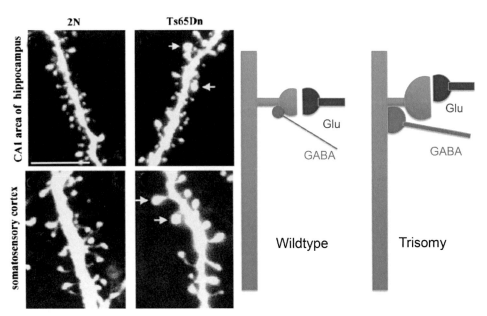

FIGURE 4   Changes in spine morphology in Down syndrome include a markedly reduced number of spines that are abnormally enlarged. These images were taken in the hippocampus and somatosensory cortex of the Td65Dn mouse model of Down syndrome compared with normal (2N) control mice. The microscopic images on the left were reproduced with permission from Ref. 3. In addition to their changed morphology, GABAergic inhibitory synapses are located at the base of the spines (as indicated in the drawing, right), where their inhibitory effect is therefore much larger, as indicated in the drawing by a larger inhibitory terminal (red) in the Trisomy case to illustrate the increase in inhibitory tone compared to Wildtype.

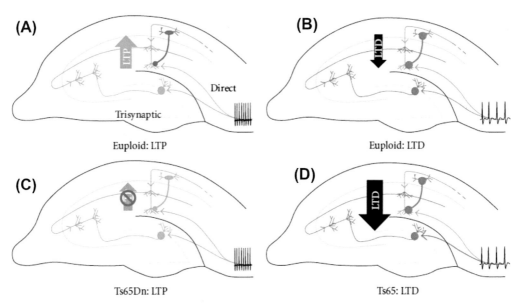

FIGURE 5   Down syndrome leads to changes in the cellular circuits underlying learning and memory. Specifically, a reduction of long-term potentiation (LTP) and amplification of long-term depression (LTD) is seen in hippocampal slices from Ts65Dn mice used as model for Down syndrome in humans. Schematic of two main pathways through hippocampus arriving from the entorhinal cortex: temporoammonic (TA)—direct to CA1 distal dendrites; trisynaptic pathway from dentate gyrus (DG) through CA3 to proximal CA1 dendrites. (A) In euploid hippocampi, high-frequency inputs induce LTP in CA1 resulting in enhanced suppression of inputs from TA by feed-forward inhibition arising from interneurons in stratum oriens. (B) Low-frequency inputs depress the trisynaptic pathway releasing distal CA1 dendrites from feed-forward inhibition and allowing information to flow through the TA pathway. (C) In Ts65Dn hippocampi, aberrant LTP in CA1 results in diminished feed-forward inhibition during high-frequency activity allowing TA inputs to become superimposed on those flowing through the trisynaptic pathway. (D) Enhanced LTD would be expected to facilitate flow of low-frequency information through the direct TA pathway in Ts65Dn mice. *Figure and Legend reproduced with permission from Ref. 4.*

in DS, and the transcription factors Olig1 and Olig2. GIRK2 is activated by the G protein–coupled GABA-B receptor, whereupon the opening of its K^+ pore stabilizes the resting potential. GIRK2 is overexpressed in human DS tissue and in mouse models of DS, and this overexpression causes enhanced neuronal inhibition. This particularly affects the input to the hippocampus and explains impairment in hippocampal-dependent learning (Figure 5). The transcription factors Olig1 and Olig2 regulate the neurogenesis of GABAergic interneurons. Their enhanced expression causes an overabundance of inhibitory neurons, which could be reversed by genetic restoration of just two copies of Olig1 and Olig2.[5]

Several proteins are involved in synaptic vesicle release or recycling. For example, DYRK1A is a dendritic protein associated with the GTPase dynamin that is involved in synaptic vesicle recycling and membrane trafficking. It modulates c-AMP response element binding protein, a c-AMP response element involved in synaptic plasticity. Overexpression of *DYRK1A* causes impaired spatial learning, making this gene a likely contributor to the cognitive defect in DS. It interacts with synaptojanin 1, a phosphatase that is required for synaptic vesicle endocytosis. Change in expression of *DYRK1A* impairs the retrieval of synaptic proteins and causes spatial memory deficits in mice. Synaptojanin 1 is further regulated by the Down syndrome critical region gene 1 (*DSCR1*), which dephosphorylates synaptojanin via the calcium-sensitive phosphatase calcineurin.

Several affected genes give rise to proteins involved in the regulation of cortical development and extension of neuronal dendrites. For example, *DOPEY2* regulates cortical cell density, and its 1.5-fold increase is likely to contribute to abnormal cortical lamination in patients with DS. *SIM2* is a transcriptional regulator that controls sonic hedgehog expression, which is involved in cell growth and differentiation. This overexpression causes mild impairment of learning and reduced sensitivity to pain.[6] The TPRD protein signals via the RhoA small GTPase to regulate neurite growth. Its overexpression in DS inhibits neurite extension, contributing to the aberrant cortical lamination.

Two overexpressed proteins directly link DS to neurodegenerative diseases and are likely responsible for premature aging and dementia in DS. Notably, the *APP* gene that gives rise to the APP protein, central to the pathology of AD, is located on chromosome 21 and hence is triplicated in DS, as is the Cu–Zn superoxide dismutase (SOD1), an enzyme of critical importance in the inactivation of oxygen free radicals. *SOD1* mutations contribute to the motor neuron degeneration underlying amyotrophic lateral sclerosis (Chapter 6). SOD1 overexpression causes impaired LTP and reduced protein clearance via the ubiquitin proteasome system.

While the above genes primarily affect neuronal proteins, chromosomal triplication also includes the secreted glial Ca^{2+} binding protein S100β, often associated with blood vessels. S100β is amplified in DS and AD, and, in transgenic mice, overexpression causes changes in dendritic spine density in the hippocampus, along with impairments in spatial learning, suggesting that an abnormal glial–neuronal interaction also contributes to the cognitive symptoms of DS.

Taken together, these genetic changes emerge as a common denominator explaining that the physiological abnormalities present in humans and mouse models are changes in synaptic structure and function. In DS these changes tend to shift the balance of excitatory and inhibitory activity toward exaggerated inhibition. The degree to which synaptic function is impaired, and the particular synapses involved, vary from individual to individual, which may explain the large range of the severity of symptomatic presentation. However, the convergence on a single functional compartment, the synapse, allows for a more focused approach in future treatments of this common developmental defect.

4.3 Emerging Treatments

Following the discovery of the chromosomal triplication causing DS in 1959, relatively little effort was put toward developing disease-modifying therapies. Given the complexity of the disease, it had been assumed that reversing the disease or meaningfully affecting its symptoms would not be feasible. This nihilism has been replaced by tacit optimism following preclinical studies with recently developed mouse models. Most excitingly, when the correct copy number of Olig1 and Olig1 is restored, the hippocampus expresses the correct number of inhibitory interneurons and essentially all cognitive symptoms were permanently corrected.[5] Similarly, memory function in Ts65Dn mice was corrected using three different GABAa receptor inhibitors (picrotoxin, bilobalide, and pentylenetetrazole).[7] These studies suggest, quite unexpectedly, that this complex and profound cognitive phenotype, long assumed to be permanent, may instead be completely reversible, providing hope that clinical trials may similarly find profound and lasting improvement in affected humans. A recently developed GABA-A receptor blocker (RG1662) has already been shown to be safe in a phase I study and is now being examined in a multicenter clinical study. In addition, GIRK2 channels, which are targeted by GABAb receptors, can be blocked with fluoxetine (Prozac), a common drug used to treat depression in adults. This drug could be readily applied to the treatment of DS if future clinical studies support this use. It is particularly exciting that these discoveries in neuroscience laboratories have resulted in renewed interest by pharmaceutical companies in a disease for which hope had been all but lost.

5. FRAGILE X SYNDROME

First described in 1943, FXS is the most common genetic form of intellectual disability and autism. It affects 1 in 3600 males and 1 in 8000 females. Affected individuals have a chromosomal expansion on the X chromosome containing a large number (>200) of trinucleotide CGG repeats. This expansion causes methylation and transcriptional repression of the adjacent fragile X mental retardation 1 gene (*FMR1*), which encodes the FMRP. FMRP is an RNA binding protein that inhibits

RNA translation into proteins and is typically associated with polyribosomes, the sites of protein synthesis.

Individuals with FXS have abnormal facial features, including a large jaw and forehead, ears, and testes. They are anxious, hypersensitive, impulsive, and suffer from attention problems akin to attention deficit/hyperactivity disorder. Most individuals with FXS have delayed language development and intellectual disabilities ranging from mild to severe. Autistic features such as shyness, poor eye contact, hand flapping, and hand biting are common. Two-thirds of all patients with FXS meet the diagnosis of ASD. Sleep problems and seizures are common comorbidities. Because females express normal *FMR1* in the nonmutated chromosome in each cell, they produce variable amounts of normal FMRP, and therefore their symptoms are typically much milder. The pathological hallmark of fragile X is the presence of numerous abnormal spines with long elongated shafts and small heads (Figure 6).

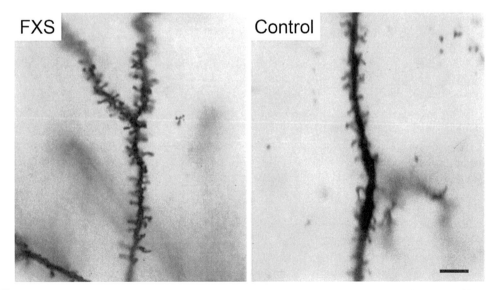

FIGURE 6 Abnormalities in synaptic spines in Fragile X syndrome (FXS). There is an increased number of long, thin dendritic spines and increased spine density in FXS compared with control. *Reproduced with permission from Ref. 8.*

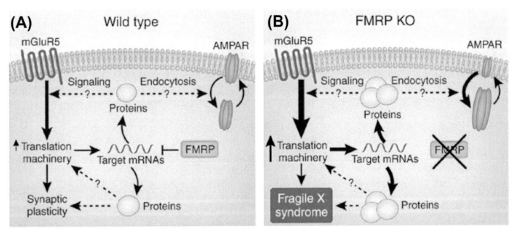

FIGURE 7   Role of fragile X mental retardation protein (FMRP) in activity-dependent regulation of protein synthesis. (A) Metabotropic glutamate receptor (mGluR) 5 signaling in wild-type mice activates the translation machinery and induces specific protein synthesis–dependent forms of synaptic plasticity. Some of the mGluR5-regulated messenger RNAs (mRNAs) are translationally suppressed by FMRP (B) In FMRP knockout mice (FMRP KO) or patients with FXS, FMRP-targeted mRNAs are translated excessively and mGluR5 signaling is enhanced causing excessive translation of mRNAs into proteins. *From Ref. 9.*

5.1 Neurobiology of FXS

Significant insight into the disease mechanism has been gained from the ability to generate transgenic animal models, including mouse, fish, and fruit fly, which mimic salient features of the disease. Most importantly, the knockout of the *FMR1* gene in transgenic mice replicates many human symptoms and has allowed the study of the underlying disease biology, providing an evolving but fairly comprehensive understanding of the disease mechanisms (Figure 7). Studies using this mouse model of disease revealed an overall enhancement of LTD, a cellular process underlying learning (Figure 8). This form of learning involves mGluRs, particularly mGluR5, which couple activity-dependent Glu release to changes in the translation of proteins involved in LTD, and FMRP acts as a negative (feedback) regulator of this activity.

mGluR5 belongs to the group 1 mGluRs and couples via a G protein (Gq) to phospholipase C. It in turn signals via extracellular signal–regulated kinase and the mammalian target of rapamycin (mTOR) to regulate protein translation at

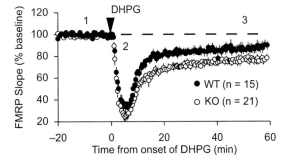

FIGURE 8   Enhancement of long-term depression by the metabotropic glutamate receptor 5 agonist DHPG. A comparison of brain slices from mice where the *FMRP* gene has been eliminated to mimic the human disease (knockout, KO) compared with wild-type (WT) controls FMRP, Fragile-X mental retardation protein. *From Ref. 10.*

polyribosomes. mGluR5 translates neuronal activity to enhanced protein synthesis. FMRP acts as a translational regulator suppressing protein biosynthesis at these ribosomes, and its absence causes protein biosynthesis to run amuck, causing aberrantly enhanced LTD due to dysregulated protein biosynthesis. mGluR–mediated LTD is insensitive to changes in intracellular Ca^{2+} and instead requires signaling via Homer → Pike-L → phosphoinositide

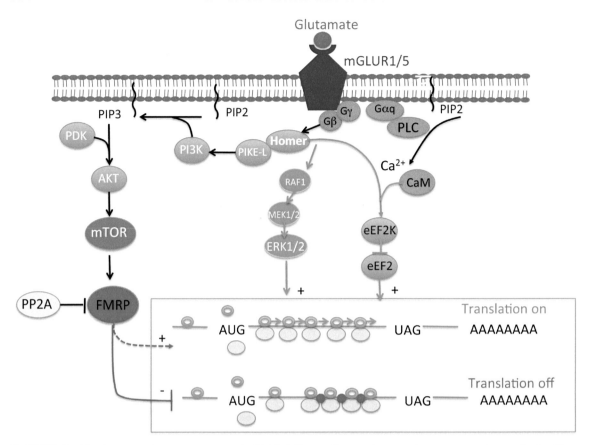

FIGURE 9 Metabotropic glutamate receptor (mGluR) 1/5 signaling pathways relevant to protein synthesis. Gluta-mate binding to mGluRs activates three main pathways, each indicated by different colors, that couple the receptors to translational regulation: The phospholipase C (PLC)/calcium-calmodulin pathway (orange ovals), the mammalian target of rapamycin (mTOR) pathway (blue ovals), and the extracellular signal–regulated kinase (ERK) pathway (green ovals). mGluR1/5 may also inhibit fragile X mental retardation protein (FMRP; red oval) function to regulate translation through a fourth pathway requiring stimulation of protein phosphatase 2A (PP2A; yellow oval). Arrows with a + indicate a positive consequence on downstream gene transcription. Box indicated active translation in the absence of FMRP, yet translation turned off by the binding of FMRP to the mRNA on polyribosomes; arrows with a − an inhibitory effect. [Ca^{2+}]i, calcium release from intracellular stores; CaM, calmodulin; Gαq, Gβ, Gγ, heterotrimeric G proteins; PIP3, inositol-1,4,5-triphosphate; PIP2, Phosphatidylinositol 4,5-bisphosphate; PI3K, Phosphatidylinositol-4,5-bisphosphate 3-kinase; phosphoinositides; Raptor, regulatory-associated protein of mTOR. *Adapted From Ref. 11.*

3-kinase, which in turn activates the mTOR extracellular signal–regulated kinase pathway (Figure 9). Protein targets affected by the mGluR5–FMRP balance include, among others, the AMPA receptor, Kc3.1b K$^+$ channels, the synaptic scaffold protein Homer that couples mGLUR5 activity to mTOR, the signaling proteins GSK3 and the extra-cellular matrix-degrading enzyme matrix metallo-proteinase-9. This list is likely to be much longer

because many additional proteins targets have yet to be identified. Together, changes in these proteins give rise to abnormal spines with altered synaptic function. Importantly, the loss of FMRP causes enhanced LTD (Figure 8) via alterations in mGluR5 signaling that lead to excessive internal-ization of AMPA receptors.

Support for a critical role of the hypothesized imbalance of mGluR5 and FMRP in generating

FXS symptoms comes from studies of transgenic mice in which the amount of mGLUR5 protein was reduced by 50%. Crossing *FMR1* knockout mice with mice that carry a heterozygous deletion of mGLUR5 led to expression of only one intact mGLUR5 allele. All FXS symptoms except enlarged testes were reversed in these animals.[12]

Obviously, neither FMRP nor mGLUR5 expression can be directly altered in a patient, and no drugs that directly alter FMRP function yet exist. However, an alternative strategy may be to reduce mGLUR5 function or its downstream targets using pharmacological inhibitors. A potent although short-lived allosteric mGLUR5 blocker, MPEP (2-methyl-6-(phenylethyl)-pyridine), crossed the blood–brain barrier and was able to acutely correct the behavioral changes in *FMR1* knockout animals. Furthermore, MPEP even normalized the abnormally increased protein expression in the hippocampus.

These experiments raise the important question of whether there exists a therapeutic developmental time period during which a correction of the imbalanced mGLUR5–FMRP signaling must be restored to reverse symptoms. In a worst-case scenario one may expect that once abnormal spines are established they cannot be corrected. In a best-case scenario the structural and functional changes present in FXS remain plastic throughout life. The already mentioned genetic studies of mice, in which the mGLUR5 protein was transgenically reduced by 50%, yielding almost complete reversal of symptoms, fail to address this question because the intervention is present throughout pre- and postnatal life. Similarly, the many pharmacological studies targeting mGLUR5 were unable to chronically suppress mGLUR5 signaling because the drugs wear off quickly. The recent development of CTEP, a specific mGLUR5 inhibitor with an exceptionally long half-life (18 h), has answered this important question with results, which are exciting.[13] When treatment of mice was started at 5–6 weeks, a time equivalent to adolescence in humans, essentially all FXS symptoms were restored to near normal, suggesting that the phenotype remains reversible,

at least throughout childhood. This suggests that this disease may ultimately be effectively treatable, or perhaps even curable, in humans.

5.2 Emerging Treatments

Known disease-modifying treatments are currently not available, and therefore treatments are largely supportive and focused on comorbidities. In light of the successes in correcting the FXS symptoms in mouse models by targeting mGLUR5, however, pilot clinical studies have begun to pursue this strategy in patients. Fortunately, one mGLUR5 inhibitor, fenobam, was already approved as a safe anxiolytic and showed promise in an open label, phase II study. More potent novel drugs targeting mGLUR5, including STX107, AFQ056, and RO4917523, are now at various stages of clinical testing.[14]

An alternative approach to changing the activity of mGLUR5 is limiting the presynaptic release of Glu. This can be accomplished using the GABA agonist baclofen, a strategy that was effective in FMR1 knockout mice; it restores spine morphology, reverses AMPA internalization, and corrects many of the cognitive symptoms. This prompted the use of the active baclofen enantiomer arbaclofen in clinical studies of patients with FXS. Although successful in a phase II, double-blind, placebo-controlled study, a subsequent phase III study showed no improvement above placebo. Another GABA agonist approved to aid the treatment of alcohol withdrawal, acamprosate, showed promise in a small, open-label study in children with FXS, and a larger-scale, placebo-controlled study is now on its way.

Preclinical studies with FMR1 transgenic mice showing promising reversal of symptoms by targeting pathways downstream of mGLUR5, prompted some trials using already approved drugs. For example, minocycline is an antibiotic that inhibits matrix metalloproteinase-9, and in animal models the drug normalized dendritic spine morphology. In a small, placebo-controlled study it improved some but not all symptoms in children 5–17 years of age. Lithium is a known mood stabilizer used in patients with bipolar disorder and depression. It targets GSK3, an enzyme that is upregulated in FXS. A small, 15-patient study found that lithium significantly improves the behavior of patients with FXS.

These examples are elegant illustrations of how basic research can drive translational medicine. The principal findings resulted from basic studies of learning and memory, which identified a cellular defect in a mouse model that subsequently elucidated a complex disease mechanism that is now pharmacologically altered. The monogenetic nature of the disease allowed for the development of a robust, clinically relevant mouse model of disease with excellent predictive value. As we have learned throughout this book, inadequate preclinical models are often a major obstacle for translational science.

6. RETT SYNDROME

RTT[15] is a rare, X-linked developmental disorder that almost exclusively affects girls. It was first described in 1966 by the Austrian pediatrician Andreas Rett, who chronicled 22 girls with a characteristic pattern of symptoms later known as RTT. Following an initial 6–18 months of normal development, during which affected girls reach typical milestones, including walking and speaking a few words, they suddenly stagnate and ultimately regress. They lose purposeful hand movements, which are often replaced by a characteristic hand-wringing or hand flapping. In addition, individuals lose language and become irritable and self-abusive. Autistic features develop, including lack of eye contact, unresponsiveness to social cues, and hypersensitivity to sound. By age five, social contact and eye contact improve; however, language does not. Cognitive functions are severely underdeveloped, and seizures are a common and potentially severe comorbidity. A gradual loss of motor control requires many patients to use a wheelchair for mobility by their teenage years. In addition to these neurological symptoms, patients exhibit variable autonomic abnormalities including severe constipation and characteristic breathing abnormalities. While life expectancy is reduced, particularly because respiratory abnormalities increase the risk of respiratory distress, a few individuals have reached their 60s and 70s. Given the fact that the disease has been followed clinically for only 50 years, comprehensive long-term outcomes are not yet available.

Consistent with a profound loss of cognitive function, the brain of affected individuals is typically reduced in volume, mainly in the frontal and prefrontal cortex. There is no sign of degeneration; instead the decreased brain volume is primarily due to the lack of postnatal growth. At the cellular level cortical neurons are smaller and densely packed, yet have simpler dendrites and immature spines. The substantia nigra shows reduced pigmentation, suggesting a reduction in dopamine. Abnormalities for essentially all neurotransmitter systems have been reported; reduced levels of acetylcholine are the most consistently reported.[16]

The incidence of RTT is 1 in 10,000, and it is most frequently (95%) caused by one of many spontaneously occurring (de novo) mutations in the methyl-CpG-binding protein 2 (MeCP2). Over 300 different missense, nonsense, and frameshift mutations have been described, each causing loss of function and mainly affecting the paternally derived allele. Because of the inactivation of one of the X chromosomes, there is a 50% likelihood that a cell expresses either the maternal or paternal gene and, consequently, at the tissue level *MECP2* expression is a mosaic, with roughly 50% wild-type and 50% mutated *MECP2*.

Since males only have one X chromosome, the same *MECP2* mutations that cause RTT in females are present in all cells and are either embryonically lethal or cause death in early infancy, although rare cases in which unknown genetic modifiers overcome this infantile lethality have been reported.

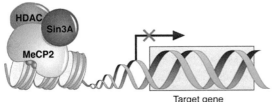

FIGURE 10 Schematic illustration of MeCP2 interaction with the DNA causing gene silencing. HDAC, histone deacetylase. *Reproduced from Ref. 17.*

The *MECP2* gene encompasses four exons that encode two isoforms of MECP2, a longer and more abundant MeCP2α and a shorter MeCP2β. Both are nuclear proteins that bind to methylated CpG dinucleotide sequences on the chromosomes via a methyl-CpG-binding domain, whereupon the transcriptional repressor site recruits histone deacetylases that induce a compaction of chromatin, making it inaccessible for transcription (Figure 10). Note that MeCP2 can both activate and repress gene expression.

6.1 Neurobiology of RTT

MeCP2 is highly expressed in the brain in both neurons and glia, and neuronal expression increases during postnatal development. A number of genes regulated by MeCP2 have been identified. The most important among them is *BDNF*, a member of the neurotrophin family that plays an important role in synapse development and function in brain regions involved with learning and memory. The *BDNF* gene is constitutively repressed by MeCP2 binding to one of its four promoters. In the normal brain, *BDNF* is transcribed in an activity-dependent manner whereby neuronal depolarization allows Ca^{2+} influx via L-type Ca^{2+} channels. This causes CaMKII-mediated phosphorylation of MeCP2 at serine residue 421, dislocating it from the BDNF promoter and thereby activating transcription. Loss of MeCP2 therefore impairs activity-dependent release of BDNF, and deletion of the *BDNF* gene

causes many of the same symptoms as genetic deletion of MECP2 in mice. Other less well-understood gene targets include the neuronal transcription factor *DLX5*, the serum glucocorticoid-inducible kinase 1 (*Sgk1*), the ubiquitin ligase *UBE3A*, the β3-subunit of the GABA-A receptor (*GABRB3*), and the neuropeptide corticotrophin-releasing hormone (*CRH*). The latter may explain some of the anxiety-related symptoms in RTT.

Mouse Models of RTT

Given the monogenetic nature of the disease, it is possible to reproduce in transgenic mouse models. Three different approaches that either deleted the entire gene, truncated the protein, or overexpressed an extra copy of MECP2 have been used. All three exhibit variable traits of the human disease, including seizures, motor dysfunction, ataxia, and hind limb clasping. Mice with MECP2 deletion develop normally for 3–6 weeks, at which time males begin to develop motor abnormalities, ataxia, tremor, hind limb clasping, and abnormal breathing. Brain development is slowed and brain weight reduced by up to one-third. Animals die by 10 weeks of age. For these mutants it is common to study males, since the females develop much milder deficits much later in life. It is surprising that many of the symptoms of RTT can be induced by overexpression of MECP2. This suggests that a proper balance of MeCP2 activity is required for normal brain function, with too little and too much being detrimental.

The study of these mouse mutants have allowed for the neurophysiological changes that may explain some of the neurological symptoms to be deciphered. The most significant changes are related to synaptic structure and function. More specifically, MECP2 null mice show a 46% reduction in the number of glutamatergic synapses and, as a consequence, show impairment in synaptic plasticity, notably a reduction in LTP.[18] In addition, GABAergic neurons in the forebrain show a 30–40% reduction in GABA release, attributable to a decrease in GABA content in synaptic vesicles.[19]

Among the most important lessons learned from mouse models of RTT is the ability to essentially completely reverse the disease phenotype by reintroducing wild-type MeCP2 in animals that were already symptomatic.[20] This suggests that the affected neurons and their signaling pathways are not permanently damaged and provides hope that disease reversal may one day be feasible in humans as well. How to do that remains somewhat enigmatic. Gene therapy approaches using viral vectors are advancing and will one day allow gene expression to be restored to some extent. In the case of RTT, however, one would have to target only those 50% of neurons that have the mutated MECP2 because overexpression of an extra copy in those cells bearing wild-type MeCP2 would be deleterious as well. Another strategy would be to capitalize on the downstream targets, of which BDNF is a prime example. Mimetics to BDNF that bind to the TrkB receptor have been synthesized and show promise in mouse models.

Role of Glia in RTT

Following the discovery of mutated *MECP2* as causing disease, it was assumed that the only affected cells are neurons. However, the protein is lost in any cell that harbors the mutated gene. In the brain, *MECP2* mutations in glia indeed seem to contribute to disease. Studies of RTT mouse mutants revealed, rather unexpectedly, that rescuing the MECP2 deficit only in glial cells is sufficient to reverse most of the RTT symptoms. When transgenic mice deficient in MeCP2 were selectively induced to express the wild-type MECP2 only in astrocytes, the dendritic morphology of neurons normalized, their presynaptic levels of Glu increased, their respiratory abnormalities improved, and the mice showed a decreased anxiety-like phenotype.[21] Most importantly, the animals lived for over 7.5 months compared to only 10 weeks for untreated littermates. How the expression

of intact MeCP2 only in astrocytes can reverse these neuronal phenotypes remains a mystery, yet the data clearly suggest a noncell autonomous function of MeCP2.

Other studies examining the role of microglial cells in RTT led to equally exciting discoveries. Microglial cells derive from bone marrow and invade the brain in early embryologic life to become the resident immune cells of the brain, where they function like macrophages to clear the brain of debris. Inflammation or brain trauma can induce bone marrow–derived cells to invade the adult nervous system, where they form new microglia. Grafting bone marrow from wild-type MECP2 mice into MECP2 knockout animals almost completely arrested the disease symptoms, and the transplant recipients lived for 1 year compared with only 8 weeks for the animals that received a transplant from a mutant mouse. While the microglial cells lacking MeCP2 showed impaired phagocytic activity, normal clearing of debris occurred through the implanted bone marrow–derived microglial cells.[22] This suggests that normal MeCP2-expressing microglial cells ameliorate disease by contributing to clearance of debris, a function that requires proteins that are under the transcriptional control of MeCP2. Which proteins specifically function in this regard remains to be shown. However, when taken together with other findings suggesting that microglial cells from MECP2 knockout animals release toxic concentrations of Glu via gap–junction hemichannels[23,24] and actively kill neurons by Glu excitotoxicity, the argument can be made that microglia cells become unsupportive and hostile if they lose functional MeCP2, as is the case in RTT.

6.2 Emerging Treatments

The findings above suggest an interesting potential therapeutic strategy.[25] A patient's bone marrow could be harvested followed by the typical myeloablative conditioning that is done in the treatment of leukemia. The harvested myeloid cells from the patient could then be sorted for wild-type and mutant MeCP2 cells, and only those with roughly 50% of the wild type could be grafted back into the patient. Aided by strategies for transiently reducing the blood–brain barrier, this may allow marrow-derived myeloid cells to seed healthy microglial cells into the patient's brain. As shown in transgenic mice, this may be sufficient to reverse disease symptoms. This strategy seems to be much more in reach than replacement of MECP2 mutant genes in every affected neuron throughout the brain.

An alternative strategy is to restore sufficient BDNF in the brain to overcome MeCP2 suppression of the endogenous gene. This can be accomplished by enhancing Glu signaling using the AMPAkine CX546 or with a BDNF mimetic, LM22A-4, that directly activates the same BDNF (TrkB) receptor.[26] Yet another strategy to overcome the inhibition of BDNF synthesis is by activating its downstream targets (Akt, phosphoinositide 3-kinase, and mitogen-activated protein kinase), which can be achieved using a recombinant insulin-like growth factor 1 (IGF-1) shown to reverse the phenotype of MECP2 knockout in mice[27]; this is now being studied in two clinical trials: one using full-length IGF-1 (clinicaltrials.gov identifier NCT01777542) and one using the IGF-1 tripeptide NNZ2566 (clinicaltrials.gov identifier NCT01703533). The recent discovery that statin drugs such as lovastatin and fluvastatin[28] can reverse the disease may provide a quick path to clinical trials as well.

Other Conditions with MECP2 Mutations

While RTT is the disease most prominently linked to MECP2, other rare neurological conditions also present with MeCP2 dysfunction. These include severe mental retardation with epilepsy, Angelman-like syndrome, and some forms of autism. Why these diseases present with different symptoms remains unknown.

7. AUTISM SPECTRUM DISORDER

ASD is a broad group of conditions historically defined by a triad of deficits in social interactions, impaired communication, and restricted interest or repetitive behavior. In its latest revision the Diagnostic and Statistical Manual of Mental Disorders, Fifth Edition, reduced this triad to a dyad: difficulties in social communication and social interaction and restricted and repetitive behavior, interest, or activities. This new definition removes language development per se in favor of recognizing its importance with regard to social communication, which involves many nonverbal forms of communication as well. Chief among those is one of the earliest forms of social communication, called joint attention, whereby a child directs attention toward an object, person, or animal by finger pointing or eye movements. The ability to share their own interest with other people is absent in autism and greatly isolates autistic individuals, who, in groups, engage in "parallel play" rather than participate in group activities. They also generally do not engage in pretend play.

Although instruments to reliably assess autism symptoms have been developed, the lack of biomarkers and the highly heterogeneous population, along with divergent presentation at different ages, make accurate diagnosis challenging. In addition, the spectrum of deficits in each arena varies so widely that autistic individuals may differ vastly in the degree of disability faced. Language can be normal, completely absent, or be reduced to echolalia or parrot-like repetition of phrases. Similarly, cognition is variably affected, ranging from severely impaired to above average. The latter defines high-functioning autistics who suffer from Asperger syndrome. As exemplified by Dustin Hoffman in the movie *Rain Man*, people with Asperger syndrome have asocial behavior, lack eye contact, and exhibit an inability to empathize, yet they have an unusual affinity toward numbers and can remember nearly endless lists of facts, numbers, and names.

Among the social deficits, the inability to show empathy and "read" other people's emotions is often highlighted as a key deficit. Often termed "theory of mind," it suggests that autistic individuals have difficulty putting themselves into the shoes of others, and hence cannot relive the emotions of others around them, which deprives the individual of important societal insight as a way of developing deeper friendships. Interestingly, autism-like traits can be observed in the general population, where they often present singly rather than in combination. For example, 59% of children who exhibited social impairments showed no other autistic behavior,[29] whereas 32% also presented with communication difficulties. The independent presence of single autistic traits suggests that autism is indeed a spectrum, not only within affected individuals but also a continuum for the general population.

Autism is typically diagnosed by the age of three, and recent studies suggest that the inability to make eye contact by 6 months of age serves as a sensitive measure for an early autism diagnosis.[30] The incidence of ASD diagnoses has skyrocketed in recent years; the latest numbers suggest that ~1 in 50 boys and ~1 in 200 girls are born with ASD in the United States, suggesting either improved detection or a truly worrisome increase in disease incidence. The fact that males are four to five times more likely to be diagnosed with autism suggests either a protective effect of being female or a hazardous effect of being male. Approximately 45% of autistic individuals have intellectual disability, and 32% show disease progression with loss of previously acquired skills. Strikingly, most people with autism (80%) carry one or many comorbidities ranging from motor abnormalities, anxiety, and sleep disorders to epilepsy (Table 2).

7.1 Risk Factors

Many risk factors have been suggested but only a few have held up to scientific study. Among those few is advanced grandparental age.

TABLE 2   Common Comorbidities Observed in Individuals with Autism Spectrum Disorders

Comorbidity	Prevalence (%)
Anxiety	42–56
Attention-deficit hyperactivity disorder	28–44
Bipolar depression	12–70
Fragile X syndrome	2–6
Gastrointestinal problems	9–70
Immune disorders	<39
Intellectual disability	45
Motor abnormalities	79
Obsessive compulsive disorder	7–24
Seizures	8–30
Sensitivity to sensory stimuli	High
Self-injury	<50
Sleep disorder	50–80
Tuberous sclerosis	1.3

Table generated with data from Ref. 31.

Fathering a child after age 50 increases the risk of having an autistic grandchild by 1.7-fold,[32] suggesting an increased risk of germ-line mutations occuring in sperm with age. As with schizophrenia, living in urban areas increases autism risk by a factor of 2. Exposure to environmental factors, particularly toxins such as mercury, in early childhood or exposure to teratogens such as thalidomide or valproate in utero have been implicated. For example, children born to mothers who used valproate to control seizures are seven times more likely to develop an ASD. One risk that continues to receive widespread media attention is the measles, mumps, and rubella (MMR) vaccine, which until 2001 contained the mercury derivative thiomersal as a preservative. However, extensive studies have ruled the MMR vaccine to be safe without any link to autism. Such a link was originally reported in *The Lancet* in 1998 but was subsequently retracted as fraudulent. In February 2012 the Cochrane Library reviewed scientific studies involving approximately 14,700,000 children and found no credible evidence of the involvement of MMR with autism.[33]

True *genetic risk* factors[34] have thus far been elusive, and a monogenetic cause for autism seems unlikely. However, heritability is high, with a concordance rate of >80% for identical twins and an overall increase in incidence of up to 8% if a sibling has autism. In all likelihood numerous genetic risks must cooperate with environmental risk factors to cause the disease. In the absence of any true autism genes, geneticist have increasingly focused their attention on monogenetic behavioral and intellectual disabilities that commonly show autistic behavior. Notably, ~30% of individuals with FXS and 7% of those with DS show autistic features, and up to 50% of patients with tuberous sclerosis complex present with autism because of mutations in *TSC1* or *TSC2*,. As already discussed, essentially all individuals with RTT show at least transient autistic features during ages 3–5 and have therefore traditionally been included in the category of ASDs. Conversely, 1% of people diagnosed with autism have mutations in *MECP2*, and another 1.3% in *TSC1/2*. The tumor suppressor gene *PTEN* is implicated in a number of disorders, including tuberous sclerosis complex, and it, too, is mutated in ~1% of autism cases. These data indicate that autism is a complex behavioral presentation that can be caused by any number of alterations, akin to a fever being a behavioral response to numerous underlying causes.

7.2 Neurobiology of ASD

While 50 years ago autism was attributed to the emotional coldness of mothers, today the neural underpinning of the disease is indisputable, yet remarkably little is known about the neurobiology of ASD. As with FXS and DS, the deficits are complex and involve numerous signaling pathways. Converging evidence from genetics, histology, and functional imaging point to a core defect in autism being preferentially located in the frontal lobe and affecting primarily synaptic function. This in turn leads to changes in network activity that underlie the complex process of social communicative behavior, particularly involving the prefrontal cortex, superior temporal sulcus, temporoparietal junction, amygdale, and fusiform gyrus. These areas, which are involved in perception and social cognition, consistently show reduced activity in autism. Additional dysfunction affects the mirror system, namely, those neuronal circuits activated equally when we perform a task or observe someone else performing a task.

Individuals with ASD have an ~15% increase in brain weight and up to ~80% more neurons in areas of the frontal lobe.[35] The synaptic density of layer II pyramidal cells is increased in the frontal, temporal, and parietal lobes.[36] Those individuals with the largest increase in spine density have the largest cognitive deficit. Therefore ASD presents with a developmental synaptic overgrowth, graphically illustrated in Figure 11, particularly affecting the frontal lobes, which would be expected to cause enhanced local activity. This is in contrast to the reduced network activity seen on functional magnetic resonance imaging (MRI), which is believed to be due to reduced network connectivity between connected regions in the brain. This notion is indeed supported by imaging studies using a combination of functional MRI and diffusion tensor MRI, which specifically images nerve fibers connecting brain regions (Figure 12). This leads to the emerging concept that ASD is associated with regional hyperexcitability and long-range hypoconnectivity.

The functional activity maps in Figure 12 were generated in controls and individuals with ASD who were performing a cognitive task that probed the strength of their "theory of mind." Each individual viewed a series of comic strips, such as the one shown in the top panel of Figure 13, and asked to select the most

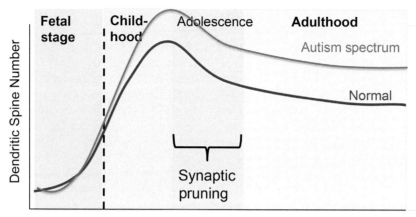

FIGURE 11   Synaptic overgrowth in autism spectrum disorder from fetal stages to adulthood. Initial overproduction of neurons and spines is not corrected by synaptic pruning, resulting in an increased density of synaptic spines throughout life.

appropriate ending. One tested physical causality (e.g., showing a glass falling from a table); the other (shown in the figure) tested intentional causality. Individuals with ASD were equally capable of solving physical causality tasks but were less able to solve the intentional causality task. Image analysis of the brain regions activated during these tasks reveal a significant degree of hypoconnectivity within the studied brain circuits. Interestingly, these connectivity maps were able to selectively identify people with ASD with 96% accuracy.[37]

Within each local circuit, changes in synaptic function are responsible for the macroscopic network changes. Table 3 lists a number of genes that are implicated in autism, many of which are related to synaptic structure and function. These include the synaptic cell adhesion molecules neurexin-1 and its endogenous ligands neuroligins-3 and -4, as well as *SHANK3*, a binding partner for the neuroligins involved in presynaptic organization of glutamatergic synapses. These synaptic proteins are currently among the "molecular frontrunners" in explaining synaptic deficits in ASD.

Initially believed to be mere adhesion molecules involved in synapse development and maintenance, it is now clear that neurexins and neuroligins are dynamically involved in signaling between the pre- and postsynaptic membranes

(Figure 13),[38] and even subtle changes in their expression and function may cause altered cognitive function. Of note, rare mutations that affect neurexin-1 and neuroligin-3 and -4 have been found in patients with ASD. Moreover, transgenic animals in which these genes were altered present with deficient synaptic function and severe autistic features. Therefore it is likely that ASD and probably other diseases with cognitive impairments involve alterations in the neurexin–neuroligin *trans*-synaptic signaling axes.

How the interaction between neurexin and neuroligins alter synaptic function is still being elucidated. However, it seems that multiple signaling axes are involved. An important one involves the postsynaptic scaffolding protein SHANK3 (Figure 14), which binds indirectly to neuroligin via PSD95, which in turn has multiple interactions with the actin cytoskeleton of the spine and various kinases that can regulate postsynaptic proteins via phosphorylation. As with neurexin and neuroligins, numerous point mutations have been found in the *SHANK3* gene in patients with ASD. These three genes now provide an opportunity to generate transgenic animals, most of which show autistic features, allowing much more detailed studies of the role of these proteins in disease while also providing much-needed animal models to study ASD.

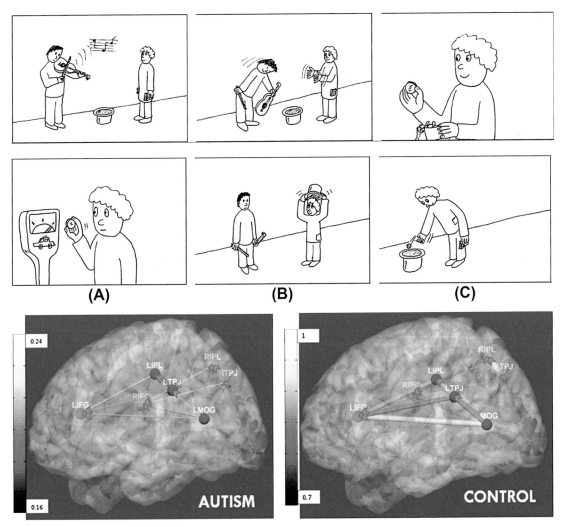

FIGURE 12 Effective connectivity in autism spectrum disorder (ASD) assessed during physical and intentional causality tasks. *Top,* comic strip illustrating an intentional causality task, where the individual has to select the most logical ending to the vignette from the three possibilities shown (A–C). Successful completion of this task requires theory-of-mind and is deficient in ASD. Eighteen regions of interest were assessed during imaging, and connectivity was established using Granger causality method for all participants. There was a significant degree of hypoconnectivity in these regions in patients with autism, as shown by the decreased color intensity and thickness of lines. *Figure courtesy of Dr Rajesh Kana, University of Alabama at Birmingham, based on data in Ref. 37.*

Animal Models of Disease

Given the genotypic and phenotypic complexity, establishing a mouse model that replicates all aspects of human ASD is a challenge. However, the relatively rare mutations in the neuroligin–neurexin signaling axes (discussed above) and their interacting partners, particularly the Shank proteins Shank1–3,[40] have allowed the generation of animals that display a range of features variably associated with ASD, including

TABLE 3 Selection of Known or Putative Autism Genes

Protein Name	Function	Gene
Neuroligin 3	Synaptic structure function	NLGN3X
Neuroligin 4	Synaptic structure function	NLGN4X
Neurexin 1	Neuroligin receptor	NRXN1
Shank3	Postsynaptic organization	SHANK3
NCAM	Neural cell adhesion molecule	NRCAM
MeCP2	Transcriptional repressor	MECP2
Ubiquitin protein ligase E3	Protein degradation/trafficking	UBE3A
Reelin	Neuronal migration	RELN
Sodium channel NAG	Action potential	SCN7A
Na$^+$/H$^+$ exchanger	pH regulation	SLC9A9
GABA-B receptor	Neuronal inhibition	GABAB3/5
Oxytocin receptor	Love/trust hormone receptor	OXTR

Data to generate the table sourced from Ref. 34.

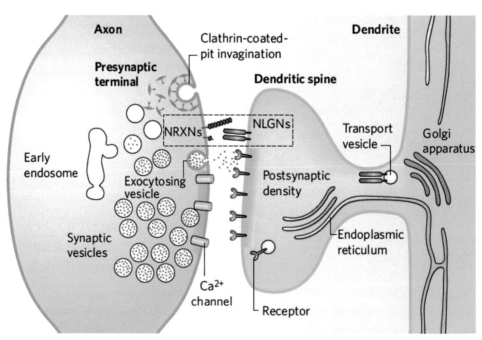

FIGURE 13 Synaptic adhesion proteins suspected to participate in autism spectrum disorder. The structure of an excitatory synapse is shown, indicating the putative locations of neurexins (NRXNs) and neuroligins (NLGNs) in the synapse. *From Ref. 38.*

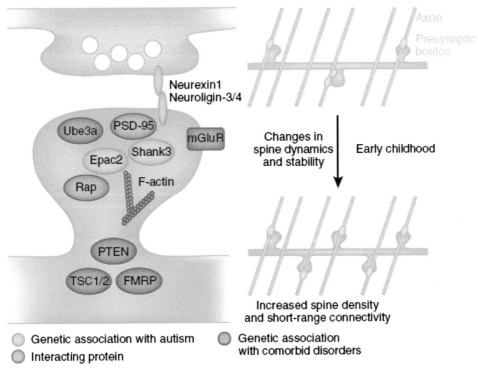

FIGURE 14   Synaptic proteins that may contribute to autism spectrum disorder (ASD). Proteins with genetic associations with ASD and comorbid disorders participate in pathways that regulate spine morphogenesis. Their disruption may alter spine dynamics and stability, leading to an increase in spine density and increased connectivity with nearby axons (blue lines) during early childhood. *From Ref. 39.*

deficits in social interactions, anxiety, and variably impaired cognitive function. Thus far, the main purpose of such animals was to gain a better understanding of cellular and molecular changes that may explain autistic phenotypes, such as changes in spine morphology and number, changes in glutamatergic and GABAergic synaptic transmission, altered transmitter receptor trafficking, and changes in LTP and LTD.

In the absence of drugs to treat ASD, these animals will continue to refine our neurobiological understanding yet hopefully will soon enable meaningful drug screens in an effort to identify pharmacological interventions for ASD.

In addition to glutamate, serotonin is increased and GABA is decreased in ASD; the latter explains seizures as a comorbidity. Further attention is focused on oxytocin and vasopressin, neuromodulators that may function as enhancers of social interactions.

7.3 Emerging Treatments

There is currently no pharmacological treatment for autism per se, only for some of its comorbidities (Table 2). While effective in altering unwanted behaviors in some children, the use of antipsychotics and antidepressants may be hazardous because the brain is in an active state of development. Effective therapy relies on the early introduction of behavioral interventions that aim to develop intelligence, communication, and socialization. A variety of different therapies are offered, typically by child psychologists, and some have been confirmed as effective through rigorous scientific study. Over 100 clinical trials are ongoing, most of which assess behavioral therapies, diagnoses, or the treatment of comorbidities. Of the neuroscience-oriented studies, the same drugs used for other cognitive diseases, particularly RTT, are being evaluated and include the Glu receptor antagonist memantine, the GABA-A receptor agonist acamprosate, and IGF-1. In light of the autistic features seen in patients with tuberous sclerosis, the mTOR inhibitor rapamycin, more commonly used in cancer, is also being evaluated. Most of these studies are in early stages, but recent neuroscience discoveries should instill some optimism that more targeted treatment strategies will soon emerge.

Prognosis

Autistic individuals have an almost 3-fold increase in mortality largely attributable to the many co-morbidities. In spite of our best efforts to reduce the disease burden on the individual through therapy, up to 75% of autistic individuals cannot live independently, typically do not work, and fail to integrate socially. The biggest challenge remains the transition into adulthood, at which time structure and outside help provided by schools disappears, leaving families to fill the void, often unsuccessfully.

8. COMMON DISEASE MECHANISM

As different as the four diseases described in this chapter appear clinically, they do share a number of common traits. (1) At the most macroscopic level, they each present with complex behavioral deficits that can be explained only by dysfunction in extensive interconnected neural networks. (2) They all show autistic behavior and intellectual disabilities to some extent. (3)

At the cellular level they present with changes in synaptic structure and function and hence could be called "synaptopathies," or diseases of synapses. All four seem to manifest altered GABAergic neurotransmission. (4) The functional defects can be explained only through changes that affect many genes and the proteins they encode. Interestingly, however, these proteins need not necessarily be mutated, as illustrated in DS and FXS, where a simple change in their number or a disequilibrium causes disease. (5) Yet another commonality, demonstrated for all but autism, is the complete reversibility of the disease long after birth. This is highly unexpected and most exciting. Who would have thought that a complex behavioral phenotype can be reversed by restoring the proper number of proteins expressed in synaptic circuits?

9. CHALLENGES AND OPPORTUNITIES

Few intellectual disabilities have a known genetic cause, yet three of the four conditions discussed here provide a window into genetic susceptibility of the nervous system to disease. Of these, FXS and RTT are among the most tragic intellectual disabilities, yet at the same time, studies of them provide unprecedented hope for this group of disorders—not because of experienced clinical successes but because the mouse models of disease suggest near complete reversibility of the disease by rectifying the underlying genetic deficits. This is exciting and unexpected and provides "proof of principle" that disease reversion is possible. I argue that the likelihood of curing RTT and FXS in the coming decades may well be greater than our ability to conquer AD, epilepsy, or ALS.

In spite of this optimism, patients currently living with any one of these conditions suffer tremendously and have a very low likelihood of independent living. The vast majority rely on family members to care for them for life. In spite of changes in our sensitivity to disabilities, intellectual disabilities continue to be associated with significant stigma.

The incidence of autism is rising at an alarming rate. Is this a true epidemic? The disease came seemingly out of nowhere, entering the stage barely 60 years ago, and it has since taken the world by storm. Is it possible that a man-made environmental toxin triggers the disease? While an association to the trace amounts of mercury found in childhood vaccines has been ruled out as causing disease, it does little to dissuade from the notion that mercury or other toxins found in the air, water, and soil may contribute to disease in genetically predisposed individuals. Given the alarming epidemiological data, increasing efforts to probe every conceivable link to environmental exposure seems imperative, before it is too late.

Acknowledgments

This chapter was kindly reviewed by Dr Alan K. Percy, Professor of Pediatric Neurology & Director, Eunice Kennedy Shriver Center for Intellectual Developmental Disabilities, University of Alabama, Birmingham.

References

1. Marin-Padilla M. Pyramidal cell abnormalities in the motor cortex of a child with Down's syndrome. A Golgi study. *J Comp Neurol*. 1976;167(1):63–81.
2. Rachidi M, Lopes C. Mental retardation in Down syndrome: from gene dosage imbalance to molecular and cellular mechanisms. *Neurosci Res*. December 2007;59(4):349–369.
3. Belichenko PV, Masliah E, Kleschevnikov AM, et al. Synaptic structural abnormalities in the Ts65Dn mouse model of Down Syndrome. *J Comp Neurol*. December 13, 2004;480(3):281–298.
4. Cramer N, Galdzicki Z. From abnormal hippocampal synaptic plasticity in down syndrome mouse models to cognitive disability in down syndrome. *Neural Plast*. 2012;2012:101542.
5. Chakrabarti L, Best TK, Cramer NP, et al. Olig1 and Olig2 triplication causes developmental brain defects in Down syndrome. *Nat Neurosci*. August 2010;13(8):927–934.
6. Chrast R, Scott HS, Madani R, et al. Mice trisomic for a bacterial artificial chromosome with the single-minded 2 gene (Sim2) show phenotypes similar to some of those present in the partial trisomy 16 mouse models of Down syndrome. *Hum Mol Genet*. July 22, 2000;9(12):1853–1864.
7. Fernandez F, Morishita W, Zuniga E, et al. Pharmacotherapy for cognitive impairment in a mouse model of Down syndrome. *Nat Neurosci*. April 2007;10(4):411–413.
8. Comery TA, Harris JB, Willems PJ, et al. Abnormal dendritic spines in fragile X knockout mice: maturation and pruning deficits. *Proc Natl Acad Sci*. May 13, 1997;94(10):5401–5404.
9. Bassell GJ, Gross C. Reducing glutamate signaling pays off in fragile X. *Nat Med*. 2008;14(3):249–250. [03//print].
10. Huber KM, Gallagher SM, Warren ST, Bear MF. Altered synaptic plasticity in a mouse model of fragile X mental retardation. *Proc Natl Acad Sci USA*. May 28, 2002;99(11):7746–7750.
11. Bhakar AL, Dolen G, Bear MF. The pathophysiology of fragile X (and what it teaches us about synapses). *Annu Rev Neurosci*. 2012;35:417–443.
12. Dölen G, Osterweil E, Rao BSS, et al. Correction of fragile X syndrome in mice. *Neuron*. December 20, 2007;56(6):955–962.
13. Michalon A, Sidorov M, Ballard TM, et al. Chronic pharmacological mGlu5 inhibition corrects fragile X in adult mice. *Neuron*. April 12, 2012;74(1):49–56.
14. Berry-Kravis E. Mechanism-Based Treatments in neurodevelopmental disorders: fragile X syndrome. *Pediatr Neurol*. April 2014;50(4):297–302.
15. Chahrour M, Zoghbi HY. The story of Rett syndrome: from clinic to neurobiology. *Neuron*. November 8, 2007;56(3):422–437.
16. Weng SM, Bailey ME, Cobb SR. Rett syndrome: from bed to bench. *Pediatr neonatol*. December 2011;52(6):309–316.
17. Chahrour M, Jung SY, Shaw C, et al. MeCP2, a key contributor to neurological disease, activates and represses transcription. *Science*. May 30, 2008;320(5880):1224–1229.
18. Chao HT, Zoghbi HY, Rosenmund C. MeCP2 controls excitatory synaptic strength by regulating glutamatergic synapse number. *Neuron*. October 4, 2007;56(1):58–65.
19. Chao HT, Chen H, Samaco RC, et al. Dysfunction in GABA signalling mediates autism-like stereotypies and Rett syndrome phenotypes. *Nature*. November 11, 2010;468(7321):263–269.
20. Guy J, Gan J, Selfridge J, Cobb S, Bird A. Reversal of neurological defects in a mouse model of Rett syndrome. *Science*. February 23, 2007;315(5815):1143–1147.
21. Lioy DT, Garg SK, Monaghan CE, et al. A role for glia in the progression of Rett's syndrome. *Nature*. July 28, 2011;475(7357):497–500.

22. Derecki NC, Cronk JC, Lu Z, et al. Wild-type microglia arrest pathology in a mouse model of Rett syndrome. *Nature*. April 5, 2012;484(7392):105–109.

23. Maezawa I, Jin LW. Rett syndrome microglia damage dendrites and synapses by the elevated release of glutamate. *J Neurosci*. April 14, 2010;30(15):5346–5356.

24. Abbracchio MP, Saffrey MJ, Höpker V, Burnstock G. Modulation of astroglial cell proliferation by analogues of adenosine and ATP in primary cultures of rat striatum. *Neuroscience*. 1994;59:67–76.

25. Derecki NC, Cronk JC, Kipnis J. The role of microglia in brain maintenance: implications for Rett syndrome. *Trends Immunol*. March 2013;34(3):144–150.

26. Li W, Pozzo-Miller L. BDNF deregulation in Rett syndrome. *Neuropharmacology*. January 2014;76(Pt C):737–746.

27. Tropea D, Giacometti E, Wilson NR, et al. Partial reversal of Rett Syndrome-like symptoms in MeCP2 mutant mice. *Proc Natl Acad Sci*. February 10, 2009;106(6):2029–2034.

28. Buchovecky CM, Turley SD, Brown HM, et al. A suppressor screen in Mecp2 mutant mice implicates cholesterol metabolism in Rett syndrome. *Nat Genet*. September 2013;45(9):1013–1020.

29. Happe F, Ronald A, Plomin R. Time to give up on a single explanation for autism. *Nat Neurosci*. October 2006;9(10):1218–1220.

30. Jones W, Klin A. Attention to eyes is present but in decline in 2-6-month-old infants later diagnosed with autism. *Nature*. December 19, 2013;504(7480):427–431.

31. Lai MC, Lombardo MV, Baron-Cohen S. Autism. *Lancet*. March 8, 2014;383(9920):896–910.

32. Frans EM, Sandin S, Reichenberg A, et al. Autism risk across generations: a population-based study of advancing grandpaternal and paternal age. *JAMA Psychiatry*. May 2013;70(5):516–521.

33. Demicheli V, Rivetti A, Debalini MG, Di Pietrantonj C. Vaccines for measles, mumps and rubella in children. *Cochrane Database Syst Rev*. 2012;2:Cd004407.

34. Miles JH. Autism spectrum disorders–a genetics review. Genetics in medicine. *Off J Am Coll Med Genet*. April 2011;13(4):278–294.

35. Courchesne E, Mouton PR, Calhoun ME, et al. Neuron number and size in prefrontal cortex of children with autism. *JAMA*. November 9, 2011;306(18):2001–2010.

36. Hutsler JJ, Zhang H. Increased dendritic spine densities on cortical projection neurons in autism spectrum disorders. *Brain Res*. January 14, 2010;1309:83–94.

37. Deshpande G, Libero LE, Sreenivasan KR, Deshpande HD, Kana RK. Identification of neural connectivity signatures of autism using machine learning. *Front Hum Neurosci*. 2013;7:670.

38. Sudhof TC. Neuroligins and neurexins link synaptic function to cognitive disease. *Nature*. October 16, 2008;455(7215):903–911.

39. Penzes P, Cahill ME, Jones KA, VanLeeuwen JE, Woolfrey KM. Dendritic spine pathology in neuropsychiatric disorders. *Nat Neurosci*. March 2011;14(3):285–293.

40. Jiang YH, Ehlers MD. Modeling autism by SHANK gene mutations in mice. *Neuron*. April 10, 2013;78(1):8–27.

General Readings Used as Source

1. Volpe JJ. *Neurology of the Newborn*. 4th ed. Saunders; 2001.

2. *Adams and Victor's, Principles of Neurology*. 9th ed. McGraw-Hill; 2009. [Chapter 38].

3. Rachidi M, Lopes C. Mental retardation in Down syndrome: from gene dosage imbalance to molecular and cellular mechanisms. *Neurosci Res*. December 2007;59(4):349–369.

4. Krueger DD, Bear MF. Toward fulfilling the promise of molecular medicine in fragile X syndrome. *Annu Rev Med*. 2011;62:411–429.

5. Bhakar AL, Dolen G, Bear MF. The pathophysiology of fragile X (and what it teaches us about synapses). *Annu Rev Neurosci*. 2012;35:417–443.

6. Chahrour M, Zoghbi HY. The story of Rett syndrome: from clinic to neurobiology. *Neuron*. November 8, 2007;56(3):422–437.

7. Lai MC, Lombardo MV, Baron-Cohen S. Autism. *Lancet*. March 8, 2014;383(9920):896–910.

8. Percy AK. Rett syndrome: exploring the autism link. *Arch Neurol*. August 2011;68(8):985–989.

Suggested Papers or Journal Club Assignments

Clinical Papers

1. Jones W, Klin A. Attention to eyes is present but in decline in 2-6-month-old infants later diagnosed with autism. *Nature*. December 19, 2013;504(7480):427–431.

2. Cuddapah VA, Pillai RB, Shekar KV, et al. Methyl-CpG-binding protein 2 (MECP2) mutation type is associated with disease severity in Rett syndrome. *J Med Genet*. 2014;51(3):152–158.

3. Nelson LD, Siddarth P, Kepe V, et al. Positron emission tomography of brain beta-amyloid and tau levels in adults with Down syndrome. *Arch Neurol*. 2011;68(6):768–774.

4. Huang W, Luo S, Ou J, et al. Correlation between FMR1 expression and clinical phenotype in discordant dichorionic-diamniotic monozygotic twin sisters with the fragile X mutation. *J Med Genet.* 2014;51(3):159–164.

Basic Papers

1. Adorno M, Sikandar S, Mitra SS, et al. Usp16 contributes to somatic stem-cell defects in Down's syndrome. *Nature.* September 19, 2013;501(7467):380–384.
2. Fernandez F, Morishita W, Zuniga E, et al. Pharmacotherapy for cognitive impairment in a mouse model of Down syndrome. *Nat Neurosci.* April 2007;10(4):411–413.
3. Chakrabarti L, Best TK, Cramer NP, et al. Olig1 and Olig2 triplication causes developmental brain defects in Down syndrome. *Nat Neurosci.* August 2010;13(8):927–934.

4. Michalon A, Sidorov M, Ballard TM, et al. Chronic pharmacological mGlu5 inhibition corrects fragile X in adult mice. *Neuron.* April 12, 2012;74(1):49–56.
5. Dölen G, Osterweil E, Rao BSS, et al. Correction of fragile X syndrome in mice. *Neuron.* December 20, 2007;56(6):955–962.
6. Lioy DT, Garg SK, Monaghan CE, et al. A role for glia in the progression of Rett's syndrome. *Nature.* July 28, 2011;475(7357):497–500.
7. Derecki NC, Cronk JC, Lu Z, et al. Wild-type microglia arrest pathology in a mouse model of Rett syndrome. *Nature.* April 5, 2012;484(7392):105–109.
8. Tabuchi K, Blundell J, Etherton MR, et al. A neuroligin-3 mutation implicated in autism increases inhibitory synaptic transmission in mice. *Science.* October 5, 2007;318(5847):71–76.

PART V

NEUROPSYCHIATRIC ILLNESSES

Mood Disorders and Depression

Harald Sontheimer

1. CASE STORY

Sarah did not know how it had happened, but here she was in a psychiatric hospital with her mother crying at her bedside and her father just staring at her with a bewildered grimace. She was supposed to finish college in 2 months, but she just could not hold it together any longer. In hindsight, her symptoms may have been lingering for a long time, but she never assumed that they constituted a real illness. She blamed her prolonged periods of sadness and emptiness on the loss of her familiar environment and the separation from her long-term friends. But she forced herself to work through this since it had been her choice to move out of state to attend an expensive, small liberal

Diseases of the Nervous System
http://dx.doi.org/10.1016/B978-0-12-800244-5.00012-4

arts college and to compete in cross-country running while pursuing a major in art history.

Although Sarah was excited and thrilled during rush, once inducted, her interest in sorority life soon waned. For most of her freshman year she found little joy in community events and weekend parties. Even among all these girls her age she felt strangely isolated and alone. She was able to hide her emotions well and often masked her feelings by drinking. On some weekends she did not remember where she had spent the previous evening. At the end of her freshman year, her mood improved and she became excited to go back home, where she would work as a lifeguard at the YMCA, a job she held throughout high school. The first few weeks, as she kept busy visiting with all of her old high school friends, she was elated and her social calendar was full. But as quickly as her sadness had vanished, it returned again. The following year, her mood was on a roller coaster, with periods of profound sadness and weeks of outright elation, during which she took on extra volunteer work, even becoming the treasurer of her sorority. However, overall, sadness predominated, and even when she felt on top of the world, she was typically unable to complete any of the tasks she took on. Her grades gradually slipped from high B's to C's, and her cross-country performances now placed her out of contention to even participate in regular competitions. She had no idea why she was feeling such apathy. During her junior year, Sarah remembered having suicidal thoughts and even researched strategies to take her own life with the least amount of pain. She started smoking cigarettes and occasionally marijuana, drinking excessive amounts of coffee, and using Adderall regularly to stay up at night to complete her assignments. No matter how hard she worked she was seemingly never able to catch up. Initially, she was excited about her thesis work, for which she elected to study the influence of Michelangelo on modern sculpture, and even contemplated a trip to Florence to visit Michelangelo's famous statue of David at the Accademia Gallery. Yet her excitement quickly turned into profound anxiety, even panic. Needless to say, her planned excursion to Italy never materialized. She could not organize her thoughts and her thesis made little progress. She failed to attend classes regularly, started sleeping at odd hours, and was socially withdrawn and barely able to maintain ordinary conversations. On the day of her final thesis presentation, as she walked toward the classroom, she experienced a breakdown. Dropping her bag in front of her classmates, she ran out of the building. Student Services found her collapsed in her dorm, crying inconsolably, and took her to the hospital where she was tentatively diagnosed with bipolar disorder, depressed.

2. HISTORY

Mood disorders are among the earliest diseases described in the literature,[1] dating back to the pre-Hippocrates era. The Greek scholars recognized two abnormal mood states and called them mania and melancholy. Melancholia derives from the Greek "black bile," in reference to the humoral theory prevailing at the time. This theory suggested that diseases were the result of an imbalance in the four bodily humors, of which black bile was one, and thus melancholia was seen as an imbalance in black bile. The term "mania" was first used by Homer and others in early Greek mythology, and typically refers to rage, with the closest Greek word, "manos," referring to an excessive relaxation of the mind.

Other early references to melancholy define it as one of the personality traits. These four traits, which include choleric (irritable), phlegmatic (calm), melancholic (gloomy), and sanguine (optimistic), all stem from the humoral theory as the basis for different types of personality. Importantly, since Hippocrates' days, and even before the presence of any scientific evidence, mood disorders have been recognized as being biological in nature and the direct result of brain illnesses.

"The people ought to know that the brain is the sole origin of pleasure and joys, laughter sadness and worry….through the brain we become insane, enraged, we develop anxiety and fears…We suffer all those mentioned above through the brain when it is ill…." (translated from Hippocrates "On the Sacred Disease"[1]).

The first reference to bipolar disorder as a distinct disease entity, with melancholy and mania defining opposite emotional phases, already appeared in the first century AD. Aretaeus of Cappadocia, a medical scholar, described mania and melancholia as having a common etiology, with melancholia being the beginning of mania and its phenomenological counterpart. In the *Canon of Medicine*, an eleventh century medical text written by the Persian physician Avicenna, major depressive symptoms were expanded to include anxiety, phobias, and suspicions of other symptoms. Thereafter, our historical knowledge darkened during medieval times, to resurface again in the 1800s when a number of French, British, and German scholars began to provide an increasingly refined picture that separated schizophrenia, then called dementia precox, from various forms of depression. In 1845, Wilhelm Griesinger introduced seasonal affective disorder to the overall spectrum of depression, and bipolar disorder made its entry into the medical literature as "manic-depressive insanity" in a textbook by Emil Kraeplin in 1896. Recognized as the "godfather of modern psychiatry," he provided extensive medical descriptions of many psychiatric conditions and recognized both the manic and depressed phases as opposite ends of the same disease occurring without any intellectual deterioration. This key observation distinguishes bipolar disorder from schizophrenia.

In spite of the universal recognition of depression as a disorder of brain function, treatments remained elusive until the accidental discovery that lithium salt suppresses the manic symptoms of bipolar disorder. At the time, this mineral was used to treat rheumatic gout, and many mineral springs contain lithium. After some people showed improvement in their mood after drinking water from these springs, the waters developed a reputation as mood stabilizers. Indeed, one Texan spring was even given the name "crazy water," as it had apparently cured depression and other psychiatric illnesses. The first prescription of lithium bromide salt to treat mania dates to 1871 (Bellevue Medical College in New York), and in 1894 it was recommended by a Danish psychiatrist to treat melancholic depression. In the ensuing decades, lithium was used at the discretion of physicians without further scientific study.

Acceptance of lithium as a specific antidepressant followed a 1949 Australian study by John Cade that showed remarkable success as a treatment for psychotic episodes, revealing that over 70% of patients derived significant benefit from lithium salts. Lithium quickly became the treatment of choice for depressive illnesses, and remains the most efficacious drug to treat mania and reduce its recurrence. Indeed, many experts consider lithium the most effective drug in all of psychiatry.

In parallel with the emergence of lithium as an antidepressant, several seemingly more specific drugs began to emerge, also through serendipitous discoveries. For example, the antimycobacterial agent iproniazid was developed to treat tuberculosis, yet it revealed unexpected psychoactive side effects whereby even terminally ill patients became more cheerful and active on iproniazid. The drug was subsequently shown to slow the breakdown of norepinephrine, serotonin, and dopamine by inhibiting the enzyme monoamine oxidase (MAO). Iproniazid not only gave rise to the first truly effective antidepressant drug, but founded a whole family of related MAO inhibitors (MAOIs) commonly used for a number of diseases including depression, schizophrenia, and Parkinson disease.

The discovery of the tricyclic antidepressant drugs also occurred serendipitously around this time when the search for an antipsychotic drug to treat schizophrenia produced imipramine (Tofranil). It induced euphoria rather than containing it, and naturally would have been a poor choice to treat schizophrenia, but was an obvious one for depression. A number of related chemicals that share a three-ring structure are now called "tricyclic" antidepressants and include Tofranil and Anafranil, which are still used today. They all inhibit the removal of norepinephrine and serotonin from synaptic terminals.

Following a refined neurobiological understanding of depression, centered on the imbalance of serotonin and norepinephrine in the brain, came the strategic development of selective serotonin reuptake inhibitors (SSRIs), with the first, fluoxetine (Prozac), reaching the market in 1987. Unfortunately, while these drugs have benefit in severely depressed individuals, they are no more effective than placebo in mild to moderately depressed patients,[2] where they present with significant side effects. These are most frequently prescribed by non-specialists, contributing to their overuse.

3. CLINICAL PRESENTATION/ DIAGNOSIS/EPIDEMIOLOGY

Depression is not the same as sadness or unhappiness. It is a profound emotional state in which a person feels worthless, desperate, and hopeless, to the point of losing energy and libido, and may become capable of harming himself or herself, or of harming others. Depression can be a reaction to life's vicissitudes, such as the loss of a loved one or job, or other anxiety-provoking situations where the response is more akin to grief. Different from transient provoked depression, major depressive illnesses are persistent and typically occur without a known provocation. It is common to distinguish the following four depressive illnesses: (1) grief reaction; (2) secondary depression as a result of neurological disease; (3) clinical (unipolar) depression; and (4) bipolar (manic–depressive) disorder.

Common symptoms shared among all forms of depression include sadness, hopelessness, guilt, fatigue, irritability, change in appetite, and difficulty sleeping. A cardinal feature is the general lack of interest in anything pleasurable, which in medical terms is often called anhedonia. Individuals suffering from depression tend to get caught up in their own condition to the point that they give predictable and narrowly focused answers related to their sadness. A major depressive disorder is clinically diagnosed by the presence of at least five of the nine symptoms listed in Table 1 from the Diagnostic and Statistical Manual of Mental Disorders, 4th. Edition (DSM-IV), and requires an individual to exhibit a chronically depressed mood for a period lasting at least 2 weeks.

3.1 Grieving

Grieving is a natural process aimed at resolving a mental state and typically lasts up to 12 weeks, though it can last longer in some instances, such as for mothers who have lost a child. During this time, it is normal for an individual to be unable to carry out functions of daily life; if he or she is able to do so, it is often without much interest or affect.

TABLE 1   Nine Criteria Used to Diagnose Major Depressive Disorder, of Which Five or More Must Be Present for a Positive Diagnosis

Symptom
Depressed mood/irritability most of the day, nearly every day
Decreased interest or pleasure in most activities
Significant weight change (5%) or change in appetite
Change in sleep: Insomnia or hypersomnia
Change in activity
Fatigue or loss of energy
Feelings of worthlessness, excessive, or inappropriate guilt
Diminished ability to think or concentrate, or more indecisiveness
Thoughts of death or suicide, or having a suicide plan

3.2 Secondary Depression

Secondary depression is commonly associated with illnesses that cause severe chronic pain, such as head or spinal injury, cancer, stroke, multiple sclerosis, Parkinson disease, Alzheimer disease, or brain tumors. Secondary depression is also common following a heart attack and can occur in conjunction with drug use, for example, corticosteroids or beta blockers. Secondary depression has the potential to resolve spontaneously, particularly if the precipitating causes resolve. However, in many instances, chronic disease may be associated with depression.

3.3 Clinical Depression

Clinical (unipolar) depression often begins in early adulthood, although it is generally more prevalent later in life. It may recur episodically, and each episode increases the likelihood of recurrence. Depression is twice as common in women as in men, and incidence increases with age for either sex. In some patients, particularly women, recurrence follows a seasonal pattern called seasonal affective disorder and is associated with fatigue, carbohydrate craving, weight gain, and insomnia. This condition is worse if one lives far from the equator, although it responds positively to sun exposure. These features are suggestive of a role for vitamin D in stabilization of mood[3] or an influence of circadian rhythms. Unipolar depression is often associated with neuroendocrine changes, including increased cortisol and corticotropin-releasing hormone. Treatment requires early administration of antidepressants, ideally in conjunction with psychotherapy. Unfortunately, current antidepressants require weeks to become effective and close to 50% of patients discontinue drug treatment if no improvement is experienced within the first 4 weeks. As discussed later in the chapter, clinical trials question the long-term effectiveness of current antidepressants.

3.4 Bipolar Depression

Bipolar depression is characterized by unexplained mood swings that alternate between euphoria (mania) and depression.[4] Mania is typically associated with inflated self-esteem and grandiosity, with severe mania producing paranoia and delusions (psychosis). During the manic state, patients are hyperactive and have lots of enthusiasm and great expectations, yet typically fail to carry out their plans. These episodes can last from several weeks to 12 months; some patients cycle so rapidly that they may experience four or more manic episodes in a given year. Before a diagnosis of bipolar disorder can be substantiated, mood swings must be present for at least 2 years. It is not uncommon for patients to go through multiple depressive episodes before experiencing their first mania. Not surprisingly, individuals with bipolar disorder are significantly impaired in the workplace and, consequently, bipolar disorder takes a tremendous toll on productivity and personal success. Indeed, bipolar disorder is the leading cause of disability among young adults throughout the developed world. Bipolar disorder affects men and women equally and typically begins between ages 20 and 30, approximately 15 years earlier than unipolar disorder. Persons with bipolar disorder often smoke, drink, and abuse recreational drugs. Genetic predisposition is strong, with an 80% concordance for identical twins. The drug of choice for the treatment is lithium carbonate, which is 80% effective during the manic phase and begins to show effectiveness within 1–2 weeks of use. It is relatively safe and cheap. Moreover, chronic use can prevent future manic episodes. Valproate and olanzapine are additional treatments for acute mania. The antiepileptic drug lamotrigine can aid during the depressed phase.

Changes in cognitive function are frequently associated with depression and can have substantial effects on cognitive performance. Impairments in episodic memories, including verbal memory deficits, are particularly common in severe depression. Patients also show reduced mental flexibility and attention as well as overall psychomotor slowing. These cognitive effects may all results from hippocampal atrophy, which can be seen on magnetic resonance imaging in patients with major depressive illness. Whether hippocampal changes are the cause or consequence of depression, however, is unknown.

A number of **risk factors for depression** are known or suspected, but only in rare cases is the link to disease strong. Such examples include endocrine abnormalities that result in changes in cortisol. This can be caused by benign pituitary cancers (Cushing syndrome) with increased cortisol, or by adrenal insufficiency (Addison disease) with abnormally low cortisol. An insufficient thyroid function may similarly present with depressive symptoms, and a number of drugs have known depressive side effects. For example, recombinant interferon for treatment of hepatitis C and other conditions is significantly linked to depression.[5] The acne drug isoretinoin (Accutane) has long been linked to depressive illness, yet recent studies argue that this side effect is patient specific and the drug may alleviate rather than cause depression in some patients.[6] Recreational use of cannabis has also been suggested as an inducement for mania, with repeated cannabis use during adolescence possibly increasing the likelihood of developing depressive illness later in life.[7] However, our overall understanding of cannabis and its relationship to psychiatric illnesses is quite poor, and one must judge these studies in the context of societal biases.

Stress is one of the most frequently mentioned risk factors for depression, yet it is a poorly defined environmental condition that is variably present in most people's lives. Clearly, growing up under conditions of food scarcity, social isolation, or continued exposure to violence undoubtedly creates anxiety and may trigger helplessness and depression, although clinical data suggests that stress alone increases depression risk only marginally.[8] As discussed later in the chapter, in animal models stress induces heritable changes in DNA that may alter the expression of proteins involved in mental health.

Prevalence of depression is very high, estimated on the order of 20% in total over an individual's lifetime. Unipolar depression is most common at 10–15%, while prevalence is estimated to be 5% in men and 9% in women for manic–depressive disorder. There is an undeniable element of genetic predisposition, with first-degree relatives of an affected individual having a 14–25% chance of developing depression. Prevalence for bipolar disorder among identical twins is 72%. Although these data suggest an underlying genetic factor, our understanding of depression-associated genes is still poor. A major concern with depression is its association with suicide. Approximately 30,000 suicides occur each year in the US, making it the eighth-leading cause of death, and it is believed that 50% of all suicides are committed by individuals suffering from depression. Suicide among adolescents and young adults is the second-leading cause of death, clearly a shocking statistic.

4. DISEASE MECHANISM/CAUSE/ BASIC SCIENCE

Our current understanding of the neurobiology of depressive illnesses is largely based on the effect of antidepressant drugs and functional human imaging studies. Together, they suggest that depressive illness is associated with region-specific neuronal cell loss, with retraction of dendrites causing persistent changes in synaptic activity. These changes lead to chronic imbalances in neurotransmission, particularly for transmitters that regulate reward, affect, and emotion.

The neural circuits specifically implicated in depression include the prefrontal cortex, the hippocampus, and the limbic system, structures that are tasked with regulation of emotion, reward seeking, motivation, and executive function. Important limbic structures include the ventral tegmental area (VTA), nucleus accumbens, locus coeruleus, thalamus, hippocampus, and amygdala. The amygdala is generally associated with attaching the emotional valence to an experience, and functional imaging studies show a strong amygdala activation associated with a feeling of sadness in both depressed and nondepressed individuals. Patients with chronic depression show sustained enhancement of amygdala activity, and disruption of such activity using an implanted deep-brain stimulator can reduce depression in these patients. The prefrontal cortex receives additional modulatory dopaminergic input from the VTA, serotonergic input from the dorsal raphe nucleus, and noradrenergic input from the locus coeruleus (Figure 1). These neuromodulators are all monoamine transmitters and provide emotional salience to processed information. Because of this, depression has historically been assumed to be a disease of monoamines. This was largely based on the finding that drugs that enhance serotonin and noradrenaline concentrations by inhibiting their removal have antidepressive effects, whereas drugs that reduce monoamine

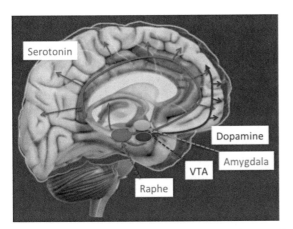

FIGURE 1 Neurotransmitter systems that regulate mood and happiness include particularly dopamine released from the amygdala and ventral tegmental area (VTA) as well as serotonin released by the raphe nucleus. *Image was modified using 3D brain, Cold Spring Harbor DNA Learning Center.*

production induce depression. This *monoamine hypothesis* is consistent with the action of the widely prescribed antidepressant drugs like the MAO inhibitor tranylcypromine, or SSRIs such as fluoxetine (Prozac). However, it would be too simplistic to call depression simply an imbalance of these monoamines. The aforementioned antidepressant drugs change the concentration of monoamines in the brain very quickly, on a time scale of minutes, yet their antidepressant effect is slow, requiring many weeks to change a patient's symptoms. Therefore, these drugs must persistently alter the connectivity within neural circuits that drive emotional valence, and must alter complex signaling cascades that are involved in structural and functional changes in the networks that regulate emotions. Pathways shown to be regulated by chronic antidepressant treatments are illustrated in Figure 2.

4.1 Structural Changes

Structural changes that accompany depression can be observed both histologically and through functional magnetic resonance imaging (fMRI) studies. fMRI studies show an overall decrease in gray matter volume in the prefrontal cortex and hippocampus, two of the brain regions involved in emotional cognition. Postmortem histology confirms a reduction in size of pyramidal neurons and a loss of GABAergic interneurons, along with loss of astrocytes and oligodendrocytes in the prefrontal cortex.[9] However, since these studies examined patients that had already received a number of antidepressive drugs over their lifetime, it is possible that these cellular changes are a consequence of drug treatment, or the result of depression itself, rather than the cause of disease. Since similar changes can be seen in mice and nonhuman primates exposed to chronic stress, it is likely that the cell loss and decreased complexity of their processes are disease related. This data suggests that depression may be considered a "mild neurodegenerative" disorder.[10] Interestingly, unlike the prefrontal cortex, the amygdala shows an increase in the

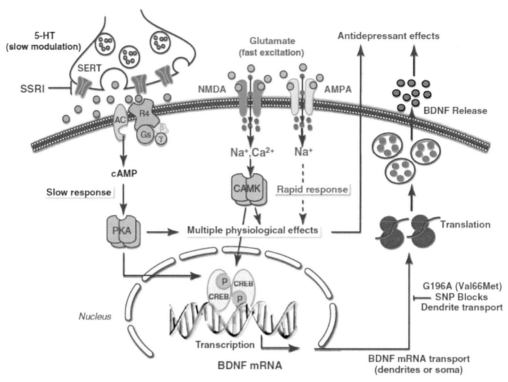

FIGURE 2   Signaling cascades important for understanding depression. Blockade of serotonin (5-HT) reuptake by selective serotonin reuptake inhibitors (SSRIs) causes activation of G-protein-coupled receptors leading to enhanced cAMP signaling, which in turns leads to transcriptional changes via the cAMP–protein kinase A (PKA)–cAMP response element binding (CREB) pathway. Chronic use of SSRIs alters expression of the transcription factor CREB and the Ca^{2+} binding protein p11. As a result, a number of genes are transcribed, including neurotropic factors such as BDNF. Rapid responses act via Glu receptors and Ca^{2+}-dependent signaling mechanisms such as the Ca^{2+} calmodulin-dependent protein kinase (CAMK). *Reproduced with permission from Ref. 10.*

size of neuronal cell bodies, with more complex dendritic branching, which possibly explains the increased amygdala activity measured in depressed individuals by functional imaging.

Hippocampal atrophy, frequently seen in patients with unipolar depression, may not be due to degeneration of existing neurons, but may instead result from a loss of hippocampal neurogenesis and synaptogenesis. The birth of new neurons is highly regulated and has been implicated in hippocampal memory function. Stress hormones are known to reduce neurotrophins and enhance hippocampal excitotoxicity. The major argument supporting a role for hippocampal neurogenesis in depression comes from studies that ablated

neurogenesis by hippocampal irradiation. While this did not induce depressive symptoms in mice, the loss of neurogenesis completely eliminated the effect of antidepressant drugs in mouse models of depression. The notion that neurogenesis may contribute to depression is also attractive, since hippocampal atrophy may readily explain the cognitive impairment of patients with depression.

4.2 Cellular and Network Changes

Much of our current understanding of the functional changes that underlie depressive disorders at the cellular level comes from studies investigating the mechanism of action in drugs that are most effective in stabilizing mood, particularly lithium and the anticonvulsive drugs valproate, carbamazepine, and lamotrigine.[11] These drugs treat primarily bipolar disorder, and findings from such studies therefore primarily inform about the etiology of bipolar disorder. While lithium was originally assumed to alter transmembrane ionic gradients, it is now clear that its activity in depression involves changes of synaptic structure and function in the hippocampus, specifically in the trisynaptic circuit, illustrated in Figure 3. This neuronal circuit comprises the dentate gyrus (DG) of the hippocampus, which contains the subgranular zone, a site of adult neurogenesis involved in learning. Neurons in the DG are glutamatergic and project their axons via the mossy fiber pathway to the pyramidal cells in the CA3 region of the hippocampus. They also synapse onto inhibitory interneurons in this region. The CA3 neurons, in turn, project their axons via the Schaffer collaterals to the CA1 pyramidal cells. These are the main output cells of the hippocampus and their axons innervate the medial prefrontal cortex, the amygdala, striatum, and hypothalamus, essentially all the regions involved in regulation of stress and mood.

Lithium directly alters the excitability of CA1 pyramidal neurons and enhances their excitatory output. This is due to a strengthening of synaptic activity on both pre- and post-synaptic sites. On the postsynaptic site, lithium prolongs the opening of α-Amino-3-hydroxy-5-methyl-4-isoxazolepropionic acid (AMPA) glutamate receptors; on the presynaptic site, it enhances brain-derived neurotropic factor (BDNF) release, which is known to enhance long-term potentiation. BDNF and its Tropomyosin receptor kinase B (TrkB) receptor are abundantly expressed throughout the limbic pathways in the adult brain, where BDNF is released in an activity-dependent manner from presynaptic terminals. In addition to its effects on long-term potentiation (LTP), a cellular form of learning, BDNF also induces synaptogenesis

and synapse stability. Experimentally induced stress in mice decreases BDNF levels, yet chronic treatment with antidepressant drugs can restore BDNF-mediated signaling. These findings suggest an important role for BDNF as a modulator for chronic stress and depression, and suggest a mechanism by which lithium may exert its mood-stabilizing effects.

Finally, lithium also affects neurogenesis in the DG, and the therapeutic benefit of the drug is lost if neurogenesis is prevented. This suggests that hippocampal neurogenesis plays an important role in both the genesis and treatment of depression, yet how these new neurons contribute to emotional well-being is entirely unknown. The most likely drug activity is via upregulation of BDNF release and increase in the expression of the neuroprotective, anti-apoptotic molecule B cell lymphoma/leukemia-2 (Bcl-2). However, it should be noted that animal models of depression may fail to adequately mimic the depression experienced in people. Whether hippocampal neurogenesis occurs and is altered in depressed humans is still unknown.

Curiously, the above studies on neurogenesis and depression also found changes in vascular endothelial growth factor (VEGF), a molecule more typically implicated in the formation of new blood vessels during angiogenesis. We typically think about new vessel formation in the context of brain tumors, where treatment strategies try to starve a tumor by blocking VEGF signaling. These observed increases in VEGF suggest the possibility of changes to the vasculature, possibly necessitated by the birth of additional neurons important in regulation of mood.

4.3 Neuroendocrine Changes

Depression is often triggered by stress and is likely a result of chronic stress at some point in life. It is well established that physical and psychological stress increases serum glucocorticoids such as cortisol. This hormone is part of

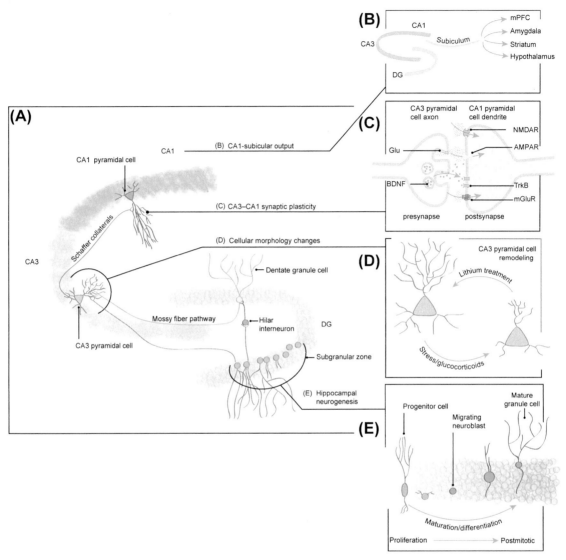

FIGURE 3 The trisynaptic hippocampal network, comprised of dentate gyrus, CA3, and CA1 pyramidal cells, in depression and as the site of action for mood stabilizing drugs. (A) The schematized trisynaptic circuit encompasses the dentate gyrus (DG, blue) that also contains the germinal subgranular zone (SGZ). DG cells (green) project excitatory terminals onto CA3 pyramidal cells and GABAergic interneurons. The output from the CA3 pyramidal cells forms the Schaffer collaterals that innervate the CA1 pyramidal neurons. (B) CA1 pyramidal cells provide the major output of the hippocampus to the medial prefrontal cortex (mPFC), amygdala, striatum, and hypothalamus. (C) Synaptic strength may be altered through pre- or postsyanptic changes involving neurotropic factors such as BDNF. (D) Mood-stabilizing drugs may restore the retraction of dendrites that occurs with stress, or (E) by promoting adult neurogenesis. *Reproduced with permission from Ref. 11.*

the "fight or flight" response that activates a state of alertness in the body. Cortisol is released from the adrenal cortex in response to adrenocorticotropic hormone from the pituitary, which itself is stimulated by corticotropin-releasing hormone produced in the hypothalamus. This pathway defines the so-called hypothalamic–pituitary–adrenal (HPA) axis schematically illustrated in Figure 4. The HPA axis is regulated by numerous inputs from the pleasure and reward circuitries illustrated in Figure 1, namely the amygdala, midbrain, cortex, and hippocampus. Changes in HPA activity, in turn, lead to changes in serum cortisol, which is a biomarker for stress. HPA hyperactivity presents with increased serum cortisol and characterizes individuals with severe mood disorders who often have enlarged pituitary and adrenal glands. Moreover, patients with pathologically increased serum cortisol levels, for example, as a result of tumor in the adrenal gland (Cushing syndrome), consistently show depressive symptoms. This places the HPA axis centrally into the mood control system, providing not only a serum biomarker but also potential targets for intervention. In support of this notion, selective elimination of the forebrain glucocorticoid receptor (GR) through conditional deletion of the GR allele produced mice with enhanced basal serum glucocorticoid levels and with behavior that mimics human depression. Interestingly, GRs themselves are regulated by chronic treatment with some (tricyclic, lithium) but not all (Celexa, Prozac) antidepressant drugs. Not surprisingly, therefore, several GR antagonists are now being evaluated in clinical trials to treat depression.

4.4 Genetics

Genetic association studies have thus far failed to uncover consistent and strong candidate genes for depression, yet a genetic basis for major depression (MD) is indisputable based on familial aggregation.[12] Most importantly, studies of twins show a 37% heritability for MD, and the odds ratio for developing MD if a

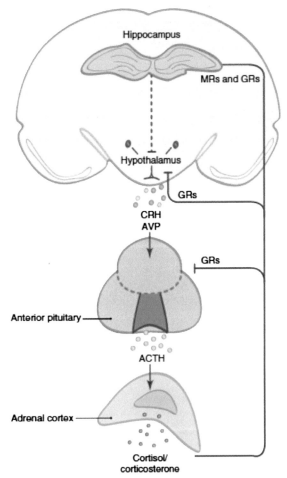

FIGURE 4 The hypothalamic-pituitary-adrenal (HPA) axis in depression. The hippocampus, along with the limbic system, controls the release of corticotropin-releasing hormone (CRH) and arginine vasopressin (AVP), which are secreted into the blood stream. They cause the release of adrenocorticotropic hormone (ACTH) from the pituitary, which in turn stimulates the production and release of cortisol from the adrenal gland. Cortisol feeds back as the stress hormone acting via glucocorticoid receptor (GR) and melanocortin receptors (MR) throughout the body and brain. *Reproduced with permission from Ref. 11.*

first-degree relative has depression is increased by a factor of 2.84. Females are more likely to develop MD, and heritance is somewhat higher in females than in males.

Since attempts to find disease-causing genes have thus far failed, research has turned to interrogate genetic susceptibility through genome-wide association studies (GWAS). These studies look for shared variations (polymorphisms) in the genome at a single nucleotide level. These studies have been similarly disappointing and thus far have not identified any definitive gene loci that confer susceptibility to MD. However, it has been argued that the nine studies that have been carried out thus far and which collectively examined data from over 17,000 affected individuals and over 21,000 control subjects, while large in size, were still underpowered to discover significant differences.

Candidate gene approaches, which make an "educated guess" regarding genes potentially involved in disease prior to examining its expression in affected and unaffected individuals, have identified almost 200 genes, of which 7 appeared significant through meta-analysis of multiple studies. These include genes for the serotonin transporter SLC6A4, the dopamine transporter SLC6A3, apolipoprotein E, the dopamine receptor DRD4, the beta subunit of the G-protein receptor GNB3, the serotonin receptor HTR1A, and the methylenetetrahydrofolate reductase enzyme MTHFR. On average, each of these genes increased the odds ratios for developing disease only marginally (~1.35) and none of them surfaced in the unbiased GWAS studies. This suggests that these genes may all be false positives, leaving us with no genetic leads at the moment. It has been argued that depression may simply not be regarded as a single disease entity but instead as a rather complex behavioral manifestation that can be triggered by numerous underlying causes, each involving very different signaling pathways. If that argument is true, different families may present with very different genetic mutations that are seemingly unrelated yet may each cause depressive illness.

It should be mentioned that one of the suspected genes involved in depression, the serotonin transporter SLC6A4, tasked with the reuptake of serotonin and the drug target of fluoxetine (Prozac), has a genetic isoform that enhances depression risk. More specifically, this transporter exists as a long and short variant. The short variant has a 50% reduced expression and capacity for serotonin uptake, and individuals with two alleles of this short variant have more depressive symptoms and suicidal tendencies.[13] Individuals with the long variant, on the other hand, have an improved response to fluoxetine.

4.5 Animal Models to Study Depression

The most reliable animal models for neurological disease introduce known genetic mutations into transgenic mice. Since we lack such genes for depression, this approach is not yet feasible, but animal models are essential for studying the biology of depression and for exploring potential new treatments. Is it possible that depression is a uniquely human phenotype? Can any of the depressed emotional symptoms be reproduced in animal models? It is likely that sadness, guilt, or suicidal behavior do not exist in animals. However, other depressive features such as helplessness, loss of pleasure, disinterest in sex, and changes in sleep and appetite can be reliably induced in animals and have yielded rather interesting and unexpected insight into disease mechanisms.[14]

The forced swim and tail suspension tests are models of **helplessness or entrapment**. As the test names imply, rodents are placed in water or suspended from their tail without any escape routes. These animals will ultimately capitulate their struggle, and the time that it takes for them to realize their entrapment can be quantified. Repeated trials cause animals to give up more quickly, and ultimately the animal will make little attempt to escape the entrapment. This behavior can be reversed using antidepressant drugs, and thus these tests have high predictive validity.

Another common model to mimic depression is **learned helplessness**. For unknown reasons this model works better in rats than in mice, and essentially entails placing a rat in an environment where it cannot escape a repeated electric foot shock. If the rat is later presented with an opportunity to escape by simply moving to a different spot in the cage, it will no longer even try, but instead will adopt a state of helplessness. Antidepressant drugs can attenuate this learned helplessness.

A **hormonal stress model** takes advantage of the above-discussed role of GRs in depression. Elimination of the GR in forebrain neurons through conditional deletion of the GR allele results in mice exhibiting depressive symptoms associated with enhanced basal serum glucocorticoid levels.

One of the most widely used depression models is called **chronic social defeat**. Here, mild-mannered mice are paired with more aggressive mice for several weeks. After being bullied for an extended period of time, the docile animals become hierarchical surrogates exhibiting anxious and withdrawn behavior. They no longer enjoy pleasurable activities such as consuming sweet treats or having sex, suggestive of an anhedonic-like phenotype. Typically, these animals overeat and become obese. Many of these behaviors are remarkably similar to those of humans with depression and, most importantly, these behavioral changes are persistent but can be reversed using antidepressant drugs.

The use of chronic social defeat has allowed researchers to identify molecular changes that begin to explain some of the behavioral changes at a molecular level. Most importantly, animals subjected to chronic social defeat show epigenetic changes in the structure of DNA, whereby many genes now bear methylation marks that prevent them from being transcribed into proteins. This particularly affects gene pathways involving signals mediating the activity of norepinephrine, serotonin, and dopamine. These epigenetic changes have been quite appropriately called "molecular scars."[15] Studies on social defeat in mice also revealed that not all mice are equally susceptible. Roughly one-third are resilient and never develop the defeated

behavior and also do not show the "molecular scars." So what might distinguish mice that are susceptible to social defeat versus those who are not, and can we extrapolate from that a differential susceptibility of humans to future development of depression?

4.6 The Environment and Depression Susceptibility

To examine whether depression is inducible and heritable, researchers[16] have studied the effect of upbringing on the resilience to depression later in live. Rats that showed good mothering, characterized by repeatedly licking their pups, were compared to passive mothers showing little interest in their offspring. Pups from the caring mothers were less anxious and produced lower concentrations of stress hormones. Importantly, the anxious pups showed methylation marks that turned off the genes producing the GRs in the hippocampus involved in producing corticosterone, the rodent equivalent of cortisol. As a result of the reduced number of receptors, these animals showed a heightened stress response. These relatively simple animal studies suggest that a nurturing environment causes epigenetic changes, essentially turning genes on and off, which make animals persistently more or less anxious (Figure 5). While one may find the behavioral changes intuitive and expected, the molecular changes observed are quite exciting, particularly since these epigenetic marks are reversible and drugs already exist to regulate them. These findings may ultimately inform future treatment of depression.

These environmental effects on gene expression are consistent with the "nature versus nurture" discussion common in biology, psychology, and neuroscience, and explains the impact of environmental influence. Quite stunningly, however, recent research suggests that these epigenetic molecular scars may even be inherited. This suggests that social defeat, or raising pups

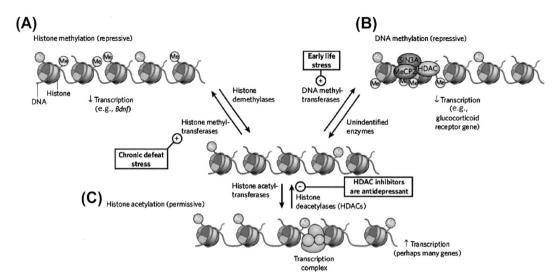

FIGURE 5   Epigenetic regulation of depression. Experiences can alter the accessibility of DNA for transcription by changes in the association of DNA with histones (A) or by methylation of the promoter sites on DNA (B). Both can be modulated using drugs that change the methyl groups that silence DNA or ethyl groups that activate it (C). *Reproduced with permission from Ref. 17.*

or children in neglect, will result in their own offspring being more susceptible to depression. This, of course, has tremendous implications for humans, as it suggests that the environment in which children are raised today can affect their own children born decades later through the passing down of these "molecular scars."

This rings true to the ideas of Jean-Baptiste Lamarck, who claimed 200 years ago that "soft inheritance" may allow certain physical traits such as a muscular body to be passed on to the following generation. Though this possibility was long dismissed, it can now be supported through epigenetic mechanisms that allow for such changes to occur without any change in the encoding genes themselves. As a cautionary note, however, studies in which sperm from defeated fathers was used to impregnate mice through in vitro fertilization showed no evidence that the depression was propagated to the offspring via the father's sperm.[18] This suggests that the trait of the father was spread behaviorally rather than through the gene line. For mothers, this may be very different, as their offspring inherit a number of proteins and even mitochondrial DNA contained in the mother's eggs that may allow for the direct inheritance of epigenetic modifications as well.

From a therapeutic point of view, these studies suggest that drugs that keep genes coated with acetyl groups instead of the methyl groups attached via epigenetic modulation may prevent the silencing of some important genes implicated in depressive behavior. This could in theory be accomplished through a group of drugs that inhibit histone deacetylases (which remove acetyl groups from histones, Figure 5). Unfortunately these drugs still lack specificity for the nervous system, but improved drugs are likely to emerge and may prove therapeutic.

5. TREATMENT/STANDARD OF CARE/CLINICAL MANAGEMENT

The most common treatment options for depression encompass psychological treatments (psychotherapy) and drug therapy, either alone

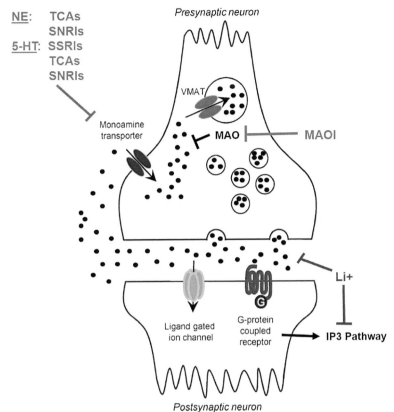

NE: TCAs
    SNRIs
5-HT: SSRIs
    TCAs
    SNRIs

Presynaptic neuron

VMAT

Monoamine
transporter

MAO — MAOI

Ligand gated
ion channel

G-protein
coupled
receptor

IP3 Pathway

Li+

Postsynaptic neuron

FIGURE 6 Targets of available drugs to treat depression. Monoamine oxidase inhibitors (MAOIs) inhibit the breakdown of monoamine transmitter by inhibiting the enzyme monoamine oxidase. Lithium inhibits presynaptic transmitter release as well as postsynaptic signaling cascades. The various transporter inhibitors such as selective serotonin reuptake inhibitors prolong the activity of the transmitter at the synapse. Abbreviations used: TCAs, tricyclic antidepressants; SNRIs, selective norepinephrine reuptake inhibitors; SSRIs, selective serotonin reuptake inhibitors; VMAT, vesicular monoamine transporter; IP3, inositol trisphosphate. *Modified after Ref. 19.*

or in combination. Less common, albeit effective, options are electroconvulsive therapy or surgical implantation of a deep-brain stimulator.

The American Psychiatric Association recommends antidepressant drugs as the first line of treatment for moderate to severe depression, ideally in conjunction with psychotherapy. Treatment of mild should employ psychotherapy prior to use of antidepressant drugs.

All currently available antidepressant drugs are based on the monoamine hypothesis, extensively discussed above, which is derived from the serendipitous finding that changes in monoamine concentrations in the forebrain and limbic system alter mood. Generally speaking, any increase in serotonin and noradrenaline enhances mood; any decrease lowers mood. Today's antidepressant drugs are more refined than the drugs used historically, often offering better side effect profiles, but in principle their mode of action is unchanged. Their neuronal targets are schematically illustrated in Figure 6.

Treatment approaches are very different for unipolar and bipolar disorder, as is their relative effectiveness, which speaks to the notion that these two conditions represent different diseases as opposed to merely two phenotypic variants. For example, psychotherapy has been shown to be equally as effective as antidepressant drugs in treating unipolar depression, though it has not yet been shown to be effective in bipolar disorder and therefore is not considered standard of care. With regard to antidepressant medications (Table 2), only two of the 12 available drugs that are recommended in unipolar depression are also approved

TABLE 2   Currently Used Pharmacological Treatments for Depressive Disorders

Drug Name	Chemical	Class	Indication	
Tofranil	Imipramine	Tricyclic	Depression	1st Generation
Elavil	Amitriptyline	Tricyclic	Depression	
Azilect	Rasagiline	MAOI	Depression	
Nardil	Phenelzine	MAOI	Depression	
Parnate	Tranylcypromine	MAOI	Depression	
Seprenyl	Selegiline	MAOI	Depression	
Prozac	Fluoxetine	SSRI	Depression	2nd Generation
Celexa	Citalopram	SSRI	Depression	
Zoloft	Sertraline	SSRI	Depression	
Luvox	Fluvoxamine	SSRI	Depression	
Paxil	Paroxetine	SSRI	Depression	
Lexapro	Escitalopram	SSRI	Depression	
Effexor	Venlafaxine	SNRI	Depression	
Savella	Milnacipan	SNRI	Depression	
Abilify	Aripiprazole	antipsychotic	Bipolar	
Risperdal	Risperdone	antipsychotic	Bipolar	
Zyprexa	Olanzapine	antipsychotic	Bipolar/Depression	
Seroquel	Quietapine	antipsychotic	Bipolar/Depression	
Geodon	Ziprasidone	antipsychotic	Bipolar	
Tegertol	Carbamazepine	antiepileptic	Bipolar	
Depakote	Divalproex	antiepileptic	Bipolar	
Lamigtal	Lamotrigine	antiepileptic	Bipolar	
Eskalith	Lithium	Salt	Bipolar	

for treatment of bipolar disorder. Indeed, essentially all others have failed to show any effect greater than that of a placebo in clinical trials.

5.1 Psychological Treatments for Clinical Depression (Psychotherapy)

There are a number of different variants of psychotherapy that, in general, all have the same goal: namely, to enhance positive thoughts and decrease negative ones.

The psychotherapy historically used is *cognitive behavioral therapy*, which is built on the premise that dysfunctional beliefs established during development lay dormant for many years until they are brought out by a life stress or traumatic event. The goal of therapy is to identify these dormant negative thoughts and extinguish them. This approach works well in unipolar depression and is as effective as antidepressant drugs, but provides no added benefit.

A variation on this approach is *interpersonal therapy*, based on the assumption that depressions arise from unresolved interpersonal conflicts, which need to be identified and resolved. Although different in approach, this works equally as well as cognitive behavioral therapy.

The most recent variant of psychotherapy is called *problem solving therapy* and operates on the premise that by learning problem-solving skills, a person with depression is better equipped to cope with his or her life situation and can resolve some of his or her specific stressors. Clinical studies suggest that this approach is the most effective form of psychotherapy and, importantly, the only one providing long-lasting improvements.

5.2 Drugs to Treat Major Depression

Antidepressant drugs are typically the first strategy to treat unipolar disease, and each of the 12 drugs listed in Table 2 is equally likely to be effective. Figure 6 shows their targets in relationship to the various sites of synaptic function. Often the drugs take many weeks to show effectiveness, such that many affected individuals become impatient and stop taking their medication. As a result, only 30–40% of patients ultimately show remission of symptoms, leaving well over half of the affected population in dire need of help.

Antidepressant drugs can be grouped into categories or generations. First-generation drugs include the tricyclic antidepressants (TCAs), such as imipramine or amitriptyline, and the MAOIs such as rasagiline and selegiline. Both increase the availability of serotonin by inhibiting either its reuptake or its enzymatic catabolism. Given their long history of use, cheap generic versions are available for all of them. The second-generation drugs include the SSRIs that are aggressively marketed as Prozac, Celexa, and Zoloft, and serotonin–norepinephrine reuptake inhibitors such as Savella. These are the most widely prescribed and advertised antidepressant drugs currently.

However, clinical studies show them to be no more efficacious than the first-generation drugs, and neither is significantly better than a placebo.

The principal side effects of TCAs are sedation and anticholinergic effects such as constipation, dry mouth, urinary hesitancy, and blurred vision. The SSRIs are better with regard to these side effects, but pose a greater risk for gastrointestinal problems and insomnia. All the SSRIs can impair sexual function, resulting in diminished libido, impotence, or difficulty in achieving orgasm. Such sexual dysfunction and weight gain are the most common reasons why patients discontinue their medication.

5.3 Treatment of Bipolar Disorder

The currently approved drugs for bipolar disorder include many of the antipsychotics also used to treat schizophrenia, several antiepileptic drugs, and lithium (Table 2). These drugs all target the manic phase of disease, with the exception of quetiapine and olanzapine, which are also approved for the treatment of depression. An increased risk of weight gain and other metabolic abnormalities is the major a concern with the antipsychotic drugs.

While none of the first- and second-generation antidepressant medications are approved for bipolar disorder, many of them are prescribed nevertheless. Note, however, that in clinical trials, none of them is superior to placebo in treating depressive symptoms. It is fair to say that there is an overuse of such drugs that is not supported by clinical or scientific evidence.

Lithium[20] has been the choice for treatment of bipolar disorder for many decades. While it only became a mainstay in treatment in the 1950s, its first use dates back to 1870. It is effective in reducing manic symptoms and in delaying their recurrence. The therapeutic dosage window is relatively narrow, and differs from patient to patient; therefore, careful blood monitoring is initially necessary to assure an ideal, patient-specific dose. Once established, the drug is quite safe, cheap, and effective. Indeed, with the exception of electroconvulsive therapy, lithium may be the single most effective treatment in psychiatry.

Lithium is generally believed to act via phosphorylation of glycogen-serine kinase 3 (GSK3). GSK3, in turn, has numerous downstream targets that include β-catenin and several Glu receptors, as well as the transcription factor cAMP response element binding (CREB), implicated in hippocampal neurogenesis. Which of these is most important in the drug's therapeutic effectiveness is unclear. Some of the potential signaling axes through which lithium may

provide relief from mania and protection from depression are illustrated in Figure 7.

5.4 Placebo

A comprehensive discussion of the placebo effect can be found in Chapter 15, Bench-Bedside Translation. In light of the fact that depression provides the strongest evidence for the power of the placebo effect, mention of this effect is warranted here. Generally speaking, all the antidepressant drugs discussed earlier improve depressive symptoms in about ~40–50% of patients. However, inert placebos work almost as well, with only a small difference in effect

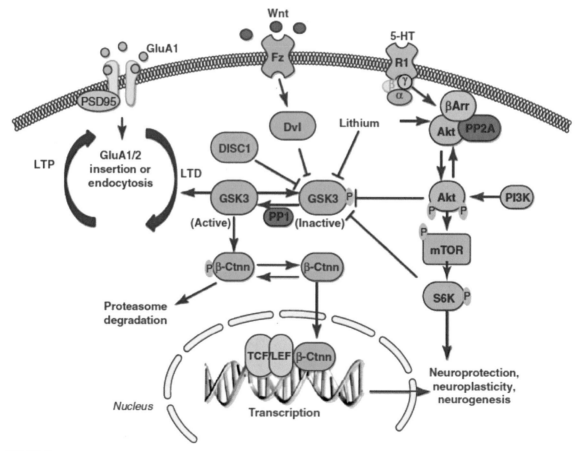

FIGURE 7　Glycogen-serine kinase 3 (GSK3), the target of lithium and its signaling pathways implicated in depressive disorders. *Reproduced with permission from Ref. 10.*

size. In fact, effects of antidepressant drugs are minimal or nonexistent, on average, in patients with mild or moderate severity of depression.[2] Indeed, if a patient receives psychotherapy alone or in conjunction with a placebo, the effect is essentially identical to that of antidepressant drugs. This is not to say that these drugs do not work. Instead it shows the power of the placebo effect, whereby a person desires to get better and, if placed in the right environment, will achieve significant relief without an active ingredient. Interestingly, the placebo effect has steadily increased over the past decades, possibly because patients who listen to commercials expect these drugs to be efficacious. Thus, when participating in clinical studies, these patients expect improvements, which, of course, also affects those who are in the placebo group. From what we know about the brain's reward system and the interaction of the limbic system with the HPA axis, it is not surprising that mentation can actually alter brain chemistry, particularly the levels of monoamines. Consequently, any condition that involves dopamine, serotonin, or noradrenaline is much more affected by the placebo effect than those that are not. Given the significant side effects of the currently available TCAs and SSRIs, which cause a large number of patients to stop taking their medication, one must seriously reevaluate current treatment strategies. Some professionals argue for expanded use of placebos and for psychotherapy to make a comeback.

5.5 Electroconvulsive Therapy

Electroconvulsive therapy (ECT) is at least as effective as medication, but its use is reserved for treatment-resistant cases and delusional depressions. As discussed in greater detail in Chapter 13 on schizophrenia, electroconvulsive therapy aims to disrupt oscillatory gamma activity in the forebrain. Side effects include acute and chronic memory impairments. A less invasive treatment is through transcranial magnetic stimulation (TMS), which is approved for treatment-resistant

depression and has been shown to have efficacy in several controlled trials. Vagus nerve stimulation (VNS) has also recently been approved for treatment-resistant depression, but its degree of efficacy is controversial. Deep-brain stimulation (DBS) is another treatment that is being used experimentally in treatment-resistant cases. What all of these treatments have in common is that they aim at disrupting neuronal networks in the brain. ECT, TMS, and VNS do so noninvasively while DBS requires neurosurgical implantation of stimulating electrodes. How and why these approaches work is largely unknown and their relative utility over antidepressant drugs is an area of active investigation.

6. EXPERIMENTAL APPROACHES/ CLINICAL TRIALS

In light of the ineffectiveness of current treatments for many patients suffering from depression or bipolar disorder, it is not surprising that numerous clinical studies are exploring more effective treatments, ranging from psychotherapy to new drugs or drug combinations to nutritional supplements such as omega-3 fatty acids. At present, over 2000 clinical trials are registered on clinicaltrials.com, yet few are at a stage where there is broad excitement that a new breakthrough may be in reach. An exception may be the surprise emergence of the local anesthetic, ketamine, also used as an illicit street drug named "Special K."

Ketamine is an NMDA receptor (NMDA-R) antagonist more typically used as an anesthetic sedative or analgesic. It also causes a dissociative state and is therefore often abused as a recreational hallucinogenic. It belongs to a group of psychedelic drugs that include Lysergic acid diethylamide (LSD), related to the ergot fungus, psilocybin from mushrooms, and mescaline, contained in peyote cacti. Like ketamine, these drugs all date back to the 1950s and 1960s. Their recreational use became associated with rebellious behavior in the 1970s, whereupon they were declared Schedule I controlled substances, limiting their further medical or investigational use. Only ketamine is used widely (as an anesthetic, primarily in veterinary use), although it is the drug of choice for acute traumatic shock. Its effects as a hallucinogenic and anesthetic are short lived (<60 min). Therefore, it was surprising to discover that even at low doses (one-third of the sedative dose), ketamine provides quick (<2 h) and lasting relief (7 days) to severely depressed individuals. Importantly, ketamine is also effective in treating suicidal thoughts. Unfortunately, in spite of dramatic improvements for severe depression following ketamine treatment, patients typically relapse after 1 week. It is therefore important to better understand the mechanism of drug action in order to selectively target downstream signaling pathways for longer-lasting relief.

Over 40 clinical trials are currently examining the potential usefulness of ketamine in severe treatment-resistant depression, in bipolar disorder, and in suicide prevention. In addition, the biological activity and specific brain regions activated by ketamine are under intense investigation. Imaging data (Figure 8) using radiolabeled glucose in conjunction with positron emission tomography (PET) suggests that ketamine enhances activity in the prefrontal cortex and the limbic structures, including the basal ganglia and thalamus.

At a cellular level, ketamine binds quite specifically to the PCP site on the NMDA receptor. However, ketamine increases, rather than decreases, glutamate levels in the brain. This paradoxical action of an NMDA-R blocker can be explained by assuming that at low doses, as

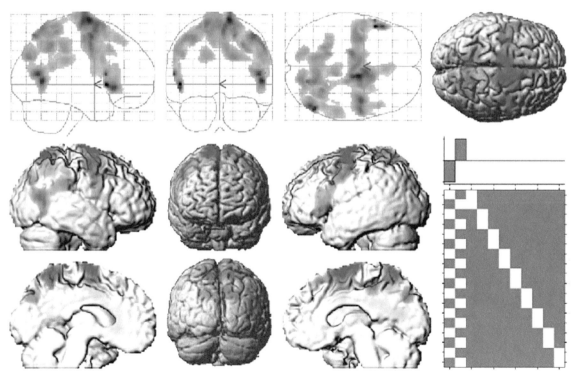

FIGURE 8 ^{18}Fluorodeoxyglucose positron emission tomography (PET) imaging in a volunteer using ketamine shows enhanced glucose utilization (indicated by red color) in frontal, temporal and parietal cortex. *Reproduced with permission from Ref. 24.*

used for the treatment of depression, ketamine preferentially blocks NMDA-Rs on GABAergic interneurons (Figure 9). This in turn reduces their inhibitory control over the prefrontal cortex. The resulting increase in glutamate then activates NMDA and AMPA-R in these prefrontal cortical neurons, causing long-lasting structural changes in synapses and cellular connectivity. Indeed, ketamine has been shown to change synaptic density and induce hippocampal neurogenesis. This may involve transcriptional changes in BDNF and GSK3 since ketamine[22] and lithium[23] are both ineffective in treating depressive symptoms in mice lacking GSK3.

7. CHALLENGES AND OPPORTUNITIES

Close to 20% of the population will go through an episode of depression during their lifetime, and for approximately 8% of the population this will permanently affect their life. With suicide being the eighth-deadliest "disease" in the US, and with 50% of all suicide victims suffering from depression, it is fair to say that this is a serious challenge to our national health. Current treatments are inadequate and largely based on an antiquated disease mechanism derived from the action of serendipitously discovered agents that alter the brain's concentration of monoamine transmitters. No drug more effective than lithium salt has been produced for treating biploar depression since the 1950s, and the repeated failure of drug companies to develop more effective drugs has brought us to near capitulation in drug discovery.

It is incumbent to take a fresh look at cognitive disorders and begin to understand mechanisms of disease before embarking on new drug development efforts. The recent surprise finding that ketamine provides fast relief through a series of mechanisms not previously considered should serve as inspiration that additional basic research has the potential to open novel avenues for drug discovery. Also, the findings that hippocampal volume decreases in depression and that hippocampal neurogenesis is essential in mood-stabilizing effects by lithium provide concrete areas and targets for investigation. Unquestionably, preclinical data points toward persistent

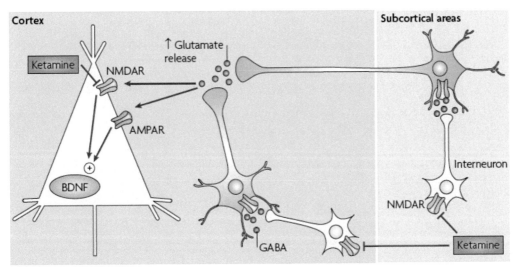

FIGURE 9 Hypothesized action of ketamine explaining how a glutamate receptor antagonist may still cause a paradoxical increase in glutamate tone. In the treatment of depression, low concentrations of ketamine predominantly affect NMDA-Rs on GABAergic interneurons. These in turn reduce their inhibitory activity on the prefrontal cortex, resulting in a net increase in gluatamatergic tone. *Reproduced with permission from Ref. 21.*

structural and functional changes in the hippocampus and areas it projects to in the prefrontal cortex. In light of the unquestionable heritability of depression, a concerted effort must be made to gain a better understanding of susceptibility genes and their interaction with environmental triggers. Unfortunately, animal models remain a challenge as the complexities of depressive disorders, particularly bipolar disorder, are not fully recapitulated in any current model used for drug discovery research. Clearly, depression may be the most complex illness in all of neurology and psychiatry, and may be the most difficult to understand. Given the magnitude of the problem, complacency is not an option.

Acknowledgments

This chapter was kindly reviewed by Dr Richard Shelton, the Charles B. Ireland Professor & Vice Chair for Research and Head of the Mood Disorders Research Center at the University of Alabama at Birmingham.

References

1. Angst J, Marneros A. Bipolarity from ancient to modern times: conception, birth and rebirth. *J Affective Disord.* 2001;67(1–3):3–19.
2. Fournier JC, DeRubeis RJ, Hollon SD, et al. Antidepressant drug effects and depression severity: a patient-level meta-analysis. *JAMA.* 2010;303(1):47–53.
3. Gracious BL, Finucane TL, Friedman-Campbell M, Messing S, Parkhurst MN. Vitamin D deficiency and psychotic features in mentally ill adolescents: a cross-sectional study. *BMC Psychiatry.* 2012;12:38.
4. Frye MA. Clinical practice. Bipolar disorder–a focus on depression. *N Engl J Med.* 2011;364(1):51–59.
5. Loftis JM, Patterson AL, Wilhelm CJ, et al. Vulnerability to somatic symptoms of depression during interferon-alpha therapy for hepatitis C: a 16-week prospective study. *J Psychosom Res.* 2013;74(1):57–63.
6. Wolverton SE, Harper JC. Important controversies associated with isotretinoin therapy for acne. *Am J Clin Dermatol.* 2013;14(2):71–76.
7. Rubino T, Zamberletti E, Parolaro D. Adolescent exposure to cannabis as a risk factor for psychiatric disorders. *J Psychopharmacol.* 2012;26(1):177–188 (Oxford, England).
8. Kendler KS, Karkowski LM, Prescott CA. Causal relationship between stressful life events and the onset of major depression. *Am J Psychiatry.* 1999;156(6):837–841.
9. Rajkowska G, O'Dwyer G, Teleki Z, Stockmeier CA, Miguel-Hidalgo JJ. GABAergic neurons immunoreactive for calcium binding proteins are reduced in the prefrontal cortex in major depression. *Neuropsychopharmacology.* 2007;32(2):471–482.
10. Duman RS, Voleti B. Signaling pathways underlying the pathophysiology and treatment of depression: novel mechanisms for rapid-acting agents. *Trends Neurosci.* 2012;35(1):47–56.
11. Schloesser RJ, Martinowich K, Manji HK. Mood-stabilizing drugs: mechanisms of action. *Trends Neurosci.* 2012;35(1):36–46.
12. Flint J, Kendler KS. The genetics of major depression. *Neuron.* 2014;81(3):484–503.
13. Ebmeier KP, Donaghey C, Steele JD. Recent developments and current controversies in depression. *Lancet.* 2006;367(9505):153–167.
14. Krishnan V, Nestler EJ. Animal models of depression: molecular perspectives. *Curr Top Behav Neurosci.* 2011;7:121–147.
15. Nestler EJ. Hidden switches in the mind. *Sci Am.* 2011; 305(6):76–83.
16. Zhang TY, Labonte B, Wen XL, Turecki G, Meaney MJ. Epigenetic mechanisms for the early environmental regulation of hippocampal glucocorticoid receptor gene expression in rodents and humans. *Neuropsychopharmacology.* 2013;38(1):111–123.
17. Krishnan V, Nestler EJ. The molecular neurobiology of depression. *Nature.* 2008;455(7215):894–902.
18. Dietz DM, Nestler EJ. From father to offspring: paternal transmission of depressive-like behaviors. *Neuropsychopharmacology.* 2012;37(1):311–312.
19. Andersen J, Kristensen AS, Bang-Andersen B, Stromgaard K. Recent advances in the understanding of the interaction of antidepressant drugs with serotonin and norepinephrine transporters. *Chem Commun.* 2009;25:3677–3692 (Cambridge, England).
20. Shorter E. The history of lithium therapy. *Bipolar Disord.* 2009;11(Suppl. 2):4–9.
21. Vollenweider FX, Kometer M. The neurobiology of psychedelic drugs: implications for the treatment of mood disorders. *Nat Rev Neurosci.* 2010;11(9):642–651.
22. Beurel E, Song L, Jope RS. Inhibition of glycogen synthase kinase-3 is necessary for the rapid antidepressant effect of ketamine in mice. *Mol Psychiatry.* 2011;16(11):1068–1070.
23. O'Brien WT, Huang J, Buccafusca R, et al. Glycogen synthase kinase-3 is essential for beta-arrestin-2 complex formation and lithium-sensitive behaviors in mice. *J Clin Invest.* 2011;121(9):3756–3762.
24. Langsjo JW, Salmi E, Kaisti KK, et al. Effects of subanesthetic ketamine on regional cerebral glucose metabolism in humans. *Anesthesiology.* 2004;100(5): 1065–1071.

General Readings Used as Source

1. Reus VI. Chapter 391. Mental disorders. In: Longo DL, Fauci AS, Kasper DL, Hauser SL, Jameson J, Loscalzo J. eds. *Harrison's Principles of Internal Medicine*, 18e. New York, McGraw-Hill; 2012.
2. Ebmeier KP, Donaghey C, Steele JD. Recent developments and current controversies in depression. *Lancet*. January 14, 2006;367(9505):153–167.
3. Krishnan V, Nestler EJ. The molecular neurobiology of depression. *Nature*. 2008;455(7215):894–902.

Suggested Papers or Journal Club Assignments

Clinical Papers

1. Mayberg HS, Lozano AM, Voon V, et al. Deep brain stimulation for treatment-resistant depression. *Neuron*. March 3, 2005;45(5):651–660.
2. Moncrieff J, Wessely S, Hardy R. Active placebos versus antidepressants for depression. *Cochrane Database Syst Rev*. 2004;1:Cd003012.
3. Naudet F, Millet B, Charlier P, Reymann JM, Maria AS, Falissard B. Which placebo to cure depression? A thought-provoking network meta-analysis. *BMC Med*. 2013;11:230.

Basic Papers

1. van Hasselt FN, Cornelisse S, Zhang TY, et al. Adult hippocampal glucocorticoid receptor expression and dentate synaptic plasticity correlate with maternal care received by individuals early in life. *Hippocampus*. February 2012;22(2):255–266.
2. Ota KT, Liu RJ, Voleti B, et al. REDD1 is essential for stress-induced synaptic loss and depressive behavior. *Nat Med*. April 13, 2014;20(5):531–535.
3. Han K, Holder Jr JL, Schaaf CP, et al. SHANK3 overexpression causes manic-like behaviour with unique pharmacogenetic properties. *Nature*. November 7, 2013;503(7474):72–77.

CHAPTER
13

Schizophrenia

Harald Sontheimer

1. CASE STORY

I am visiting my friend John at Bryce Psychiatric Hospital. I still cannot believe how this all unfolded. We had been roommates at Ole Miss for the past year. I thought we were a pretty good match and shared similar interests. We both liked tinkering with things. He was pursuing a mechanical engineering major while I was still undeclared. Like me, John was quiet and shy; one

Diseases of the Nervous System
http://dx.doi.org/10.1016/B978-0-12-800244-5.00013-6

would probably call us introverted, but among the engineering students we were pretty typical. Neither of us took any interest in Greek life and we didn't really do much with other students. Most evenings we just hung out in our room watching Netflix movies or listening to loud music. After Christmas break John seemed a bit different, more absent-minded than usual and somewhat expressionless. At dinner he wasn't very talkative and he often just stared into space. Then he would nervously look around as if someone were approaching him. I didn't make much of it as first. He also started complaining about noises from adjacent rooms or the floor above, yet I never heard anything unusual. He began waking up at night and pacing endlessly up and down the room. I heard him toss and turn for hours on end. One night he literally woke me every hour. I asked him to stop since I had a quiz in the morning. He started shouting. He said he was hearing voices yelling at him, saying that he was good for nothing and he should just go to hell. After class I insisted that John meet with a student counselor, which he did. He claimed that he felt much better after counseling and that he realized the stress had been getting to him. Initially John did seem better but I noticed him getting up and leaving the room almost every night. I had no idea what he was up to. He said he was taking long walks that calmed him down. Upon returning one night he was panting, totally out of breath as though he had been running sprints. "Guy, help me! I can't stop those voices. They are driving me crazy. They are chasing me. They are here to get me! Help!" I called campus safety immediately, who took John to the emergency room at Baptist Memorial Hospital. I didn't see or hear from John until his parents came by to pick up his stuff. I learned that he had been taken to a psychiatric hospital in Alabama and would be there for some time. I have visited him once every month for the past year. The last time John seemed fine at first and we talked about him coming back and making up his lost classes. Then seemingly out of nowhere, he started screaming. "I have

to get inside, they are here to shoot me." Once that passed, he switched topics almost randomly in conversation without a clear train of thought. On the way out I ran into his mother, who was in tears. "Schizophrenia, just like his grandfather," she said. "Dad took his life when I was 5, and John is just as my mother described my dad."

2. HISTORY

Throughout history, madness and insanity have received little differentiation in terms of such behavior. In antiquity, madness was not viewed as an illness but rather as a divine punishment or possession by demons. This view was first challenged by the Greek physician Hippocrates (460–377 BC), who not only attributed mad behavior to abnormal brain function but even suggested that it was a treatable condition: "Only from the brain springs our pleasures, our feelings of happiness, laughter and jokes, our pain, our sorrows and tears. . . . This same organ that makes us mad or confused, inspires us with fear and anxiety…" (Hippocrates, *The Holy Disease*). He suggested that rebalancing the four humors through diet, bloodletting, or purgatives should cure the illness. Unfortunately, this visionary perspective was largely neglected until the 1950s, when antipsychotic drugs would permit treatment of psychotic patients. In the meantime, countless people who suffered from psychiatric illnesses were killed as heretics or witches during the Middle Ages or on eugenic grounds in Nazi Germany. If not killed, they were institutionalized for life and often found themselves strapped to a hospital bed with little contact with the outside world.

Although ancient literature describes psychotic individuals, it is difficult to ascertain whether these people indeed may have suffered from schizophrenia. The earliest accurate portrayals of the illness date to 1809, when two psychiatrists independently published schizophrenia cases that they ascribed to "premature dementia." In 1871 Karl Ludwig Kahlbaum and Erwald Hecker categorized different forms of madness, including catatonia and paranoia. In 1896 Kraeplin introduced the term *dementia praecox*, emphasizing a dementia that was associated with psychosis and manic-depressive episodes, a term that was later replaced by *schizophrenia*, derived from the Greek roots for "split mind."

Throughout the first half of the twentieth century, psychiatry was influenced by the psychoanalytic theories of Sigmund Freud, who claimed that schizophrenia resulted from a weak ego and the apparent unconscious inability of a child to form a strong bond with a parent of the opposite sex. He advocated treatment through rebirthing to restore the impaired parent–child bond. Since no medications were available to contain the psychotic behavior, it was quite common for schizophrenia patients to wind up in closed, state-operated psychiatric wards.

The recognition that schizophrenia is indeed an organic disease of the brain came with the introduction of effective neuroleptic drugs, starting with chlorpromazine in 1954 and followed by haloperidol in 1958. These drugs target dopamine receptors and, consequently, schizophrenia became a "dopamine disorder." Importantly, the scientific community gradually recognized schizophrenia as a disease of neurochemical origin. The notion that a neurochemical imbalance is a major contributing factor to schizophrenia is alive and well today, but with the inclusion of other neurotransmitter systems, including monoamines, glutamate (Glu), and γ-aminobutyric acid (GABA). Unfortunately, the success of antipsychotic drugs directed much of the research focus in the late twentieth century on understanding drugs rather than disease etiology. In stark contrast, the recent identification of mutations in genes that are fundamentally involved in neuronal development and that enhance the susceptibility for disease has made a compelling case to consider schizophrenia a neurodevelopmental disorder that begins well before a patient has his/her first psychotic break.

Over the centuries, literature and art have described the struggle of patients affected by schizophrenia; an example is shown in Figure 1. This painting, entitled "The Scream," by Edvard Munch (1863–1944), depicts one of his psychotic episodes. The 1994 movie "A Beautiful Mind" captured the heart-wrenching struggle with schizophrenia of the Nobel Prize-wining mathematician John Nash.

3. CLINICAL PRESENTATION/ DIAGNOSIS/EPIDEMIOLOGY

Schizophrenia is a complex syndrome that is defined by a collection of symptoms that are dominated by psychosis. However, symptoms also include deficits in thinking, behavior, and affect and are typically grouped into (1) positive

FIGURE 1   "The Scream," by Edvard Munch, a Norwegian painter who suffered from schizophrenia. http://brainz.org/10-great-painters-who-were-mentally-disturbed/.

symptoms such as hallucinations, delusions, or paranoia; (2) negative symptoms that include a loss of motivation, apathy, asocial behavior, loss of affect, and poor use and understanding of speech and language; and (3) cognitive symptoms such as impaired working memory, dissociated thought processes, and impaired executive function.

A typical patient is in his or her early 20s, a smoker, and often somewhat unkempt in appearance and exhibits odd social behavior. The person has a reduced ability to sense pleasure, frequently showing signs of anhedonia. Reduced eye contact and muted facial expression are common. In conversation one may find little substantive thought content, and topics may change multiple times during the conversation. Patients with schizophrenia may be hypersensitive to sounds or lights and have a distorted view of themselves with regard to the world around them, often feeling like detached observers. They may even claim that their body belongs to someone else. Patients typically pay little attention to their appearance, hygiene, and immediate living environment.

Schizophrenia is the most severe neuropsychiatric illness and probably the most severe disorder with unknown cause. It affects men and women equally, and, with a lifetime prevalence of 1%, it is a common illness. Approximately 2.2 million patients currently live with schizophrenia in the United States, and about 300,000 patients are hospitalized. These patients fill more hospital beds than those with any other disease. Since the illness typically begins during late adolescence or early adulthood (18–25 years of age), it exacts a tremendous toll on the individual, the family, and society. Many patients are unable to work and are disabled without income for life. Not surprisingly, over 200,000 patients with schizophrenia are homeless, and these account for one-third of the entire homeless population in the United States. Unfortunately, these patients are often outside the reach of a doctor and not medicated.

Because of an absence of biomarkers, a schizophrenia diagnosis is based entirely on the assessment of symptoms by a trained psychiatrist who bases his or her judgment on a number of features that are described in the *Diagnostic and Statistical Manual of Mental Disorders* (DSM, now in its fifth revision). A typical examination involves an extensive interview with the patient and ideally includes immediate family members or relatives. It seeks to assess the illness from the patient's own perspective, particularly the disease history and evolution of symptoms. The physician looks for neurovegetative signs affecting sleep, appetite, or sex drive. Family and personal history must be carefully evaluated, in particular to establish a timeline regarding changes that may have occurred. Prior drug treatment, as well as alcohol, nicotine, and drug use, will be questioned. A mental status examination is administered and includes appearance, affect, mood, speech, cognition, judgment, and insight.

Most patients first seek medical help after they have suffered a psychotic episode characterized by delusions and/or hallucinations. In such cases the person is overcome by panic and a feeling of helplessness with auditory hallucinations; (s)he may feel haunted, possessed, or pursued by evil spirits or manipulated and controlled by the Federal Bureau of Investigation. It is likely that the disease and its symptoms began many years earlier, with subtle signs of social, cognitive, and motor deficits during childhood and leading to anxiety, moodiness, withdrawal, and social isolation by adolescence. This is often called a prodromal (forerunner) phase, during which precipitating factors such as life stresses or drug use may ultimately lead to a psychotic attack. In spite of the seemingly violent behavior exhibited during a psychotic attack, schizophrenia does not usually produce violent behavior.

In the past based on its presentation schizophrenia was subclassified into disorganized, catatonic, paranoid, and residual. This classification has been abandoned with the recent adoption of the DSM-V, and schizophrenia is now regarded as a disorder with a spectrum of behavioral presentations that likely share common underlying disease cause(s) and mechanism(s). Schizophrenia is easily misdiagnosed, particularly if diagnosis is primarily based on psychotic behavior, which can result from illicit drug use alone. The use of amphetamines and cocaine can easily mimic positive symptoms. Also, brain injury through trauma, stroke, or malignancies can cause psychotic behavior, but these are readily excluded through brain magnetic resonance imaging (MRI) or computed tomography (CT).

There is considerable overlap between the negative symptoms of schizophrenia and depression,[1] such as an overall lack of energy, social withdrawal, and loss of interest in pleasurable activities (anhedonia). Indeed, 60–75% of patients with schizophrenia report depression, and it has been argued that depression may be a prodromal symptom that can precede the first psychotic attack by many years. Therefore depression is an important comorbidity that must be treated independent of the psychosis because it may be the single greatest risk for suicide in patients with schizophrenia.

A number of external risk factors have been reported, but most increase risk only moderately, and some are based on equivocal scientific evidence. There has been a fascination with epidemiological studies showing that babies born in winter have a significantly increased risk of developing schizophrenia over their lifetime compared with those born in summer. This led to numerous studies examining a potential link to maternal viral infections, which are more common in the winter. Maternal malnutrition, exposure to cytokines, and perinatal injury are additional prenatal exposure risks. Growing up in an urban environment, being part of a minority group, and cannabis use are additional confirmed risk factors.[2]

As in most psychiatric conditions, schizophrenia is highly heritable, although no monogenetic causes have been identified. Within families the risk of developing schizophrenia ranges from ~6% for a first-degree relative to 50% for homozygotic twins.

Current therapy is quite effective in containing positive symptoms but largely ineffective for both negative and cognitive symptoms. While some older literature suggests that 15–30% of patients can be medically cured, it is more accurate to refer to this as disease management rather than a cure. Long-term remissions with treatment approach 50%, leaving approximately half of all patients with schizophrenia dependent on life-long supervision by a caregiver. In spite of much-improved antipsychotic drugs, 15% of all patients are permanently hospitalized. Shockingly, schizophrenia accounts for more lost life-years than cancer and heart disease. Overall, life expectance is reduced on average by 25 years, in part because of unhealthy lifestyles leading to heart disease and respiratory complications resulting in a significantly enhanced risk for suicide. Sadly, approximately 10% of patients with schizophrenia take their own life.

4. DISEASE MECHANISM/CAUSE/ BASIC SCIENCE

Schizophrenia is a functional disorder that, in spite of a clear behavioral phenotype, does not show unequivocal biological or histological disease markers. Many patients with schizophrenia have a reduced cortical thickness on MRI, particularly affecting the cortical gray matter, and as a result the ventricles appear somewhat enlarged. An extreme example, illustrated in Figure 2, compares MRIs of identical twins, one affected by schizophrenia and one unaffected, and visibly makes this case. Though not specific for schizophrenia, it signifies an underlying degenerative pathology similar to that observed in frank neurodegenerative diseases. Surprisingly, however, there does not seem to be a reduction in neuronal numbers; instead, it is neuropile, which encompasses glial cells and the

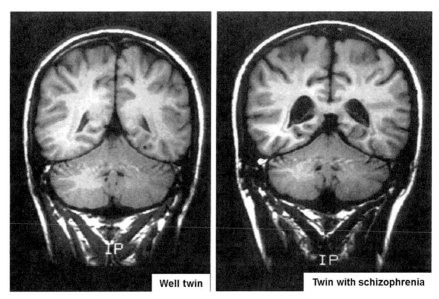

Well twin

Twin with schizophrenia

FIGURE 2   Magnetic resonance images from 28-year-old identical male twins. The one with schizophrenia (right) has markedly enlarged posterior ventricles, suggesting the loss of brain tissue associated with the disease. *Courtesy of Drs E. Fuller Torrey and Daniel Weinberger.*

dendritic neuronal processes, that is reduced in volume. This suggests a decrease in the complexity of neurites and their dendritic spines.

4.1 Imaging

Structural changes are not always as dramatic as in the example shown in Figure 2. Functional imaging studies,[3] however, using either positron emission tomography (PET) or functional MRI, are sensitive enough to detect more modest reductions in functional and/or metabolic activity in a number of brain regions that are involved in executive function, episodic memory, emotional regulation, and social recognition. Figure 3 schematically summarizes the most affected brain regions. The most well-researched region is the dorsolateral prefrontal cortex (DLPFC), which is tasked with executive function, task initiation, motivational drive, and working memory. Reduced activation also has been reported for the amygdala and the prefrontal cortex, which explains the flattened affect displayed by individuals with schizophrenia.

The past 10 years have led to an explosion of functional imaging studies of schizophrenia, allowing for an assessment of neurotransmitter systems (single photon emission CT and PET) as well as brain connectivity (diffusion tensor imaging). Particularly insightful were studies using the dopamine precursor F-Dopa, adopted from imaging of Parkinson's disease, with which the presynaptic concentrations of dopamine can be inferred. In addition, single photon emission CT imaging allows a more direct measurement of GABA and Glu. Dopamine release can be indirectly measured as the displacement of a radio-labeled substrate from its receptor. These approaches have significantly transformed our understanding of the disease. Most important, they have allowed the role of different transmitter systems in disease to be questioned, as discussed in some detail below.

A number of molecular markers are useful on autopsy tissue because they cannot yet be

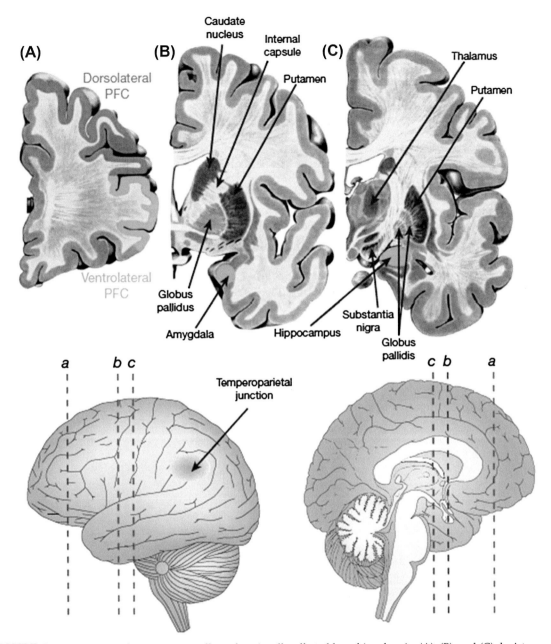

FIGURE 3   Brain regions that are structurally or functionally affected by schizophrenia. (A), (B), and (C) depict coronal brain sections corresponding to the planes illustrated in the schema below as *a, b* and *c*. Enhanced activity is indicated by red colors, reduced activity by blue colors. *From Ref. 3.*

studied noninvasively. Consistent findings include a reduction in the glutamic acid decarboxylase, the enzyme responsible for the synthesis of GABA, particularly the 67-kD form (GAD67); a decrease in the alpha-1 subunit of the GABAa receptor; and an overall ~50% decrease in GABAergic synapses. There also seems to be decreased dopaminergic innervation in the prefrontal cortex, particularly because of reduced projections from the thalamus. In addition, Glu receptor subunits, particularly those forming N-methyl-D-aspartate (NMDA) receptors (NMDA-Rs), are decreased in postmortem tissue. This suggests that schizophrenia is a multisystemic disease involving not only multiple brain regions but also multiple neurotransmitter systems.

4.2 Genetics

As with most neurological diseases, genetic causes of schizophrenia have been difficult to delineate. However, the more closely one is related to an individual with schizophrenia, the more likely one is to develop it as well. Identical twins have a 50% likelihood of both being affected by schizophrenia, which is reduced to 10% in fraternal twins and 6.6% in first-degree relatives. If both parents are affected, their children carry a 40% disease risk. These numbers do not support a simple Mendelian heritability and rule out a monogenetic cause of disease, yet they support a strong polygenetic genetic risk for disease.

A number of schizophrenia susceptibility genes have been identified on chromosome 22. These include deleted in schizophrenia complex 1 (*DISC1*), catechol-*O*-methyltransferase (*COMT*), neuregulin, and its receptor *ERBB4*. Interestingly, neuregulin and the ERBB4 receptor are involved in the orderly development of the cortex, in particular the differentiation of parvalbumin-expressing GABAergic interneurons. Dysfunction of the neuregulin–ERBB4 pathway could readily explain the reduced dendritic arborization and the thinning of the cortical gray matter frequently observed on MRI. Similarly, *DISC1*, which was first described in a single affected family in Iceland, functions in neuronal maturation and process outgrowth. However, none of these genes alone is sufficient to cause disease. When human mutations were introduced into transgenic mice, a number of structural and behavioral phenotypes resulted, yet none was sufficient to model schizophrenia, if that is even possible (further discussed in Section 4.5).

By far the strongest known genetic risk factor for schizophrenia is a hemizygous deletion (affecting only one chromosome) of a 3-megabase segment on the long arm of chromosome 22q11.2 that wipes out up to 60 genes (Figure 4). The result causes a complex

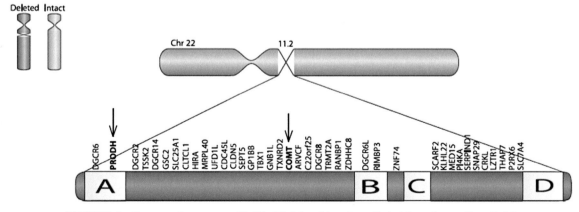

FIGURE 4   Genes on chromosome 22 (Chr22) deleted in velocardiofacial syndrome. *From Ref. 5.*

disease called **velocadiofacial** or **DiGeorge syndrome**.[4] This autosomal dominant deletion is the most common chromosomal deletion in humans and affects about 1 in 4000 newborns. The primary features are subtle to severe facial abnormalities and heart defects. It causes a number of neuropsychiatric conditions including autism, attention deficit hyperactivity disorder, and mood and anxiety disorders.[5] Importantly, about a quarter of these children develop schizophrenia that is indistinguishable from idiopathic forms of the illness. These account for 1–2% of all schizophrenia cases. The chromosomal segment that is deleted affects about 60 genes. Among these are two microRNAs called *DGC8* (DiGeorge syndrome critical region 8) and *miR185*. microRNAs are small, 21- to 25-nucleotide, noncoding RNAs that bind target sites in messenger RNAs (mRNAs) and regulate protein expression by repressing mRNA translation or inducing a breakdown of the mRNA. Each microRNA can control hundreds of RNAs, and *DGC8* and *miR185* seem to regulate dendritic and spine development via different pathways. *miR185* is a regulator of the endoplasmic reticulum Ca^{2+}-ATPase (SERCA2) that modulates intracellular Ca^{2+}, which is important in synaptic plasticity. Its targets also include Cdc42, which controls actin polymerization, and RhoA, a GTPase that destabilizes the actin cytoskeleton, causing a reduction in dendritic branches. In light of the observed loss of dendritic complexity in the DLPFC, these target genes readily explain mechanistically how the chromosomal deletion causes the characteristic schizophrenia pathology. Also included in the deleted region is the gene for catechol-*O*-methyl transferase (*COMT*; Figure 4), an enzyme involved in the postsynaptic degradation of catecholamines such as dopamine. Reduced expression of *COMT* due to loss of one allele would result in reduced dopamine clearance and therefore enhanced dopamine concentrations at affected synapses, which could directly contribute to psychotic behavior since blockade of dopamine receptors inhibits psychosis.

To broaden the search for susceptibility genes possibly linked to schizophrenia, a number of genome-wide association studies (GWASs) have been conducted, with disappointing results. Although they identified over 40 candidate genes, their effect size, or the degree to which their expression alters susceptibility to develop disease, is very small. As we have seen with other neurological diseases (e.g., multiple sclerosis), GWAS findings suggest that schizophrenia must be a polygenic disease where mutations in multiple genes, each with limited penetration, compound to increase disease risk.

4.3 Neurotransmitters

Shortly after the discovery of dopamine as a neurotransmitter in the 1950s by Carlsson and Lindqvist and an accidental discovery of the first antipsychotic drugs operating as D2 antagonists, schizophrenia became a disease of dopamine signaling. Since then, however, just about every neurotransmitter system has been implicated in schizophrenia, with current emphasis on GABA, Glu, dopamine, and serotonin, each of which has its own "hypothesis," further discussed below, yet in all likelihood all are probably interconnected. As already alluded to, in most instances either the therapeutic efficacy of antipsychotic drugs or the mimicry of disease by blockade of neurotransmitter receptors sets the stage. Specifically, antipsychotic drugs, such as haloperidol, act primarily as dopamine D2 receptor antagonists, yet the second generation of antipsychotics, such as clozapine and risperidone, target serotonergic receptors as well. The "glutamate hypothesis" owes its birth to the fact that silencing NMDA receptors with antagonists such as ketamine and phencyclidine (PCP) induces some of the typical schizophrenia symptoms, including hallucinations and psychosis, in normal individuals. Before further discussion of the individual transmitters, it is important to emphasize that inferring a disease mechanism from the activity of drugs, many of which may have "off-target" effects, is potentially dangerous. Fortunately, most of the corroborating data now suggests changes in expression of the neurotransmitters, receptors, transporters, or enzymes involved in their synthesis.

4.4 The Dopamine Hypothesis

The dopamine hypothesis posits that enhanced dopamine function is responsible for the cardinal symptom(s) in schizophrenia[2]. It originated from the accidental discovery that D2 receptor antagonists reduce psychotic symptoms in patients with schizophrenia and that drugs that increase dopamine concentrations, such as amphetamine and cocaine, can induce psychotic behavior. However, there is little evidence for changes in the D2 receptors or dopamine transporters in schizophrenia. Instead, molecular imaging studies, as well as radiotracer studies, now suggest that changes in dopamine signaling are almost entirely caused by an increased capacity to synthesize presynaptic dopamine and load it into vesicles for release, resulting in an overall increase in baseline dopamine concentration. This is not just the case in schizophrenia but also in patients who have psychosis linked to temporal lobe epilepsy. Dopamine release is greater in patients who are acutely psychotic compared with patients between psychotic breaks. Hence increased dopamine metabolism is linked to psychotic behavior specifically, but not to the other symptoms of schizophrenia, explaining the selective antipsychotic effect of D2 antagonists. We also know that the increased capacity to produce and release dopamine seems to develop from early childhood, possibly even before birth, and can be the result of various insults, including complications during childbirth, caesarian delivery, prenatal exposure, or infection. In addition, some of the genetic risk factors include enzymes involved in dopamine synthesis, such as the already mentioned COMT that is deleted in velocardiofacial syndrome.

How might dopamine cause psychosis? Dopamine is the principal transmitter in the reward pathway, specifically in the so-called mesolimbic pathway that encompasses the ventral tegmental area (VTA) of the midbrain and projects into the nucleus accumbens, amygdala, hippocampus, and frontal cortex (Figure 5). The nucleus accumbens contains medium spiny neurons, which receive dopaminergic input from the VTA. These neurons also receive glutamatergic input from the hippocampus, amygdala, and prefrontal cortex. Their main output is GABAergic projections to the ventral pallidum of the basal ganglia, which in turn projects to the thalamus and prefrontal cortex. The nucleus accumbens plays an important role in assigning emotions, pleasure, fear, and aggression and it reinforces learning by attaching an expectation of reward to behavior, so-called wanting, whereby dopamine release occurs in anticipation of reward. Dopamine functions to help discriminate whether we find a stimulus worthy of working for; in other words, dopamine attaches "motivational salience" to the stimulus.

In the context of schizophrenia, one may envision that the assignment of motivational salience to a sensory input or experience is perturbed. Improper dopamine concentrations may attach a wrong salience to objects, people, or actions, forming a biased cognitive schema. Experiences or people that are repeatedly assigned the wrong salience may be interpreted as threatening. Moreover, dopamine may impair the ability to distinguish internal from external stimuli, misassigning internal signals (voices) to external ones. Hearing threatening voices may be explained this way. This model is clearly speculative but may explain the selective association of dopamine with psychosis. What is clear from this discussion is that changes in dopamine release, rather than its receptor, underlie psychosis, and therefore dopamine antipsychotic drugs act downstream from their actual cause.

4.5 The Glutamate Hypothesis

The glutamate hypothesis suggests that reduced function (hypofunction) of NMDA

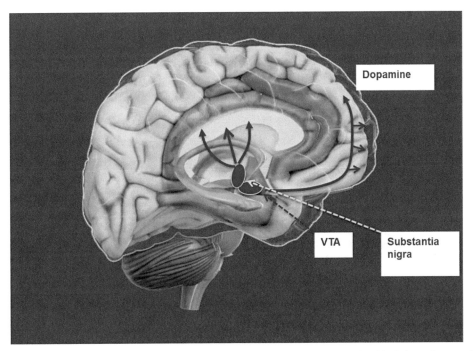

FIGURE 5   Dopamine pathways in the brain. Dopaminergic projections (schematized in blue) from the vental tegmental area (VTA) project to the nucleus accumbens, amygdala, hippocampus and frontal cortex. *Image was modified using 3D brain, Cold Spring Harbor DNA Learning Center.*

receptors, particularly in the frontal cortex, is responsible for the development of both positive and negative symptoms of disease.[6] This hypothesis originated from the finding that PCP and ketamine, which each block NMDA receptors, mimic both positive and negative symptoms of disease. Moreover, when given to patients with schizophrenia, ketamine also elicits positive and negative symptoms.[7] Numerous animal studies show activation of glutamatergic pathways in response to PCP. In addition, postmortem studies show a region-specific loss of certain NMDA receptor and AMPA receptor subunits, which would predict altered synaptic functioning in cortical and subcortical brain regions. The tools for study of postmortem tissue are constrained by the specificity of antibodies and the relative instability of proteins or mRNA that are being probed, both of which rapidly degrade after death. However, structural findings are gaining increasing support from functional, noninvasive,

spectroscopic MRI, where Glu concentrations and changes thereof can be measured in patients with schizophrenia and volunteers treated with ketamine. These studies have confirmed region-specific changes in the activity of glutamatergic pathways. These could best be summarized as showing slightly elevated Glu concentrations in untreated patients with early stage disease and decreased concentrations after treatment.[8]

What makes the Glu hypothesis attractive is the fact that Glu is involved in brain regions that are responsible for both positive and negative symptoms. However, the fact that 90% of all excitatory pathways in the brain are glutamatergic also makes it more difficult to precisely delineate the deficit in disease. Figure 6 shows a schematized glutamatergic synapse that illustrates not only the complexity of the transmitter receptors per se, encompassing multiple types (NMDA, AMPA, kainate), but also subtypes by subunit composition. In addition, trafficking,

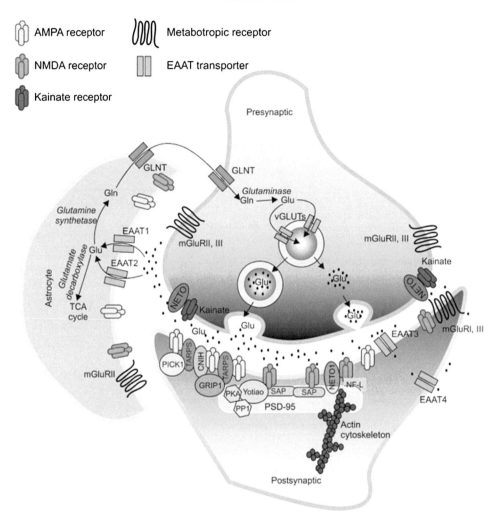

FIGURE 6   The tripartite glutamate synapse composed of pre- and postsynaptic neurons and regulatory astrocytes, along with many sites of regulation by auxiliary proteins. *From Ref. 6.*

stability, and phosphorylation regulate receptors as well as coagonists such as glycine or D-serine, which are largely produced by astrocytes that control the synaptic concentration of Glu to some extent. Therefore a tremendous number of proteins could fail in schizophrenia, making them each potential drug targets for regulating glutamatergic signaling. Many of these reach way beyond the typical Glu receptor drugs, and some of the resulting drug trials are discussed in Sections 5 and 6. Any treatment targeted at

Glu receptor function hold promise to address the full spectrum of symptoms as opposed to merely the negative symptoms treated by currently used antipsychotic drugs.[9]

Importantly, the Glu and dopamine hypotheses are not really in conflict if one considers their convergence in the striatum. Here, presynaptic dopamine release is indirectly controlled by the Glu receptor, which regulates the intrinsic GABAergic neurons in the striatum and frontal cortex. Naturally, therefore, any

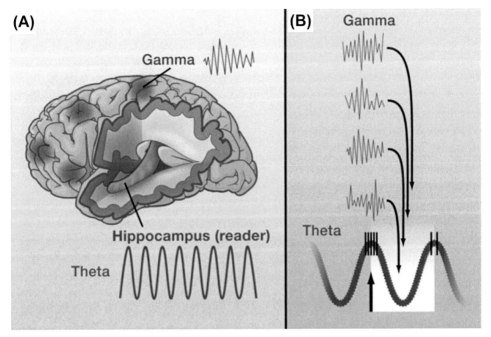

FIGURE 7   Gamma oscillation and schizophrenia. (A) Gamma oscillations generated throughout the cortical regions are involved in working memory and are ultimately read by the hippocampus. Here, much slower theta oscillation synchronizes information received from widespread neocortical areas (reflected by transient gamma oscillations illustrated in red hotspots). (B) The hippocampal theta oscillations (white area) imposes the timing during which neocortical circuits are read. They represent a readiness phase for the hippocampus. It is hypothesized that any cortical gamma oscillations that coincide in this readiness phase and are in phase are bound together as a narrative. Erroneous binding of such information may occur in patients with schizophrenia. *From Ref. 10.*

change in Glu receptor function is likely to alter dopamine release. This can be indirectly demonstrated in humans by imaging dopamine receptor availability with a PET ligand such as ^{123}I-iodobenzamide before and after administration of ketamine.[8] Such studies indeed show a decrease in binding of the PET ligand, suggesting that a block of the NMDA receptor by ketamine causes an increase in dopamine release, which then occupies the receptors.

4.6 The GABA Hypothesis, Network Changes, and Gamma Oscillations in Schizophrenia

The idea that GABAergic neurons may be pivotal in the etiology of schizophrenia originated with the finding that glutamic acid decarboxylase (GAD67), the enzyme that converts Glu to GABA, is decreased by 23–30% in GABAergic interneurons in the DLPFC of patients with schizophrenia. Overall, the density of GABAergic interneurons is reduced in the prefrontal cortex.[11]

Another consistent finding among patients with schizophrenia is a decrease in an electroencephalographic (EEG) signal called gamma oscillations.[10, 12]

Gamma oscillations (Figure 7) result from the rhythmic interaction of excitatory glutamatergic pyramidal cells with inhibitory GABAergic interneurons. Excitation drives the upswing and inhibits the downswing of activity. The frequency with which neurons depolarize and hyperpolarize in synchrony determines the oscillation frequency, which is typically between 30 and 200 Hz. As with the slower theta (6-10 Hz), alpha (8–12 Hz) and beta (12–30 Hz) waves we discussed for epilepsy, gamma oscillations can be recorded on the skull using surface electrodes, provided that there is a synchronous discharge of thousands of cells.

The excitatory pyramidal cells in layer 3 of the DLPFC, which receive input on their extensive dendritic trees from various parts of the cortex and thalamus, are responsible for

gamma oscillations. Their activity is modulated by two different interneurons: GABAergic chandelier cells, which have their synapses on the axon just next to the axon initial segment where action potentials are generated, and basket cells, which have their GABAergic synapses on the cell body (Figure 8). Both typically exert an inhibitory influence, but GABA can become excitatory under certain conditions. We have already encountered

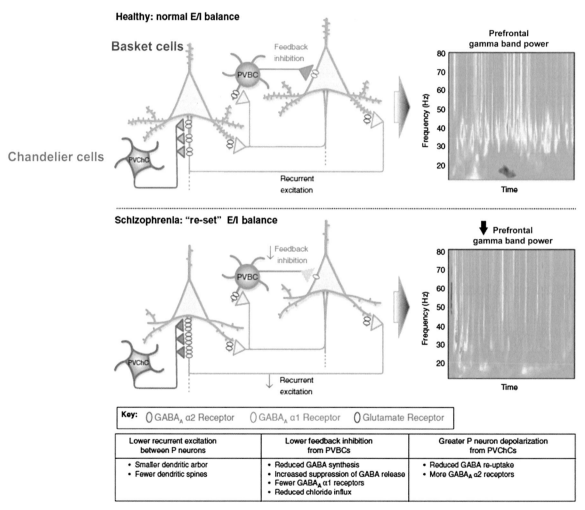

Lower recurrent excitation between P neurons	Lower feedback inhibition from PVBCs	Greater P neuron depolarization from PVChCs
• Smaller dendritic arbor • Fewer dendritic spines	• Reduced GABA synthesis • Increased suppression of GABA release • Fewer GABA$_A$ α1 receptors • Reduced chloride influx	• Reduced GABA re-uptake • More GABA$_A$ α2 receptors

FIGURE 8 Changes in the excitation–inhibition (E/I) balance underlying altered gamma oscillations in schizophrenia. The overall effect of changes in GABAergic inhibition results in a decrease in the power or synchrony of gamma oscillations in patients with schizophrenia. GABA, γ-aminobutyric acid; PVBC, Parvalbumin-expressing Basket Cells; PVChC, Parvalbumin-expressing Chandelier Cells. *From Ref. 11.*

this in epilepsy, where the loss of the neuronal KCC2 transporter causes the cell to accumulate Cl^-, and when GABA gated Cl^- channels open, Cl^- leaves the cells and depolarizes rather than enters and hyperpolarizes. GABA essentially becomes excitatory. Indeed, evidence for an excitatory action of GABA in schizophrenia is now emerging, where GABA itself may change. The most consistent finding from autopsy tissues is a significant decrease in GAD67, the enzyme responsible for synthesizing GABA. In addition, the pyramidal cells of the DLPFC have much smaller cell bodies and fewer dendritic spines and therefore fewer opportunities to receive excitatory inputs. Finally, the duration of the hyperpolarization mediated by GABA receptors differs based on their molecular composition; receptors containing alpha 1 are faster than those containing alpha 2. These changes result in an altered excitation–inhibition balance and a change in gamma oscillation frequency or amplitude measurable in schizophrenia.

So why bother with gamma oscillations? Functionally, gamma oscillations are believed to be the substrate for coordinating or "binding" information flow in the brain.[11] Singer and colleagues first described them in the visual system, where they serve to bind different parts of a picture. Since then, gamma waves have emerged as the substrate for working memory and can be seen in various brain regions involved in learning and memory, including the hippocampus and prefrontal cortex. Working memory refers to the task of accessing different memory traces in parallel and binding them into a cohesive narrative or experience. That is, visual, auditory, spatial, and emotional memories are bound together. For example, as I enter a bakery, my prefrontal cortex must process the smells, images, noises, and language and identify faces and location to produce the experience.

This occurs in the prefrontal cortex, in particular the DLPFC, where gamma oscillations thread together memory, emotions, thoughts, and experiences into a comprehensive picture.

In addition, cortical information may be read out by the hippocampus that binds synchronous gamma activity during certain readiness phases imposed by the hippocampuses intrinsic slower theta waves (Figure 7). Throughout the cortex the amplitude of gamma oscillations increases with an increasing working memory load, and different frequency patterns indicate parallel narratives where memories are being bound and organized together. In patients with schizophrenia the amplitude of these oscillations is reduced (Figure 8), suggesting that they have a deficit in binding the various elements that form the narrative of their cognitive task. If our understanding of gamma oscillations is correct, they may be one functional substrate of cognition, and consequently they may serve as a biomarker for disease. Interestingly, changes in gamma oscillation can be observed long before the first psychotic break.[11]

If the deficit in gamma oscillation is not just correlative but causes the cognitive deficit and incoherence of thought, one might argue that this electrical signal should be a target for treatment. The delicate excitation–inhibition balance in the DLPFC circuitry has a number of cellular and molecular substrates through which one may restore the excitation–inhibition balance and thereby proper binding or working memory (Figure 8). Indeed, it is believed that the often successful transcranial magnetic stimulation and electroshock therapies (discussed in section 6) do just that. Moreover, recent studies also suggest that benzodiazepines that enhance GABA receptor function can increase gamma power and reverse memory deficits in patients with schizophrenia.[13] Finally, the long-used antipsychotic drug haloperidol can reduce gamma oscillations in healthy adults.[14]

4.7 Synaptic Changes in Schizophrenia

The thinning of the cortex visible on MRI has prompted researchers to examine more

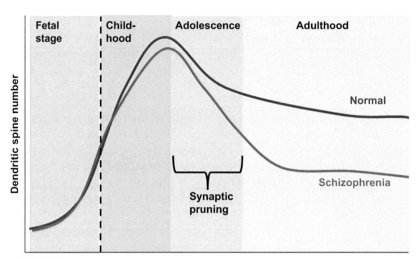

FIGURE 9   Schematic of spine development throughout life. *Adapted from Ref. 15.*

specifically which neurons are primarily affected and what cellular changes are most prominent. Examinations of postmortem tissues singled out three brain regions in which neuronal changes are most pronounced, namely, the above-mentioned DLPFC, the auditory portion of the superior temporal gyrus, and the subiculum and CA3 regions of the hippocampus. In all these regions, pyramidal cell bodies are smaller in size and their dendritic processes have significantly fewer spines. Since spines are the sites for glutamatergic input, one would therefore expect an overall reduction in excitatory activity within these regions. These structural changes may explain the reported hypoactivity of the Glu system and may also explain why ketamine and amphetamine, which each suppress glutamatergic activity in normal individuals, cause symptoms that mimic some schizophrenia symptoms.

But how do these spine changes come about? Might they be related to the genetic changes reported? It is now well accepted that a large excess of synapses is formed during early childhood. Subsequently, during childhood and adolescence, synapses are pruned back in an activity- and experience-dependent way,

reinforcing those synapses that are most active (Figure 9). In schizophrenia, genetic defects are likely responsible for overpruning, resulting in a markedly reduced number of synapses. Genes that have been identified as potential schizophrenia susceptibility genes include the already mentioned *DISC1*, *NRG1*, and *ERB4*, as well as two postsynaptic proteins, Kalirin-7 and PSD-95. As illustrated schematically in Figure 10, all these proteins are involved in different aspects of synapse maintenance, both structural and functional. Nrg1 is coreleased by presynaptic terminals with neurotransmitters and acts as a growth factor binding to its postsynaptic ERB4 receptor, which is tyrosine kinase. Its overexpression increases spine density and excitatory neurotransmission; conversely, its loss does the opposite. PSD-95 is a chaperone protein that directly interacts with NMDA receptors and organizes the postsynaptic density. Its expression is reduced in the schizophrenic cortex. The function of DISC1 is not entirely known, but loss of DISC1 also reduces spine density. It seems to interact with kalirin-7, a signaling protein that interacts with RAC, which is required to stabilize the F-actin cytoskeleton of the spine. Kalirin-7 is reduced specifically in the DLPFC

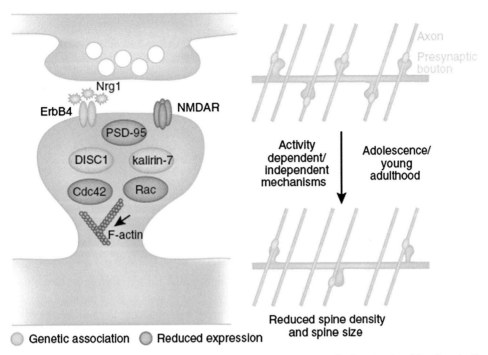

FIGURE 10   Pre- and postsynaptic molecules proposed to contribute to spine dysfunction in schizophrenia. *From Ref. 15.*

of the schizophrenic cortex. The emerging picture suggests that schizophrenia is a disease of dendritic spines, where a number of genetic defects in a variety of proteins that are involved in synapse maturation and stability results in a cortex where synaptic spine density is greatly reduced in certain regions. Since spine development and synaptogenesis have a unique time course during development, this emerging disease picture also explains the gradual development of schizophrenia and its onset in late adolescence (Figure 9).

4.8 White Matter Changes in Schizophrenia

Conspicuously, the prodromal period, that is, the time period preceding a psychotic break, during which schizophrenia develops, is congruent with a time window during which extensive white matter changes occur. Indeed, one of the last parts of the brain to become myelinated are the frontal lobes, also the site primarily implicated in many schizophrenia symptoms. Evidence of a defect in myelination, derived initially from structural MRI, indicates a small overall reduction in white matter volume.[16] Stereological analysis of postmortem tissue supports a 27% decrease in the number of oligodendrocytes in the frontal lobe of schizophrenic brains, and histological analysis also indicates impaired myelin compactness with signs of swelling, suggestive of dysfunctional white matter. However, the important question is whether this results in altered signal conduction along these white matter tracts. The relatively recent introduction of diffusion tensor MRI, which specifically permits the inference of connectivity from the directed diffusion of water along fiber tracts, allows at least an indirect functional assessment. Such studies are beginning to show region-specific differences suggestive of reduced myelination and in functional connectivity, for

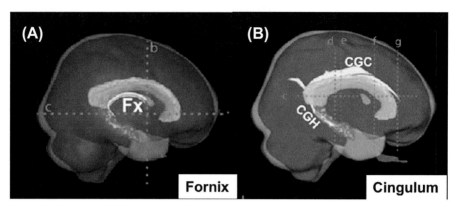

FIGURE 11   Two major white matter projections are the fornix (yellow in (A)), which connects the hippocampus (pink), thalamus, and nucleus accumbens, and the cingulum (yellow in (B)), which connects the prefrontal cortex, premotor areas, association cortex in the parietal and occipital lobes, parahippocampus, and thalamus. The corpus callosum is indicated in green. *From Ref. 17.*

example, between the fornix and cingulum (Figure 11), two prominent white matter tracts that function as "information highways" within the brain.[17]

The fornix connects the hippocampus, thalamus, and nucleus accumbens; the cingulum connects prefrontal cortex, premotor areas, association cortex in parietal and occipital lobe the parahippocampus and thalamus. Another indication that myelin dysfunction may contribute to hypoconnectivity comes from microarray analysis of postmortem tissues from schizophrenic brains. Of 6500 genes examined, only 7 had significantly reduced expression, and 6 of these were myelin related and included myelin-associated glycoprotein, which was also discussed as a major disease target in multiple sclerosis. It should be emphasized, however, that a decrease in myelination and even the number of oligodendrocytes might be a consequence rather than cause of schizophrenia. More specifically, the greatly reduced spine density and therefore synaptic activity that results in less activity may not require an equal number of axons to carry this signal.

4.9 Schizophrenia as a Neurodevelopmental Disorder

If one considers the limited overall knowledge of the etiology of schizophrenia, one can readily make the argument that it is a developmental disorder, as schematically illustrated in Figure 12. The genes that increase susceptibility are those involved in fetal brain development and the pruning of neural circuitries in early childhood. Maternal malnutrition and viral or cytokine exposure are most likely to provide an insult at this stage, affecting proliferation, migration, and arborization. This sets up a vulnerability, which during development into adolescence goes through what some describe as a latent period. Hormonal changes during puberty may be the final trigger causing the first psychotic attack. Similar to the current model of epileptogenesis, where subtle subclinical changes that ultimately trigger seizures and epilepsy develop over time, it is likely that subclinical changes that lead to a schizophrenic brain occur throughout early life. This is by no means a universally accepted point of view but one that can be substantiated experimentally.

4.10 Animal Models

Given the subjective nature of its symptoms and a near absence of biomarkers, few diseases may be as challenging to model in animals as schizophrenia. However, animal models are essential for advancing knowledge and developing new therapies. In spite of the obvious

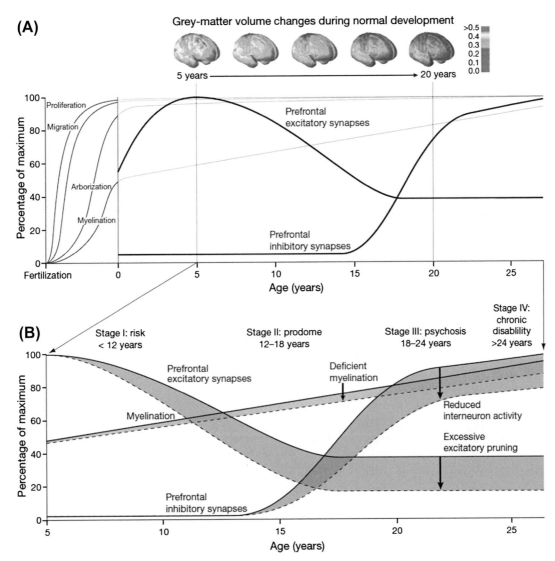

FIGURE 12 Developmental presentation of schizophrenia. The graphs show the time periods during which cell proliferation, migration, myelination, and arborization occur. (A) The combined effects of pruning of the neuronal arbor and myelin deposition are thought to account for the progressive reduction of grey-matter volume observed with longitudinal neuroimaging. (B) The trajectory in children developing schizophrenia could include reduced elaboration of inhibitory pathways and excessive pruning of excitatory pathways leading to altered excitatory–inhibitory balance in the prefrontal cortex. Reduced myelination would alter connectivity. Based on this data one may define 4 phases of disease, namely the risk stage, the prodromal stage, a psychosis phase and the chronic disability phase. Early detection of a child in the prodromal phase may allow to prevent progression into the psychosis phase and hence development of disease. *From Ref. 18.*

heritability of disease, producing a mouse model that replicates schizophrenia based on any of the known mutations has thus far been impossible. The closest model comes from the replication of a microdeletion on chromosome 22q11.2 that produces the already mentioned velocardiofacial syndrome, or the null mutation of the two targeted microRNAs that is responsible for up to 2% of all schizophrenia cases. Indeed, Dgcr8 deficiency caused deficient dendritic complexity and changes in short-term plasticity. In addition, such mice have reduced hippocampal neurogenesis. While some of these models may permit the study of cognitive function, the greatest challenges are models of psychosis. Can mice have hallucinations and delusions? If so, how do they present? Mice have been used to examine each of the proposed neurotransmitter hypotheses, and each presented with a defined set of behavioral deficits. For example, repeated treatment of mice with amphetamine induces hyperactivity, disrupts memory as judged from maze tests, and impairs novel object recognition. PCP treatment induces similar impairments. Manipulation of individual schizophrenia-associated genes such as *DISC1* and neuregulin-1 shows changes in cortical development associated with subtle behavioral symptoms. Based on the already discussed potential prenatal viral link to disease, animal models using viral infection of fetuses have been able to show behavioral and neural abnormalities. Yet given the weak evidence of a viral cause in human disease, how valid are such animal models? Similarly, the glutamate hypothesis of schizophrenia is largely based on the hallucinogenic effects of the positive NMDA receptor modulators PCP and ketamine. When mice with a reduced number of NMDA receptors were generated, they showed deficits in mating and nest building and were less interested in novel objects. However interesting these behaviors are, how much should we trust treatments and disease mechanisms built around such animals? It may simply be premature to pursue animal model studies of schizophrenia. Instead, we may focus on studying cellular models of disease, which can be readily derived from stem cells isolated and expanded from patients with schizophrenia. These models may give insight into molecular mechanisms of disease from which more applicable animal models may derive in the future. This approach yielded insight regarding glutathione deficiency in schizophrenia and its consequences.[19]

5. TREATMENT/STANDARD OF CARE/CLINICAL MANAGEMENT

The principal goal of treatment is to prevent psychotic episodes and to improve negative thoughts and apathy. It is now clear that patients treated quickly after their first psychotic episode have better outcomes than those who remain untreated for a considerable length of time and suffer repeated psychotic breaks. In many patients hospitalization is initially unavoidable because the psychosis prevents them from functioning at even the most basic level and caring for themselves so that they require the constant presence of a caregiver. Hospitalization also ensures that the patient adheres to the scheduled regimen of antipsychotic drugs until the acute phase has passed. There are numerous antipsychotic drugs (Table 1) that control psychotic behavior, including hallucinations and delusions, quite effectively. The overall success rate in treating a first psychotic episode using an antipsychotic drug is 70%, with almost immediate improvement and full remission within a few weeks. Unfortunately, all of these drugs have unwanted side effects such as weight gain, slowness of motion, and loss of libido, and up to 80% of patients do not comply with continued treatment, stop taking their medication, and relapse. Most of these drugs affect the dopamine system to some extent, particularly D2 receptors. Drugs are divided into the typical neuroleptics, which are often also called first-generation drugs and include the D2 receptor antagonists chlorpromazine and haloperidol, and atypical (new) antipsychotics such as clozapine, risperidone, and olanzapine. These atypical antipsychotics not only work through dopaminergic receptors but also affect serotonergic, noradrenergic, and histaminergic neurons to some extent.

TABLE 1   Currently Available Antipsychotic Drugs

Class	Chemical	Drug Name	Target	Cost in 2014 (year*)
Typical	Chlorpromazine	Thorazine	D2, α-adren	$1680
	Haloperidol	Haldol	D2-4, 5HT, α-adren	$170
	Perphenazine	Gen	D1-4, 5HT, α-adren Hist	$1680
	Fluphenazine	Gen	D2, α-adren	$135
Atypical	Clozapine	Clozaril	D2, NMDA, 5-HT, α-Adren, mAcH, Hist	$2400
	Risperidone	Risperadal	D2, 5HT, Hist, α-adren	$500/$6000
	Olanzapine	Zyprexa	D1, D2, 5HT, α-adren, mAcH, Hist	$250/$7000
	Quetiapine	Seroquel	D1-4, 5HT, α-adren, mAcH, Hist	$900/$7200
	Ziprasidone	Geodon	D1-5, 5HT, α-adren	$3600/$8000
	Aripiprazole	Abilify	D1/2, α/β-adren, Hist	$9600
	Paliperidone	Invega	D4, α−adren, Hist	$7300

D1–5, dopamine receptors; 5HT, serotonin; α/β−adren, adrenergic receptors; Hist, histamine; mAcH, muscarinic acetylcholine; NMDA, N-methyl-D-aspartate. *Cost name brand/generic.*

Because dopamine plays an important role in the control of movement by the midbrain, specifically through D1 and D2 receptors in the striatum, dopamine receptor antagonists have the expected side effect of reducing the initiation of movement, akin to the bradykinesia seen in Parkinson's disease. These so-called pyramidal side effects must be considered in the treatment of psychosis with drugs that target D2 receptors. Typically, anticholinergic drugs are coadministered to increase the availability of acetylcholine in the muscle endplate, thereby reducing bradykinesia. However, the motivation for synthesizing second-generation atypical antipsychotic drugs was to reduce or eliminate these pyramidal side effects. Indeed, they do have a lower affinity for striatal D2 receptors.

Moreover, they provide relief from psychosis in 50% of patients who were initially unresponsive to classical antipsychotic drugs. As a consequence, and illustrated in Figure 13, second-generation antipsychotic drugs that are aggressively marketed in TV commercials under the names Seroquel or Abilify have essentially replaced the classical drugs in clinical use. However, when their relative effectiveness was compared in a large, multicenter, randomized, double-blind trial,[21] the improved efficacy of the second-generation drugs was called into question. This trial found that the older and much cheaper drugs work just as well as the newer, expensive drugs that promise a better side effect profile. The side effects of all of them resulted in 64–82% of patients stopping their medication. Although not by a huge margin, olanzapine had the lowest discontinuation rate. But the claim of reduced side effects in second-generation antipsychotics was questionable. Patients taking olanzapine (Zyprexa) experienced more weight gain. Also, some of the second-generation drugs show potentially life-threatening side effects, most notably cardiovascular complications from the metabolic syndrome.[22] It is important to state that none of these drugs effectively controlled the negative symptoms of the disease, including flat affect, social withdrawal, and confusion.

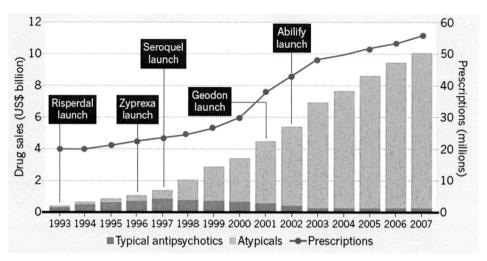

FIGURE 13   The prescription of second-generation "atypical" antipsychotic drugs increased between 1993 and 2007. The worldwide market for schizophrenia drugs has grown 10-fold since the introduction of atypical antipsychotics such as risperdal. Annual prescriptions have also increased dramatically. *From Ref. 20.*

Further critical evaluation of 38 randomized clinical trials with 7323 participants further tampered the enthusiasm for the overall effectiveness of both first- and second-generation antipsychotics when compared with a placebo; 41% of patients responded to the drug and 24% to placebo. While significant, this 17% increase in response rate is a disappointing effect size. Interestingly, the same study also found that haloperidol is just as effective in reducing negative and depressive symptoms as any of the second-generation drugs.[23] Finally this meta-analysis also unexpectedly revealed that across patients the effectiveness of these drugs decreases over time.

Taken together, these drug trials have dealt a huge setback to patients and to the drug companies manufacturing these drugs. Importantly, these studies also suggest that, unlike initially assumed, second-generation drugs also act in an antipsychotic fashion entirely through the inhibition of D2 receptors, with little if any role for 5-HT2A activity. At least the available data do not bear out any of the proclaimed movement-tempering serotonin effects. In all likelihood this is because the drugs were screened against models that only detect D2 effects.

This leaves us in a real crisis; 80% of patients discontinue the only drugs that stop psychosis and receive no help coping with confusion and negative emotions, whereas drug companies have all but given up developing better drugs.

6. EXPERIMENTAL APPROACHES/ CLINICAL TRIALS

In the long term, 60% of patients will recover to the point that they live at home and integrate socially; 30% remain helpless and 10% remain institutionalized. Consequently, the need for novel, more effective interventions is huge; not surprisingly, there are currently over 400 clinical trials investigating various aspects of schizophrenia, ranging from gluten-free diet to electroconvulsive therapy. A few examples are discussed below.

6.1 Novels Allosteric Modulators of NMDA Signaling

Since a hypofunction of NMDA-R is undoubtedly involved in schizophrenia symptoms,

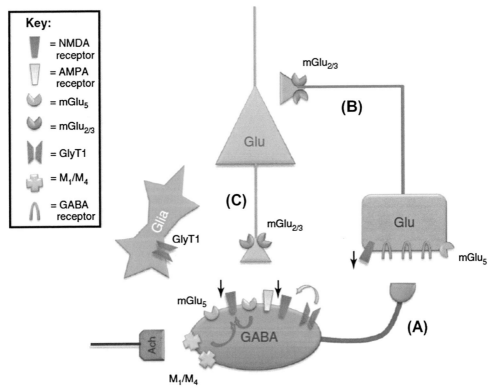

FIGURE 14   Potential interventions that can indirectly enhance N-methyl-D-aspartate (NMDA) receptor function by activating muscarinic acetylcholine receptors or metabotropic glutamate receptors (mGluRs) or by altering the concentration of the NMDA receptor coagonist glycine. *From Ref. 9.*

significant efforts to augment its function are underway.[9] The NMDA-R itself is a difficult pharmacological target because it is so widely used throughout the nervous system. For this reason, most drugs that target it in stroke have failed. However, NMDA activity can be indirectly (allosterically) modulated in at least three different ways, each of which has yielded compounds that are being actively investigated in clinical trials. More specifically, these include metabotropic Glu receptors on pyramidal cells and interneurons, muscarinic acetylcholine receptors on GABAergic interneurons, and the glycine transporter GlyT1 on astrocytes (Figure 14).

The metabotropic glutamate receptor (mGluR) 5 interacts physically and functionally with NMDA receptors and enhance NMDA activity in the striatum, hippocampus, and subthalamic nucleus. mGluR2 is found in the limbic areas of the forebrain, where it functions as an inhibitory modulator. The hypoactivity of NMDA receptors is hypothesized to cause decreased activity of GABAergic interneurons, which in turn causes disinhibited glutamatergic output in the prefrontal cortex. Activation of glutamate receptor 2 is believed to dampen this activity, thereby reducing the amount of Glu being released in the prefrontal cortex. A drug that acts as an mGluR2-specific agonist (LY404039) after conversion from a prodrug (LY2140023) has advanced to clinical trials for schizophrenia, and a recent report suggests benefits equal to, but not superior to, current second-generation antipsychotic drugs.[24]

M1 muscarinic receptors colocalize and potentiate NMDA currents in hippocampal pyramidal cells. In light of the reduced working memory in patients with Alzheimer's disease, xanomeline, the agonist of M1 receptors, has been used in Alzheimer's disease, where it showed significant cognitive enhancements. Since postmortem studies also suggest a reduction in the density of muscarinic M1 and M4 cholinergic receptors in the prefrontal cortex and hippocampus, the drug was given to patients with schizophrenia, in whom it significantly enhanced verbal learning and short-term memory but also presented some significant peripheral side effects that preclude it widespread use.[25] However, these results demonstrate that M1 and M4 receptors are viable targets for further drug design.

Yet another modulatory site on the NMDA receptor that has gained increasing attention is the glycine coagonist site (Figure 15). Only when occupied by either glycine or D-serine is the NMDA receptor fully functioning. It should be noted that glycine and D-serine concentrations are reduced in the cerebrospinal fluid of patients with schizophrenia, arguing for a contribution of this pathway to disease. D-serine is the more potent of the two and is made in

the brain by conversion of the L to the D form by the serine racemase, which is primarily found in astrocytes. D-serine is degraded via the D-amino acid oxydase (DAAO), an enzyme that is elevated two-fold in schizophrenia. Interestingly, the gene encoding DAAO is a known schizophrenia susceptibility gene.[26] Logically, one should be able to overcome the hypoactivity of the NMDA receptor by giving patients high concentrations of glycine or D-serine. Indeed, the coadministration of glycine or D-serine with second-generation antipsychotics yielded some positive outcomes in clinical studies. However, a buildup of glycine is typically prevented by uptake into astrocytes via the transporter GlyT1. Therefore drug companies have begun to develop specific GlyT1 transport inhibitors. One of these, RG1678, is based on a naturally occurring substance, N-methyl-glycine (sarcosine), and in early clinical testing caused significant improvement in the negative symptoms of schizophrenia, with an acceptable side effect profile.[27]

6.2 Glutathione

Changes in tissue antioxidant status are commonly associated with nervous system diseases, particularly in conditions where mitochondrial injury is present, as in Parkinson's disease, or when inflammatory cytokines produce reactive oxygen species, as is the case in seemingly all neurological conditions. The most prominent cellular defense is glutathione (GSH), an endogenously produced, highly reactive tripeptide that readily reduces reactive oxygen species and other reactive molecules. GSH is generally assumed to attenuate inflammation and neuronal injury. GSH concentrations are consistently reduced by up to 50% in postmortem tissue and cerebrospinal fluid samples from patients with schizophrenia, most likely as a result of genetic changes in the GSH synthesis pathway.[19] Interestingly, the NMDA receptor has a redox regulation site through which GSH strongly enhances

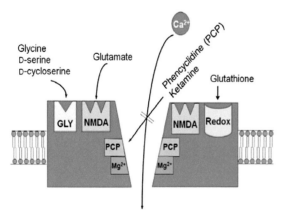

FIGURE 15 Various modulation sites on the N-methyl-D-aspartate (NMDA) receptor could be used to develop improved drugs to treat schizophrenia. Gly, glycine; PCP, phencyclidine. *From Ref. 8.*

NMDA activity[28] (Figure 15). Since we discussed at great length that schizophrenia presents with hypoactivity of the NMDA receptor, it seems logical that GSH should be an ideal drug to augment and restore normal NMDA function through modulation at the redox site. As a seemingly innocuous antioxidant molecule, GSH is offered as a nutritional supplement in many health stores and online. However, GSH does not permeate the blood–brain barrier well, and hence these supplements would be ineffective for nervous system disorders. However, GSH is produced in multiple steps from its precursor molecules Glu, cysteine, and glycine, with cysteine being rate-limiting. Administration of N-acetyl-cysteine, which readily passes the blood–brain barrier and cellular membranes, markedly increases GSH production by increasing the availability of cysteine. When coadministered in conjunction with the antipsychotic risperidone it enhanced the antipsychotic effects in a placebo-controlled phase II study. Two ongoing studies are currently examining the potential direct effects of N-acetyl-cysteine alone on cognitive function and as a general neuroprotective agent to prevent cortical thinning.

6.3 Nicotine

It has long been recognized that the vast majority of patients with schizophrenia smoke (40–90%), eclipsing the general population by a significant margin (15–25%). Moreover, they smoke more cigarettes each day and use ones with higher nicotine content. Epidemiological data also show that the majority of patients with schizophrenia smoke well before their first episode and, when one analyzes a large epidemiological sample of individuals without known disease, smoking rates are inversely proportional with progression to schizophrenia. In other words, smokers were less likely to develop disease. Nicotine binds to two receptors in the brain. The most abundant is a4b2, which has a high affinity and is believed to confer drug

seeking or addiction. Less abundant a7 homomeric receptors have a lower affinity that is not associated with addiction, yet it is the one implicated in the well-established cognitive enhancement effects seen in normal individuals as well. GWASs have reported copy number variations in the gene locus encoding the a7 nicotinic acetylcholine receptor, and reduced expression of a7 is seen in the postmortem brains of patients with schizophrenia. The a7 receptors are in a prime location to regulate cognitive function as well as mood. They are found on presynaptic terminals of dopaminergic, glutamatergic, and GABAergic neurons in the VTA, subthalamic nucleus, and nucleus accumbens. Activation of the receptor is therefore compatible with all of the neurotransmitter systems implicated in disease. Most important, the fact that nicotine enhances cognition is well established, and hence a selective agonist for a7 receptors would be the ideal drug. It turns out that such drugs exist. These are so-called allosteric modulators, or partial agonists, which enhance the opening of the receptor only in the presence of the natural ligand. Two of these, DMXB-A and TC-5619, have been evaluated in limited clinical studies of schizophrenia, where they reduced the negative symptoms of disease. While early, these are exciting findings; the reduced expression of a7 is augmented by drugs that make the remaining receptors work harder by staying open longer. This is quite an elegant strategy, although whether these or related drugs hold up in further clinical studies remains to be seen.[29] As of this writing, there were four trials recruiting patients to evaluate a7 modulators.

6.4 Electroconvulsive Therapy

Used since the 1930s, electroconvulsive therapy, often called electroshock therapy, induces focal seizures to relieve psychiatric symptoms. Initially based on the unsubstantiated hypothesis of the Hungarian psychiatrist Ladislas Meduna, electroshock therapy was widely used to treat all forms of psychiatric illnesses from the 1940s through the 1960s. It consists of placing two electrodes on each side of the forehead and applying square wave currents of sufficient intensity to evoke a seizure. The procedure is done under anesthesia and EEG can assess the seizure threshold. It produces acute cognitive side effects that typically disappear within minutes. It was largely displaced as treatment for schizophrenia by the introduction of pharmacological antipsychotic drugs. Electroshock therapy is still in use for treatment of severe depression and is making a comeback as a treatment of last resort in patients with catatonic, unresponsive, pharmacoresistant schizophrenia.[30] It may also be used in conjunction with antipsychotic drugs, and several active clinical trials are evaluating its effectiveness. A meta-analysis of existing data suggests that it is effective in a subgroup of patients with schizophrenia, namely, catatonic patients at the first psychotic episode and suicidal patients.[30] The mechanism by which electroconvulsive therapy alters a patient's behavior is not well understood.

6.5 Transcranial Magnetic Stimulation Therapy

Transcranial magnetic stimulation (TMS) therapy is different way in which to noninvasively change neuronal activity by applying a strong magnetic field to alter neuronal ion channel function. Furthermore, there is better mechanistic understanding of how it may work to enhance cognitive performance in affected individuals. Using simultaneous EEG measurements with TMS, it was possible to show that TMS specifically alters the gamma wave activity that characterizes the DLPFC, as discussed above. With the right stimulation parameters, gamma activity could be restored to normal, with patients showing improved neurocognitive performance on working memory tasks.[31] These are early pilot studies, but 27 studies currently investigate the effectiveness of TMS alone or in conjunction with other treatment modalities.

6.6 Vitamin D

The use of dietary changes or supplements to treat disease is widespread, and schizophrenia is no exception. A gluten-free diet, vitamins B_6, B_{12}, and E, and omega-3 fish oils are just a few examples. Most of these have unknown benefits and few ever give rise to rigorous clinical studies. As of 2014 there were three clinical studies examining the potential benefit of vitamin D supplementation in schizophrenia. The motivation for these trials is very indirect and possibly unjustified. The idea that vitamin D may help in schizophrenia originated with the peculiar finding that dark-skinned individuals carry a higher risk for schizophrenia than fair-skinned people. Synthesis of vitamin D, being largely generated in the skin, is reduced in highly pigmented skin. Urban living also carries an increased risk and is often associated with reduced vitamin D concentrations simply because people do not get out into the sun as much. Finally, the seasonal correlation of schizophrenia risk by month of birth suggests the highest risk in babies born during the winter months, with little sun exposure. So, one can generate a scenario, albeit a tentative one, in which all these situations involve vitamin D as the common denominator. Unfortunately, these ongoing studies do not address the potential benefit of prenatal vitamin D supplementation, possibly the most important modifiable time window in disease susceptibility.

6.7 Early Detection and Prevention

The idea that schizophrenia develops slowly during childhood has stimulated a search for early warning signs with which one could identify children who are likely to develop schizophrenia later in life and thus consider treating them through various forms of psychotherapy and family therapy to prevent disease onset. From 2003 to 2008, the North American Prodrome Longitudinal Study searched for such early features of schizophrenia to define a prodromal stage (*prodrome* is from the Latin term meaning "running ahead") through a questionnaire that looked for social and school troubles, fragmented thought, and peculiar emotions, among others. The published results of this study[32] showed an 80% accuracy of identifying at-risk children using this approach. However, it also raised concerns regarding the 20% who were falsely identified and therefore exposed to significant psychological harm and societal stigma. Furthermore, in spite of the accuracy of early detection, whether early interventions can prevent the onset of disease remains to be seen. The National Institutes of Health is currently supporting a second, expanded study following over 1000 individuals at multiple sites.

7. CHALLENGES AND OPPORTUNITIES

Schizophrenia is clearly the most disabling brain disorder in young people and tragically robs young adults of a socially integrated, fulfilling, and productive life. In the absence of biomarkers, our understanding of disease etiology and risks for susceptibility are poor. The ineffectiveness of available medication to consistently manage both positive and negative symptoms is tragic, and with pharmaceutical companies' waning interest in developing more effective drugs, the outlook is bleak. Yet this disease represents a health epidemic of major proportion, affecting well over two million people in the United States alone, a far greater number than all neurological diseases discussed in this book except stroke and Alzheimer's disease. Recent, exciting neurobiological discoveries suggesting that schizophrenia is most likely a disease of synaptic spines being overpruned during development offers hope that new treatment strategies will emerge. One can envision drugs that regulate activity-dependent synaptic pruning, the stability of neurotransmitter receptors, or even

the spine cytoskeleton being targeted. Most important, however, this research suggests that the disease is a neurodevelopmental disorder that begins years before the first psychotic break occurs. This prodromal time period may offer a window of opportunity to intervene and limit the overpruning of synapses and thereby prevent disease onset. I remain hopeful that these discoveries will truly be a game-changer and make us think outside the current box, which largely defines schizophrenia only as a disease of dopamine, serotonin, or glutamate neurotransmitter systems. One of the greatest challenges is the stigma that continues to be associated with schizophrenia. Active efforts must be undertaken to educate the public that schizophrenia is a neurodevelopmental disease that can affect anyone and that it is not restricted to people living on the fringes of society. Not only will this generate further support toward research, it may also change the willingness of affected individuals to seek professional help early, when intervention offers the most promise.

Acknowledgments

This chapter was kindly reviewed by Dr. James H. Meador-Woodruff, Heman E. Drummond Professor and Chair, Department of Psychiatry, University of Alabama at Birmingham.

References

1. Mulholland C. The symptom of depression in schizophrenia and its management. *Adv Psychiatr Treat.* 2000;6(3):169–177.
2. Howes OD, Murray RM. Schizophrenia: an integrated sociodevelopmental-cognitive model. *Lancet.* 2013.
3. Meyer-Lindenberg A. From maps to mechanisms through neuroimaging of schizophrenia. *Nature.* 2010;468(7321):194–202.
4. Forstner AJ, Degenhardt F, Schratt G, Nothen MM. MicroRNAs as the cause of schizophrenia in 22q11.2 deletion carriers, and possible implications for idiopathic disease: a mini-review. *Front Mol Neurosci.* 2013;6:47.

5. Jonas RK, Montojo CA, Bearden CE. The 22q11.2 deletion syndrome as a window into complex neuropsychiatric disorders over the lifespan. *Biol Psychiatry*. 2014;75(5):351–360.

6. Rubio MD, Drummond JB, Meador-Woodruff JH. Glutamate receptor abnormalities in Schizophrenia: implications for innovative treatments. *Biomol ther*. 2012;20(1):1–18.

7. Tamminga CA, Lahti AC, Medoff DR, Gao XM, Holcomb HH. Evaluating glutamatergic transmission in schizophrenia. *Ann N Y Acad Sci*. 2003;1003:113–118.

8. Poels EM, Kegeles LS, Kantrowitz JT, et al. Imaging glutamate in schizophrenia: review of findings and implications for drug discovery. *Mol Psychiatry*. 2014;19(1):20–29.

9. Field JR, Walker AG, Conn PJ. Targeting glutamate synapses in schizophrenia. *Trends Mol Med*. 2011;17(12):689–698.

10. Buzsaki G. Neural syntax: cell assemblies, synapsembles, and readers. Neuron. 2010;68(3):362–385.

11. Lewis DA, Curley AA, Glausier JR, Volk DW. Cortical parvalbumin interneurons and cognitive dysfunction in schizophrenia. *Trends Neurosci*. 2012;35(1):57–67.

12. Williams S, Boksa P. Gamma oscillations and schizophrenia. *J Psychiatry Neurosci: JPN*. 2010;35(2):75–77.

13. Lewis DA, Cho RY, Carter CS, et al. Subunit-selective modulation of GABA type A receptor neurotransmission and cognition in schizophrenia. *Am J Psychiatry*. 2008;165(12):1585–1593.

14. Ahveninen J, Kahkonen S, Tiitinen H, et al. Suppression of transient 40-Hz auditory response by haloperidol suggests modulation of human selective attention by dopamine D2 receptors. *Neurosci Lett*. 2000;292(1):29–32.

15. Penzes P, Cahill ME, Jones KA, VanLeeuwen JE, Woolfrey KM. Dendritic spine pathology in neuropsychiatric disorders. *Nat Neurosci*. 2011;14(3):285–293.

16. Davis KL, Stewart DG, Friedman JI, et al. White matter changes in schizophrenia: evidence for myelin-related dysfunction. *Arch Gen Psychiatry*. 2003;60(5):443–456.

17. Abdul-Rahman MF, Qiu A, Sim K. Regionally specific white matter disruptions of fornix and cingulum in schizophrenia. *PloS one*. 2011;6(4):e18652.

18. Insel TR. Rethinking schizophrenia. *Nature*. 2010; 468(7321):187–193.

19. Gysin R, Kraftsik R, Sandell J, et al. Impaired glutathione synthesis in schizophrenia: convergent genetic and functional evidence. *Proc Natl Acad Sci USA*. 2007;104(42):16621–16626.

20. Abbott A. Schizophrenia: the drug deadlock. *Nature*. 2010;468(7321):158–159.

21. Lieberman JA, Stroup TS, McEvoy JP, et al. Effectiveness of antipsychotic drugs in patients with chronic schizophrenia. *N Engl J Med*. 2005;353(12):1209–1223.

22. de Hart M, Schreurs V, Vancampfort D, van Winkel R. Metabolic syndrome in people with schizophrenia: a review. *World Psychiatry: Off J World Psychiatr Assoc (WPA)*. 2009;8(1):15–22.

23. Leucht S, Arbter D, Engel RR, Kissling W, Davis JM. How effective are second-generation antipsychotic drugs? A meta-analysis of placebo-controlled trials. *Mol Psychiatry*. 2009;14(4):429–447.

24. Adams DH, Kinon BJ, Baygani S, et al. A long-term, phase 2, multicenter, randomized, open-label, comparative safety study of pomaglumetad methionil (LY2140023 monohydrate) versus atypical antipsychotic standard of care in patients with schizophrenia. *BMC Psychiatry*. 2013;13(1):143.

25. Shekhar A, Potter WZ, Lightfoot J, et al. Selective muscarinic receptor agonist xanomeline as a novel treatment approach for schizophrenia. *Am J Psychiatry*. 2008;165(8):1033–1039.

26. Lin CH, Lane HY, Tsai GE. Glutamate signaling in the pathophysiology and therapy of schizophrenia. *Pharmacol, Biochem, Behav*. 2012;100(4):665–677.

27. Javitt DC. Glycine transport inhibitors in the treatment of schizophrenia. *Handb Exp Pharmacol*. 2012;213:367–399.

28. Kohr G, Eckardt S, Luddens H, Monyer H, Seeburg PH. NMDA receptor channels: subunit-specific potentiation by reducing agents. *Neuron*. 1994;12(5):1031–1040.

29. Young JW, Geyer MA. Evaluating the role of the alpha-7 nicotinic acetylcholine receptor in the pathophysiology and treatment of schizophrenia. *Biochem Pharmacol*. 2013;86(8):1122–1132.

30. Pompili M, Lester D, Dominici G, et al. Indications for electroconvulsive treatment in schizophrenia: a systematic review. *Schizophr Res*. 2013;146(1–3):1–9.

31. Farzan F, Barr MS, Sun Y, Fitzgerald PB, Daskalakis ZJ. Transcranial magnetic stimulation on the modulation of gamma oscillations in schizophrenia. *Ann N Y Acad Sci*. 2012;1265:25–35.

32. Cannon TD, Cadenhead K, Cornblatt B, et al. Prediction of psychosis in youth at high clinical risk: a multisite longitudinal study in North America. *Arch Gen Psychiatry*. 2008;65(1):28–37.

General Readings Used as Source

1. Papadakis Maxine A, McPhee Stephen J, Rabow Michael W, eds. *Current Medical Diagnosis and Treatment*. McGraw-Hill Education; 2014.

2. Harrison's Principals of Internal Medicine, Chapter 390, Biology of Psychiatric Disorders. Robert O. Messing; John H. Rubenstein; Eric J. Nestler.

3. Harrison's Principals of Internal Medicine, Chapter 391, Mental Disorders, Victor I. Reus.

4. Insel TR. Rethinking Schizophrenia. *Nature*. 2010; 468(7321):187–193.

5. Howes OD, Murray RM. Schizophrenia: an integrated sociodevelopmental-cognitive model. *Lancet*. 2013.

6. National Institute for Mental Health (www.NIMH.org).

Suggested Papers or Journal Club Assignments

Clinical Papers

1. Cannon TD, et al. Prediction of psychosis in youth at high clinical risk: a multisite longitudinal study in North America. *Arch Gen Psychiatry*. 2008;65(1):28–37.
2. Lewis DA, et al. Subunit-selective modulation of GABA type A receptor neurotransmission and cognition in Schizophrenia. *Am J Psychiatry*. 2008;165(12):1585–1593.
3. Lieberman JA, et al. Effectiveness of antipsychotic drugs in patients with chronic Schizophrenia. *N Engl J Med*. 2005;353(12):1209–1223; assign together with the commentary paper: Abbott A. Schizophrenia: The drug deadlock. Nature. 2010;468(7321):158–159.
4. Aberg KA, McClay JL, Nerella S, et al. Methylome-wide association study of schizophrenia: identifying blood biomarker signatures of environmental insults. *JAMA Psychiatry*. March 1, 2014;71(3):255–264.
5. Baker JT, Holmes AJ, Masters GA, et al. Disruption of cortical association networks in schizophrenia and psychotic bipolar disorder. *JAMA Psychiatry*. February 2014;71(2):109–118.

Basic Papers

1. Schobel SA, et al. Imaging patients with psychosis and a mouse model establishes a spreading pattern of hippocampal dysfunction and implicates glutamate as a driver. *Neuron*. 2013;78(1):81–93.
2. Brans RG, et al. Heritability of changes in brain volume over time in twin pairs discordant for Schizophrenia. *Arch Gen Psychiatry*. 2008;65(11):1259–1268.
3. Farzan F, et al. Transcranial magnetic stimulation on the modulation of gamma oscillations in schizophrenia. *Ann N Y Acad Sci*. 2012;1265:25–35.
4. Gysin R, et al. Impaired glutathione synthesis in schizophrenia: convergent genetic and functional evidence. *Proc Natl Acad Sci USA*. 2007;104(42):16621–16626.

COMMON CONCEPTS IN NEUROLOGICAL AND NEUROPSYCHIATRIC ILLNESSES

14

Shared Mechanisms of Disease

Harald Sontheimer

Diseases of the Nervous System
http://dx.doi.org/10.1016/B978-0-12-800244-5.00014-8

1. INTRODUCTION

By now, the reader of this book and probably every student of neurological illness will have heard some hypothesized disease mechanisms multiple times, in seemingly very different diseases. For example, glutamate toxicity appears to be involved in just about any disease where neurons die, be that as a result of stroke, genetic mutation in amyotrophic lateral sclerosis (ALS), or brain cancer. Similarly, the accumulation and aggregation of proteins, either inside or outside of cells, seem to occur quite abundantly in neurological disease. Even when genes are identified that unequivocally cause disease, as is the case with mutated superoxide dismutase 1 (SOD1) causing ALS, or with mutations in the methylcytosine-binding protein 2 (MECP2) gene underlying Rett syndrome, the resulting pathology is often complex and not readily explained by the mutated gene, and sometimes is not understood at all. Indeed, most of the animal models used, even those that accurately model the disease, are poor predictors for subsequent treatment of patients.

This chapter will attempt to synthesize commonalities that bridge multiple neurological disorders with the hope that common mechanisms may teach us important principles. Some of these findings, however, may be exaggerated, in part because scientists have selectively studied certain mechanisms across the spectrum of diseases, thereby overlooking important ones that are not shared. Nevertheless, a critical examination allows us to synthesize at least some true conceptual similarities that are shared by many diseases of the nervous system. A summary overview that identifies some of the hallmarks of disease, separating the disease groups as static, primary and secondary progressive neurodegeneration, developmental illnesses, and neuropsychiatric disease, is provided in Table 1.

2. NEURONAL DEATH

At first glance, neuronal death appears to be the most defining characteristic of most neurological illnesses. Cell death may occur almost instantaneously following a stroke or traumatic insult, or with a delayed onset that defines primary progressive neurodegenerative diseases, including Alzheimer disease (AD), Parkinson disease (PD), Huntington disease (HD), and ALS. Neuronal death also characterizes what we classified as secondary neurodegenerative diseases, namely brain tumors, multiple sclerosis (MS), and nervous system infections. Neuronal death may be indirect and concealed, and therefore is a less prominent attribute in some disorders. For example, the white matter loss that characterizes MS is associated with axonal dysfunction and ultimately causes axonal death as the loss of myelin removes important metabolic and trophic support. Other diseases, too, present with less conspicuous white matter loss, as in schizophrenia and ALS, and therefore we must surmise that axons may die in these diseases as well.

The question then becomes to what extent neuronal loss is the cause or the consequence of disease. This is more than a rhetorical question, as it guides future therapeutic strategies. Clearly, the neurological symptoms of a disease are directly related to neuronal loss, and, in fact, the symptoms often allow us to directly identify the affected neuronal cell population. For example, the motor weaknesses and gait imbalances in poliomyelitis and ALS selectively affect the motor neurons in the cortex and spinal cord and present with hyperreflexia, while the gait imbalances in neurosyphilis are due to sensory neuron loss in the spinal cord and present with areflexia.

But let us consider a bacterial infection of the meninges, meningitis, which initially presents with muscle stiffness of the neck and may lead to severe headaches or even coma. In the end stage of disease, neuronal

TABLE 1 Common Hallmarks of Neurological Diseases

Hallmark	Acute/Static			Progressive Neurodegenerative				Secondary Neurodegenerative			Development	Neuropsychiatric	
	Stroke	Trauma	Epilepsy	AD	PD	ALS	HD	MS	Glioma	Infection	Down/Rett/Fragile-X	Mood	Schizophrenia
Neuronal death	Yes-nonspecific	Yes-nonspecific	minor	Yes-diffuse	Yes	Yes-motor	Yes-motor	secondary	Yes-nonspecific	Yes-nonspecific	Yes	?	?
White matter involvement	Yes-specific	Yes-specific	?	Yes	some	Yes-specific	Yes-specific	Primary	Yes-nonspecific	Non-specific	variable	?	Yes
Glutamate toxicity	Yes	Yes	some	?	Yes	Yes	Yes	Yes	Yes	Yes	Yes	?	Yes
Protein aggregates	no	no	no	Plaques/Tangles	Lewy Bodies	Inclusion bodies	Inclusion bodies	no	no	no	rare	no	no
Mitochondrial dysfunction	secondary	secondary	secondary	secondary	primary	primary	primary	?		secondary		?	
Known genetic cause	none	none	Poly-genetic	Poly-genetic	Poly-genetic	Poly-genetic	single	?	Poly-genetic	none	Often single gene	?	Poly-genetic
Rare familial forms	no	no	yes	yes	yes	yes	Fully penetrant	?	Yes/NF1	none	no	?	?
Cell autonomous	no	no	no	?	?	no	maybe	yes	yes	no	no	?	no
Non-cell autonomous	yes	yes	yes	?	?	yes	astrocyte				Microglia astrocytes	?	
Impaired BDNF	Yes	Yes	Yes	Yes	Yes	Yes	Yes	?	No	No	Yes	Yes	Yes
Epigenetic	No	No	Yes	?	?	Yes	?	?	Yes	Yes	Yes	Yes	Yes
Vascular abnormalities	Yes	Yes	Probably	Yes	?	?	?	?	Yes	Yes	?	?	?
Inflammation	Major	Major	Minor	some	some	Major	Minor	Major	Major	Major	Some	Yes	Yes
Animal models/therapeutic value	Several/very poor	Several/variable	Many/variable	Several/poor	Several/poor	Several/poor	Several/poor	Several/poor	Yes/poor	Variable/good	Yes/some promising	Inadequate	Emerging

cell death is common, yet at no point in the disease are neurons anything but innocent bystanders. We assume the same to be the case in MS, where neurons may never be the direct target of activated lymphocytes, but eventually become victims. The same case could be made for cerebrovascular infarct and central nervous system (CNS) trauma, both of which affect brain tissue in a rather nondiscriminant fashion, yet, owing to the unique dependence of neurons on oxidative metabolism, they are the most vulnerable cell population and are the first to die as energy supply becomes limited. To take this reasoning even further, we learned that in some neurodegenerative or neurodevelopmental diseases, for example ALS or Rett syndrome, introducing the mutated, disease-causing genes into astrocytes or microglia is sufficient to cause disease even when the neurons continue to make the wild-type protein. Here too, neurons ultimately die, but once again they appear to be collateral damage rather than the principal cell targeted by disease. So while neuronal cell loss is a common feature of neurological disease, and occurs almost without exception at some stage of the illness, one must not make the mistake of focusing too heavily on mechanisms of neuronal cell death to understand the underlying disease mechanisms. Instead, we must accept the fact that neuronal death may occur in many different ways as a final stage of disease, and is mechanistically a consequence of disease rather than its cause.

3. GLUTAMATE TOXICITY

One must also be careful about being overly fixated on any singular molecular cause of neuronal cell death, particularly if death is caused by an extrinsic chemical such as glutamate (Glu). Without question, Glu excitotoxicity is one of the most common threads, appearing in essentially all neurological disorders. Not only does excessive Glu kill neurons, it can also kill oligodendrocytes and thereby cause white matter injury as well. Astrocytes, by contrast, are unaffected by even very high glutamate concentrations.

Glu toxicity was first proposed from extensive studies on the health effect of monosodium glutamate, which is used as a food-flavoring agent. It was found to kill neurons in infant monkey brain regions, particularly those not protected by the blood–brain barrier (BBB).[1] Within the following two decades, Glu emerged as a major culprit in numerous disorders including stroke, HD, and ALS.[2] Indeed, we have since learned that ~90% of all neurons in the brain express Glu receptors and all of those neurons are, in principle, susceptible to Glu toxicity.

3.1 Why is Glutamate Toxic?

As we discussed in Chapter 4 (Aging and Alzheimer disease), glutamate excitotoxicity derives from the peculiar properties of the Glu receptors involved in learning. One may argue that we entered into an "evolutionary bargain" where we accepted this vulnerability to gain a mechanism for cells to learn that two stimuli or pieces of information are related. More specifically, neurons utilize two different types of Glu receptors (Figure 1). Fast signaling occurs via AMPA/kainate receptors that primarily flux Na^+ into the cell, thereby depolarizing the postsynaptic cell upon ligand binding. The second type of Glu receptor is the NMDA receptor (NMDA-R), which fluxes both Na^+ and Ca^{2+}. In fact, the primary role of NMDA receptors is to flux Ca^{2+}, and thus their permeability for Ca^{2+} is about 5-fold greater than that for Na^+. Importantly, NMDA-Rs are usually inactive, since they are blocked by intracellular Mg^{2+} ions that are lodged inside the permeation pore. Only when the cell becomes depolarized does this Mg^{2+} ion temporarily dislodge, allowing Ca^{2+} and Na^+ to enter the cell.

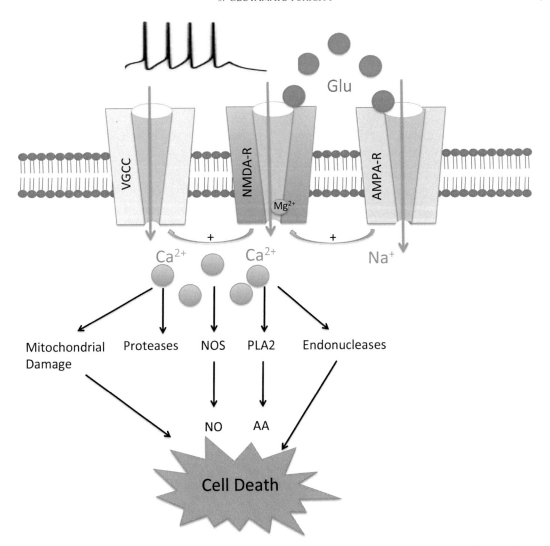

FIGURE 1 Glutamate excitotoxicity. Glu release causes activation of AMPA-Rs. The resulting Na^+ influx depolarizes the membrane to remove the Mg^{2+} block from the NMDA-R, allowing Ca^{2+} influx. It also activates voltage-gated Ca^{2+} channels (VGCC). Sustained depolarization causes uncontrolled Ca^{2+} influx, activating phospholipase A2 (PLA2) that forms arachidonic acid (AA), nitric oxide synthase (NOS), which produces NO, endonucleases, and proteases, and causes mitochondrial damage. Collectively, these molecules create a toxic intracellular milieu that causes excitotoxic cell death that is commonly observed in neurological disorders.

If a postsynaptic terminal receives coincident signals paired less than 50 ms apart, the first stimulus will depolarize the cell, thereby relieving the Mg^{2+} block, and the second signal will now trigger a Ca^{2+} influx. This leads to activation of Ca^{2+}/calmodulin-dependent protein kinase II (CaMKII) and induces long-term changes in receptor number on the cell surface. For the next minutes to hours, this causes a potentiation of all signals that travel through this synapse, a process that we call long-term potentiation (LTP) and generally accept as an important element of learning at the cellular level. As in many cellular processes, Ca^{2+} acts as the second messenger that activates Ca^{2+}-dependent kinases, which in turn cause changes in gene transcription that lead to permanent synaptic changes. Of fundamental importance is a close temporal and spatial pairing of the stimuli, as otherwise the Mg^{2+} quickly resumes its place blocking the NMDA channel.

This seemingly elegant design has one major flaw. Should a neuron become depolarized for reasons unrelated to signal transmission and learning, such as during energy failure resulting from a stroke, this too will remove the Mg^{2+} block that keeps the Ca^{2+} from getting into the cell. This happens most notably in stroke and trauma, where inadequate blood supply of glucose results in insufficient ATP production in neurons. Energy failure then causes the neuron to gradually lose its negative resting potential, and eventually opens the flood gates for uncontrolled Ca^{2+} entry via NMDA receptors. Defective mitochondria, as present in PD and ALS, can similarly cause neuronal depolarization, as can injury- or seizure-associated accumulations of extracellular K^+. Unfortunately, any nonspecific depolarization of the neuron potentially has the same outcome, resulting in an overload of the cell with Ca^{2+}.

3.2 Role of Ca^{2+}

The physiological range for Ca^{2+} to act as a second messenger in the neuronal cytoplasm is about 100–1000 nM. Above this range, Ca^{2+} activates destructive enzymes such as proteases, phospholipases, endonucleases, and caspases, which begin to cleave proteins, DNA, and membrane lipids (Figure 1). Also, upon Ca^{2+} entry, mitochondria open a large transition pore through which they release cytochrome C, which in turn initiates the apoptotic

death cascade. Since extracellular Ca^{2+} concentration is ~1 mM under normal conditions, cells must maintain a steep, 1000–10,000-fold concentration gradient for Ca^{2+} across the plasma membrane. This gradient is maintained through several energy-dependent Ca^{2+} transport systems. Any loss of energy, such as that seen during an ischemic stroke, puts the cells at immediate risk for a pathological rise in intracellular Ca^{2+} due to the loss of this tightly maintained gradient. Of course, this can also occur with prolonged exposure of neurons to elevated concentrations of glutamate. Even a change as small as an excess of only 50–100 μM glutamate for just 2 min is sufficient to cause essentially irreversible neuronal death 24 h later.

Note that any chronic depolarization of neurons causes uncontrolled influx of Ca^{2+} into their presynaptic terminals via both voltage-gated Ca^{2+} channels and NMDA-Rs. The additional massive release of synaptic Glu when the presynaptic terminal is depolarized then creates a "perfect storm" scenario in which neurons die as a consequence of the Glu they themselves release.

Other cells can also contribute to Glu toxicity. For example, we learned in Chapter 9 that brain tumors release massive amounts of glutamate and deliberately kill neurons to create room for tumor growth. In addition, whenever inflammation occurs and activated microglia are present, microglia release glutamate as well.

Although astrocytes are not subject to Glu toxicity, Glu readily kills oligodendrocytes and this appears to involve the same principal mechanisms. As we learned in Chapter 8, oligodendrocytes express NMDA-Rs in their myelin, as well as a subtype of AMPA receptor that is also Ca^{2+} permeable. During normal activity, axons release glutamate as a physiological stimulus to maintain the myelin sheath. Pathological dysregulation of glutamate then becomes toxic to the myelin.

3.3 Astrocytic Glutamate Transport

Because Glu has tremendous toxic potential, neurons are typically protected from uncontrolled rises in extracellular Glu. This is accomplished by Glu transporters expressed on astrocytic processes that surround glutamatergic synapses, illustrated in Figure 2, where the astrocytic processes are colored in blue.[3] Any spillage of Glu from the synaptic cleft binds to one of two excitatory amino acid transporters expressed by astrocytes. Excitatory amino acid transporter 2 (EAAT2) is the most abundant Glu transporter in the mature brain and, in fact, is the most abundant of all brain proteins.[4] EAAT2 has a high affinity for Glu (~12 µM) and imports Glu into astrocytes by harnessing the Na^+ gradient as an energy source. This gradient is electrogenic; three Na^+ ions and one H^+ ion are transported into the cell along with each Glu^- molecule. During this process, one K^+ ion is also counter-transported out of the cell (Figure 3), creating a situation in which there is a net effect of two extra positive charges entering the cell (four positive and one negative charge entering, minus one positive charge leaving, yields two positive charges entering). These two extra positive charges facilitate Glu uptake into the cell, as the transporter can now effectively harness both the chemical and electrical gradient. It is worth mentioning, however, that the transporter operates very slowly, transporting only 14 ions/sec (compared to other ion channels that flux $>10^6$ ions/sec) and therefore has a very low capacity. To compensate for this limited capacity, the density of EAAT2 Glu transporters on astrocytes is extraordinarily high, approximately 1000–10,000 transporters/$µm^2$, such that binding of Glu to transporters will be sufficient to capture all extra Glu being released from synapses, even if no transport occurs.[3] The system is designed so that under normal conditions no spillage of Glu should occur, unless, of course, astrocytes fail to express a sufficient number of them. Therefore it stands to reason that in any disease where we observe Glu toxicity, astrocytes must be compromised in their ability to neutralize this threat. This is actually the case in many neurological diseases, including stroke, trauma, MS, ALS, PD, HD, and glioma, where a loss of EAAT2 expression or function has been demonstrated in patient

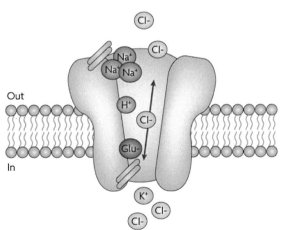

FIGURE 3 Schematic of the astrocytic EAAT2 transporter. Transport of each negatively charged Glu is coupled to three Na^+ and one H^+ into the cell, with counter-transport of one K^+ for a net charge difference of two positive charges. This charge adds an electrical gradient to the chemical gradient, allowing more effective Glu uptake. It also makes it more difficult for the transporter to run in reverse. *Reproduced with permission from Ref. 3.*

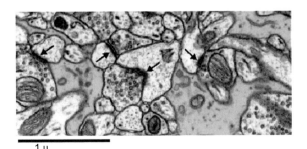

1 µ

FIGURE 2 Ensheathment of glutamatergic synapse (arrow) by astrocytes (blue) in the hippocampus shown by electron microscopy. Astrocytic processes are strategically placed to prevent spillage of glutamate from the synaptic cleft. *Modified with permission from Ref. 5 (Figure 1).*

tissue and/or animal models of disease. In fact, in ALS, motor neuron death can be attenuated by experimentally restoring EAAT2 expression using ceftriaxone, an antibiotic that acts as a transcriptional regulator of EAAT2 and enhances its expression in astrocytes up to 3-fold.[6] This drug has shown promise in a clinical trial for ALS[7] (ClinicalTrials.gov, NCT00349622), and ceftriaxone may have the potential to be more broadly used to mitigate glutamate excitotoxicity in other diseases.

3.4 Extrasynaptic NMDA Receptors

The role of NMDA-Rs in learning and memory, as well as glutamate toxicity, is more complex than was described above. NMDA-Rs are expressed not only on postsynaptic terminals but also at extrasynaptic sites located at a distance from the next synapse[8] (Figure 4).

These receptors can sense Glu spilling from adjacent synapses if astrocytes do not capture all the Glu. The extrasynaptic NMDA-Rs have a different subunit composition from synaptic ones. All NMDA-Rs are heterotetramers (composed of four different subunits) that share two obligatory GluN1 subunits, typically with a combination of two additional GluN2 subunits. These come in two isoforms, GluN2A and GluN2B. GluN2A is the most common subunit of NMDA-R at synapses, whereas GluN2B is commonly found at extrasynaptic sites. These subunits both promote different signaling cascades, which in turn exert different effects on the cell. Synaptic GluN2A-containing receptors enhance LTP, the cellular equivalent of learning, and activate cellular signals that promote cell survival. By contrast, activation of extrasynaptic GluN2B-containing receptors contribute to long-term depression (LTD), the cellular equivalent of

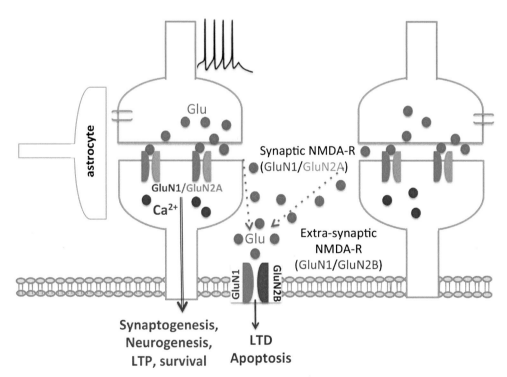

FIGURE 4   Extrasynaptic NMDA receptors containing the GluN2B subunit activate apoptotic cell death pathways whereas synaptic receptors containing the GluN2A subunit are largely prosurvival.

forgetting, and also activate cell death pathways. For unknown reasons, the number of GluN2B-containing extrasynaptic receptors increases in neurological disease. This has now been demonstrated for AD, HD, epilepsy, and acute insults such as stroke and trauma.[8] Extrasynaptic receptors become activated as neuronal Glu release increases, such as when neurons depolarize under energy failure or during hyperexcitability.

The most direct way to inhibit Glu toxicity under such circumstances would be via inhibition of Ca^{2+} influx through NMDA-Rs, ideally by specific inhibition of extrasynaptic NMDA-Rs. In principle, this should be feasible with drugs such as memantine, a use-dependent blocker of NMDA-R that barely affects NMDA currents at physiological Glu concentrations but becomes more and more effective as Glu concentrations increase into a pathological range. A number of other high-affinity blockers, such as ketamine and MK801, or modulators of the NMDA-R, such as D-serine and glycine, exist. Unfortunately all attempts with this approach have failed across all the diseases discussed here. Indeed, at least two decades of stroke research have unsuccessfully pursued the neuronal NMDA-R as therapeutic target, with little to no success. In only two diseases were modest improvements seen with drugs that directly or indirectly attenuate Glu toxicity. These are riluzole, which delays motor neuron death in ALS, and memantine, which brings modest cognitive improvements in AD patients.[9] These same drugs, however, have universally failed when used in other diseases.

I would argue that while Glu excitotoxicity is among the most common hallmarks of neurological disease, it is also the least specific and may be the final "nail in the coffin." The lack of success in attenuating excitotoxicity in human disease strongly suggests that we must look upstream to understand each unique cause of disease. One must consider the possibility that different challenges to a cell's health may each present with a similar dysregulation in Glu signaling; if true, one must therefore resist the temptation to assume that there should be additional commonalities between diseases that present with Glu toxicity.

4. PROTEIN AGGREGATES AND PRION-LIKE SPREAD OF DISEASE

All of the classical neurodegenerative diseases are histopathologically defined by intracellular or extracellular protein aggregates. Examples of the most common aggregates observed in AD, PD, ALS, HD, and prion disease are illustrated in Figure 5. For each disease, the protein that causes the aggregate is different. However, a shared commonality is that accumulating proteins differ in their tertiary structure so as to be misfolded and insoluble. Most typically, the tertiary structure changes from an alpha helix to a beta-sheet conformation, which makes the proteins sticky and hydrophobic. Seemingly small changes in the primary structure of a protein can have a profound effect on its tertiary structure. For example, under normal conditions, the cleavage of beta-amyloid causes water-soluble nontoxic amyloid. Yet, in the case of AD, a difference in only two or three amino acids affects beta-amyloid cleavage at the 42 position and causes toxic amyloid plaques. In HD, up to 35 polyglutamines produce a normal huntingtin protein that is highly water soluble, while 36–39 can cause disease in some individuals, and 40 or more repeats guarantee protein inclusions and disease in every affected individual.

Another commonality is that protein aggregates tend to develop well before a person becomes symptomatic. In AD, for example, non-invasive imaging allows detection of amyloid deposits up to 15 years prior to disease onset. Not surprisingly, it is therefore widely assumed that protein aggregates cause disease, a notion that is heavily debated and is far from certain. The only example where protein inclusions have been shown to be necessary and sufficient for disease is in HD, where the removal of deposits

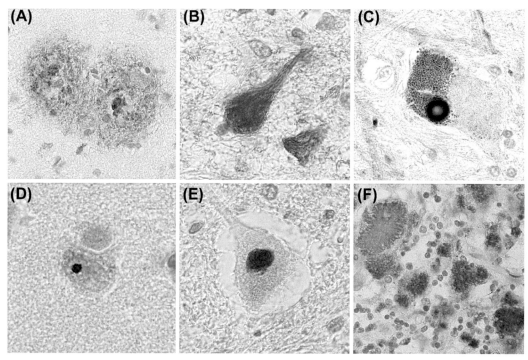

FIGURE 5   Protein aggregates in neurodegenerative disease. (A) Senile amyloid plaques in the neocortex of a patient with Alzheimer disease. (B) Neurofibrillary tangles in the hippocampus of a patient with frontotemporal dementia with parkinsonism (FTDP-17). (C) Lewy body in the substantia nigra of a patient with Parkinson disease. (D) Intranuclear polyglutamine inclusion in the neocortex of a patient with Huntington disease. (E) Ubiquitinylated inclusion in spinal cord motor neuron of patient with ALS. (F) Protease-resistant PrP prion protein in the cerebellum of a patient with Creutzfeldt–Jakob disease. *Reproduced with permission from Ref. 10.*

after the disease is already fully manifest causes a nearly complete reversal of symptoms.[11]

While we assume that only the mutated, hyperphosphorylated, or misfolded proteins accumulate, it would be a mistake to assume that their loss of function is directly responsible for disease. Instead, one must consider that protein aggregates also contain numerous other proteins, including wild-type protein, transcription factors, DNA/RNA regulatory proteins, chaperone proteins involved in protein folding and trafficking, or proteins that confer resistance to cellular stress. It is likely that the sequestration of these supportive proteins by protein aggregates contributes to the functional losses observed in neurons as these proteins are prevented from participating in their normal function.

Protein aggregates may resemble temporary or permanent storage sites for junk proteins in the cell. In all likelihood the aggregates per se are not doing irreversible harm, but result from an inability of the cell to remove junk proteins in a timely fashion. Therefore, a problem in protein homeostasis or proteostasis exists where the production of new protein is in disequilibrium with removal of old or nonfunctioning protein. This leads to a "toxic gain of function" attributed to inclusion bodies but largely explained by a lack of participation of proteins that are caught in these aggregates. Should commonalities prevail, the removal of the protein aggregates therefore should restore neuronal health, and this has been conclusively demonstrated only for HD. Note that protein aggregates do not form in secondary neurodegenerative disorders or acute insults, and are also absent in neuropsychiatric illnesses.

4.1 Proteostasis

Understanding what leads to the formation of aggregates in the presymptomatic stages of disease holds tremendous promise and potential in preventing disease.[12] This concept serves as the rationale behind a large National Institutes of Health-supported AD prevention trial that is enrolling 100 controls and 100 subjects who harbor the presenilin-1 (PSEN1) E280A mutation that causes familial early onset of AD by age 44. The participant's cognitive status will be assessed prior to disease onset and throughout continuous treatment with the monoclonal antibody crenezumab (Genentech) to clear beta-amyloid before it can form aggregates.[13]

The notion that impaired proteostasis is at least responsible for intracellular protein accumulation is very attractive. A network of proteins called the preotestasis network maintains an orderly balance of protein synthesis and degradation. Such proteins include molecular chaperones that help newly synthesized polypetides to assume the correct three-dimensional shape, i.e., tertiary and quaternary structure. It also involves proteins that dispose of old, nonfunctional, mutated, or misfolded proteins. Such proteins are degraded by two parallel systems. Small peptides and proteins are tagged by polyubiquitin marks attached by ubiquitin ligases and transported to the proteasome, an intracellular complex that proteolytically cleaves proteins into their amino acid constituents (see also Box 5.1, Chapter 5). Protein aggregates that are too large to fit into the core of the proteasome aggregate into autophagosomes, which then fuse with the cell's lysosomes, where proteins are enzymatically degraded (see Figure 7.9, Chapter 7). If this equilibrium is disrupted by either increased supply or reduced disposal, junk proteins accumulate and give rise to inclusion bodies.

4.2 Prion-Like Spread of Misfolded Protein

Essentially all diseases that present with protein aggregates show considerable local spread within tissue. For example, in ALS it is typical for adjacent motor neurons to become gradually affected by disease. In AD, disease spreads gradually from the hippocampus to the entorhinal cortex, then encroaches on the frontal lobe and spreads toward the parietal brain. This propagation of disease is consistent with a toxic agent moving from cell to cell. Inclusion bodies are very stable beta-sheet structures, similar to mutated prions, which cause mad cow or Creutzfeldt–Jakob disease, which we discussed in Chapter 10. Mutated prions propagate by imposing their shape on wild-type prion proteins in their vicinity, causing them to assume the same shape. It has been postulated that neurodegenerative disease may also spread within the nervous system in a prion-like fashion, as misfolded mutated proteins coax wild-type proteins in adjacent cells to assume their misfolded state by acting as a template. At least for AD, a prion-like spread of disease has been demonstrated in mice by inoculating mice that overexpress a mutant form of APP with Aβ from brain homogenates into only one brain hemisphere. This was sufficient to cause widespread Aβ deposits in both hemispheres consistent with a prion-like mode of propagation.[14] If true across all diseases with misfolded proteins, this would be a frightening proposition, as prions have been shown to be resistant to just about any chemical intervention. What raises doubt for the generalization of this mechanism to all diseases with protein aggregates is the fact that at least in HD, the inclusions dissolve and disease reverses when the production of new mutated huntingtin protein is experimentally stopped. If the mutated huntingtin protein behaved as prion, it could not be disposed of by the proteasome. Nevertheless, a prion-like spread of neurodegeneration cannot yet be excluded for AD, PD, and ALS.

5. MITOCHONDRIAL DYSFUNCTION

The brain is the organ with the highest relative energy use. Weighing only 2% of the body's weight, the adult brain uses 20% of all calories consumed. The reason for this incredible energy use is the sustained high neuronal activity. Action potentials (AP) require maintenance of ionic gradients across the axonal membrane. For Ca^{2+} to function as second messenger, it needs to be kept at a concentration 10,000-fold below the extracellular concentration by pumping it out of the cell or into organelles. In addition, neurotransmitters need to be recycled and packaged into synaptic vesicle. These processes all directly or indirectly consume energy in the form of ATP. In most organs, ATP can be generated glycolytically in the cytoplasm or through aerobic metabolism in the mitochondria. The former process is wasteful, as it only captures 3 mol of ATP per 1 mol glucose. Hence, the brain utilizes the more efficient oxidative metabolism exclusively, yielding 36 mol ATP per 1 mol glucose. Oxidative phosphorylation exclusively occurs in the mitochondria, and these organelles are found abundantly throughout most neurons. The mitochondrial tricarboxylic acid (TCA) cycle, also known as citric acid or Krebs cycle, not only provides effective aerobic energy production but also generates some important amino acid precursors for the synthesis of transmitters and proteins. Oxidative phosphorylation utilizes glucose and oxygen delivered via the cerebral vasculature. Failure to provide either oxygen or glucose or both causes a rapid loss of ATP, depleting the cells' energy stores within less than 120 s. This occurs in stroke, where loss of blood flow causes a rapid cessation of energy production. As is evident from the sudden onset of symptoms in stroke, neuronal function ceases very quickly. Moreover, if the energy disruption persists, glutamate excitotoxic death, discussed above, will ensue. Even in the continued presence of energy substrates, ATP production by the mitochondria can become impaired, causing a similar loss of neuronal function.

The possibility that mitochondrial dysfunction may contribute to neurological disease first surfaced in studies of HD and PD.[15] In PD, mitochondrial poisons such as the pesticide rotenone, which inhibits complex I of the respiratory chain, cause Parkinsonian movements. Similarly, an early animal model of HD used the mitochondrial poisons 3-nitropropionic acid or malonate to elicit Huntington-like motor dysfunction in mice. The idea emerged that environmental toxins may exert deleterious function by inhibiting mitochondrial function. However, much stronger evidence for concrete disease-causing contribution of mitochondrial dysfunction came from genetic studies.

5.1 Genetic Mitochondrial Defects in Parkinson Disease

Although familial forms of PD are uncommon, several genes have been identified that cause rare familial forms of PD. Two of these (Table 1) are directly related to mitochondrial health, namely Parkin and PTEN-induced kinase 1 (PINK1). Mutations in either gene cause recessive early-onset PD. Parkin encodes an E3 ligase that attaches ubiquitin residues to proteins or organelles, marking them for degradation. PINK1 encodes a protein that is expressed on the surface of mitochondria, where it interacts with Parkin to provide continuous surveillance of mitochondrial health. In normal mitochondria, PINK1–Parkin interactions signal a healthy organelle. As mitochondria become sick, their membrane potential, which indirectly harnesses the power to generate ATP, deteriorates. This is sensed by PINK1, which then phosphorylates Parkin, causing it to mark the mitochondrium for degradation (Figure 5.16, Chapter 5). When either gene is mutated, however, this surveillance fails and the neuron begins to accumulate faulty mitochondria that are unable to generate ATP effectively. This leads to an increased release of a highly reactive oxygen species (ROS) that can harm DNA and cellular proteins. Importantly, mitochondria harbor their own mitochondrial DNA to produce most essential proteins required for the synthesis of ATP. The presence of oxygen radicals puts this mitochondrial DNA at risk to produce faulty enzymes, jeopardizing the role of the mitochondria in energy production.

5.2 Mitochondrial Defects in Other Neurodegenerative Diseases

Defective mitochondrial enzymes are commonly observed in PD and HD neurons and even in fibroblast or muscle cells from patients. As a result, the mitochondria become very inefficient in converting glucose to ATP. In HD, the formation of intracellular inclusion bodies impairs the transport of mitochondria along the axon toward sites of high energy demand. This is also the case in AD, where intracellular hyperphosphorylation of tau destabilizes the microtubules that form the guiderails of axonal transport. As a result, mitochondria can no longer be effectively transported along the axon.

In HD, mitochondria often look noticeably sick. They are swollen, containing vacuoles and disturbed cisternae. Inclusions of mutated huntingtin protein can be found lodged under the outer mitochondrial membrane, where they impair the function of the mitochondrium. This causes measurable decreases in activity of key enzymes in the TCA cycle and complex II–IV activity. Owing to the ineffective use of glucose,

HD patients show increased compensatory caloric intake.

Similar changes in the activity of mitochondrial enzymes are found in mitochondria isolated from AD patients. In such cases, pyruvate dehydrogenase and isocitrate dehydrogenase are both impaired, and the expression of complexes I and IV is reduced. Increased glucose uptake is seen in patients with AD by positron emission tomography (PET) imaging and assumed to be a compensatory response.

As already mentioned above, any impairment in mitochondrial function has the potential to generate reactive oxygen radicals, which can damage mitochondrial DNA. These, in turn, generate faulty enzymes that further contribute to the ill health of the mitochondria. In ALS, a common mutation of both familial and sporadic disease affects the SOD1 gene, which is essential for the mitochondrial detoxification of oxygen radicals produced in the context of oxidative phosphorylation. Mutant SOD1 causes swelling of the mitochondria, impairs respiratory complexes, disrupts Ca^{2+} homeostasis, and leads to markedly reduced ATP production.

Mitochondrial dysfunction appears uniquely critical in diseases involving neurons that have very high AP discharge rates. This is particularly the case in the midbrain, where movement control pathways fire at rates of tens of APs/s, and some neurons have pacemaker activity, such as in the substantia nigra. These cells are literally on the edge, with a very close balance needed to be able to sustain the energy demand with the energy they produce. Even a small decrement in efficiency of their mitochondria would quickly impair their function.

While PD, HD, AD, and ALS are the diseases in which mitochondrial dysfunction is most strongly implicated in disease, this appears to be a much more universal pattern, and likely also contributes to acute and nonprogressive conditions such as stroke, trauma, and epilepsy. To what extent mitochondrial dysfunction is a consequence of disease or contributes to its cause remains to be elucidated. Thus far, a strong causal role has only been demonstrated in familial PD that presents with Parkin or PINK1 mutations.

5.3 Impaired Redox Status

Even in normal cells, mitochondrial respiration is a major cause for unpaired ROS. The most prevalent mitochondrially produced ROS is the superoxide radical. Even in a normal cell, the mitochondria do not work flawlessly and thus up to 2% of all electrons that pass through the respiratory chain are prematurely reduced. Since these ROS are highly reactive, the cells try to capture and neutralize them. This is done by antioxidant molecules such as vitamin E and glutathione, and by enzymes such as SOD that degrade the superoxide radical to H_2O_2 and water. SOD1 mutations are found in up to 7% of patients with ALS, making it a leading genetic cause of disease, and the only genetic cause that we know to affect redox status in neurological disease. That said, there is not a single disease that does not implicate a challenged redox environment as a contributing factor to disease, be that stroke or trauma, frank degeneration or infection. It is important to note, however, that while under these conditions an imbalance in the production and neutralization of ROS may contribute to cell damage, it is likely to be not causal of disease but rather an aggravating circumstance.

5.4 Improving Mitochondrial Function to Ameliorate Disease

To compensate for mitochondrial dysfunction, several strategies have been pursued in animal models and human clinical trials.[16] First, phosphocreatine is a high-energy substrate that is utilized by nerve and muscle cells to generate ATP. Phosphocreatine supplementation extends the life expectancy of mice carrying mutated

SOD1. Nutritional supplementation, explored in a pilot clinical trial for PD, failed to show improvements. However, high doses of creatine monohydrate slowed cortical atrophy in HD patients, and this trial was recently expanded to a multicenter phase III study (clinical trial registration #NCT00712426).

Another attempt to improve overall mitochondrial health is through supplementation with CoQ10, which is an endogenous substrate for the respiratory chain. It is often sold as an antioxidant dietary supplement and has shown neuroprotective qualities in various mouse models of neurodegenerative disease. Ongoing clinical studies are evaluating CoQ10 in PD, HD, and ALS. As is often the case, small early pilot studies show promising results, yet when expanded to larger patient cohorts the effect disappears. At present there is no strong evidence that CoQ10 supplementation improves outcome for any of the neurodegenerative diseases.

Finally, dietary supplements with cellular antioxidants including glutathione, vitamins C and E, omega 3 fatty acids, and beta-carotene have undergone experimentation in many diseases, yet none have proven to be effective. Given the ready access to these nutritional supplements, they continue to be widely used by patients nonetheless.

6. IN SPITE OF OBVIOUS DISEASE HERITABILITY, GENETIC CAUSES OFTEN REMAIN ELUSIVE

The majority of nervous system disorders, with possibly the only exceptions being stroke, trauma, and infections, are suspected to have genetic causes. However, in only a few instances have mutations in single genes been shown to cause disease. Examples discussed in this book include HD, Rett syndrome, and fragile X syndrome. It is much more common for disease-associated genes to characterize just a small subset of individuals within

TABLE 2   Rare Mutations that Cause Familial Forms of Neurodegenerative Disease

Gene	Disease	Prevalence	Inheritance
Parkin	Parkinson	<1%	Recessive
PINK1	Parkinson	<1%	Recessive
α-Synuclein	Parkinson	<1%	Dominant
LRRK2	Parkinson	<1%	Dominant
APP	Alzheimer	<5%	Dominant
PSEN1/2	Alzheimer	<5%	Dominant
TDP-43	ALS/FTD	2%	Dominant
SOD-1	ALS	7%	Dominant
FUS	ALS	2%	Dominant

APP: Amyloid precursor protein; FUS: Fused in sarcoma ; SOD-1: Superoxide dismutase; TDP-43: Transactive response DNA-binding protein 43; PSEN1/2: Presenilin; LRRK2: Leucine-rich repeat kinase 2; PINK1: PTEN-induced putative kinase 1.

affected families, yet these familial forms of disease are phenotypically often indistinguishable from sporadic disease. Therefore, we assume that familial and sporadic diseases are by and large identical and that eventually other genetic alterations will be discovered that explain the disease. Such is the case for three of the classic neurodegenerative diseases, ALS, AD, and PD. In each instance, studies of gene mutations in rare familial diseases have informed us about disease mechanisms and have allowed us to model the disease in animals by introducing similar mutations. This approach has also yielded model systems in which to test therapeutic drugs. Table 2 lists some prominent examples of genes that give rise to familial disease in a small percentage of affected individuals.

On the other side of the spectrum are diseases such as depression and schizophrenia, where any attempt to identify disease-causing genes, combinations of genes, or even susceptibility genes has thus far provided a complicated and largely unsuccessful result.

Nevertheless, we know from epidemiological data that these diseases "run in families" and that disease risk increases quite significantly if

any other family member also has the disease. With schizophrenia, the risk of a sibling of an affected individual developing schizophrenia is 10%. Even for multiple sclerosis, where we understand very little about the genetics of the disease and overall disease etiology, sibling studies suggest significant heritability, with a 33% chance for identical twins to develop disease and still a 5% change in risk if a sibling suffers from MS.

6.1 Gene–Environment Interactions

These diseases teach us that genetic factors alone are, in most cases, insufficient to cause disease. Instead, genes generate disease susceptibility that only manifests under the right (or, shall I say, *wrong*) conditions at the wrong time. Moreover, most diseases are polygenic, probably involving a number of susceptibility genes and possibly a combination of environmental factors. Note that by environment we are not only implying a person's physical environment but also the cellular environment as depicted in Figure 6, which includes nutritional status, redox environment, temperature, cellular stressors, and many other unknown factors.

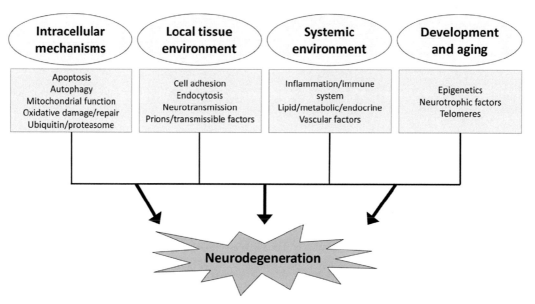

FIGURE 6 Environmental influences and gene expression interactions. Conceptual model of candidate pathways contributing to neurodegeneration. Candidate pathways influencing the balance of neuronal survival and degeneration are displayed within broader functional groups based on their major site or mode of action (intracellular mechanisms, local tissue environment influences, systemic influences, and mechanisms related to neurodevelopment and aging). The pathways and overarching functional groups in this model are highly related and can have overlapping or interacting components that can collectively modulate neurodegenerative processes. *Reproduced with permission from Ref. 17.*

6.2 Monogenetic Diseases

For only three diseases discussed in this book do we know a single gene that alone is responsible for a disease. These examples are HD, Rett syndrome, and fragile X syndrome. In HD, a microsatellite CAG expansion on chromosome 4 gives rise to a mutated protein (huntingtin) that causes progressive neurodegeneration. In Rett syndrome, mutations in the methyl-CpG-binding protein 2 (MeCP2) gene that regulates chromatin compaction alters the transcription of numerous downstream signals, including brain-derived neurotrophic factor (BDNF), thereby causing a progressive X-linked neurodegeneration. In fragile X, a microsatellite (CGG) expansion causes methylation and transcriptional repression of the fragile X mental retardation gene 1 (FMR1), which encodes an RNA-binding protein inhibiting RNA translation into proteins. Note that in two of these

diseases, Rett syndrome and fragile X, the affected gene and protein per se are entirely normal and only their quantity is altered. In HD, the mutation causes abnormal folding and protein inclusions, similar to those seen in PD and ALS.

Genes for the above diseases were identified through linkage analysis long before whole genome sequencing became available. As we learned for HD, large families spanning multiple generations of affected individuals were analyzed to find common regions on the genome. The chromosomal stretches were then narrowed down to smaller and smaller segments to eventually identify a single gene. This approach only works for diseases caused by a single gene.

Through a heroic effort, and involvement of many laboratories, a large-scale cloning effort completed the sequencing of the entire human genome in 2003. So why, in spite of the much-heralded ability to sequence a human's entire genome, have we been unable to define a genetic cause for most of these disorders? For starters, we do not have whole genome sequencing data for many people yet. This is still a time-consuming and costly endeavor. It will probably be feasible and affordable within another decade or two, yet even then it is not likely to become a standard diagnostic procedure. To put it in perspective, an analysis must be done for two copies of 23 chromosomes harboring 25,000 genes encoded by 3.2 billion base pairs. Of these, 99.9% will be identical for most individuals. Among the remainder, many differences will be irrelevant. Others will affect genes of still unknown function. The cloning of the first human genome took 13 years and cost approximately $3 billion. Even if it were feasible to obtain every person's full genomic sequence, the primary obstacles remain for bioinformatics to identify meaningful differences and for biologists to figure out which of the changes cause proteins to show altered function in disease. We are certainly still trying to find the needle in the haystack.

6.3 Polygenetic Diseases and Genome-Wide Association Study

To deal with current constraints regarding the ability to sequence and align entire genomes, the most common gene mining approach used to date is called a genome-wide association study (GWAS). Through this approach, the assumption is made that certain small genomic changes, namely single nucleotide polymorphisms (SNPs, *pronounced snip*), occur randomly in our genome, albeit with a relatively low frequency. Typically, such changes do not alter the amino acid and protein that is encoded, since the nucleotide code is highly redundant. For example GTT, GTC, GTA, and GTG each encode for the same amino acid (valine). In this example, two of the three nucleotides can differ and yet still produce the same amino acid and will therefore not change the resulting incorporated protein. GWAS typically compare two groups of people, those with disease and those without (Figure 7). They search for changes that occur nonrandomly and therefore are more likely in people affected by the same disease compared to controls. If an SNP can be identified to be more common in a patient with disease than in a control group, we define the SNP to be associated with disease, and it will mark a region on the genome that appears to influence the disease risk. Such a search can be done without having a gene candidate in mind, or by restricting the search to a genomic region that encodes for a candidate gene or protein suspected of contributing to the disease.

But even using GWAS, the degree of genetic variability in the population requires us to analyze huge data sets. For diseases that are common and homogeneous in their presentation, it should ultimately be possible to identify the underlying susceptibility genes and factors, unless they involve networks of many genes. Indeed, with enhanced bioinformatics capabilities, GWAS data can now be analyzed to search for changes in genes that are within a cellular pathway or network.[17] Nevertheless, for rare diseases, such as ALS, and heterogeneous diseases, such as depression or autism, gene mining through GWAS may still be a futile effort. In depression, for example, 17,000 cases were insufficient to yield statistical power. Hence, the fact that we have failed to identify new gene candidates for these diseases is largely explained by technical issues rather than by the absence of a genetic disease cause.[18]

It is important to recognize that an absence of a clear monogenetic cause of disease makes it difficult to establish a good, highly reproducible animal model of disease in which new therapeutics can be studied.

7. EPIGENETICS

If the polygenetic nature of neurological disease were not complex enough, we now have to contend with epigenetic changes as well. This emerging field provides important insight that explains the above-mentioned interactions of

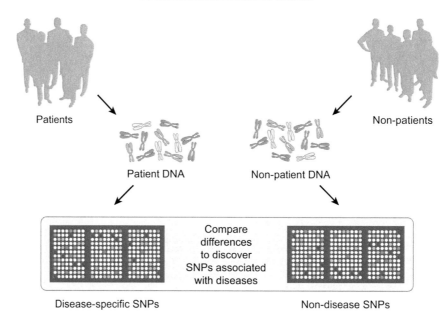

FIGURE 7   Genome-wide association studies. Scientists compare DNA from thousands of patients to DNA from thousands of control subjects, looking for differences ("SNPs" or single nucleotide polymorphisms) between the two groups. © *Pasieka, Science Photo Library.*

genes with the environment.[19,20] Importantly, epigenetic mechanisms can confer lasting, even trans-generational environmental risk or protection for developing disease even if a genetic risk is present. In its simplest form, epigenetics refers to persistent but reversible, heritable changes in gene expression without alteration of the DNA itself. Therefore, the gene(s) of interest are not directly affected, but the ability of the gene to be transcribed into protein is regulated via a number of modifications at the level of chromatin or DNA. These modifications are called epigenetic "marks," and the two most common marks are schematically illustrated in Figure 8.

First, a common epigenetic regulation involves changes in the interaction of chromatin with histones. Histones serve to compact chromatin and conserve space, as they allow for 146 bp of DNA to be tightly wrapped around each histone octamer like a spool of thread (Figure 8(A)). The histone tails protruding from this spool interact with the chromatin. Modification to

these tails by acetylation, methylation, or phosphorylation (Figure 8(B)) determines whether the chromatin becomes accessible for transcription by RNA-polymerases or not. For example, acetylation, or attachment of an acetyl group to a lysine residue on the histone, causes the chromatin to relax. This makes the DNA accessible for transcription, and thus acetylation results in transcriptional activation. Acetylation is carried out by a group of enzymes called histone acetyltransferases (HATs). Their activity is antagonized by histone deacetylases (HDACs), which cause transcriptional repression.

Another common epigenetic mechanism to regulate transcription is DNA methylation (Figure 8(C)). This occurs at specific cytosine residues in the vicinity of guanines, which are encoded by CpG dinucleotides, often called CpG islands. DNA methylation typically causes the methylated gene to be silenced, and therefore protein expression is reduced. DNA methylation is catalyzed by a group of enzymes called DNA

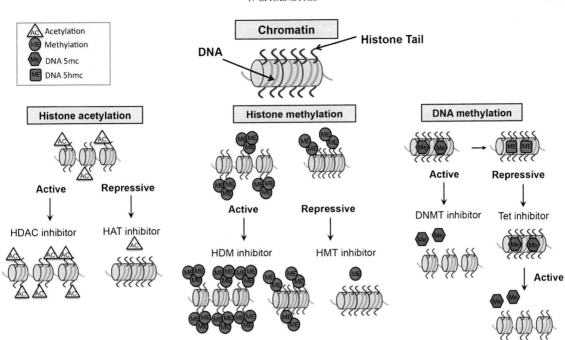

FIGURE 8 Schematic representation of the most common epigenetic marks that regulate transcription. DNA is condensed within the nucleus through interactions with histones, and this DNA–protein complex is referred to as chromatin. Two copies each of the histones H2A, H2B, H3, and H4 assemble to form a histone octamer, around which 146 bp of genomic double-stranded DNA are wrapped. The N-terminal tail of a histone contains many sites for epigenetic marking via histone acetylation, methylation, and phosphorylation. For example, acetylation of H3, shown as the addition of yellow triangles to the tails, results in a relaxed chromatin state that promotes gene transcription, whereas methylation (shown via red circles) can either promote or repress gene transcription. Methylation of DNA is another method of epigenetic marking of the genome, where a methyl group (shown as red diamonds) is transferred to cytosines in genomic regions in and around gene promoters that are rich in cytosine guanine nucleotides (CpG islands). The addition of methyl groups at gene promoters is generally linked to transcriptional repression. Inhibiting histone acetyltransferases (HATs) and histone methyltransferases (HMTs) prevents gene transcription while inhibiting histone deacetylases (HDACs) promotes gene transcription and histone demethylases (HDMs) promotes or prevents gene transcription depending on the form of histone methylation or degree of methylation. Image courtesy of Farah Lubin, Ph.D., Department of Neurobiology, University of Alabama Birmingham.

methyltransferases (DNMTs). Humans have five different DNMTs, which recognize the cytosines to be methylated via one of several proteins that contain a methyl-CpG-binding domain (MBD). One of these MBDs is the MECP2 gene mutated in Rett syndrome, discussed further below.

Epigenetic changes are surfacing as important genetic regulators in normal biology but also in disease, as simplistically illustrated in Figure 9. Normally the gene of interest is in an unmethylated state and can be transcribed into protein. In disease, the affected gene is hypomethylated,

thereby repressing its transcription. This scenario is described below for brain tumors. However, different scenarios are possible, such as when proteins involved in applying the epigenetic marks themselves become mutated and cause disease, best exemplified by Rett syndrome, discussed in Chapter 11. In most instances, however, the actual epigenetic signaling pathway operating in neurological disease has not been elucidated, nor has it been shown that epigenetic changes are a consequence rather than cause of disease. Remember, however, that this is a very young field of study

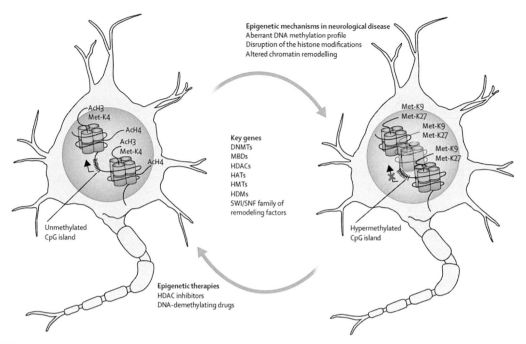

FIGURE 9 Epigenetics and neurological disease: In healthy neurons or glia (left), expression of the mRNA of a gene occurs in the presence of an unmethylated promoter CpG island and a set of histone modifications associated that cause an open "chromatin" conformation (e.g., hyperacetylation and methylation of lysine 4 of histone H3). Blue cylinders indicate octamers of histones, consisting of histones H2A, H2B, H3, and H4. These form the nucleosomes, and the double strand of DNA is wrapped around them. In neurological disease (right), a combination of selection and targeted disruption of the DNA methylation and histone-modifier proteins disrupts the epigenetic circumstances in the cell. The aberrant epigenetic inactivation of the disease-associated genes ("closed" chromatin conformation) is associated with dense hypermethylation of the CpG island promoter and the appearance of repressive histone modifications. Epigenetic drugs such as DNA-demethylating drugs and HDAC inhibitors can partially rescue the distorted epigenetic processes and restore gene expression of the neuronal or glial gene by removing chemical modifications (e.g., DNA methylation) and inducing the presence of modifications (e.g., histone acetylation). Ac=acetylation. DNMT=DNA methyltransferase. HAT=histone acetyltransferase. HDAC=histone deacetylase. HDM=histone demethylase. HMT=histone methyltransferase. MBD=methyl-CpG-binding domain protein. Met-K4=methylation of lysine 4. Met-K9=methylation of lysine 9. Met-K27=methylation of lysine 27. SWI/SNF=switching/sucrose nonfermenting chromatin-remodeling complex. *Reproduced with permission from Ref. 20.*

that was, until recently, focused almost exclusively on cancer. The few examples below should illustrate that epigenetic mechanisms may very well explain some of the gene–environment interactions commonly observed in polygenetic, complex neurological illnesses. They certainly already explain the fact that identical twins typically do not show identical diseases. In spite of both twins carrying the same DNA, the actual transcription of DNA into protein can differ absolutely due to regulation by epigenetic marks.

7.1 Epigenetics and Rett Syndrome

Rett syndrome is a severe, X-linked developmental intellectual disability caused by sporadic mutations in the MECP2 gene. Following a short period of seemingly normal development, a number of neurological symptoms appear that include stereotypic hand-wringing motions, motor abnormalities including toe walking, breathing abnormalities, autonomic dysfunction, autistic features, and loss of language.

Mutations in MECP2 cause a transcriptional dysregulation of many proteins involved in normal dendritic and spine development, including BDNF. While none of the affected proteins themselves are mutated and are therefore functional, their relative abundance is altered in such an unfavorable way that it leads to disorderly brain development, ultimately causing disease and developmental regression.

7.2 Epigenetics and Brain Tumors

As we discussed in Chapter 9, brain tumors are almost universally fatal and typically do not respond to chemotherapy. This is in part due to the fact that chemotherapeutic drugs tend to damage DNA, yet cancer cells have acquired DNA repair enzymes that effectively fix these damages just as quickly as they are inflicted. One of the most promising current chemotherapeutics for brain tumors is the DNA alkylating agent, temozolomide. While it was found to be completely ineffective in the majority of glioma patients, it provides very favorable outcomes in a subgroup. These 20% of patients carry a hypermethylated, and therefore inactivated, DNA repair enzyme, the O-6-methyl-guanine-DNA methyltransferase (MGMT),[21] tasked with repairing the DNA damage inflicted by temozolomide. It is now possible to screen for the methylation status of the gene encoding for MGMT, and its silencing predicts a favorable temozolomide response whereby the tumor can less effectively escape the DNA damage inflicted by the drug. Note that all patients have the same MGMT gene, yet only a small minority have MGMT epigenetically silenced through methylation.

7.3 Epigenetics and Epilepsy

Epilepsy is a very heterogeneous disease, but is commonly associated with unprovoked recurring seizures. It is generally assumed that an imbalance of GABAergic inhibition and glutamatergic excitation is responsible for seizures, and the general therapeutic strategy is to rectify this balance. One of the most effective drugs to treat seizures, sodium valproate (or valproic acid), has been assumed to show its anticonvulsant properties by enhancing GABAergic inhibition. Surprisingly, however, the therapeutic effect of valproate, while immediate, increases with prolonged use. This is difficult to explain on the basis of GABAergic signaling alone. The solution to the mystery may be an unexpected activity of valproate, which is also a potent HDAC inhibitor. As HDAC activity causes transcriptional repression, inhibition of this enzyme relieves this transcriptional activation, possibly affecting many genes. Consistent with a predominantly epigenetic effect of valproate is the finding that hippocampal neurogenesis, commonly blamed for the cognitive changes associated with epilepsy,[22] is suppressed by valproate. This also opens the possibility that epilepsy is associated with altered epigenetic marks, particularly those affecting the histone association of chromatin, possibly due to aberrant HAT or HDAC function. Valproate is also effective as a mood stabilizer in certain neuropsychiatric conditions such as posttraumatic stress disorder, and may similarly affect epigenetic marks that contribute to aberrant gene expression in these conditions.

7.4 Epigenetic Role in Anxiety and Depression

Our understanding of neuropsychiatric illnesses, particularly depression, is considerably less developed than that of neurodegeneration, epilepsy, and brain tumors. In Chapter 12 we highlight the complex nature of depressive disorders that, in spite of a near absence of known genetic risk factors, are still quite heritable. Depression seems to run in families, with an ~15% chance of a family member developing depression if any first-degree relative has also been affected. Is it possible that affected individuals share epigenetic marks inherited from their parents that predispose the family to depression? Given several related studies on the heritability of maternal stress and childhood neglect, this seems quite plausible.[23]

The notion that stress can cause trans-generational changes in behavior was first demonstrated in rats. Offspring of "bad" rat mothers that rarely licked or groomed their pups showed significantly elevated anxiety later in life than pups raised by "good" caring mothers. The latter not only showed lower anxiety but also had decreased expression of the glucocorticoid "stress" receptors that respond to the adrenally released stress hormone corticosterone (equivalent to human cortisol). The reduced expression of corticosterone receptors was the result of gene silencing by DNA hypermethylation.[24] While increased anxiety in mistreated or neglected offspring may be somewhat intuitive and expected, it was surprising to note that the epigenetic marks responsible for the differences in anxiety were passed on for several generations. Therefore, maternal neglect can have lasting effects not only on the mother's direct offspring, but also on future generations of her grandchildren as well.

While in the above example the epigenetic marks were the result of a life experience, namely stress or neglect, one must consider the possibility that similar epigenetic marks may arise by chance. As a consequence, some of us may be more resilient to stress than others and some may be more likely to develop depression. These differences in epigenetic marks may well explain why monozygotic twins often differ dramatically in their resilience to stress or their susceptibility to depression.

We are only beginning to understand the profound effects that epigenetic changes may have on human mood and behavior and how life experiences may cause heritable changes in our emotional state without ever causing direct changes to a person's DNA. The potential consequences are both fascinating and frightening.

They certainly open a new avenue for therapeutic intervention that may reversibly turn gene expression on and off simply by altering epigenetic marks.

8. NON-CELL-AUTONOMOUS MECHANISMS

Since Theodor Schwann's nineteenth-century discovery that cells are independent biological entities that function autonomously, we assumed that diseases, too, present in a cell-autonomous fashion. That is, we expected that neurological disorders resulted from the death or functional change of a defined population of neurons. Hence, these neurons alone are responsible for the disease. As an example, once mutations in the SOD1 gene were found to cause 20% of familial ALS, we assumed that SOD1 causes selective motor neuron loss by impairing the biology of this population of cells and this in turn causes the characteristic motor weakness. This is a reasonable assumption that we now know not to be true.

The gold standard for establishing a single gene as disease causing has been the introduction of the mutated gene into the germline of transgenic mice. This yielded valuable disease models for ALS, where various point mutations in the SOD1 gene produced motor neuron death and shortened life span comparable to human ALS. Similarly, germline expression of mutated tau causes hyperphosphorylation of the tau protein and replicates important aspects of AD. Likewise, the expression of the mutated MECP2 gene responsible for Rett syndrome largely mimics disease in mice. However, we have largely neglected the fact that in each of these instances, the mutated genes are expressed in essentially all cell types of body and brain. Could the mutated genes affect other cells that may contribute to disease? Could these non-neuronal cells possibly be equally or even more important in disease etiology? This indeed appears to be quite commonly the case. As a result we must radically change our viewpoint from a cell-autonomous to non-cell-autonomous cause of disease. Instead of mutations affecting just a single neuronal population, we must consider their effects on all cell types in the brain and elsewhere in the body as well. This is particularly true for the glial support cells, astrocytes, oligodendrocytes, and Schwann cells, as well as microglia acting as innate immune cells of the brain. In addition, even other seemingly unrelated cells, such as the endothelial cells of the cerebral blood vessels or bone marrow-derived macrophages, may contribute to certain neurological diseases in a non-cell-autonomous fashion.

8.1 Microglia and Disease

In ALS, approximately 20% of all familial cases are caused by mutated SOD1, presumably through toxic protein aggregates, mitochondrial impairment, and possible transcriptional regulation. Overexpression of mutated SOD1 fails to produce disease in mice, unless the mutated SOD1 is also found in microglial cells. Interestingly, selective ablation of mutated SOD1 only in microglial cells (or astrocytes) while leaving the wild-type protein in neurons significantly attenuates the severity of disease and slows its time course.

An even more striking example is Rett syndrome. As discussed in the previous paragraph, the transcriptional regulator MeCP2 changes the expression of a number of downstream genes, including BDNF, causing a progressive degeneration of cognitive function, language, and ambulation, along with peculiar changes in breathing in mice. These phenotypes are accurately reproduced by germline mutation of the MECP2 gene in mice. However, only correcting the mutated gene in microglial cells while leaving the mutated gene in neurons is sufficient to rescue mice from the disease.[25]

Given our broadening understanding of microglial cells in synaptic development, maintenance, and function, their role in neurological disease should not be that surprising after all.[26] It is now clear that microglial cells release factors that are involved in synapse formation and maintenance of synaptic connectivity. During development, pruning of synapses adjusts the number of synapses to match the activity between two cells. It turns out that microglial cells are the effector cells that execute this activity-dependent pruning.[27] They do so in a classic complement-dependent fashion, whereby microglial cells produce C3 proteases that lyse synapses tagged with the complement protein C1q. Consequently, impaired synaptic pruning by microglial cells may potentially alter synaptic connectivity in neurodevelopmental disorders like Rett syndrome and perhaps other disorders as well. Indeed, a lack of C1q expression on neurons is sufficient to generate epileptic seizures due to impaired developmental synaptic pruning by microglia.[28]

8.2 Astrocytes and Disease

For ALS, HD, and Rett syndrome, astrocytes appear to play important roles in disease. In ALS, disease is less severe when the mutated SOD1 protein is corrected in astrocytes. In addition, there appears to be a defect in astrocytic glutamate transport, as the principal transporter EAAT2 shows reduced expression in autopsy tissue from ALS patients and is similarly absent in mouse models of ALS. Yet, simply restoring normal EAAT2 expression in astrocytes can attenuate disease severity and prolong the life of mutant mice, suggesting that a breakdown of astrocyte Glu homeostasis plays an important role in ALS.

In Rett syndrome, the astrocytic MECP2 mutation appears to contribute significantly to disease. When MECP2 function was selectively restored in astrocytes, the dendritic morphology of neurons normalized, presynaptic levels of Glu increased, and even the respiratory abnormalities improved.[29] Most importantly, the animals lived for over 7.5 months compared to only 10 weeks. How one can explain the astrocytic protection from these diseases, however, remains unclear.

One important role of astrocytes is the regulation of extracellular K^+, and this typically occurs through diffusional uptake of K^+ via the Kir4.1 K^+ channel encoded by the KCNJ10 gene. Changes in $[K^+]$ occur with every AP during the repolarization phase, when K^+ leaves the cell. Accumulation of extracellular K^+ is most pronounced in the vicinity of neurons that fire at high frequency. In HD, astrocytic Kir4.1 expression is significantly reduced in the striatum, which contains medium spiny neurons that fire constantly at high frequency. As a result, extracellular K^+ is higher than normal. Restoring normal Kir4.1 expression in astrocytes through viral delivery of the gene is sufficient to attenuate motor dysfunction and prolong life in a mouse model of HD.[30]

While these examples illustrate a co-dependence of neurons on support cells, with somewhat defined sharing of responsibility, we do not yet understand the complex interaction between different populations of brain cells. Instead, it is clear that in addition to finding genes that cause disease, we must also ask in which cell types these genes function. We must expect that the same gene mutation may affect neurons, glia, and vascular cells differently, and that the sum of these effects are needed to explain the disease phenotype. This notion is commonly referred to as "non-cell-autonomous" mechanisms of disease.

9. INFLAMMATION

An inflammatory response is a common characteristic shared among all nervous system disorders. Inflammation is normally a transient process that helps wound clearance and repair, yet in some neurodegenerative diseases inflammation can become a persistent pathological response. Brain inflammation primarily engages microglia and astrocytes, which together form the brain's innate immune system, but occasionally blood-derived immune cells, macrophages, and lymphocytes contribute as well.

Along with macrophages, microglia share the capability to engulf cells and debris, and are thus often called the resident macrophages of the brain. However, their lineages diverge early in embryonic life when yolk-derived primitive precursor cells give rise to microglia that take residence in the developing nervous system long before the formation of the brain's vasculature.[31] In their unactivated state, microglial cells are ramified and process-bearing in appearance and constantly patrol the brain for evidence of damage or injury. Once microglia sense injury, they become activated and change their appearance to an ameboid shape (Figure 10). In this activated state, microglial cells are tasked with clearing debris and initiate a healing response by secreting proinflammatory molecules, including TNFα, interferon γ, interleukin 1β, nitric oxide (NO), and ROS. As injury resolves, microglial cells release anti-inflammatory factors that facilitate healing such as insulin-like growth factor 1 and interleukins 4 and 10. Microglia express complement proteins and can present antigens to blood-borne immune cells for the production of antibodies. They are even capable of eliminating entire neurons that are weakened, yet still functioning, in a process called "phagoptosis."[33]

Astrocytes play an important role in containing sites of injury through the formation of a scar. Akin to the tenacious tissue that seals the skin after an insult, astrocytes seek to compartmentalize injured and normal tissue. At the scar, they appear morphologically transformed, with thickened intertwined processes, and are called reactive astrocytes. Like microglial cells, they participate in the inflammatory response and may indeed regulate the microglial response through the release of cytokines and chemokines.

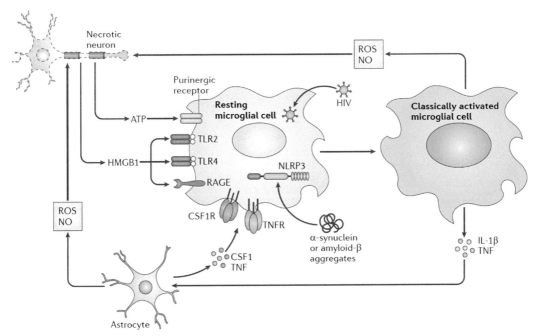

FIGURE 10 Microglial activation via exposure to environmental stimuli is detected by various receptors. A number of disease-associated factors can activate microglia through pattern recognition and purinergic receptors to establish a classical activated microglial cell phenotype. Such factors include HIV infection, damage-associated molecular patterns (DAMPs, such as high-mobility group box 1 protein [HMGB1], histones, and ATP), and neurodegenerative disease-specific protein aggregates (such as α-synuclein or amyloid-β aggregates). Proinflammatory mediators produced by classically activated microglia activate astrocytes, and the products released by activated microglia and astrocytes may exert neurotoxic effects. Activated astrocytes also release cytokines—including colony-stimulating factor 1 (CSF1) and tumor necrosis factor (TNF)—that further induce the activation and proliferation of microglia. Communication between microglia and astrocytes may therefore amplify proinflammatory signals initially sensed by microglia and thereby contribute to the pathology of neurodegenerative disease. IL-1β, interleukin-1β; NLRP3, NOD-, LRR-, and pyrin domain-containing 3; NO, nitric oxide; RAGE, receptor for advanced glycation end products; ROS, reactive oxygen species; TLR, Toll-like receptor. *Reproduced with permission from Ref. 32.*

9.1 Microglial Activation in Primary Neurodegenerative Disorders

While acute inflammation is common in trauma, stroke, and infections, neurodegenerative diseases are typically associated with persistent, low levels of inflammation that presents similarly for all classical neurodegenerative diseases, AD, PD, HD, and ALS.[32,34] Common among them is a progressive accumulation of protein aggregates, often preceding disease onset. These aggregates are perceived as foreign substances by microglial cells and astrocytes, which recognize these molecules through the family of Toll-like receptors (TLRs). These are cell-surface receptors that detect structurally conserved molecules typically produced by microbes. Different subtypes of TLRs have well-defined substrate specificity. For example, bacterial lipopolysaccharides are recognized by TLR4, viral DNA is recognized by TLR3, while TLR1 and TLR2 recognize bacterial peptides. Protein deposits including amyloid plaques are primarily detected via TLR4 (Figure 10).

Microglia and astrocytes sense neuronal injury through purinergic P2X7 receptors that are activated by ATP released from injured or dying neurons. In response to either TLR or purinergic activation, microglia, and to a lesser extent astrocytes, release proinflammatory molecules (TNFα, IL-1β, MCP-1, NO, ROS). These recruit additional microglial cells, astrocytes, and even invading macrophages and T cells to the disease or injury site. Astrocytes can increase the number of microglial cells by stimulating their local proliferation through the release of colony-stimulating factor 1 (CSF1), a growth factor for microglial cells.

In autopsy tissue from AD patients, activated microglia and reactive astrocytes surround amyloid plaques and stain positive for many proinflammatory mediators, including MCP-1, TNFα, IL-1β, and IL-6. Microglial cells detect the presence of Aβ plaques via TLR4 receptors, causing plaque removal by phagocytosis. This suggests that microglial cells make every effort to clear the potentially toxic protein aggregates, and under these conditions, microglia may well be called neuroprotective microglia (Figure 11). However, the clearance of Aβ ultimately fails and plaque burden increases, resulting in a sustained activation of microglia and astrocytes by the plaque. These chronically activated microglial cells now produce ROS molecules, particularly toxic superoxide radicals (OO^-) and hydrogen peroxide (H_2O_2). Superoxide radicals are generated by NADPH oxidase, a membrane-bound enzyme that transfers electrons to molecular oxygen.[35] Superoxide radicals kill bacteria and fungi and are released together with hydrogen peroxide in a "respiratory burst" to kill pathogens. In so doing, these chronically activated, ROS-producing microglial cells also harm adjacent neurons, and can therefore be regarded as neurotoxic microglia.

Abundant evidence suggests that neurotoxic microglia cells contribute to neurodegenerative disease (see schematic in Figure 11). For example, the expression and activity of NADPH oxidase are upregulated in AD and in PD, where its activity contributes to the loss of dopaminergic neurons. In sporadic and familial ALS, the mutation of SOD1, found in 7% of patients, increases the activity of NADPH oxidase,[36] and the increased production of ROS is a likely contributor to the progressive motor neuron death.

As neurons die, they release ATP, which activates P2X7 purinergic receptors on microglial cells, keeping them in their ROS-producing neurotoxic state. Pharmacological inhibition or genetic ablation of the P2X7 receptor suppresses the neurotoxic response of microglial cells,[37] potentially providing a novel avenue to specifically suppress the activation of neurotoxic microglial cells.

Similar microglial activation can be seen in essentially all CNS insults ranging from infection to trauma. In each of these conditions, neuroprotective and neurotoxic properties have surfaced, and it is likely that both operate at the same time. The eventual outcome, i.e., whether neuroprotective or neurotoxic properties prevail, is determined by the various receptor systems on microglial cells that become activated. One important commonality to remember is that microglial cells are one of the most important sources of toxic ROS molecules in the nervous system. Their ROS production is normally beneficial, as it is required for fighting pathogens. It is their chronic activation, in the absence of foreign pathogens, that turns them from friend to foe in neurological disorders.

9.2 Astrocytes and Inflammation

Although astrocytes are not immune cells, they too become activated after injury and during disease, visible by a change in their overall appearance and by an increase in the expression of GFAP. These morphological changes are typically called reactive gliosis and are often associated with the formation of a scar, a physical barrier that seals off a site of injury (Chapter 2, Box 3). The mechanical containment of an injury site by a physical barrier may be considered one of the important contributions of astrocytes to injury and chronic disease. Such scars can be transient or permanent, depending on the insult, and can also be beneficial or detrimental. In spinal cord trauma, for example, the scar has been suggested to be an impediment to axonal regrowth,

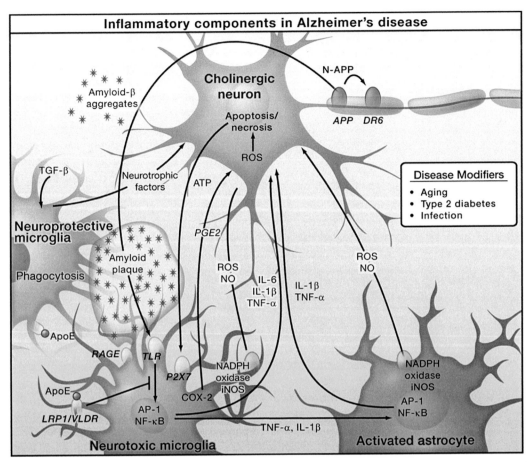

FIGURE 11 Inflammation in AD. Amyloid-β peptide, produced by cleavage of amyloid precursor protein (APP), forms aggregates that activate microglia, in part by signaling through Toll-like receptors (TLRs) and RAGE. These receptors activate the transcription factors NF-κB and AP-1, which in turn induce the production of reactive oxygen species (ROS) and drive the expression of inflammatory mediators such as cytokines. These inflammatory factors act directly on cholinergic neurons and also stimulate astrocytes, which amplify proinflammatory signals to induce neurotoxic effects. Apoptosis and necrosis of neurons result in release of ATP, which further activates microglia through the purinergic P2X7 receptor. Microglia can also play protective roles by mediating clearance of Aβ through ApoE-dependent and ApoE-independent mechanisms. Cholinergic neurons in the basal forebrain, the neurons that are primarily affected in AD, are presumed to be important targets of inflammation-induced toxicity, but other types of neurons, such as glutaminergic and GABAergic neurons, may also be affected. *Reproduced with permission from Ref. 34.*

yet suppression of scarring increases the size of the lesion.[38] In mesial temporal lobe epilepsy, glial scarring appears to contribute to epileptic seizures, which do not stop until the scar is surgically removed. Importantly, astrocytes also contribute to inflammation through the release of the very same cytokines and chemokines, both proinflammatory and anti-inflammatory, that we already discussed for microglial cells above. Astrocytes use these signals to amplify the initial inflammatory response and recruit additional microglial cells and astrocytes to a site of injury or disease. Reactive astrocytes express major histocompatibility complex class II and can present antigens such as plaque fragments to invading immune cells. Overall, however, their role is most similar to that of an orchestra conductor: fine tuning the immune response though the release of modulator signals, assisting in the removal of glutamate and potassium, clearing edema, and protecting the integrity of the vasculature.

9.3 Blood-Derived Cells and Inflammation

T cells are major contributors to inflammation. Subpopulations of T cells are specifically recruited to sites of injury and invade the CNS. T cells recognize foreign antigens and become activated. In turn, they recruit cytotoxic T cells through the release of inflammatory cytokines. They can also attract B cells to the brain. B cells then generate antibodies that can induce complement-dependent lysis of proteins or cells. In the normal brain, blood-borne immune cells are absent, and their presence in the cerebrospinal fluid is an indication of brain inflammation and disease. Acute injuries that breach the BBB allow the rapid entry of blood-borne immune cells, many of which contribute to the resolution of the injury. However, in many neurological disorders, infections, and neurodegenerative diseases, the integrity of the BBB is lost and lymphocytes and macrophages enter the brain. In multiple sclerosis, the entry of auto-activated T cells is believed to cause disease. In stroke and CNS trauma, the entry of blood cells is an inevitable and uncontrolled by-product, yet in most other disease conditions, the entry of blood-borne immune cells is viewed as disease ameliorating, with bacterial meningitis being an excellent example.

9.4 Directing the Immune Response to Treat Disease

A universal component shared among all neurological disorders is the very presence of an immune response. This is characterized by the activation of the innate immune cells of the brain, both microglia and astrocytes. Both sense foreign particles via Toll-like receptors and dying cells via purinergic receptors. They jointly try to kill pathogens and engulf and remove debris. Astrocytes seal off the acute lesion and release cytokines and chemokines that guide the process, with different molecules operating during different phases of the immunological response. The uncontrolled respiratory burst causing the release of toxic ROS molecules to fight pathogens causes neurons to die, most likely as bystanders rather than targets.

Given how commonly inflammation accompanies nervous system diseases, it is easy to envision how inflammation contributes in a negative way to disease. However, this would be overly simplistic. As with systemic inflammatory responses, one must assume that the tissue repair ultimately serves the purpose of healing rather than destruction. Abnormal inflammation, as is the case in the autoimmune attack of the myelin by activated T and B cells in multiple sclerosis, is a different matter. Here, inflammation is a major pathological component central to disease etiology. Otherwise, however, the universally observed inflammatory response provides an opportunity to ameliorate disease by instructing the innate immune system to be reparative rather than destructive. Thus far, our knowledge is insufficient to do so, but this will certainly change with future research. At present, anti-inflammatory approaches have been attempted across the spectrum of neurological diseases, but by and large they have not been successful. Among the more promising examples is the antibiotic minocycline, which attenuates microglia activation. It is being explored in active clinical trials for just about every neurological disease ranging from bipolar depression to ALS (see www.clinicaltrial.gov).

10. VASCULAR ABNORMALITIES

We have already repeatedly emphasized the brain's reliance on constant delivery of oxygen and glucose to provide sufficient energy substrates to sustain brain function. Even brief or focal disruptions can cause irreversible cell death, with stroke and trauma being examples in which vascular occlusions or ruptures directly cause disease. However, more subtle changes in the cerebral vasculature also have the potential to significantly contribute to disease. We must consider several different functional disruptions that may occur singly or in combination, including ischemia, vascular breach and edema, and breach of the BBB with entry of toxic molecules and cells.[39]

Unlike the vasculature of peripheral organs, the cerebral vasculature is completely impermeable to cells and water-soluble molecules, thereby separating the blood from the cerebrospinal fluid of the brain through the so-called BBB, which we discussed more extensively in Box 2.3, Chapter 2. A major role of the BBB is to keep immune cells from entering the brain and restrict entry of potential toxic molecules such as glutamate or albumin, which are toxic to neurons.

The integrity of the BBB depends on the continuous presence of pericytes and/or astrocytes on the vascular endothelial cells that form the vessel walls. Loss of either of these cells has been shown to decrease the expression of tight junction proteins (TJPs) that glue adjacent endothelial cells together. The loss of TJPs in turn opens microscopic spaces in the vessel wall through which molecules and even blood cells can enter. Protein deposits such as amyloid on blood vessels, or the release of inflammatory molecules and ROS from neurons and microglia in the context of disease, each weaken the BBB. Once the BBB breaks down, toxic molecules such as albumin, thrombin, and plasmin enter the brain. Plasmin degrade the laminin basement membrane, further accelerating vascular degeneration. Ultimately, red blood cells enter through microbleeds, allowing hemoglobin and iron, which are both toxic to neurons, into the brain. These various vascular dysfunctions are schematically illustrated in Figure 12. With a breach of the BBB, serum enters the brain,

causing vasogenic edema or swelling of the brain parenchyma. As the integrity of the vessel walls fails, focal ischemia results, thereby depleting energy substrates. This energy depletion affects both neurons and endothelial cells, as each requires ATP to maintain ion and amino acid transporters.

The above scenario plays out to some degree in nearly all neurological illnesses, albeit in some diseases more visibly than in others. A breach of the BBB following stroke or trauma is obvious. However, bacterial or viral infections similarly cause inflammation that induces transient opening of the endothelial BBB. As a result, fluid enters the brain, causing tissue swelling or edema. This can also be observed around brain tumors, where newly formed blood vessels lack tight junctions and thus fluid can enter the peritumoral tissue. Moreover, gliomas associate with the existing brain vasculature, thereby causing a focal breach of the BBB through which immune cells and blood-borne molecules such as glutamate can enter.[40]

Vascular abnormalities are also prominent in AD, vascular dementia, ALS, and MS,[39] where the expression of TJPs that form the BBB decreases. This may be in part due to the activity of extracellular matrix-degrading enzymes such as metalloproteinases (MMPs), which are employed in tissue remodeling. The TJPs are substrates for proteolytic cleavage by MMPs.

An important consideration is the fact that pericyte numbers decrease with age in the normal brain, and consequently their trophic support for the vasculature and the integrity of the BBB wanes. Age alone, therefore, contributes to progressive vascular changes, visible by a reduction in blood flow that can be detected in normal aging individuals by PET.[41] Interestingly, people carrying the ε4 allele of the APOE4 apolipoprotein, the major risk factor for AD, show enhanced regional decline in cerebral blood flow compared to people without this allele.[42] Amyloid deposits are often found in close association with blood vessels, and may directly contribute

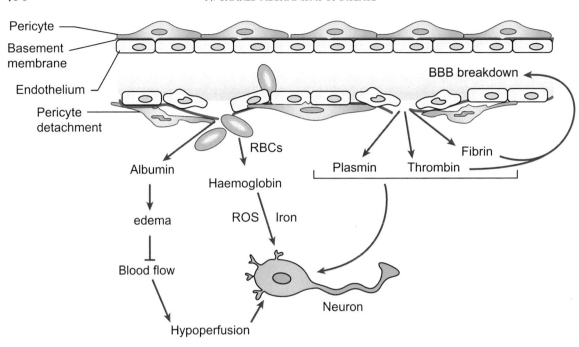

FIGURE 12 Vascular dysfunction and its contribution to neuronal injury and disease. Blood–brain barrier (BBB) breakdown caused by pericyte detachment leads to leakage of serum proteins and focal microhemorrhages, with extravasation of red blood cells (RBCs). RBCs release hemoglobin, which is a source of iron. In turn, this metal catalyzes the formation of toxic reactive oxygen species (ROS) that mediate neuronal injury. Albumin promotes the development of vasogenic edema, contributing to hypoperfusion and hypoxia of the nervous tissue, which aggravates neuronal injury. A defective BBB allows several potentially vasculotoxic and neurotoxic proteins (for example, thrombin, fibrin and plasmin, and glutamate) to enter the brain. *Reproduced with permission from Ref. 39.*

to further impairments in cerebral blood flow. Indeed, noninvasive imaging of cerebral blood flow suggests that amyloid burden correlates with the degree of reduction in cerebral blood flow in presymptomatic patients carrying disease-causing PS1 or APP mutations.

To what extent the above age- or pathology-related changes in the vasculature are contributing to disease or are the consequence of disease remains controversial. Clearly, we must consider vascular health as an absolute requirement for healthy aging, and strategies to accomplish this require a better understanding of the brain–vascular interactions. We must particularly improve our understanding of the role of astrocytes and pericytes in maintaining a healthy endothelial vessel wall with intact TJPs. This may provide an untapped opportunity for the future development of drugs that could be beneficial across all neurological illnesses.

11. BRAIN-DERIVED NEUROTROPHIC FACTOR

Following the accidental discovery that nerve growth factor (NGF) stimulates the outgrowth of sympathetic sensory nerve fibers in the 1950s, over 50 additional growth factors have been described in the nervous system. They can be divided into five families, namely the neurotrophins, cytokines, transforming growth factor-β, fibroblast growth factor, and insulin-like

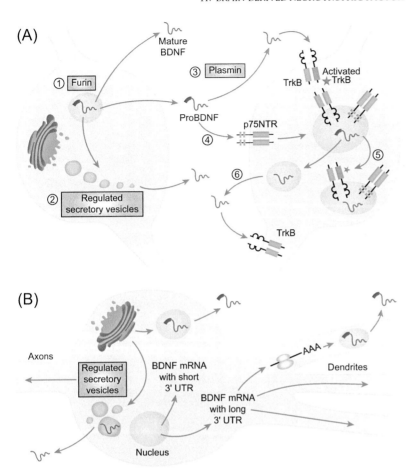

FIGURE 13 BDNF secretion in the CNS. (A) ProBDNF may be processed to mature BDNF by several cellular mechanisms. ProBDNF can be cleaved within the endoplasmic reticulum by furin (1) and in regulated secretory vesicles by proconvertase enzymes (2). If proBDNF reaches the extracellular milieu, it can be processed by plasmin, and the mature BDNF produced can then activate cell-surface TrkB receptors (3). Alternatively, extracellular proBDNF can bind p75NTR and become endocytosed and then cleaved to produce mature BDNF that either activates TrkB within endosomes (5) or is recycled to the cell surface (6). (B) The site of BDNF translation within the neuron may determine the form of BDNF released. BDNF mRNA with a short 3′ UTR accumulates in the neuronal soma, whereas BDNF mRNA with a long 3′ UTR is trafficked to dendrites. The soma supports BDNF cleavage within the Golgi, but the majority of dendrites lack Golgi elements necessary for processing of proBDNF, and therefore proBDNF may be the predominant form released. *Reproduced with permission from Ref. 44.*

growth factor families. Most growth factors show a degree of cell type specificity, and most operate throughout an individual's life span. With regard to diseases of the nervous system, BDNF appears to play a particularly prominent role in many of the nervous system disorders, and therefore deserves some expanded discussion.[43]

Normal Brain-Derived Neurotrophic Factor Signaling

Identified in 1982, BDNF is a member of the neurotrophin family, which also includes NGF, neurotrophin-3 (NT3), and neurotrophin-4 (NT4). BDNF is synthesized and released as a 247 amino acid precursor molecule. Following an initial

cleavage by proteolytic enzymes such as furin, the pro-BDNF is packaged into secretory vesicles and released into the extracellular space (Figure 13). After its release, the pro-BDNF is further proteolytically cleaved by plasmin or MMPs, giving rise to the mature BDNF. Mature BDNF binds with high affinity and specificity to the tropomyosin-related kinase type B (TRKB) receptor (Figure 14). The TRKB receptor is a tyrosine kinase that mediates its activity by phosphorylation of tyrosine residues on downstream signaling molecules, eventually activating the external receptor kinases ERK1/2 or AKT. Both mature and proBDNF also bind with low affinity to the p75 receptor, a member of the tumor necrosis factor family. The function of p75, which is not a kinase, is less well

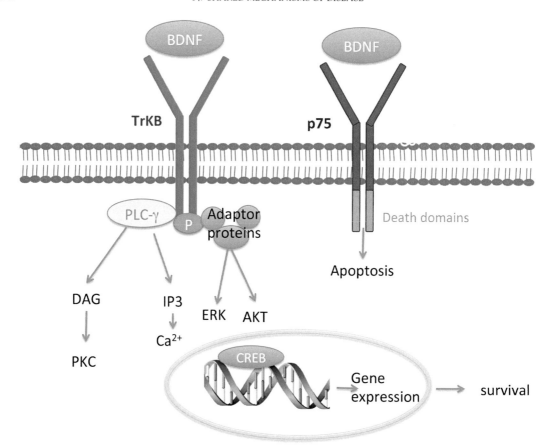

FIGURE 14   Receptors for brain-derived neurotrophic factor (BDNF). BDNF binds with high specificity to the tropomyosin-related kinase receptor type B (TrKB) and to the low-affinity neurotrophin receptor p75. "Mature" BDNF binds with greatest affinity to TRKB, whereas proBDNF binds with a higher affinity to p75. TRKB predominantly supports neuronal survival and expression of several functional genes, including extracellular signal-regulated kinases (ERKs) AKT and cyclic AMP-responsive element-binding protein (CREB). By contrast, separate p75 activation results in proapoptotic signaling. BDNF binding to the transmembrane TRKB leads to dimerization and autophosphorylation of tyrosine sites adjoining the cytoplasmic carboxy-terminal domain; this in turn activates several adaptor proteins that ultimately activate AKT, thus enhancing cell survival. Activation of PLCγ1 results in the generation of inositol-1,4,5-triphosphate (InsP$_3$) and diacylglycerol (DAG), which results in the mobilization of calcium stores and activation of calcium-dependent protein kinases that influence synaptic plasticity.

understood. Some research suggests that it may be a decoy receptor that sequesters BDNF if there is too much present. Importantly unlike TRKB, p75 is promiscuous and binds all of the neurotrophins including NGF, NT3, and NT4.

BDNF may be described as a reward factor. BDNF is released in an activity-dependent fashion; the more active a neuron, the more BDNF is synthesized and released. On adjacent cells it enhances dendritic branching and induces growth of spines. Continuous signaling assures the survival of postsynaptic neurons. The presence of BDNF has also been shown to alter synaptic plasticity; most importantly, BDNF enhances LTP, the cellular substrate for learning.

BDNF plays a major role in Rett syndrome, Huntington disease, Parkinson disease, Alzheimer disease, and ALS, and in every case the loss of BDNF causes a reduction in neural activity, LTP, and even cell death. Moreover, BDNF signaling is important in acute injury and trauma, and genetic changes suggest that aberrant BDNF signaling may contribute to depression.

Brain-Derived Neurotrophic Factor and Rett Syndrome

BDNF appears to be the major disease target in Rett syndrome, which is caused by de novo mutations in the MeCP2 protein. The MeCP2 protein regulates gene expression by binding to methylated CpG dinucleotide sequences on chromosomes. After binding to the transcriptional repressor site, MeCP2 recruits HDACs that induce a compaction of chromatin. Chromatin compaction makes the gene inaccessible for transcription. This process particularly affects the BDNF gene, which becomes constitutively repressed by MeCP2 binding to one of its four promoters. In the normal brain, BDNF is transcribed in an activity-dependent manner, whereby neuronal depolarization allows Ca^{2+} influx via L-type Ca^{2+} channels. This Ca^{2+} influx causes CaMKII-mediated phosphorylation of MeCP2 at serine residue 421, dislocating it from the BDNF promoter, thereby activating transcription. Functional loss of MeCP2 in Rett syndrome therefore impairs the activity-dependent release of BDNF, impairing its effect on learning and dendritic development. Indeed, one of the pathological features of Rett syndrome is a reduced number of abnormal, dysmorphic spines, a phenotype that can be completely rescued by restoring normal BNDF release.[45]

Brain-Derived Neurotrophic Factor in Huntington Disease

The main disease pathology of HD is the progressive loss of medium spiny neurons in the striatum. These glutamatergic neurons are part of the indirect movement control pathway that fine-tunes voluntary movements, and their loss causes the abnormal chorea movements of affected patients. Striatal neurons rely on the continuous supply of BDNF they receive from layer 5 cortical motor neurons. In these motor neurons, wild-type huntingtin protein promotes the transcription of BDNF by binding to exon II of the BDNF promoter via REST, a transcriptional repressor. This causes the activation of the BDNF promoter and stimulates BDNF transcription. The synthesized BDNF is then axonally transported through the corticostriatal pathway and released onto the striatal neurons as a trophic factor. In HD patients, the mutated huntintin protein is unable to promote BDNF transcription, resulting in a loss of BDNF to be released at the cortiocostriatal synapses. Moreover, protein inclusion bodies disrupt the axonal transport of BDNF to the target synapse. In turn, the striatal neurons involved in controlling voluntary movements are deprived of BDNF and are gradually lost, causing the characteristic movement symptoms.

Brain-Derived Neurotrophic Factor in Parkinson Disease

In PD, the neurons most affected by a loss of BDNF are the same striatal medium spiny neurons mentioned above for HD, but the dopaminergic neurons in the substantia nigra are lost as well. BDNF protein expression is reduced in the midbrain of patients with PD. In monkeys, PD-like symptoms can be induced by infusion of the chemical 1-methyl-4-phenylpyridium, which causes a lesion of the striatum. Subsequent infusion of BDNF rescues both the anatomical and behavioral effects, suggesting that treatments that restore normal BDNF signaling may ameliorate disease in PD patients.

Brain-Derived Neurotrophic Factor in Alzheimer's Disease

In AD, a progressive loss of cholinergic neurons throughout the cortex underlies the relentless dementia. The synaptic loss in the hippocampus impairs both short- and long-term memory, particularly spatial memory, which is localized in the hippocampus. BDNF is normally produced by neurons in the entorhinal cortex immediately adjacent to the hippocampus, and is then transported to the hippocampus and released in an activity-dependent manner. A reduction in BDNF can be seen in the entorhinal cortex and the hippocampus in Alzheimer disease. The infusion of BDNF into the hippocampus rescues hippocampal-dependent learning and memory in mouse and monkey models of AD. Ultimately, clinical studies are needed to examine whether an infusion of BDNF or a BDNF mimetic is feasible and efficacious in patients suffering from AD.

Brain-Derived Neurotrophic Factor and Amyotrophic Lateral Sclerosis

ALS is characterized by the progressive and selective loss of upper and lower motor neurons in the cortex and spinal cord, causing progressive muscle weakness and paralysis. Approximately 6% of familial cases of ALS present with mutations in the RNA/DNA binding protein fused in sarcoma (FUS). Mutated FUS leads to a splicing defect of the mRNA for BDNF, causing impaired TRKB-mediated BDNF signaling,[46] which likely contributes to the neuronal cell death. Unfortunately, thus far, attempts to supply exogenous BDNF to ALS patients in clinical studies have failed, most likely because BDNF does not reach its target neurons in sufficient quantity.

Brain-Derived Neurotrophic Factor and Depression

Depression is among the least well-understood neuropsychiatric disorders. However, one of the few clues we have regarding genetic causes of depression is a SNP in the BDNF gene, which makes individuals more prone to develop depression. More specifically, if the normal valine at position 66 of the BDNF gene is substituted with a methionine, the resulting Met allele (Val-66Met) is less efficiently packaged into secretory vesicles and therefore causes a reduction in activity-dependent BDNF release. Individuals with at least one Met allele have a 50% reduction in the wild-type allele and are at increased risk to develop depression or related affective disorders, suggesting a strong causative role of BDNF signaling in depression. BDNF and TrkB are abundantly expressed throughout the limbic pathways, where activity-dependent BDNF release is believed to contribute to a regulation of emotion. Interestingly, the most effective drug to treat bipolar depression, lithium, has been shown to exert some of its therapeutic benefit via a stimulation of BDNF release from presynaptic terminals. Stress decreases BDNF levels, yet chronic treatment with antidepressant drugs can restore BDNF-mediated signaling, suggesting an additional important role for BDNF as modulator for chronic stress and depression. Note that there is a rich literature on the mood-stabilizing effect of exercise, which suggests that this effect is at least partially mediated via increased release in BDNF.

Structural Effects

In addition to these neurodegenerative and neuropsychiatric diseases, acute insults such as stroke and CNS trauma, too, may involve disruption of BDNF signaling or may at least benefit from BDNF supplementation. In the case of spinal cord injury, for example, animal models clearly demonstrate that BDNF enhances axonal sprouting and regeneration. In CNS trauma and stroke, a recovery of function by sprouting processes and enhancing synaptogenesis will certainly benefit recovery. In mouse models of ischemia, infusion of BDNF has been shown to reduce infarct size and speed up recovery of function.

Feasibility of Brain-Derived Neurotrophic Factor Therapy

While the cellular mechanisms whereby a deficit in BDNF is somewhat disease specific, a bridging theme is that inadequate BDNF signaling consistently causes impaired neuronal signaling and, commonly, progressive cell death. Therefore, it seems that a strategy that could employ the supplementation of BDNF from an exogenous source would be almost universally beneficial. This idea, as mentioned above, is supported by studies in mice and nonhuman primates, but has yet to be done in human subjects. Several major challenges exist. First, it is difficult to produce sufficient BDNF in the test tube for treatment. One approach taken to overcome this issue is the production of mimetics, which are artificial peptides that are chemically different from BDNF but bind to the TrKB receptor. Ideally, mimetics have some enhanced properties; for example, they are more diffusible or have a higher affinity for the TrKB receptor. One of these mimetics, LM22A-4, directly activates the same receptor as BDNF (TrkB) and is being pursued to treat Rett syndrome.[45]

The second challenge is the difficulty of delivering BDNF reliably and in high enough quantity to the targeted neurons, in part because the BDNF molecule diffuses poorly in tissue and does not cross the BBB. To overcome these challenges, gene delivery systems are being developed that package the BDNF gene into adeno-associated viruses, for example, and a number of clinical trials using gene delivery of BDNF are in advanced stages of planning.

As mentioned above, lithium enhances BDNF release in depression, and has been shown to cause a 30% increase in BDNF in patients with mild AD. With this rationale, and the overall safety of lithium, it is now being studied in clinical trials for AD (NCT00088387) and ALS (NCT00790582). Another strategy, pursued in Rett syndrome, is to activate the downstream targets activated by BDNF (Akt, PI3K, and MAPK), using a recombinant insulin-like growth factor 1 (IGF1) currently in clinical trials (NCT01777542).

Remember, however, that one reliable way to augment BDNF levels in the nervous system is exercise. Clearly, once disease is established, an individual may no longer be able to exercise, or exercise may no longer be able to supply sufficient BDNF. Moreover, if the BDNF gene is silenced, for example via a MECP2 mutation, it

is unlikely that exercise can solve the problem. However, much of the beneficial effect of exercise in neurological disease may in fact be attributable to the positive effects of BDNF on neuronal health. In lectures to lay audiences, I like to refer to BDNF as an endogenously produced "multivitamin" for the brain. While overly simplistic, this notion may not be far from the truth.

12. CHALLENGES AND OPPORTUNITIES

This quick round-up of common disease mechanisms makes several important observations that are beginning to inform our approach to research and may ultimately change the development of new therapeutic strategies. Most importantly, we have become less neurocentric, and instead are beginning to consider the contribution of non-neuronal cells, and particularly vascular cells, glia, and microglia as well. Neuron-targeted therapies will ultimately fail as they only treat one of the affected cell types. This complicates matters, as the same genetic mutations may have different biological consequences in neuronal and non-neuronal cells.

Mitochondrial health is taking center stage, as these organelles not only fuel all brain activity but also function as modulators of intracellular second messengers, such as Ca^{2+}, NO, and ROS. Additionally, mitochondria also function as arbiters of cell death. Development of drugs that target mitochondria may provide novel ways to protect neurons and non-neuronal cells alike.

Our understanding of inflammation has evolved. It has lost its exclusive negative connotation as we are beginning to see the many positive contributions of inflammatory cells and processes to disease. Once we learn to specifically redirect the brain's immune cells to their neuroprotective state we may be able to harness the endogenous healing powers of our body to attenuate disease.

Finally, the genetics of disease, believed to be the "holy grail" necessary for our understanding of disease, continues to frustrate us. The majority of nervous system disorders have complex poly-genetic causes, convoluted by superimposed epigenetic regulation. However, there is no doubt that given time, we will gain a handle on the molecular genetics and epigenetics that predispose us to disease. Who said it will be easy? This will take time and patience. However, such insight, particularly as regards our knowledge of epigenetic regulation of disease processes, holds tremendous potential for the development of novel therapeutics to ameliorate or even cure disease.

References

1. Olney JW, Sharpe LG. Brain lesions in an infant rhesus monkey treated with monsodium glutamate. *Science*. 1969;166(903):386–388. 329.
2. Rothstein JD. Excitotoxicity hypothesis. *Neurology*. 1996;47(4 Suppl. 2):S19–S25; discussion S26.
3. Tzingounis AV, Wadiche JI. Glutamate transporters: confining runaway excitation by shaping synaptic transmission. *Nat Rev Neurosci*. December 2007;8(12):935–947.
4. Danbolt NC. Glutamate uptake. *Prog Neurobiol*. 2001;65(1):1–105.
5. Ventura R, Harris KM. Three-dimensional relationships between hippocampal synapses and astrocytes. *J Neurosci*. August 15, 1999;19(16):6897–6906.
6. Rothstein JD, Patel S, Regan MR, et al. Beta-lactam antibiotics offer neuroprotection by increasing glutamate transporter expression. *Nature*. 2005;433(7021):73–77.
7. Berry JD, Shefner JM, Conwit R, et al. Design and initial results of a multi-phase randomized trial of ceftriaxone in amyotrophic lateral sclerosis. *PloS One*. 2013;8(4):e61177.
8. Parsons MP, Raymond LA. Extrasynaptic NMDA receptor involvement in central nervous system disorders. *Neuron*. April 16, 2014;82(2):279–293.
9. Kurz A, Grimmer T. Efficacy of memantine hydrochloride once-daily in Alzheimer's disease. *Expert Opin Pharmaco*. September 2014;15(13):1955–1960.
10. Forman MS, Trojanowski JQ, Lee VM. Neurodegenerative diseases: a decade of discoveries paves the way for therapeutic breakthroughs. *Nat Med*. October 2004;10(10):1055–1063.
11. Yamamoto A, Lucas JJ, Hen R. Reversal of neuropathology and motor dysfunction in a conditional model of Huntington's disease. *Cell*. March 31, 2000;101(1):57–66.
12. Powers ET, Balch WE. Diversity in the origins of proteostasis networks—a driver for protein function in evolution. *Nat Rev Mol Cell Biol*. April 2013;14(4):237–248.
13. Mullard A. Sting of Alzheimer's failures offset by upcoming prevention trials. *Nat Rev Drug Discov*. 2012;11(9):657–660. 09//print.
14. Stohr J, Watts JC, Mensinger ZL, et al. Purified and synthetic Alzheimer's amyloid beta (Abeta) prions. *Proc Natl Acad Sci USA*. July 3, 2012;109(27):11025–11030.
15. Chaturvedi RK, Flint Beal M. Mitochondrial diseases of the brain. *Free Radic Biol Med*. 2013;63(0):1–29. 10//.
16. Uttara B, Singh AV, Zamboni P, Mahajan RT. Oxidative stress and neurodegenerative diseases: a review of upstream and downstream antioxidant therapeutic options. *Curr Neuropharmacol*. March 2009;7(1):65–74.
17. Ramanan VK, Saykin AJ. Pathways to neurodegeneration: mechanistic insights from GWAS in Alzheimer's disease, Parkinson's disease, and related disorders. *Am J Neurodegener Dis*. 2013;2(3):145–175.
18. Flint J, Kendler KS. The genetics of major depression. *Neuron*. February 5, 2014;81(3):484–503.
19. Jiang Y, Langley B, Lubin FD, et al. Epigenetics in the nervous system. *J Neurosci*. November 12, 2008; 28(46):11753–11759.
20. Urdinguio RG, Sanchez-Mut JV, Esteller M. Epigenetic mechanisms in neurological diseases: genes, syndromes, and therapies. *Lancet Neurol*. November 2009;8(11): 1056–1072.
21. Hegi ME, Diserens AC, Gorlia T, et al. MGMT gene silencing and benefit from temozolomide in glioblastoma. *N Engl J Med*. 2005;352(10):997–1003.
22. Jessberger S, Nakashima K, Clemenson Jr GD, et al. Epigenetic modulation of seizure-induced neurogenesis and cognitive decline. *J Neurosci*. May 30, 2007;27(22):5967–5975.
23. Nestler EJ. Epigenetic mechanisms of depression. *JAMA Psychiatry*. April 1, 2014;71(4):454–456.
24. Weaver IC, Cervoni N, Champagne FA, et al. Epigenetic programming by maternal behavior. *Nat Neurosci*. August 2004;7(8):847–854.
25. Derecki NC, Cronk JC, Lu Z, et al. Wild-type microglia arrest pathology in a mouse model of Rett syndrome. *Nature*. April 5, 2012;484(7392):105–109.
26. Gomez-Nicola D, Perry VH. Microglial dynamics and role in the healthy and diseased brain: a paradigm of functional plasticity. *Neuroscientist*. April 10, 2014.
27. Stephan AH, Barres BA, Stevens B. The complement system: an unexpected role in synaptic pruning during development and disease. *Annu Rev Neurosci*. 2012;35:369–389.
28. Chu Y, Jin X, Parada I, et al. Enhanced synaptic connectivity and epilepsy in C1q knockout mice. *Proc Natl Acad Sci USA*. April 27, 2010;107(17):7975–7980.
29. Lioy DT, Garg SK, Monaghan CE, et al. A role for glia in the progression of Rett's syndrome. *Nature*. July 28, 2011;475(7357):497–500.
30. Tong X, Ao Y, Faas GC, et al. Astrocyte Kir4.1 ion channel deficits contribute to neuronal dysfunction in Huntington's disease model mice. *Nat Neurosci*. May 2014;17(5): 694–703.

31. Ginhoux F, Greter M, Leboeuf M, et al. Fate mapping analysis reveals that adult microglia derive from primitive macrophages. *Science*. November 5, 2010;330(6005):841–845.

32. Saijo K, Glass CK. Microglial cell origin and phenotypes in health and disease. *Nat Rev Immunol*. 2011;11(11): 775–787. 11.

33. Brown GC, Neher JJ. Microglial phagocytosis of live neurons. *Nat Rev Neurosci*. April 2014;15(4):209–216.

34. Glass CK, Saijo K, Winner B, Marchetto MC, Gage FH. Mechanisms underlying inflammation in neurodegeneration. *Cell*. 3/19/2010;140(6):918–934.

35. Lull ME, Block ML. Microglial activation and chronic neurodegeneration. *Neurotherapeutics: the journal of the American Society for Experimental NeuroTherapeutics*. October 2010;7(4):354–365.

36. Liu Y, Hao W, Dawson A, Liu S, Fassbender K. Expression of amyotrophic lateral sclerosis-linked SOD1 mutant increases the neurotoxic potential of microglia via TLR2. *J Biol Chem*. February 6, 2009;284(6): 3691–3699.

37. Apolloni S, Parisi C, Pesaresi MG, et al. The NADPH oxidase pathway is dysregulated by the P2X7 receptor in the SOD1-G93A microglia model of amyotrophic lateral sclerosis. *J Immunol*. May 15, 2013;190(10): 5187–5195.

38. Faulkner JR, Herrmann JE, Woo MJ, Tansey KE, Doan NB, Sofroniew MV. Reactive astrocytes protect tissue and preserve function after spinal cord injury. *J Neurosci*. 2004;24(9):2143–2155.

39. Zlokovic BV. Neurovascular pathways to neurodegeneration in Alzheimer's disease and other disorders. *Nat Rev Neurosci*. 2011;12(12):723–738.

40. Watkins S, Robel S, Kimbrough IF, Robert SM, Ellis-Davies G, Sontheimer H. Disruption of astrocyte–vascular coupling and the blood–brain barrier by invading glioma cells. *Nat Commun*. 2014;5:4196.

41. Martin AJ, Friston KJ, Colebatch JG, Frackowiak RS. Decreases in regional cerebral blood flow with normal aging. *J Cereb Blood Flow Metab*. July 1991;11(4): 684–689.

42. Thambisetty M, Beason-Held L, An Y, Kraut MA, Resnick SM. APOE epsilon4 genotype and longitudinal changes in cerebral blood flow in normal aging. *Arch Neurol*. January 2010;67(1):93–98.

43. Nagahara AH, Tuszynski MH. Potential therapeutic uses of BDNF in neurological and psychiatric disorders. *Nat Rev Drug Discov*. March 2011;10(3):209–219.

44. Barker PA. Whither proBDNF? *Nat Neurosci*. February 2009;12(2):105–106.

45. Li W, Pozzo-Miller L. BDNF deregulation in Rett syndrome. *Neuropharmacology*. January 2014;76(Pt C):737–746.

46. Qiu H, Lee S, Shang Y, et al. ALS-associated mutation FUS-R521C causes DNA damage and RNA splicing defects. *J Clin Invest*. March 3, 2014;124(3):981–999.

General Readings Used as Source

1. Saijo K, Glass CK. Microglial cell origin and phenotypes in health and disease. *Nat Rev Immunol*. 2011;11(11):775–787.

2. Forman MS, Trojanowski JQ, Lee VM. Neurodegenerative diseases: a decade of discoveries paves the way for therapeutic breakthroughs. *Nat Med*. 2004;10(10):1055–1063.

3. Nagahara AH, Tuszynski MH. Potential therapeutic uses of BDNF in neurological and psychiatric disorders. *Nat Rev Drug Discov*. 2011;10(3):209–219.

4. Parsons MP, Raymond LA. Extrasynaptic NMDA receptor involvement in central nervous system disorders. *Neuron*. 2014;82(2):279–293.

5. Ramanan VK, Saykin AJ. Pathways to neurodegeneration: mechanistic insights from GWAS in Alzheimer's disease, Parkinson's disease, and related disorders. *Am J Neurodegener Dis*. 2013;2(3):145–175.

6. Chaturvedi RK, Flint Beal M. Mitochondrial diseases of the brain. *Free Radical Bio Med*. 2013;63(0):1–29.

7. Zlokovic BV. Neurovascular pathways to neurodegeneration in Alzheimer's disease and other disorders. *Nat Rev Neurosci*. 2011;12(12):723–738.

8. Urdinguio RG, Sanchez-Mut JV, Esteller M. Epigenetic mechanisms in neurological diseases: genes, syndromes, and therapies. *Lancet Neurol*. 2009;8(11):1056–1072.

Suggested Papers or Journal Club Assignments

Clinical Papers

1. Thambisetty M, Beason-Held L, An Y, Kraut MA, Resnick SM. APOE epsilon4 genotype and longitudinal changes in cerebral blood flow in normal aging. *Arch Neurol*. 2010;67(1):93–98.

2. McDade E, Kim A, James J, et al. Cerebral perfusion alterations and cerebral amyloid in autosomal dominant Alzheimer disease. *Neurology*. 2014;83(8):710–717.

Basic Papers

1. Qiu H, Lee S, Shang Y, et al. ALS-associated mutation FUS-R521C causes DNA damage and RNA splicing defects. *J Clin Invest*. 2014;124(3):981–999.

2. Stevens B, Allen NJ, Vazquez LE, et al. The classical complement cascade mediates CNS synapse elimination. *Cell*. 2007;131(6):1164–1178.

3. Ginhoux F, Greter M, Leboeuf M, et al. Fate mapping analysis reveals that adult microglia derive from primitive macrophages. *Science*. 2010;330(6005):841–845.

4. Weaver IC, Cervoni N, Champagne FA, et al. Epigenetic programming by maternal behavior. *Nat Neurosci*. 2004;7(8):847–854.

PART VII

BENCH-TO-BEDSIDE TRANSLATION

Drug Discovery and Personalized Medicine

Harald Sontheimer

1. INTRODUCTION

Almost daily we hear news reports about promising clinical trials that will revolutionize the treatment of cancer, heart disease, stroke, Alzheimer disease, and many other dreaded diseases. We learn that commonly established clinical practices, such as routine mammograms to reduce cancer risk, are suddenly questioned. Commercials advertise cognitive intervention programs "developed by neuroscientists" that increase memory and cognitive ability, and nutritional supplements that ward off dementia and make us centenarians have become a cottage industry.

How do we make sense of all this information? What can we believe? What is utter nonsense? How much can we trust clinical studies? Why do so many—in fact most clinical trials—fail? How is it possible that drugs approved through clinical trials are suddenly withdrawn? Did investigators get it wrong in the first place? Finally, what is this buzz surrounding personalized medicine when treatments for most neurological illnesses remain ineffective?

In this chapter the reader is introduced to the process of moving a laboratory discovery into clinical trials and the various obstacles faced when bringing drugs to patients. This is written primarily for a nonmedical audience to help them understand the complexity of human clinical trials; it is not intended as a guide for clinicians to implement clinical trials.

2. HOW DID WE GET TO THIS POINT? A BRIEF HISTORY

Throughout history there have been healers in just about every culture. Based on a unique knowledge (claimed or true), they used secret potions and various concoctions to heal the sick or wounded who entrusted themselves to the healer's care. Surprisingly, well into the twentieth century, a patient was an object at the disposal and mercy of such healers, whom we now generally refer to as doctors. Those entering this honorable profession swear to the Hippocratic oath, which states that "I will prescribe regimens for the good of my patients according to my ability and *my judgment* and never do harm to anyone."

While we typically emphasize the "do no harm" phrase in the oath, just as important is the fact that to this day we defer judgment to the physician. While that may be justified in many instances, it is also what has enraged many, including me, when it comes to situations where a physician is clearly in no position to render superior judgment on the matter of treatment. Fortunately, we live in a time when we are increasingly practicing evidence-based medicine and teaching medical students to ask for evidence before using a treatment. Such evidence comes from human subject research or clinical trials.

Studies using human participants have been done by healers for thousands of years, and some of these have contributed to our anecdotal knowledge about the healing powers of certain plants or animal extracts. In Asian medicine it is the hand-me-down passage of such anecdotes that much of traditional medicine relies on. However, many of these healers got it wrong and misinterpreted their findings. To this day we take certain things for granted without much scientific evidence. As is discussed in more detail below, disease often resolves and patients spontaneously improve with time, and people generally have a strong desire to get better. If during this time they receive an inert substance, a potion or herbal tea, for example, they and their physician may erroneously associate the improvement with the healing power of the potion or tea. In other instances we take as common sense that an intervention heals. For example, in a recent study a neurosurgeon asked the question of whether there is evidence that removing brain tumors actually improves life expectancy.[1] Much to my bewilderment, it turns out that the evidence is scant and indeed does not unequivocally support the use of surgery to prolong the life of a patient with glioma. Fortunately, the use of potions, medicines, and surgical and technological approaches is now highly regulated and subjected to extensive clinical validation through human clinical trials. These allow us to systemically distinguish which treatments work and which do not. This is, however, a complicated process fraught with many challenges.

For historic purposes, let us take a look at the first documented credible clinical trial, the study by James Lind on the treatment of scurvy in seamen.[2] In 1762, Lind published a study he conducted while he served as a surgeon on the HMS Salisbury. During this voyage, many of the sailors aboard the ship fell ill to a common ailment called scurvy. We now know that scurvy is the result of a deficiency in vitamin C, which is required for the synthesis of collagen in tissues throughout the body. The sailors were fatigued, with skin sores, muscle weakness, and overall malaise. Lind picked 12 sailors that were roughly in the same state of health and moved them into a common living quarter; everyone received the same basic foods, water, and attention. He then assigned two seamen each to six treatment groups, which he supplemented with different food items. These supplements included cider, vinegar, seawater, barely water, sulfuric acid, and citrus fruit (two lemons and an orange). Within 6 days, it was clear that those receiving the citrus fruit fared the best by far; indeed, they were able to return to work. Long before the discovery of vitamins and their role in connective tissue, Lind had identified a way to heal sick seamen by simply giving them citrus fruit. While this eventually became standard practice and led to giving English sailors the nickname "limeys" (for eating limes), the findings were also challenged because others who were simply eating fresh fish or meat were equally protected. Of course we know today that meat also contains ample vitamin C, as do other food sources. However, of key importance to the success of Lind's experiment was that he (1) started the study without a preconceived notion of the outcome (bias); (2) randomized treatment groups; (3) compared subjects with similar disability under otherwise identical conditions; and (4) meticulously documented and published his findings. Such a trial is now called a "fair" trial free of bias, and the experiment by Lind is frequently used as an example for the student of clinical studies.

3. DRUG DISCOVERY: HOW ARE CANDIDATE DRUGS IDENTIFIED?

Barely a day passes without the media reporting a scientific breakthrough that will soon cure Alzheimer disease, multiple sclerosis, Parkinson disease (PD), or cancer. But as years pass, the treatments that are offered to patients affected by these illnesses have changed little, if at all. Why is there such an apparent discrepancy between the reported breakthroughs and actual medical practice? Are the scientists dishonest? Are the news media getting it all wrong? Are these just hyped up findings? The answer is neither, but it is complicated!

Let us begin by looking at what scientists actually discover and report in the scientific literature. Most neuroscientists study brain function or mechanisms of disease, and many excellent examples of exciting discoveries are described throughout this book. For example, they may report how plaques and tangles form in a mouse with Alzheimer disease and how mutations in certain proteins causes misfolding of these proteins. This misfolding leads to aggregates, which in turn irritate neurons and cause them to die. Involved at every step of the discovered pathway are molecules, typically proteins, that do not function properly. When scientists discuss their findings, they close their papers with the conclusion that restoring the function of these proteins may cure a disease. They call the identified proteins and the genes encoding them "potential therapeutic targets." These targets are, however, a far cry from a drug that can be picked up at a pharmacy. Getting from a target to a drug is a long, winding road that is described further in this chapter. However, we must assume that the scientist reporting such a therapeutic target is honest (more on that toward the end) and that this target is therefore a good starting point for drug discovery. It is the discovery of potential new targets and the promise that these targets hold that is frequently reported in the news. The distinction between the target and an actual drug is important to understanding the seeming disconnect between promise and reality.

3.1 Drug Discovery by Chance

So just how do we get from a therapeutic target to a drug? Well, let me get the bad news out first. While we are good at finding targets, we are terrible at designing and making new chemicals that are specific for a new target. In fact, we are not very good at making new biologically useful chemicals, period! I know I may be ruffling the feathers of some budding chemists reading this chapter, but the truth is that few of the drugs on a pharmacy shelf came out of the deliberate synthesis of a new drug specific for a new target. Instead, most of them were serendipitous discoveries, many even surprising the lucky person who discovered them. Consider, for example, the antidepressant iproniazid, which was originally developed as an antimycobacterial agent to treat tuberculosis. Observant physicians noticed an unexpected psychoactive side effect when even terminally ill patients became more cheerful and active. The drug was subsequently shown to slow the breakdown of norepinephrine, serotonin, and dopamine by inhibiting the enzyme monoamine oxidase (MAO) and gave rise to the first class of antidepressant drug, as well as the MAO inhibitors now widely used to treat schizophrenia and PD. Similarly, in the 1960s the antiviral

drug amantadine was commonly used in nursing homes to treat patients with flu symptoms. Yet many patients suffering from PD showed a reduction in their dyskinetic, jerky movements following treatment with amantadine. The drug was subsequently shown to block N-methyl-D-aspartate (NMDA) receptors and became the only approved neuroprotective drug for PD. Who could have anticipated that the antibiotic minocycline would not only effectively treat bacterial infections in the skin (acne) and brain (Lyme's disease) but also provide a broad spectrum of effects that range from neuroprotective and anti-inflammatory to antipsychotic. In summary, one path to drug discovery is serendipity, where unexpected clinical side effects point to a different clinical use.

3.2 New Diseases for Old Drugs

The second path involves discovery of a new mode of action for an already existing drug through laboratory experiments. Most drugs have multiple targets. Some drugs may well be considered the "Swiss army knives of the pharmacy." Glucocorticoids (e.g., cortisol) are a good example because they treat just about anything from edema to fatigue and eczema, making them among the most commonly used drugs clinically. The antibiotic minocycline is used to treat bacterial infections such as acne and Lyme disease but also has beneficial effects on neuroAIDs and depression. The antinausea medicine temozolamide also slows brain cancer. There are many similar examples, and the list grows daily. These additional, unanticipated effects of drugs are discovered largely by accident in the laboratory rather than by rational design. Most scientists routinely dismiss claims that a drug may be highly specific, knowing from experience that this holds true only until the next unspecific target for the drug is found. Consider, for example, the tricyclic Prozac. It was believed to be a specific serotonin reuptake inhibitor (SSRI) that alters mood by prolonging the action of serotonin. Little did we know that this drug would also block multiple K^+ channels, G protein–regulated channels involved in the inhibitory action of GABA-B receptors, Cl^- channels, and even acetylcholine receptors. It may even lead one to question whether the primary antidepressive action of Prozac is in fact mediated by its serotonergic activity.

3.3 Biological Drugs: Nature's Helping Hand

So are there no truly specific drugs out there? There actually are quite a few, but these are by and large not made by man but designed by nature. Many of the proteins that we wish to target in neuroscience are ion channels, receptors, transporters, or signaling molecules that have been conserved through evolution and are also found in nematodes and fruit flies. Most of these molecules are involved in signaling to some degree and have led to the evolution of specific molecules with which a predator may paralyze or kill its prey or an organism can fend off its attacker. Examples include the snake venom α-bungarotoxin, the puffer fish poison tetrodotoxin, the bee venom apamine, or the black widow spider venom α-latrotoxin. Many nervous system drugs are contained in plants, where they serve to either protect them from being devoured by animals or attract animals to distribute their seeds. Examples include the muscarinic agonist atropine, pain-killing opioids, the anti-inflammatory aspirin, and hallucinogenic cannabinoids and cocaine. Even bacteria produce highly specific nervous system chemicals, as exemplified by botulism or tetanus toxin. When one analyzes the specificity of all of these poisons for their respective targets, it is clear that nature has done a remarkable job of refining them for a perfect "Goldilocks" fit. These molecules typically have the smallest possible size with the highest affinity and specificity, all as a result of millions of years of evolutionary refinement.

Following the cloning of the first neurotransmitter and peptide receptors in the early 1980s, there was tremendous enthusiasm among chemists who believed that knowing the receptors' structure to a single-molecule resolution would soon allow the design of much more specific "designer drugs." Rational drug design seemed to be within easy reach, and drug companies began to invest huge amounts of money in this endeavor. What happened? Why is Valium still on the pharmacy shelf instead of being replaced by a better and more specific benzodiazepine? Why have we not been able to generate a perfect modulator for NMDA receptors to prevent excitotoxicity after stroke? Knowing the anatomy of a lock (i.e., the receptor) does not allow us to make a better key (i.e., the drug) after all. Following two decades of attempts to create specific designer drugs, the art of rational drug design has largely fizzled into oblivion, and many pharmaceutical companies have divested from those research endeavors.

But wait—aren't there designer drugs out there? Isn't personalized medicine all the buzz? True, there are highly specific drugs on the market for some diseases, but these are largely based on antibodies. Here, once again the design of the key to fit the lock has been left to nature as opposed to the chemist (who must hate me by now, but please be patient—I redeem your profession below). By immunizing mice with a target protein or a fraction thereof, specific antibodies that bind to the target with high affinity and specificity are made. These antibodies can be humanized and made in larger quantities as monoclonal antibodies. The resulting drugs are often referred to as "biological," and several examples in the treatment of multiple sclerosis have been discussed. Unfortunately antibodies are often not functional, meaning that even if they bind a given receptor, they may not alter its function, so repeated immunizations are required to develop the rare functional variety. The difficulty of producing these functional antibodies in large quantities also makes them very expensive. Moreover, given the fact that they are immune modulators, many have shown unanticipated side effects or have even lost their effectiveness in humans over time. Pharmaceutical companies therefore always prefer a chemically made small molecule to a biological. Given the paucity of new chemicals being successfully invented, however, biologicals have recently taken the drug market by storm. In many ways antibodies have become the successor to rational drug design.

3.4 Unbiased Screens of Chemical Libraries

A final and common method of drug discovery is through large-scale screens. Every time a chemist modifies a molecule, a new chemical is created. Over the past century, pharmaceutical companies have taken nearly every drug that has known activity in the human body and made countless modifications to them, some as little as adding or removing a hydrogen or carboxyl group. More often than not, the newly generated chemical is less active than envisioned. Thus these chemicals become part of growing chemical libraries that can be used to screen against novel targets in hopes of finding a suitable binding partner. Such screens often pull hundreds of chemicals from a library, and these "hits" then need to be tested in biological systems. The goal is to reduce the number of candidate drugs to a manageable number for further biological and behavioral studies in the laboratory. Once the ideal candidate chemical is found, chemists may once again introduce a few modifications to see if the chemical can be tweaked to be even more efficacious or specific. Through this iterative process, medicinal chemists generate new drugs that are close relatives to the starting molecule for experimental medical use. It is important to emphasize that this approach rarely produces an entirely new class of chemicals; instead it expands the variety of chemicals within a known class. SSRIs or tricyclic antidepressants are excellent examples of how this approach was used to expand the available drug arsenal to treat disease. Once a drug that is specific for a target involved in disease has been identified, it can move into the clinical trials process.

4. WHAT ARE CLINICAL TRIALS AND WHY DO THEM?

A clinical trial quite simply is a controlled experiment with voluntary human participants. Put another way, it is a planned experiment involving human participants that is designed to identify the most appropriate intervention for future patients with a given medical condition. The treatment to be studied is typically a drug, a device, or nutritional change that is being tested in an effort to positively affect a disease or health condition.

The need to examine interventions through a well-planned experiment in humans may be obvious. However, the principal reason for a clinical trial is to determine whether an intervention has a true benefit. While this sounds trivial, it is actually critically important. Unlike the preclinical animal models in which drugs are initially studied, people have very heterogenous genetics and biological makeup. Even people of the same race, age, sex, height, and weight may respond very differently to the same treatment. In addition, as we will see in the following sections, there are many ways in which results can be misinterpreted or a study design can be biased that lead to false conclusions. To begin, we need to remember that in many cases sick people tend to get better over time, even when they are not treated. While this may not be the case for all diseases, it is generally true for many illnesses. If during such a period a patient is given medical attention, medicine, or even a potion of sorts, the physician and patient may both wrongly assume that it is the treatment that made them better rather than a natural recovery as the disease runs its course. It is only when treatments are compared with proper controls that receive the same attention without the drug that we can begin to evaluate the true drug effect

TABLE 1   Phases of a Clinical Trial

- Preclinical validation in animal models, dose finding and safety studies for first-in-human drugs

- Phase 0: Pilot studies, for example, studies of bioavailability

- Phase I: Studies of drug safety in patients or nonpatient volunteers

- Phase II: Studies of drug efficacy

- Phase III: Studies of broader applicability of effect; may include comparisons with other effective treatments

=> Drug/device approval

- Phase IV: Postmarketing analysis for interaction with other drugs and long-term effects

separate from the effects of time, attention, and so on. Hence one bias we are constantly fighting is natural healing; another may be the effect of medical attention alone or being in the hands of an expert in the absence of any intervention. These interventions per se may improve the outcome. Ideally, any careful study compares a drug with an inert substance, procedure or intervention called a placebo and does so in a double-blinded way in which neither the patient nor the doctor knows whether a given patient is receiving the active drug or the inert placebo. Clinical trials typically proceed through six well-defined stages (Table 1), each with unique objectives, pitfalls, and challenges.

4.1 Preclinical Evaluation

Before a drug can enter a clinical study, extensive preclinical testing needs to be documented. Importantly, whether a drug was derived from a drug screen or made by medicinal chemistry, a clinical trial must be motivated by robust laboratory findings or an epidemiological observation providing a strong rationale for the promise the intervention opportunities in humans. Let us consider as an example the discovery of a new GABAergic drug that causes a reduction in

neuronal firing in an in vitro slice model of epilepsy. The investigator reasons that if this effect holds up in humans, it would have the potential to selectively reduce seizures in patients suffering from epilepsy. The next step would be to validate the findings in animal models of the disease, using multiple models if possible. Ideally, other investigators should also validate this study so that a consensus develops in the epilepsy field that this approach is indeed promising. With such solid rationale, a clinical trial may be proposed.

The transition from preclinical to clinical study is directly influenced by whether a drug has been previously approved for human use for other indications. If so, safety data are already available; however, if a drug is a new chemical that is slated for first human use, more extensive preclinical testing is necessary to determine drug-limiting toxicity. Administering increasing drug doses to animals determines an LD50, or the effective concentration at which 50% of the animals are killed. Toxicity studies are usually done in rodents, but the US Food and Drug Administration (FDA) may also require testing in nonhuman primates before approving a clinical study. Contract laboratories typically do this work and can certify these values. Based on the knowledge of toxicity in animals, the "human equivalent dose" (HED) can be calculated. Unfortunately, this is often calculated incorrectly in the literature. The HED scales by body surface[3] rather than body weight and thus uses the following formula:

HED (mg/kg) = Animal dose (mg/kg) × (Animal K_m/Human K_m)

K_m constants are given by the FDA: mouse = 3, rat = 6, dog = 20, child = 25, and adult = 37. For example, a ketamine dose of 200 mg/kg would likely be lethal to a mouse. For a 20-g mouse, this dose would equate to 4 mg of ketamine (200 mg/0.02 kg = 4 mg ketamine). Using the HED formula with a mouse K_m value of 3 and a human K_m value of 37, we can calculate that the HED is 16.2 mg/kg (200 mg/ kg × (3/37) = 200 mg/kg × 0.081 = 16.2 mg/kg). Therefore, a potentially lethal dose for a 60-kg human would be estimated at 972 mg (16.2 mg/ kg × 60 kg = 972 mg ketamine). To emphasize the importance of scaling HED by body surface rather than weight, consider that the potentially lethal dose for a 60-kg human would be 12,000 mg if calculated by body weight (200 mg/ kg × 60 kg = 12,000 mg ketamine). Using the more conservative HED value given by scaling to body surface helps prevent lethality caused by inadvertent overdosing during the initial stages of a clinical trial.

Once an effective dose range to be used in human studies is determined, demonstrating an adequate shelf life of the drug at various temperatures and storage conditions, and to develop an assay to assess whether the drug is still efficacious, may also be necessary. This may be a simple enzyme-linked immunosorbant assay, or it may entail more complex physiological testing in cells, tissues, or even animals. Once researchers are satisfied that the drug is safe and efficacious in animal models and have verified the drug's stability and shelf life, it is ready to enter a trial with the first patient cohort.

4.2 Clinical Phases

The first phase of testing is a pilot phase often called phase 0. This phase has a very narrow objective, namely, to establish a method for drug delivery and sufficient bioavailability. It typically enrolls just a few subjects ($n = 10$), who may in fact be nonpatient volunteers.

Safety testing begins in earnest in phase I. In this phase, 12–50 subjects are given a single or multiple drug doses. A dose escalation schema is followed, with the first cohort received the lowest starting dose, the next cohort receiving an increased dose, and so forth, until the highest prespecified dose is reached. The objective is to show, either in nonpatient volunteers or in individuals with disease, that the drug does no

harm. What side effects may be deemed acceptable versus which ones may not need to be specified up front. Naturally, drugs to treat severe and untreatable conditions, such as brain tumors, permit more significant side effects than drugs that treat more benign health conditions such as attention deficit hyperactivity disorder. The results of the phase I study are reviewed by the FDA to ensure that the primary safety end points are met with an acceptable side effect profile.

The actual efficacy testing begins during phase II. At this stage a trial should be controlled, if possible, by a comparison of the drug with a placebo group and double-blinded so that neither the investigators nor patients know whether they belong to the treatment or control group. The drug dose or range of dosages determined in the phase I study are used for treatment. To expedite the study, phase II trials typically involve multiple clinical centers, with a total study populations ranging from 100 to 500 patients. A clearly defined efficacy end point must exist, with multiple possible secondary end points. For example, a brain tumor study may look primarily at the radiological response on magnetic resonance imaging (MRI), seeking to show contraction of the tumor. Secondary end points may be progression-free survival and overall survival after the intervention. Once again, the FDA reviews these data, and it is common for the FDA to inspect the data sets periodically throughout the study if it is possible to do so.

If successful, the drug moves on to phase III, which is the final stage of efficacy testing required for drug approval. Phase III studies are large, multicenter studies that enroll >1000 patients and compare the drug with other treatments that are considered the current gold standard for a given condition. In a multiple sclerosis trial, for example, this could be a comparison to interferon-β. The end point of the phase III study is likely different from the phase II study and seeks a significantly improved outcome over other available therapies. Phase III studies may also consider combined use of a new drug with an existing drug. Successful completion of phase

III and drug approval by the FDA is required for a drug to go to market. Approval is issued only for the particular disease indication for which the drug was studied.

In the following years of widespread clinical use, some drugs undergo a phase IV study that examines unexpected drug interactions or looks at the response of patient population to the drug. For example, an epilepsy drug may be monitored for interaction with beta-blockers used in patients with heart disease.

4.3 Important Considerations

The entire plan of a study, including its objective, enrollment, primary and secondary end points, methods of collecting, handling, and analyzing data, and expected sample size needed at each stage, is defined up front and forms the clinical trial protocol. This protocol must be approved by an institutional review board (IRB) and the FDA.

Enrollment

Unlike in animal studies, where most animals are the same except for sex and age, a human study has to thoroughly consider the appropriate target population that could and should benefit from the treatment, as well as a well-matched comparison group. This is not always trivial, and use of the wrong comparator or lack of a control group may easily send a trial off course. Moreover, a balanced approach must be taken to ensure inclusion of patients of all races, both sexes, various ages, and other variables, unless a strong argument can be made for the exclusion of a certain subgroup. Typically, early clinical trials exclude children because they cannot give informed consent, unless the condition of interest is exclusively a childhood disease.

End Points

Another important consideration is meaningful end points to be studied. In animal experiments for an antiseizure medication, for example, one may measure seizure frequency determined by continuous electroencephalography recordings. This is readily feasible in laboratory animals but not in humans since we cannot expect people to be in the hospital for weeks to receive continuous monitoring. Hence we may select a self-reported seizure log as a substitute. In addition, a secondary end point may be defined. This could be cognitive improvement, which would require a battery of behavioral cognitive tests. This greatly complicates the study and introduces additional confounding factors.

Sample Size

Determining the correct sample size before beginning the study is an important component of a successful trial and is the most commonly flawed aspect of trials that are reviewed throughout this book. We often talk about a study being underpowered to show significance for an expected end point. Before each phase of the trial, the necessary population must be calculated with the help of a statistician through empirical power size calculation rather than an open-ended experimentation. The size of the study population depends on the expected effect size, with a larger effect requiring a smaller population for study. It makes intuitive sense that detecting a 5% decrease in average infarct volume after a stroke intervention is more difficult to detect than an increase in the number of patients surviving for at least 12 months, and thus would require a larger population for study. Sample size is also affected by the effect size of the control or placebo group and the expected dropout rate in the trial. Excellent example calculations are found in the literature.[4]

Patient Safety and Informed Consent

Human experimentation has, sadly, not always been conducted ethically. The violations are many, ranging from the brutal atrocities performed by Nazi doctors during World War II to the well-intended experimentation with the polio vaccine by Jonas Salk on intellectually

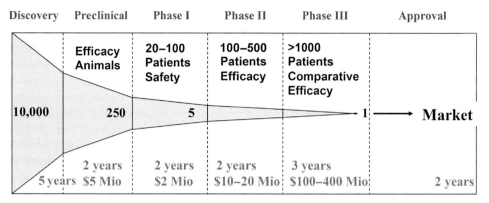

FIGURE 1   The timeline of a clinical trial along with the relative success rate, number of patients involved in each phase, and overall cost estimates. Mio = million.

disabled children living in orphanages without their consent. One objectionable study that has become a mandatory read for the scholar of clinical trials is the Tuskegee Syphilis Study, which ran from 1932 to 1972 and was sponsored by the US Department of Health. It examined the effects of untreated syphilis in 400 African American men. Researchers withheld treatment even when curative penicillin G became widely available. Most importantly, the study participants were never told that they were participating in an experiment.[5]

Several important safeguards have now been implemented to ensure the safety of those participating in clinical studies. Foremost is the requirement for voluntary participation and informed consent and the assurance that a participant can withdraw at any time for any reason or for no reason whatsoever. Moreover, an IRB reviews each trial well ahead of its approval. These boards typically include lay people from the community who scrutinize the objective of a trial. The IRB ensures that the expected benefits far outweigh the risks to participating individuals and that the study has clear conditions under which a trial will be ended. The IRB also ensures that participants are well informed and consent to the study. IRBs have become a major administrative hurdle to clinical trials and are perceived by some investigators as overreaching.

However, given the history of human experimentation, it is important to err on the side of patient benefit and safety.

Reporting and Registration

In the US all clinical trials must be registered with clinicaltrials.gov and all pertinent information publically displayed. The investigator also must ensure that the outcome of the data are published. Unfortunately, this requirement does not require publication and wide dissemination in a peer-reviewed journal, only an abbreviated summary on this Website.

Success and Cost

As of 2010 there were approximately 8650 drugs in clinical testing. As already alluded to, few of these are truly novel compounds. Many are recycled for use for a different disease indication. The entire process of moving these drugs from discovery to market, along with success, failure, and costs, is illustrated in Figure 1. This graph provides startling news. For every 10,000 drugs that enter the discovery phase, only one drug makes it to market. Most drugs that begin preclinical testing fail. For those that remain, the cost to market is enormous. As of 2005, the cost to develop a single drug was $1.2 billion, with the cost per patient ranging from $12,000 in phase III to $20,000 in a phase I trial. The whole process takes, on average, 15 years, of which the clinical testing phase occupies 6–8 years. Consider that current patent protection in the US is typically only 17 years from the time of issuance, and therefore most drug companies have only a few years to market their drug with exclusive rights before generic drugs become available. This explains the high cost of most newly developed drugs; drug companies must recoup the cost of drug development in the few years during which they can exclusively market the drug. Also, much of the development cost is the same whether it is for a large or a small patient market, making it difficult to justify developing treatments for rare diseases with a small market.

5. THE PLACEBO EFFECT

As already mentioned, a clinical trial should ideally use a placebo control group rather than a nontreatment group if possible. A placebo is typically an inert substance given to a patient in lieu of an actual medication. While it is often thought of as a "sugar pill," a placebo may be better defined as an intervention designed to simulate medical therapy, but it is not considered to be an actual treatment at the time of its use. This definition broadens a placebo to the use of devices and even sham surgery in treatment. The use of placebos in medicine was quite common throughout the eighteenth and nineteenth centuries. Indeed, in 1807 President Thomas Jefferson reported that one of the most successful physicians at the time confided that he "uses more bread pills, drops of colored water and powders of hickory ashes than all of other medicines together."[6] Until the 1950s, a placebo was considered a "humble humbug" that comforted a patient without doing any harm. Then, however, it became elevated through a study published by Beecher[7] in 1955 showing an overall 35% effectiveness of a placebo compared to an actual treatment. This, along with reports that sham surgery can provide medical relief, defined what we now call the placebo

effect, namely, clinical improvements achieved in the absence of an active intervention. Of course, in hindsight, many of the potions provided by traveling doctors and self-anointed healers throughout the centuries provided little more than a placebo effect. Today, the placebo has become the most important control in defining whether a drug or intervention is actually curative per se or if the improvements are the result of the environment and belief system in which the treatment occurs.

The placebo effect must not be dismissed as humbug. It is real, it is strong, and it is particularly applicable to neurological illnesses. Our understanding of the neurobiology underlying the placebo effect is incomplete, but it is clear that our mind strongly influences brain and body function through the release of neurotransmitters and hormones. The placebo effect has a strong psychological component and is therefore probably only effective when the patient is convinced that (s)he receives an active treatment as opposed to a placebo. Interestingly, however, patients may improve even when told that they are receiving a placebo, but whether these patients actually trusted this information or persisted in the belief that they were receiving an active treatment is unknown. While placebo treatment works for many ailments, it does differ in efficacy and is strongest for conditions with a large psychological component, such as pain, anxiety, and depression.

Functional neuroimaging studies done in conjunction with the use of placebo analgesics have been particularly insightful. These studies demonstrate that the belief that an analgesic has been applied before inflicting a painful stimulus is sufficient to suppress the activation of nociceptive pathways in the spinal cord.[8] This effect is most likely mediated through the release of endogenous opioids in the spinal cord in response to cortical activation of the descending pain control systems in the brain stem; indeed, opioid antagonists block the placebo effect. These experiments show that the placebo effect is mediated by an actual physiological change that is identical to that of an active drug, a truly remarkable observation.

One of the most pronounced placebo effects is observed in the study of antidepressant drugs. When antidepressants drugs such as Prozac were compared to placebo, they were found to be equally effective at improving mood. Importantly, just as with placebo analgesics, for patients who responded to the placebo, it affected the very same regions in the limbic brain involved in experiencing sadness as did the antidepressant drug, causing an identical change in glucose utilization imaged through positron emission tomography.[9] This once again shows a physiological effect of the placebo that is identical to that of the active drug.

These placebo studies present us with quite a dilemma. Depressed patients who are untreated stay depressed. However, 75% of those who take medication get better, but so do the majority of patients who take the placebo. In fact, the difference in effect size is only 1.8 points on a 51-point behavioral scale and would thus be considered clinically meaningless. In short, placebos are as effective as antidepressants, yet antidepressants can have dramatic side effects including weight gain, fatigue, loss of libido, and suicidal thoughts. Moreover, they can cost as much as $9000/year. Yet a physician cannot prescribe a placebo to a patient under the pretense that it is a real drug. This approach can be used only with informed consent in a clinical study. Until we find a solution to this dilemma, drug companies are enjoying >$10 billion/year in sales for drugs that may be no better than sugar pills.

Yet another consideration that contributes indirectly to the placebo effect in clinical trials is the fact that simply being part of a medical study and in the hands of a specialist has a positive effect on treatment outcome. The simple act of showing up at the doctor's office and being cared for by white-coated professional with countless diplomas decorating the office walls makes people feel better. Taken together with the placebo effect, it is clear that any clinical study must correct for these nonspecific effects to ascertain a drug's true biological activity.

Unfortunately, this typically increases the number of patients required for a study, since the effect size (difference between drug and placebo) is always reduced by these nonspecific effects. For example, in the case of antipsychotic drugs given to patients with schizophrenia, this difference is a mere 17%, with placebo being almost as effective as the actual drug (24% vs 41%).

6. WHY DO CLINICAL TRIALS FAIL?

There are roughly 50,000 clinical studies being conducted in the United States in 2014. Unfortunately, the fast majority of them will fail before a product can make it to market. Many of these failures occur during phase I or II trials, where drug safety and efficacy are examined; others fail in larger, multicenter phase III studies in which larger cohorts of patients are enrolled. This enormous failure rate is unsustainable in the long run. On average, each patient enrolled in a study costs between ~$12,000 and $20,000, and the cost of successfully moving a drug from candidate to approval is estimated at $1.2 billion. Hence drug companies are spending an exorbitant amount of money to find the few drugs that stand up to these multiple phases of clinical testing. Not surprisingly, a number of studies have examined why the failure rate is so high and have identified several flaws, most of which could be eliminated.

6.1 Irreproducible Laboratory Data

As already discussed above, drug targets are typically identified in academic laboratories. Thousands of investigators strive to find innovate new drug targets. The competition to be first to publish is fierce and is rewarded by recognition, tenure, promotions, and additional research grants. Scientists become invested in their stories, sometimes to the point of allowing bias to creep into their work. Data points may be unjustly excluded as outliers or improper scientific methods may be used to reveal statistical significance in cases where a proper analysis would discard the finding as noise. Indeed, a recent analysis of 157 articles published in the five leading neuroscience journals (*Nature*, *Science*, *Neuron*, *Nature Neuroscience*, and *Journal of Neuroscience*) made the startling discovery that 50% of these articles used the wrong method for statistical analysis, and in 66% of these cases the conclusion would not have been supported had the correct test been used.[10] Therefore it may not come as a surprise that drug companies are wary of the reproducibility of laboratory findings. As a result, they have established their own laboratories where they first attempt to reproduce the published findings and validate the targets. A recent article described that in doing so, drug companies are finding that 75–80% of scientific findings cannot be reproduced in their laboratories.[11] While they do not claim that the scientists were intentionally misreporting their findings, it nevertheless exposes a major problem that must be fixed. If we cannot trust published data, we are wasting taxpayer dollars that support such research. In 2006, *Nature* stated in an editorial that "Scientists understand that peer review *per se* provides only minimal assurance of quality and that the public conception of peer review as a stamp of authentication is far from the truth." One problem repeatedly identified is the pressure to publish, which may result in scientists failing to spend the necessary amount of time needed to understand and validate their experimental results using multiple strategies to challenge the original hypothesis. Indeed, it has become common for scientists to "prove" their hypothesis, which, of course, to the student of science, is impossible. One can only challenge a hypothesis repeatedly, and if no experiment refutes it, one gains greater confidence that it may in fact be correct. That level of confidence is often not reached in published studies.

6.2 Publication of Negative Data

Unfortunately, for the same reasons explained above, publishing "negative" data, that is, those experiments that refute a hypothesis, has become difficult. They tell a "negative" story, which may happen to be the truth but carries little appeal. Journals such as *Nature* and *Science* are drawn to newsworthy findings, such as an environmental link to autism, a gene that causes schizophrenia, or new evidence that cold fusion is possible. That precludes many valid negative stories from being distributed via the most widely read journals with the highest impact. Furthermore, it carries the danger that many of these stories published in high-impact journals may in fact be wrong. Let us assume that 20 researchers work on a possible link of lead with autism. They all conduct similar studies using mouse models, and each reports their data with a significance threshold of $P < 0.05$. Hence, for each study there is a 5% chance that their findings are wrong due to chance. It thus stands to reason that of the 20 studies, 19 (95%) reach the correct conclusion and find no such link (let us assume this to be correct for the sake of argument). Yet one study (5%) incorrectly reports a striking, but erroneous, link between lead exposure and autism. Remember, based on a threshold of significance of $P < 0.05$, we would expect that 1 in 20 studies would get it wrong, without bias, purely by chance alone. Now, of these 20 studies, which is most likely to be reported in *Nature*? This example shows that, by pure statistics and by publication bias toward positive data, we may indeed publish and disseminate incorrect findings. Such a selection bias toward "positive" findings is nothing new and was already recognized in 1959 by the statistician Theodore Sterling. Sterling found that 97% of all published psychological studies at the time found the effect that they were looking for,[12] and this result did not change when he reexamined this question in 1995. Once a wrong conclusion is published, particularly in a high-profile journal, scientists tend to crowd the field, all eager to add to their own data. Within a few years, additional erroneous studies establish a new paradigm that may be wrong and very resistant to change.

6.3 The Truth Wears Off

Even when findings have been replicated many times, and therefore the effect is assumed to be well established, the initially observed effect size has a tendency to decrease over time with repeated studies.[13] This is even the case when the same scientist in the same laboratory repeats the experiments. This troubling phenomenon is now recognized in many areas of science, from medicine to ecology, and has become subject to rigorous debate. It seems that the more we repeat studies, particularly complex ones and ones involving humans, the smaller the effect size becomes, to the point where it may eventually be lost altogether. This was recognized initially in trials of antidepressants, which lost 50% of their effect within just a few years. The gradual inclusion of a more diverse study pool, be they mice from different mothers, cells that have been passaged 10 times more often, or people of different ages and ethnicities, yields a broader diversity with more noise, thereby reducing the effect size.

6.4 Poor and Potentially Biased Trial Design

Early phase clinical studies are frequently the domain of small biotechnology companies supported by venture capital. Many are "one trick ponies" that have only one drug in the pipeline, and their success and return on investment or valuation rests entirely on positive clinical data. One way to achieve early success is to structure a clinical trial that is biased toward maximal success. This may entail having a narrow enrollment group, for example, patients of only a narrow age window with a severe and homogenous presentation of disease. Other factors that may bias toward success are a simple primary end point, a comparison group that receives no treatment rather than placebo, or even no control group at all. Let us consider as an example a study examining the effectiveness of an oncolytic virus to treat brain tumors and glioma. Since only terminally ill patients are enrolled, including a control group that receives no treatment or placebo would be unethical. Therefore, one must compare the results to historical data from other studies. However, the company may enroll patients who overall are in superior neurologic health, such as younger patients, who tend to live longer even when ill. This alone can skew survival data toward longer survival compared with that of historical control groups including all ages. The trial may therefore be deemed a success, moving into phase III, where it is likely to fail once used across the entire spectrum of glioma. Using end points that are likely to show positive results, yet may be rather meaningless, is also common. This has been common in multiple sclerosis trials in which the radiological response of drugs was used as primary outcome, yet patients did not improve clinically. Thus, while these trials were successful, they provided no functional benefit to the patient.

6.5 Interpretation Bias

Yet another challenge is overcoming the bias of experts toward their own opinion or intuition. An excellent example of such interpretation bias is the recently published phase III study of the effectiveness of mammography to reduce cancer deaths in women. A 25-year study with over 95,000 participants showed no survival benefit for women who received mammograms compared with those who do not.[14] In spite of this evidence, radiologists around the US are crying foul, since one of their primary "cash-cows" is being called into question. It seems that these physicians demonstrate a bias toward their opinion and judgment over a large, long-term rigorous study. Of course, in addition to interpretation bias, this also reveals a potential financial conflict of interest that may cloud sound judgment. While bias and financial conflict of interest are inappropriate scientific practices, this is merely one of many examples that show expert bias among individuals who were paid as consultants to voice opinions on drugs made by the very companies who paid them.

6.6 Approval of Drugs Based on Only Two Positive Trials

Finally, the current approval process by the FDA requires only two studies that show positive data in support of a drug, irrespective of how many studies were done in total, and may not show any effect at all. Indeed, in the case of antidepressants, over 80 studies failed to show any effect of SSRIs compared with placebo. However, two studies showed statistically significant improvements that were clinically meaningless. SSRIs were approved based on these two studies, justifying >$8 billion/year in sales for a class of drugs that is as clinically effective as a sugar pill.

The clinical researcher who is perhaps most critical of the current state of clinical trials is John P. A. Ioannidis, an epidemiologist at Stanford who has been examining the extent to which even large-scale clinical trials are often misinterpreted. He suggests that 90% of all published information that doctors rely on in their clinics is flawed.[15] He analyzed randomized controlled clinical trials, the holy grail of clinical studies, and found that even those cannot be trusted. He argues that just as scientists are biased toward the outcome of their studies, so are the clinicians running these trials. Moreover, as stated above for SSRIs, if only 2 of 87 studies show a positive result, only those 2 that show positive results will be published. Hence, a publication bias is perpetuated, resulting in the medical community never learning the truth about the other 85 "negative" studies. Ioannidis examined 49 of the most highly regarded research findings in medicine, including trials of coronary stents for heart disease, low-dose aspirin to control blood pressure, and hormone replacement therapy for menopause. A retest of the findings from these double blind, placebo-controlled studies showed that a striking 41% were unequivocally wrong, yet many doctors remain unaware of this inaccuracy.

7. PERSONALIZED MEDICINE

Often defined as "the right treatment (drug) for the right patient at the right time in its most effective form (concentration)" *personalized medicine* has become a buzzword in science and medicine and has generated a lot of excitement.[16] Let us begin by acknowledging that many treatments are already personalized to some extent. We each may have a certain drug dose combination that works best for us. For example, a patient with epilepsy may take multiple drugs in combination that uniquely helps their condition yet may be ineffective for others. However, personalized medicine seeks an a priori definition of an effective treatment by taking into consideration a patient's genetic makeup or other biomarkers that would predict the usefulness or applicability of a given drug or intervention. In its most extreme form, personalized medicine encompasses harvesting a patient's skin cells, reprogramming them into induced pluripotent stem cells, introducing a missing gene, such as dopamine decarboxylase (which encodes for an enzyme critical in the production of dopamine and serotonin), and then implanting these cells into the midbrain of a patient with PD. Increasingly, these approaches will become possible, and promising examples are highlighted throughout this book. The greatest challenge, however, may be validating their success through clinical trials and the fiscal challenges of delivering personalized medicine broadly.

Let us take the temozolamide (TMZ) treatment of gliomas as an example. TMZ is effective only in the 20% of all glioma patients who carry a methylated (inactivated) DNA repair enzyme; this enzyme would otherwise mitigate the effect of TMZ. A simple polymerase chain reaction can determine whether a patient stands to benefit from TMZ treatment and could spare 80% of the population the side effects of TMZ treatment. However, this test is currently not a common clinical practice. Is it because of the time required for and cost of the test? Is it pressure by the manufacturer, who would suffer an 80% reduction in sales of TMZ? Is it a lack of insistence by the insurer, who could save 80% of the unnecessary cost? Maybe it is the absence of more effective treatments, which one may argue retains hope for all to be among the 20% who stand to benefit from treatment. It is a dilemma—one that we will be facing with many other personalized medicines as well.

An even greater challenge will be mass production of effective drugs for very small markets. As we have learned, the average cost of bringing a new drug to market is $1.2 billion. Hence the manufacturer has to recoup this investment. If the cost of this investment is spread over a smaller number of treatments, each will become significantly more expensive, as we are already seeing in the cost of biologicals used to treat multiple sclerosis. One possible solution to this dilemma would entail drug companies selling their drug together with a test kit, where they stand to profit not only from the drug sale but also from screening for who will benefit and who will not.

8. CHALLENGES AND OPPORTUNITIES

Obviously, this chapter has raised many issues, ranging from the inadequacies of drug discovery to the failure of clinical trials and the unexpected prevalence of the placebo effect. It is important to stress that clinical experimentation has come a long way and is now almost consistently yielding evidence-based medicine. However, this comes with the unpleasant discovery that many drugs that were believed to be biologically active are in fact no better than a sugar pill. We need to critically assess how

we deal with such findings; currently, the most widely prescribed class of drugs, second generation antidepressants, are a prime example of drugs that could, perhaps in many cases, be effectively replaced by a sugar pill or green tea. The buzz around personalized medicine get louder by the day, yet our understanding of what it is, what it promises, and how the promises can be delivered remains less clear. Personalized medicine has academic appeal, but whether generating drugs for small groups of people will be economically feasible remains to be shown. In the end, health care costs paid through insurance is collectively borne by the entire insured population. Thus financial considerations also feed into the expanding use of biologicals. These biologicals are currently the hottest trend, yet I would call it unsustainable. As for multiple sclerosis, the typical drug cost is about $60,000 per year, exceeding the average annual family income. We simply will not be able to treat every patient with such expensive drugs, and these drugs can therefore be viewed as transitional drugs used only until we can create more viable solutions. Indeed, one must pursue viable drugs for general use; otherwise, we are on the verge of creating a multitiered health system in which only those with sufficient resources can purchase treatments for disease. This would widen our already troublesome health disparity.

One way to reduce the cost of drugs is to increase the success rate along the entire development pathway. We must ensure that drugs and drug targets are honestly reported and that negative findings receive the same prominence as positive ones. We should insist that all clinical studies be published in peer-reviewed journals, regardless of the outcome, so that physicians can accurately weigh the evidence in favor of or against the use of a new drug or procedure. Clearly, a culture change away from publication bias and back to truthful and accurate reporting of both basic and clinical findings is essential.

Finally, why not reconsider a placebo as a viable alternative treatment? I am not advocating for deceiving a patient; instead, I suggest including this option as a true alternative in the consultation of patients by their doctors.

References

1. Eyupoglu IY, Buchfelder M, Savaskan NE. Surgical resection of malignant gliomas-role in optimizing patient outcome. *Nat Rev Neurol*. 2013;9:141–151.
2. Bhatt A. Evolution of clinical research: a history before and beyond James Lind. *Perspect Clin Res*. 2010;1:6–10.
3. Reagan-Shaw S, Nihal M, Ahmad N. Dose translation from animal to human studies revisited. *FASEB J*. 2008;22:659–661.
4. Sakpal TV. Sample size estimation in clinical trial. *Perspect Clin Res*. 2010;1:67–69.
5. Markman JR, Markman M. Running an ethical trial 60 years after the Nuremberg Code. *Lancet Oncol*. 2007;8:1139–1146.
6. de Craen AJ, Kaptchuk TJ, Tijssen JG, Kleijnen J. Placebos and placebo effects in medicine: historical overview. *J R Soc Med*. 1999;92:511–515.
7. Beecher HK. The powerful placebo. *J Am Med Assoc*. 1955;159:1602–1606.
8. Eippert F, Finsterbusch J, Bingel U, Buchel C. Direct evidence for spinal cord involvement in placebo analgesia. *Science*. 2009;326:404.
9. Mayberg HS, Silva JA, Brannan SK, et al. The functional neuroanatomy of the placebo effect. *Am J Psychiatry*. 2002;159:728–737.
10. Nieuwenhuis S, Forstmann BU, Wagenmakers EJ. Erroneous analyses of interactions in neuroscience: a problem of significance. *Nat Neurosci*. 2011;14:1105–1107.
11. Prinz F, Schlange T, Asadullah K. Believe it or not: how much can we rely on published data on potential drug targets? *Nat Rev Drug Discov*. 2011;10:712.
12. Sterling TD. Publication decision and the possible effcets of inferences drawn from test of significance—or vice versa. *J Am Stat Assoc*. 1959;54:4.
13. Kirsch I, Deacon BJ, Huedo-Medina TB, et al. Initial severity and antidepressant benefits: a meta-analysis of data submitted to the Food and Drug Administration. *PLoS Med*. 2008;5:e45.
14. Miller AB, Wall C, Baines CJ, et al. Twenty five year follow-up for breast cancer incidence and mortality of the Canadian National Breast Screening Study: randomised screening trial. *BMJ (Clin Res Ed)*. 2014;348:g366.
15. Ioannidis JP. Why most published research findings are false. *PLoS Med*. 2005;2:e124.
16. Hamburg MA, Collins FS. The path to personalized medicine. *N Engl J Med*. 2010;363:301–304.

General Readings Used as Source

1. *Testing Treatments, Better Research for Better Healthcare*. The British Library; 2010, ISBN: 978-1-905177-35-6. by Pinter & Martin Ltd.
2. Kirsch Irving. *The Emperor's New Drugs: Exploring the Antidepressant Myth*. Basic Books; 2011, ISBN: 0465022006.
3. Karsh LI. A clinical trial primer: historical perspective and modern implementation. *Urol Oncol*. July–August 2012;30(suppl 4):S28–S32.
4. Ezekiel Emanuel MD, Emily Abdoler BA, Leanne Stunkel BA. *Research Ethics*. The National Institutes of Health.

Suggested Papers or Journal Club Assignments

1. Eippert F, Finsterbusch J, Bingel U, Buchel C. Direct evidence for spinal cord involvement in placebo analgesia. *Science*. October 16, 2009;326(5951):404.
2. Mayberg HS, Silva JA, Brannan SK, et al. The functional neuroanatomy of the placebo effect. *Am J Psychiatry*. May 2002;159(5):728–737.
3. Nieuwenhuis S, Forstmann BU, Wagenmakers EJ. Erroneous analyses of interactions in neuroscience: a problem of significance. *Nat Neurosci*. September 2011;14(9):1105–1107.
4. Ioannidis JP. Why most published research findings are false. *PLoS Med*. August 2005;2(8):e124.

PART VIII

NEUROSCIENCE JARGON

"Neuro"-dictionary

Harald Sontheimer

***

Absence seizures Absence seizures are short-lasting seizure episodes, during which an affected individual shows a complete absence of attention and sensation and appears to be staring into space, incognizant of his or her surroundings. The patient regains consciousness just as quickly as it was lost, and is often unaware of any preceding lapse in consciousness. An absence seizure is, at best, associated with very subtle motor signs, such as rapid eye blinking or chewing movements of the mouth. However, electroencephalography typically shows abnormal synchronous 3 Hz spike and wave activity affecting both hemispheres, suggesting that it is a generalized seizure. While they may sound relatively mild in their severity, patients with uncontrolled absence seizures may have several hundreds of these episodes per day, leading to extreme difficulty in functioning at school, work, etc.

acetylation The addition of an acetyl group (COCH^3) to a protein is called acetylation. This is most typically catalyzed by enzymes such as acetyltransferases, whereas the removal of acetyl groups is called deacetylation and is catalyzed by a group of enzymes called deacetylases. Acetylation changes the biological activity of many proteins, including histones and tubulins.

acquired immune deficiency syndrome (AIDS) AIDS is a disease caused by the human immunodeficiency virus (HIV), a retrovirus that infects cells of the body's immune system. AIDS manifests in advanced stages of untreated HIV infection. The infection of T lymphocytes by the HIV virus causes uncontrolled production of virus and loss of T lymphocytes, leaving the body unable to defend itself against pathogens. Over a period of a few weeks the patient develops fever, headache, and swollen throat and lymph nodes, as well as severe muscle pain and fatigue.

actin–myosin Actin and myosin are molecular motor proteins that use energy in the form of ATP and convert it to enable muscle contraction and cell movement.

action potential The principal electrical signal that is used for communication throughout the nervous system is called an action potential or spike. It is an all-or-none signal and typically has a constant amplitude and very short duration. It is generated near the neuronal cell body at the axon hillock and travels along the axon toward the synapse, where it causes the release of neurotransmitters. The upstroke of the action potential is mediated by rapid influx of Na$^+$ ions transiently depolarizing the membrane to +50 mV. The downstroke is due to the closure of Na$^+$ channels in concert with an efflux of K$^+$ that repolarizes the membrane back to its resting membrane potential of ~ −70 mV.

acute meningoencephalitis or primary amoebic meningoencephalitis (PAM) PAM is a rare and fatal brain disorder that is caused by the amoeboid Naegleria fowleri. This single-cell organism is commonly found in bodies of warm fresh water such as lakes or canals in the southern United States. The amoebae enter the brain via the nasal passage and, once resident in the brain, digest brain tissue. The inflammation of the meninges and brain tissue causes severe headaches, even coma. Patients typically die within 2 weeks of becoming symptomatic.

adaptive transfer Harvesting immune cells from a donor to a recipient with the goal of transferring the immune function of the donor is called adaptive transfer. In multiple sclerosis, the transfer of T lymphocytes from an affected animal to a normal control causes the recipient to develop disease, allowing for the creation of an animal model of disease.

adeno-associated virus (AAV) Adeno-associated virus (AAV) is a small virus that infects humans. It relies on co-infection with a helper virus in order to replicate and is therefore nonpathogenic and safe. AAV binds to heparin sulfate expressed on the surface of many nervous system cells and can integrate its DNA into a specific site on chromosome 19, thus permitting insertion of a foreign gene into the host DNA. AAV is used in ongoing clinical trials treating Parkinson disease, ALS, and epilepsy.

adrenaline (epinephrine) Adrenaline (epinephrine) is a monoamine neurotransmitter made by the adrenal gland and is often called the fight-or-flight hormone. It binds to a variety of alpha or beta adrenergic receptors expressed by sympathetic neurons.

affect Affect is a synonymous term for feeling or emotion.

aging Aging is the sum of all the biological, physical, and psychological changes that occur in an organism over its lifetime. Aging is a chronological process that begins at birth and ends with death.

allosteric modulators Allosteric modulators are molecules that indirectly alter the activity of a receptor through sites that are different from the ligand binding site. Examples are the glycine or serine binding site on the NMDA glutamate receptor. These differ from orthostatic modulators that bind to an agonist or antagonist binding site.

alpha-synuclein (α-synuclein) Alpha-synuclein is a protein abundantly expressed in neurons. When misfolded, alpha-synuclein proteins form aggregates called Lewy bodies, which are a pathological hallmark for Parkinson disease.

Alzheimer disease (AD) Alzheimer disease is the most common neurodegenerative disease in humans. It presents late in life with insidious forgetfulness and cognitive decline, and typically causes death within 10 years of diagnosis. Familial forms of disease with early onset are associated with mutations in genes involved in the processing of beta amyloid. Pathological hallmarks of Alzheimer disease include protein aggregates inside and outside of cells; these are called tangles and plaques, respectively.

AMPA α-*Amino-3-hydroxy-5-methyl-4-isoxazolepropionic acid* is a selective agonist of a subtype of glutamate receptor called the AMPA receptor and the chemical mimics the effect of glutamate, which is the biological agonists. AMPA receptors are responsible for the majority of fast neurotransmission in the central nervous system. It contrasts to the NMDA receptor that is primarily involved in learning and memory.

amplified developmental instability Amplified developmental instability is a scientific hypothesis that seeks to explain the biology of Down syndrome. It argues that the extra copy of chromosome 21 (trisomy 21) causes an imbalance in gene expression, with some genes being upregulated due to an extra gene copy and others being downregulated if the extra gene copy is a transcriptional suppressor. Hence, the exact outcome of this trisomy is unpredictable but imbalanced.

amyloid plaques The amyloid plaque is an extracellular protein aggregate of insoluble beta amyloid. Beta amyloid is produced from a large amyloid precursor protein (APP), which is cleaved by enzymes called secretases. If cleavage occurs at the wrong site, there is a greater abundance of amyloid peptides that have 42 amino acids, called amyloid-beta-42. These are sticky and form aggregates called amyloid plaques. Amyloid plaques are the pathological hallmark of Alzheimer disease, and are typically only visible if the brain is autopsied. However, with a radiolabeled contrast medium, it is now possible to use PET imaging to identify the presence of amyloid plaques long before a patient is symptomatic for Alzheimer disease. Whether plaques cause disease or are the consequence of disease remains controversial.

amyloid positron emission tomography (PET) Amyloid is a polypeptide of approximately 38–42 amino acids that accumulates as insoluble protein aggregates called plaques in patients with Alzheimer disease. Plaque-associated amyloid has a beta-sheet configuration, which binds a group of chemicals called thioflavins. Thioflavin T can be conjugated with a radioactive isotope such as 18-Fluor to create an imaging contrast agent that can detect the presence of amyloid plaques in patients long before the symptoms of Alzheimer disease are present. The contrast molecule is sometime referred to as Pittsburgh Compound B or PiB, because it was developed at the University of Pittsburgh. It is detected in patients through positron emission tomography (PET). With this approach, it is now possible to diagnose Alzheimer disease up to 10 years prior to disease onset. Its use, however, is controversial because currently no treatments exist, making it difficult for patients and their families to respond to a positive diagnosis. Therefore, the use of amyloid-PET is still largely restricted to research purposes.

amyloid precursor protein (APP) The APP is a large-membrane glycoprotein of unknown function. It is cleaved by secretases into soluble fragments of beta-amyloid ranging in size from 36 to 43 amino acids. Soluble beta amyloid is believed to have a role in synaptic transmission. Too much of the amyloid 42 variant, which is sticky, causes their aggregation as amyloid plaques.

amyotrophic lateral sclerosis (ALS) ALS is a progressive disease that presents with muscle weakness due to the selective loss of motor neurons in the brain and spinal cord. ALS spares sensory nerves but eventually affects essentially all muscle groups of the body. Death occurs within 1–2 years as muscles involved in breathing fail.

aneurism A congenital malformation of a blood vessel that makes it more prone to rupture. Aneurisms can be enlarged, balloon shaped and often have a thinner more fragile vessel wall.

angiogenesis A shared mechanism of most tumors is the ability to grow new blood vessels by sprouting branches from existing vessels, known as angiogenesis, to supply sufficient energy to a growing tumor.

anhedonia Anhedonia, or an inability to experience pleasure in otherwise enjoyable activities, is a cardinal feature of depression.

anterograde amnesia Amnesia is the inability to remember things, or a loss of memory. This can affect all memory, called global amnesia. When memory loss selectively affects old memories related to past events, it is called retrograde ("backwards") amnesia. When amnesia selectively affects new memories that are about to be formed, it is called anterograde ("forward-looking") amnesia. The famous case of H.M., a patient who had both of his hippocampi removed due to intractable epilepsy, serves as an excellent example of anterograde amnesia. For H.M., every day started anew, as he was completely unable to form new memories of people or events that would last to the next day. Yet, H.M. still remembered his childhood and everything he experienced prior to his surgery. He also maintained procedural memory (the ability to remember the necessary procedure to complete a task), although he had no memory of learning these new procedures. The case of H.M. taught us the importance of the hippocampus as a gateway for the formation of new memories.

antibodies Antibodies, also known as immunoglobulins, are reactive, Y-shaped molecules that are made by immune cells to recognize, bind, and destroy foreign substances, viruses, or bacteria.

antigen An antigen is any foreign substance that can stimulate immune cells to generate antibodies and trigger an immune response.

antisense oligonucleotides (ASOs) These are single-stranded short strings of 8–50 nucleic acids that bind to target RNA by traditional Watson–Crick base pairing. ASOs can be engineered for high stability, and are very effective in disrupting protein synthesis through degradation of the encoding RNAs. The end result is a gradual depletion of the targeted protein from the cell.

aphasia Aphasia derives from the Greek language and means speechless. The term is used scientifically to describe the difficulty in understanding and using language that is a consequence of brain dysfunction, such as a stroke, rather than an inability to move the mouth, larynx, and tongue.

apolipoprotein E (APOE4) Apolipoproteins are lipid-binding proteins that are synthesized in the liver but can also be produced in the brain by astrocytes and microglial cells. Apolipoproteins transport lipids such as cholesterol throughout the body, and also make it possible to transport lipids through water-based fluids, such as lymph and blood. Of the many different apolipoproteins, apolipoprotein E (APOE) has emerged as a risk factor for Alzheimer disease. More specifically, the e4 allele of the APOE4, found in 14% of the population, predicts a 4- to 15-fold heightened risk for an individual to develop Alzheimer disease.

apoptosis Also known as programmed cell death, apoptosis involves the activation of caspases, enzymes that degrade cellular proteins. Hallmarks of apoptosis include condensation of the cell cytoplasm, condensation of chromatin and breakdown of the nuclear envelope, and eventually the breakdown of cells into smaller vesicles. Unlike necrotic cell death, there is no spillage of cellular constituents into the surrounding tissue.

arborization Arborization is the branching of neuronal processes into finer and finer branches, creating a tree-like shape or arbor.

areflexia The complete absence of a reflex is called areflexia. It is often a sign of nerve damage.

Asperger syndrome A condition first described by Hans Asperger, Asperger syndrome, is characterized by a profound disinterest in people, fixation on objects, repetitive behavior, and delays in language. Asperger syndrome shows variable cardinal features of autism, and thus is on the autism spectrum. Because individuals with Asperger syndrome have normal or above-normal IQ, it is considered a high-functioning variety of autism.

astrocytes Astrocytes are the most common brain support cell in the family of glial cells, which also includes oligodendrocytes, Schwann cells, and microglia. They are star-shaped (astro = star) cells that serve many support functions, most notably the sequestration of neuronally released neurotransmitters and potassium. They extend processes onto the vasculature, through which they influence blood flow and integrity of the blood–brain barrier. Astrocytes are often identified by their expression of glial fibrillary acidic protein (GFAP) as part of their cytoskeleton.

astrocytoma An astrocytoma is a primary brain tumor believed to derive from astrocytes or glial progenitor cells. These are typically WHO grade I–III and stain variably for astrocytic markers such as GFAP.

ataxia Ataxia is an unsteady, imbalanced movement, typically resulting from poor coordination and lack of balance.

atherosclerosis Atherosclerosis is the hardening of blood vessels through deposition of plaque on the inside of the vessels. The plaque is composed of cholesterol, blood-borne proteins, lymphocytes, and macrophages. It reduces the blood flow, placing surrounding tissue at risk for a deficit in nutrients and gas supply. Plaque can grow to a thrombus that occludes the vessel, or break off in pieces called emboli that can occlude smaller vessels downstream. Atherosclerosis is promoted by low-density lipoprotein (LDL), also called bad cholesterol, making LDL a risk factor for heart attack and stroke.

atrioventricular septal defect Atrioventricular septal defect is a congenital heart defect that causes mixing of oxygenated and nonoxygenated blood. It is a common condition in children with Down syndrome and can be surgically corrected.

atypical (new) antipsychotics Atypical antipsychotic is a term used to describe a group of new drugs that affect multiple neurotransmitter receptors, including dopamine, serotonin, norepinephrine, and histamine.

autism Autism is a behavioral abnormality first described by Leo Kanner in 1943. The cardinal features of autism include profound disinterest in people, focus on objects, repetitive behavior, and often a delay in language. In addition, many autistic individuals have below-normal IQ. The presentation of autism is broad, suggesting a spectrum of behavioral abnormalities that is now called the autism spectrum.

autocrine or paracrine feedback loop An autocrine feedback loop is formed by signals that are released by a cell and can act on its own receptor. If these signals can act on adjacent cells, the loop is called paracrine feedback.

autophagosome system The autophagosome is a large vesicle with a double membrane that encloses large protein complexes or invading microorganisms. The autophagosome then fuses with a lysosome that contains proteolytic enzymes for the degradation of its content.

axolemma The axolemma is the membrane that surrounds an axon and contains the ion channels through which axon potentials are generated.

axon initial segment (AIS) The AIS is a part of the axon close to the cell body at which a high density of Na^+ channels ensures the generation of an action potential. The action potential then travels along the myelinated axon.

axonal transport Axonal transport is a cellular transport system that is responsible for moving molecules and organelles synthesized by the neuronal cell body toward the synapses, and also moves synaptic molecules back to the cell body. The axonal transport uses microtubules as a rail system along which motor proteins such as dynein and kinesin propel bound cargo in an energy-dependent fashion along the microtubules.

β-amyloid (Ab) Beta amyloid is a cleavage product of the APP, which has been cleaved by beta secretase, giving rise to a 36–42 amino acid peptide. Of these peptides, the beta-amyloid 42 is primarily responsible for the amyloid plaques that are a pathological hallmark for Alzheimer disease.

B lymphocytes B cells or B lymphocytes are immune cells that are made in the bone marrow. They are the principal cells involved in antibody production to mount a humoral immune response.

Babinski sign A neurological readout that allows a physician to determine whether a patient with muscle weakness has impaired function in the upper (cortical) or lower (spinal cord) motor neurons. Stroking the bottom of the foot normally causes the toes to curl inward. If the toes curl upwards this is abnormal and is called a Babinski sign (after the neurologist who first described this reflex). It suggests a lesion of the upper motor neurons in either the cortical or corticospinal tract. A lower motor neuron lesion would cause a failure to elicit any response at all.

basal ganglia A collection of nuclei at the interface of the forebrain, posterior forebrain, and midbrain that contain neurons involved in the fine-tuning of voluntary movements. The basal ganglia include the striatum, the subthalamic nucleus, and the susbtantia nigra. The neurons of the basal ganglia are mostly dopaminergic.

basal nucleus of Meynert This nucleus is also called the nucleus basalis. Like all brain nuclei, these are congregations of neuronal cell bodies. The nucleus basalis is located anterior to the midbrain in the basal forebrain. It contains cholinergic (acetylcholine-releasing) neurons that project to the cortex. The nucleus degenerates in Alzheimer disease, and this is believed to contribute to the cognitive decline. Pharmacological treatment of dementia seeks to counteract the deficit in cholinergic function resulting from the degeneration of the nucleus basalis.

basket cells Basket cells are a certain type of interneuron that is defined by its basket-shaped dendritic tree. Basket cells are typically inhibitory and use the neurotransmitter GABA.

Basso, Beattie, Bresnahan (BBB) scale The BBB scoring system is widely used in laboratories to rate the motor disability of an animal following injury. The scale ranges from 0 = no movement to 21 = normal movements.

beta-glucocerebrosidase (GBA) GBA is a housekeeping enzyme found in lysosomes, the organelles involved in cellular recycling of molecules. GBA breaks down large glycolipid molecules such as glucocerebrosides into sugar and fat. Mutations in GBA causes Gaucher disease, characterized by enlarged inner organs.

bias Bias is the same as prejudice. In science, it refers to an investigator favoring one outcome or interpretation over another without supporting evidence. The reasons an individual may be biased are multitudinous. Being conflicted by potential financial gains or increase in status are among them. Bias in clinical studies may affect the selection of patients to be lopsided, so as to favor a positive outcome.

biomarkers Any measurable change in the biology that correlates highly with an altered biological state, for example, a disease, is called a biomarker. The availability of a reliable biomarker makes diagnosis independent of human judgment and error.

bipolar cells Bipolar cell refers to a cell type in the retina that conveys the light signal between photoreceptors and the retinal ganglion cells that form the optic nerve.

bipolar disorder Bipolar disorder is a form of depression that is characterized by unexplained mood swings from euphoria, often called mania, to depression. During the manic state, patients are hyperactive, have lots of enthusiasm and great expectations, yet typically fail to carry out their plans.

blast injury A blast injury is a type of injury that results from the pressure waves and gradients caused by an explosive device.

blood–brain barrier (BBB) The BBB is a physical and enzymatic barrier that separates circulating blood from the cerebrospinal fluid bathing the brain. The BBB is established by tight junction proteins that glue adjacent endothelial cells together, forming an impermeable vessel wall. Hence, any molecules that need to be exchanged between blood and brain or vice versa need to be actively transported across the endothelial cell, providing a tightly regulated environment for the brain to protect it against toxins and other damaging factors.

blood oxygen level dependent functional magnetic resonance imaging (BOLD fMRI) As oxygen is released from hemoglobin in the blood into bodily tissues, the ratio of oxyhemoglobin (oxygen-bound) and deoxyhemoglobin (not oxygen-bound) decreases. These changes can be detected by magnetic resonance imaging. Since oxygen use correlates with neural function, an increase in oxygen use can be interpreted as an increase in neuronal activity. This is the principal readout used in functional magnetic resonance imaging (fMRI), commonly used to interrogate brain function.

botulism (Botox) Botulism is a disease caused by botulinum toxin (Botox), a highly toxic poison produced by the *clostridium* bacteria found abundantly in the environment. Under favorable anaerobic and acidic conditions, bacterial spores germinate and produce bacteria that release toxin into food; less commonly, spores may be found in wounds and produce toxin at the site. Botulinum toxin specifically binds to molecules involved in the release of acetylcholine in the muscle synapse, thereby causing long-lasting paralysis.

bradykinesia A slowness or poverty of movement, or of initiating movement, bradykinesia is a cardinal feature of Parkinson disease and is caused by a loss of dopamine in the midbrain. Bradykinesia is a side effect of antipsychotic drugs that antagonize dopamine receptors, as these can also affect the dopaminergic motor control pathways in the midbrain.

brain-derived neurotrophic factor (BDNF) An important growth factor produced by many neurons, BDNF is produced as a preform of the final protein that is released from neurons in an activity-dependent fashion. BDNF can be transported via the axonal transport over long distances and plays a major role in brain development, sustained neuronal health, learning, and memory.

brain-eating amoeba The amoeboid Naegleria fowleri is a single-cell organism commonly found in warm lakes or canals in the southern United States. In very rare instances, the amoebae enter the brain through the nasal passages. Once in the brain, they digest brain tissue.

c9orf72–chromosome 9p21 locus The intronic GGGGCC repeat expansion on chromosome 9 is present in up to 50% of familial ALS cases and 20% of sporadic ALS. How this chromosomal change contributes to disease is not known.

calcineurin A Ca^{2+}- dependent protein phosphatase that activates T cells, calcineurin interacts with NMDA glutamate receptors and GABA receptors in the nervous system. Calcineurin deficiency or loss of function has been implicated in schizophrenia.

Capgras syndrome Capgras syndrome is named after a French neurologist who first described this delusional disorder in which an affected individual insists that a spouse, friend, or relative is in fact an impostor. The syndrome can occur in many different diseases, including schizophrenia, neurodegenerative diseases, and brain trauma.

carbamazepine (Tegretol) Tegretol is a commonly used antiepileptic drug and mood stabilizer. It enhances GABAergic inhibition by potentiating the current flow through GABA-A receptors. It also reduces neuronal excitability by keeping Na$^+$ channels in an inactivated state.

carbidopa Carbidopa is a drug that inhibits DOPA-decarboxylases, the enzymes that convert the L-DOPA precursor to dopamine. It is used to treat Parkinson disease and prevent the premature conversion of the dopamine precursor L-DOPA (levodopa) to dopamine in the body. Carbidopa thereby allows the L-DOPA to reach the brain, since only here is its conversion to dopamine able to improve motor coordination in patients with Parkinson disease. Carbidopa is therefore administered together with levodopa as a combination therapy.

Carl Wernicke Carl Wernicke was a German physician who studied brain disorders that cause deficits in speech and language. He discovered a brain region on the left hemisphere near the posterior temporal gyrus that is required for language comprehension. This area is now called Wernicke's area. It is independent from Broca's area, which is located in the left inferior frontal gyrus and is required for articulation of language.

caspases Caspases are a family of cysteine proteases that cleave cellular proteins, resulting in the slow cellular disassembly known as apoptotic cell death.

catechol-O-methyltransferase (COMT) COMT is an enzyme responsible for the degradation of catecholamines such as dopamine and norepinephrine.

caudate nucleus The caudate nucleus is one of the three midbrain nuclei that forms the basal ganglia, along with the putamen and globus pallidus. The caudate nucleus is involved in the fine-tuning of voluntary movements.

central nervous system (CNS) The central nervous system encompasses the brain and spinal cord, which are separated from the peripheral nervous system (PNS) by a triple layer of connective tissue, the meninges. The CNS is entirely encased by the bone of the skull and spinal column and has its own fluid system, the cerebrospinal fluid.

cerebral vasculature The network of interconnected blood vessels that supply the brain with blood. Fine branches reach to within 50 μm of each neuron, assuring sustained delivery of oxygen and glucose.

cerebrospinal fluid **(CSF)** The fluid that bathes the brain and spinal cord, cerebrospinal fluid, is a cell-free ultra-filtrate of blood. CSF serves to supply nutrients, to eliminate waste products, and to cushion the brain.

cervical spine The vertebrae of the human spinal column are labeled as cervical, thoracic, lumbar, and sacral, and the most anterior segment of these is the cervical spine.

chandelier cell The chandelier cells are a type of fast-spiking GABAergic interneuron in the cortex. They express the Ca^{2+}-binding protein parvalbumin.

channelopathy Diseases or symptoms that can be attributed to changes in the expression of function of ion channels are called channelopathies. Most typically, congenital mutations in ion channel proteins alter their function, thereby impairing normal neural activity. Epilepsy is an excellent disease example, where certain forms of epilepsy can be unequivocally linked to mutations in Na^+, K^+, or Ca^{2+} channels. Another excellent example is the lung disorder cystic fibrosis, where mutations in Cl^- channels cause abnormal buildup of mucus. Instead of genetic alteration, an immune attack on ion channels may result in a channelopathy, as is the case in Lambert Eaton myotonia, where a presynaptic Ca^{2+} channel is attacked by autoantibodies, or in myasthenia gravis, where the acetylcholine-gated receptor channels in the muscle endplate are the targets of immune cells.

chlorpromazine The first drug used as a specific antipsychotic in humans, chlorpromazine, is a dopamine receptor antagonist. For unknown reasons, it is also the most effective drug to treat infections with the brain-eating amoeba Naegleria fowleri.

choline acetyltransferase (ChAT) Choline acetyltransferase is the enzyme responsible for the synthesis of the neurotransmitter acetylcholine.

chorea A dance-like movement that involves the entire body, chorea is a characteristic early motor sign in individuals with Huntington disease.

choroid plexus The choroid plexus is the membrane structure at the roof of the third and fourth ventricle. It produces the cerebrospinal fluid that bathes the brain and spinal cord.

chronic social defeat In the chronic social defeat paradigm, pairing a docile animal with a more aggressive animal will eventually result in the docile animal showing defeat, unable to make any attempt to escape the abuse by the bully. This results in depression-like phenotypes, and can be useful for animal modeling of depression.

chronic traumatic encephalopathy (CTE) Also called punch-drunk syndrome, CTE is a severe chronic inflammation of the brain as a result of repeated injuries. Patients may show signs of dementia or cognitive impairments at an early age.

cingulum The cingulum is a white matter fiber pathway that connects the prefrontal cortex, premotor cortex, associate cortex, parahippocampus, and thalamus.

clinical trials (preclinical evaluation and phases) Clinical trials are experiments with volunteer human participants that seek to study the safety and efficacy of new or existing drugs, procedures, or interventions in a controlled and supervised way. Clinical trials are conducted in phases that have distinctly different objectives. Pilot studies (phase 0) study drug administration and feasibility of procedures. Phase I studies evaluate safety and escalate drug dose to find limits to safe drug dosing. Phase II studies seek efficacy and the most efficacious dose of use for an intervention. Phase III studies expand to larger patient populations examining the drug in a blinded fashion, ideally compared with an inert placebo. Phase IV studies are postmarketing surveys to examine drug interactions in a given disease indication, and may provide evidence that a drug should not be used for persons of a certain group. For example, a pain drug may have contraindications for patients who had suffered a prior heart attack.

closed- or open-head injuries An injury to the head that penetrates the skull is called an open-head injury; one that leaves the skull intact is called a closed-head injury.

Clostridium tetani A bacterium of the genus *Clostridium*, *Clostridium tetani* produces tetanus toxin, a poisonous substance similar to botulinum toxin. This tetanus toxin can cause spastic paralysis or tetanus. Bacterial spores enter an open wound where, under appropriate conditions, they germinate, and the resulting bacteria produce tetanus toxin. The toxin is taken up by motor terminals and transported via the axonal transport to the motor nuclei of the brain. From there, the toxin is released and crosses to the adjacent presynaptic GABAergic terminals, where the toxin binds to synaptic proteins involved in transmitter release. Blockage of the release of inhibitory neurotransmitter causes tetanic muscle contractions and spastic paralysis. Note that botulinum toxin, by contrast, causes flaccid (limp) paralysis.

coenzyme Q10 The Q10 coenzyme functions as an electron carrier and antioxidant in the mitochondria. Supplementation of the diet with Q10 is being pursued in a number of diseases in which mitochondrial dysfunction is suspected.

cognitive behavioral therapy CBT is a summary term for certain nonmedicinal psychotherapies to alter mood and depression, also often called talk therapy.

compression Compression occurs when force is applied to the surface of a cell or organ, thereby exerting pressure on its constituents without compromising its integrity. Due to their high water content, brain cells are incompressible, and thus force applied to the brain typically results in an increase in intracranial pressure and cerebrospinal fluid being forced from the brain.

computed tomography (CAT) A CAT scan is an X-ray-based imaging technique that produces a three-dimensional image by taking serial X-rays. It is readily available in most medical centers and relatively inexpensive. It is preferred over an MRI to detect blood in the brain due to a hemorrhage.

constraint-induced therapy Constraint-induced therapy is a physical therapy approach used primarily to treat hemiplegia or one-sided impairment in muscle control. The therapy places the unaffected good arm or leg in a temporary cast, forcing an individual to make increased use of the affected or injured extremity. This approach has shown promise in children with cerebral palsy and adults who have suffered a stroke.

contre-coup Brain injury often causes a bruise or contusion on the opposite site of the impact. This injury is called the contre-coup, while the site of the impact is called the coup.

convulsion The repeated contraction and relaxation of muscle groups lead to an uncontrolled shaking of the body. Convulsions are a common behavioral symptom associated with certain forms of epileptic seizures, yet not all seizures present with convulsions.

cortical-mesial-temporal lobe (MTL) network The mesial-temporal lobe is an anatomical part of the cortex located in the inner aspect of the temporal lobe. It encompasses the hippocampus and para-hippocampus, and is often involved in epilepsy.

corticobulbar tract The corticobulbar tract carries information from cortical neurons to the brain stem motor nuclei of the cranial nerves that innervate "bulbar" muscles of the face and neck; these muscles are involved in facial expressions, chewing, speaking, and swallowing.

corticospinal tract The corticospinal tract is a nerve tract formed by the axons of cortical motor neurons projecting to motor neurons in the spinal cord. These neurons are called upper motor neurons, and this tract is involved in voluntary control of movement.

corticosteroids Corticosteroids include synthetic steroid hormones similar to those normally produced by the adrenal gland. These drugs are used broadly to stabilize membranes, tame inflammation, and reduce edema. Dexamethasone is one example frequently prescribed for conditions ranging from joint inflammation to brain tumors. These drugs provide short-term benefit, yet lack disease-specific mechanisms of action.

corticosterone Corticosterone is the stress hormone released from the adrenal gland in rodents. It is equivalent to cortisol in humans.

cortisol Cortisol is the stress hormone made in the adrenal gland of primates that puts our body into a fight-or-flight mode.

coup Brain injury at the site of impact with an object is called coup; injury on the opposite site is called contre-coup. Both injury sites present with contusion that is typically visible by CT or MRI scan. Injury may be the result of a moving object hitting the stationary head or vice versa.

CpG islands DNA regions in which cytosine nucleotides are found next to guanine nucleotides and are linked by a phosphate group C-p-G are called CpG islands. These cytosines can be methylated, which typically results in transcriptional repression.

cranial nerves The cranial nerves are 12 pairs of nerves that exit directly from the brain, as opposed to the spinal cord. The cranial nerves may carry sensory information, motor information, or both to various parts of the body. (See Box 1)

craniotomy A craniotomy is the surgical removal of a bone flap from the skull to gain access to the brain; it is also sometimes performed to relieve pressure after brain injury or an infection that causes edema.

cyclic AMP response element-binding protein (CREB) The cyclic AMP response element-binding protein CREB is a transcription factor that acts as an intermediate signaling molecule. Following activation of a membrane receptor, a rise in Ca^{2+} or production of cAMP causes activation of kinases such as cAMP-dependent protein kinase A (PKA). These kinases translocate to the nucleus and activate CREB, which binds to DNA and causes gene transcription. CREB plays important roles in neuronal plasticity and the formation of long-term memories.

Creutzfeldt-Jakob disease (CJD) CJD is a fatal neurodegenerative disease that is caused by sporadic mutation in the prion proteins. CJD is the equivalent of spongiform myelitis or mad cow disease in bovines. The disease propagates within the brain by misfolded prions imposing their shape on other prion proteins. Transmission among humans is rare but can occur following cornea or organ transplantation.

cytokines Cytokines are small proteins that are released in the context of infection, inflammation, or disease. Cytokines regulate immune cells, and may include chemokines, interferons, interleukins, lymphokines, and tumor necrosis factor β. They can be proinflammatory or anti-inflammatory, and are typically found in very small quantities, on the order of 10^{-12} mol.

cytoskeleton A molecular assembly of fibers and proteins that give a cell its rigidity and form, the cytoskeleton is made of microtubules, interacting proteins, and actin filaments. In the nervous system, the cytoskeleton is involved in the transport of molecules, for example, along the long axons. Cell type- specific differences in cytoskeletal proteins often allow the identification of cell types based on the selective presence of certain cytoskeletal molecules. Neuron-specific neurofilament and astrocyte-specific GFAP are two examples.

D-serine D-serine is an amino acid that is produced by serine racemase in the brain from L-serine contained as a building block of proteins. It also acts as an allosteric modulator of the neuronal NMDA glutamate receptor.

BOX 1

THE CRANIAL NERVES

The cranial nerves are 12 pairs of nerve bundles that exit directly from the brain (Figure 1). They are numbered in roman numerals with ascending numbers moving from the front (rostral) to the back (caudal) of the brain. The cranial nerves originate from sensory and/or motor nuclei in the brain stem or forebrain and carry specific sensory and motor information related to the head and neck. Except for the optic nerve (II), which is part of the central nervous systems, all cranial nerves are considered peripheral nerves. The cranial motor neurons are considered lower motor neurons and are comparable to the motor neurons in the ventral horn of the spinal cord. Similarly, the cerebral sensory nuclei that contain the cell bodies of sensory fibers are comparable to the dorsal root ganglia. The axons of cranial nerves typically have a central branch that is myelinated by oligodendrocytes and a peripheral branch that is myelinated by Schwann cells (except for the optic nerve, which is entirely central). Table 1 below lists the name, type, and function for each cranial nerve along with common symptoms experienced after a nerve lesion.

The names of the optic nerves are often remembered using mnemonics, for example:

Old **Op**ie **Oc**casionally **Tries Trig**onometry **And** Feels **V**ery **Glo**omy, **Vag**ue, **And Hypo**active.

To remember the type: **S**ome **S**ay **M**arry **M**oney, **B**ut **M**y **B**rother **S**ays **B**ig **B**rains **M**atter **M**ost (where "S" is sensory, "M" is motor, and "B" is both or mixed).

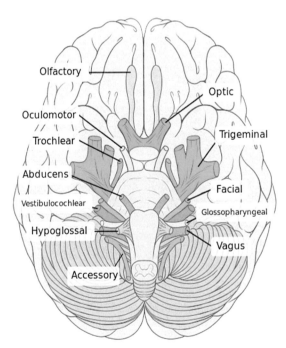

FIGURE 1   Human brain seen from below, showing exit of the 12 cranial nerve pairs. *Source: Wiki Common, Patrick J. Lynch, medical illustrator.*

TABLE 1   Cranial Nerves, Number, Name, Type, and Function, as Well as Behavioral Effects Following Lesion

#	Name	Type	Function	Lesion
I	Olfactory	Sensory	Olfaction	Loss of taste and smell
II	Optic	Sensory	Vision	Loss of vision, visual field defects
III	Oculomotor	Motor	Eye movement, pupillary light reflex	Double vision, nystagmus, loss of pupillary reflex
IV	Trochlear	Motor	Eye movements	Double vision
V	Trigeminal–3 branches V1 ophthalmic (sensory) V2 maxillary (sensory) V3 mandibular (mixed)	Mixed	Facial sensations sensory to meninges, gums, and teeth; chewing muscles	Trigeminal neuralgia, impaired facial sensation, difficulty chewing and moving mouth
VI	Abducens	Motor	Eye movements	Inability to look laterally
VII	Facial	Mixed	Facial expression ,salivary and lacrimal glands	Unilateral facial weakness loss of facial sensation
VIII	Vestibulocochlear	Sensory	Hearing and balance	Hearing loss, vertigo, loss of balance, tinnitus
IX	Glossopharyngeal	Mixed	Tongue sensation, swallowing reflex, salivary glands	Difficulty swallowing, pseudobulbar palsy
X	Vagus	Mixed	Laryngeal muscles, vocal cord and visceral innervations	Inability to speak and control blood pressure and heart rate
XI	Accessory	Motor	Stabilize head and neck	Inability to shrug and move head
XII	Hypoglossal	Motor	Tongue movement	Weakness of tongue

de- and hyperpolarize Depolarization and hyperpolarization *both refer to a* change of the membrane voltage away from the normal resting state. Depolarization moves in the positive direction, typically associated with increased neuronal activity, whereas hyperpolarization moves in the negative direction and typically reduces neuronal activity.

deafferentation Deafferentation is the loss of sensation due to injury or disease of the sensory nerve.

declarative memory Declarative memory is the ability to remember facts, events, and names. It is believed to draw narrative from different memory storage sites in the cortex and requires hippocampal activity. It contrasts to *nondeclarative memory*, used for skills we perform without awareness, such as riding a bicycle. Unlike declarative memories, nondeclarative memories can be performed even after the hippocampus is lesioned.

deep brain stimulation (DBS) DBS is a treatment strategy involving an implanted device, similar to a pacemaker, that delivers electric pulses from a stimulator via a lead wire. The electric pulses are then conveyed to a platinum electrode implanted deep inside the brain. It is now widely used to treat Parkinson disease, but is also being tested for many other neurological and psychiatric illnesses.

delusions A delusion is a wrong belief that is held with conviction even in the face of evidence to discredit the belief. It is typically associated with psychiatric illness.

dementia Dementia is a general decline in memory function that interferes with functions of daily life. It contrasts to benign forgetfulness in the elderly, which is a normal, nonpathological part of aging. However, forgetfulness can be a precursor to dementia.

dementia with Lewy bodies (DLB) Dementia with Lewy bodies (DLB) is the third-leading cause of dementia and is characterized by intracellular aggregates of alpha-synuclein protein. These aggregates are named after their discoverer, Frederick Lewy. Unlike in Alzheimer disease, DLB patients often show Parkinson-like impairment in movement.

designer drugs In the past, designer drugs were chemicals that mimicked a controlled substance, or a substance of abuse manufactured by clandestine laboratories to bypass legal restrictions on their use. These could, for example, be new performance-enhancing drugs for which athletes are not currently tested. Today, the term designer drug is often used to describe the outcome of rational drug design, whereby the knowledge of a receptor or target was used to instruct the design of a specific chemical that only binds to its target. Examples of designer drugs would be humanized monoclonal antibodies.

developmental intellectual disabilities Developmental intellectual disabilities is the umbrella term currently used for diseases developed before the age of 18 that present with alterations in cognition and/or behavior. Most commonly, the disability is already present at birth or occurs shortly thereafter.

diencephalon The posterior forebrain includes the thalamus, hypothalamus, and pineal gland, collectively referred to as the diencephalon.

diffusion-tensor MRI (DT-MRI) DT-MRI is a noninvasive imaging technique that allows visualization of tracts of nerve fibers or white matter. It takes advantage of the anisotropy of water movements, which occur preferentially along the length of the axon.

diplopia Double vision, most commonly due to impaired contraction of the extraocular muscles that position the eye balls, is called diplopia.

direct pathway The direct pathway is one of two pathways in the basal ganglia involved in control of movements. This is an excitatory pathway, thereby promoting movement; in contrast, the indirect pathway is inhibitory and suppresses movement.

disability-adjusted life year (DALY) The DALY is a metric often used by epidemiologists to express the burden of a disease on the general population. It computes the average number of years of life lost due to poor health. One DALY equals one health life year lost to disease. Importantly, it takes into consideration differences in life expectancy in different parts of the world.

DNA methyltransferases (DNMTs) DNMTs are enzymes that catalyze the methylation of DNA by adding a methyl group at CpG sites.

dopamine hypothesis The dopamine hypothesis attributes the root cause for schizophrenia to a problem in dopaminergic signaling. This old hypothesis derives from the beneficial effect achieved through typical neuroleptic drugs that block the D2 dopamine receptors.

dorsal root ganglia (DRG) The DRG is a structure located just outside the dorsal (back) side of the spinal cord and containing the cell bodies of sensory nerves. The dorsal root ganglion cells carry sensory information from the periphery toward the midbrain and cortex.

dorsolateral prefrontal cortex (DLPFC) The DLPFC is a brain region that is tasked with executive function and the initiation of tasks and is also the site of motivational drive.

double-blinded To assure that the effect ascribed to a drug or intervention is not due to the placebo effect or the desire of the patient to get well, but rather to a drug's or intervention's ability to improve health, clinical trials should test drugs in a double-blinded fashion. Here, neither the patient nor the physician knows whether the patient is receiving the active drug or intervention or the inert placebo. Only after the trial ends will the "blindfold" be removed and the outcome of each patient ascribed to either active drug or placebo. Note that we frequently talk about placebos in the context of drugs, but inert intervention can also encompass placebo surgery.

Down syndrome Down syndrome is a genetic disorder caused by a triplication of all or parts of chromosome 21. Affected individuals have dysmorphic features, including small hands and a round, flattened face with a poorly developed nasal ridge, leading to the use of the term "mongolism" until the 1950s. It was since renamed after John Landon Down, a British doctor who devoted his life to the passionate care of individuals with trisomy 21. Down syndrome is characterized by overall delayed development and impaired cognition, with a mean I.Q. of 50. Individuals with Down syndrome age prematurely and develop Alzheimer disease by age 40. Incidence increases with the age of the mother, from around 1/1000 in young mothers to 1/10 in older mothers. Since the early 1960s, prenatal screening can detect the chromosomal abnormality.

doxycycline Doxycycline is an antibiotic used to treat common infections. In neuroscience, it has become common to produce genes that have promoter elements that can be reversibly turned on or off by the presence of doxycycline, called inducible promoters. These inducible promoters can be used in transgenic mice to study the effects of altering gene transcription during only a certain stage of development or during certain conditions.

drug targets The target of a drug is typically a receptor, enzyme, ion channels, or signaling molecule involved in a biological process. By binding to its target, a drug alters the biology in which the target participates. For example, the targets of the painkiller oxycodone, a synthetic opium derivative, are opioid receptors on neurons expressed throughout the nervous system. By binding to the receptor, oxycodone changes the electrical properties of the neuron.

dura mater The meningeal coverings of the brain are the connective tissues that separate the brain tissue from the skull; the outermost of these is the dura mater. It is the most tenacious of the three layers. The innermost layer is the pia mater, and the middle layer is the arachnoid mater.

dying forward hypothesis The dying forward hypothesis was proposed to explain the progressive motor neuron death in ALS. It suggests that upper motor neurons in the cortex kill the lower motor neurons through excessive release of glutamate.

dying back hypothesis An alternative hypothesis proposed to explain the progressive motor neuron death in ALS, the dying back hypothesis suggests that the disease begins in the muscle as opposed to motor nerves. A loss of trophic factors released from the muscle to the innervating lower motor neurons causes its degeneration, which in turn causes the retrograde death of upper motor neurons.

dynein Dynein is a motor protein involved in axonal transport. It binds cargo molecules such as peptides or organelles. Dynein preferentially transports cargo toward the cell body, whereas kinesin preferentially transports cargo away from the cell body.

dystonia Involuntary sustained muscle contractions that cause the body to twist and assume unusual postures are collectively referred to as dystonia.

early-onset Alzheimer disease Most cases (95%) of Alzheimer disease (AD) are sporadic, without known cause, and develop later in life, age >65. Rare mutations in genes that process APP can cause early onset of disease, before age 65 and as early as age 35. These account for 5% of all cases.

early stages of Parkinson disease The early stages of Parkinson disease are characterized by resting tremor, postural instability, bradykinesia, and rigidity. During this stage, dopamine replacement therapy is most effective.

EC50 Half-effective dose of a drug or intervention. It is customary to express the effectiveness of a drug over a large dosing range to determine the maximal achievable change, the minimal dose to detect any change, and the midpoint (the dose at which 50% of the effect is achieved). Comparable useful values are LD50, the lethality dose at which 50% of test subjects (mice) die.

echolalia Echolalia is the repeating of phrases, sounds, or sentences without any evidence of understanding their meaning, as in a parrot talking back.

edema Swelling of tissue is referred to as edema. In the nervous system, we differentiate between cellular and vascular edema. Cellular edema is caused by water moving along with ions into cells, causing them to swell. In vascular edema, serum from the blood vessels crosses into the brain after injury or during disease and accumulates extracellularly.

electrocauterization Electrocauterization is a method to close blood vessels through injection of an electric current into a resistive metal wire or electrode in order to stop bleeding. The metal wire or electrode will heat up as the electric current is injected; as it touches a blood vessel, the heated wire or electrode will cause the vessel's walls to collapse and close.

electroconvulsive therapy (ECT) Also called electroshock therapy, ECT is a nonmedicinal treatment for mental illness that relies on passing electric currents through the brain, eliciting seizures with the goal to change brain chemistry. Its actual mechanism of action remains elusive, yet the approach is used with some success as a last resort in severe cases of mental illness.

electroencephalography (EEG) Electroencephalography is a technique that allows the detection of brain activity through electrodes applied to the surface of the skull. It is primarily used to diagnose seizure disorders such as epilepsy and localize their origin.

electromyography or electromyogram (EMG) EMG is a diagnostic technique that directly measures muscle function either superficially or through insertion of a needle into the muscle. In response to electrical stimulation of the nerve, the resulting voltage change is recorded in the muscle that drives muscle contraction. The technique is called electromyography, and the resulting record is called an electromyogram.

embolus A "floating disaster," an embolus is a piece of atherosclerotic plaque that has broken off of the plaque and floats with the blood into finer and finer vessels. Unless degraded, it can eventually occlude a vessel, causing it to cease blood flow or, worse, burst open. Embolytic strokes are caused by such a piece of plaque in circulation.

encephalitis Inflammation of the brain is called encephalitis. This can be caused by viral, bacterial, or fungal infections, or can be due to an autoimmune response.

endopetidase An endopeptidase is an enzyme that cleaves polypeptides or proteins at internal amino acids. In contrast, ectopeptidases cleave the terminal amino acids at either the N- or C-terminus.

enteroviruses Enteroviruses are a diverse group of single-stranded RNA viruses that affect humans by entering the gastrointestinal tract or airways. Important examples are poliovirus, rhinovirus, and Coxsackievirus.

ependymoma An ependymoma is a primary brain tumor arising from the ependymal cells that line the cerebral ventricles. Most commonly found in children, they can often be surgically removed and carry a favorable prognosis.

epigenetic Epigenetics refers to persistent, but reversible, heritable changes in gene expression without alteration of the DNA itself. The gene(s) of interest are not directly affected, but the ability of the gene to be transcribed into protein is regulated via a number of modifications at the level of the chromatin or individual genes. The two most common epigenetic regulations involve the compaction of chromatin around histones, which affect whether a gene can be transcribed or not, and the methylation of individual gene promoters along the chromatin, which causes transcriptional repression. Specific enzymes interact with histone tails and loosen the chromatin by applying modification groups. For example, addition of an acetyl group (acetylation) relaxes the DNA, making it accessible for transcription. Similarly, the methyl marks on the promoters are applied by a group of enzymes called DNA methyltransferases, which typically silence the gene of interest. Epigenetic changes can explain how environmental conditions can influence disease and behavior and confer susceptibility to disease or resistance without changing a person's genetic makeup.

epilepsy Epilepsy is a very common neurological disorder characterized by the spontaneous recurrence of two or more seizures. These seizures are caused by abnormal synchronous discharge of neural activity and result in a range of behavioral symptoms. Epilepsy can be the result of known or unknown inborn genetic changes, or can develop later in life as a result of a brain insult such as trauma, infections, vascular changes, or tumors. About two-thirds of patients with epilepsy can be effectively treated with a range of antiepileptic drugs, yet for the one-third of patients suffering from pharmacoresistant epilepsy, quality of life is severely affected.

epileptogenesis This refers to a time period during which the brain acquires epilepsy. It is a prodromal phase during which an individual does not show overt signs of epilepsy; no behavioral seizures are present and EEG abnormalities may not be visible. During this time, progressive changes in the interconnectivity of neural networks lead to clusters of cells that have the potential to fire in synchrony and initiate seizures. Such seizure-prone networks must be provoked by additional intrinsic or extrinsic factors to generate a seizure, but many of these factors are not yet known. Such factors probably differ from patient to patient, but may include diet, drugs, temperature, light, sounds, smells, stress, or sleep deprivation, to name just a few. Unfortunately, little is known about this period of time preceding presentation of disease, although this is an area of active research.

episodic memory Memories that capture episodes in one's life in an autobiographical fashion are referred to as episodic memories. These include places, times, and emotions within the context of life stories, such as a collection of experiences that form a narrative. For example, remembering one's wedding is an episodic memory.

Epstein–Barr virus Epstein–Barr virus is a DNA virus of the *Herpes* family that causes mononucleosis. It has been implicated as a risk factor for multiple sclerosis, yet it is among the most common viruses in humans, with an estimated presence in 90% of the population.

essential tremor Often called benign tremor, essential tremor is a high-frequency, low-amplitude shaking or movements that affect hands, arms, head, or even muscles involved in articulation of speech. This is the most common movement disorder, affecting 5% of the population, with impairments ranging from mild to substantial.

evidence-based medicine Evidence-based medicine is the term used to describe the philosophical shift in medicine whereby treatments and interventions should be based on solid evidence provided by rigorous unbiased clinical studies rather than on the personal experience of any physician or anecdotal evidence. This philosophy implies that physicians have a responsibility to stay informed on the current state of drug discovery and clinical research in their respected fields, and that they will alter their practice and prescribing behavior as evidence in support of certain drug treatments or intervention changes.

excitation–inhibition balance (E-I balance) The E-I balance between glutamatergic and GABAergic synaptic activation determines the overall excitability of a neural network, and is presumably perturbed in epilepsy and other neurological conditions.

excitotoxicity Excitotoxicity occurs when a cascade of processes that result in neuronal cell death is initiated through overactivation of NMDA-type glutamate receptors. These flux excessive amounts of Ca^{2+} into cells, which in turn activates destructive enzymes such as caspases, proteases, endonucleases, phospholipases, and others that destroy the cell and its DNA.

executive function Executive functions are important mental processes that enable planning, strategizing, organizing, paying attention, remembering details, and managing time, resources, and space.

experimental autoimmune encephalomyelitis (EAE) The inoculation of experimental animals, mainly mice, rats, or rabbits, with components of myelin can elicit an autoimmune response called EAE that mimics many aspects of multiple sclerosis (MS). Therefore, EAE has become a popular model to study MS in laboratory animals.

extracellular matrix The space surrounding brain cells is called the extracellular space and is filled with a collection of molecules called the extracellular matrix. These molecules include laminin, vitronectin, collagen, and hyaluronic acid. Extracellular matrix provides structural and biochemical support for the surrounding cells. Its gel-like consistency serves as a mechanical cushion but is also the substance through which cells signal to each other, for example through growth factors. Following injury or in disease, extracellular matrix is degraded by proteases to allow for wound healing. Afterward, new extracellular matrix is deposited by adjacent cells.

extracellular signal-regulated kinase (ERK)/mitogen-activated kinase (MAPK) Mitogen-activated kinase (MAPK) and extracellular signal-regulated kinase (ERK) are signaling molecules that add phosphate groups to proteins, thereby turning them on or off. These signals are engaged following the binding of an extracellular mitogen to a membrane receptor. An example of this is the epidermal growth factor binding to the epidermal growth factor receptor.

F-DOPA (fluorodopa) F-DOPA is an L-DOPA precursor radiolabeled with a radioactive 18-Fluor isotope. This allows for imaging of L-DOPA in the brain by positron emission tomography (PET).

fasciculations Spontaneous involuntary muscle twitches are also called fasciculations. They can occur in people without disease, but may indicate a denervation of a small muscle group by nerve damage.

Febrile seizures Convulsive seizures that are evoked by elevations in body and brain temperature, typically through fever, are referred to as febrile seizures. Febrile seizures are common in infants and young children, who typically outgrow them. They are often associated with a loss of consciousness, and typically stop once the fever is controlled.

floppy baby syndrome A loss of muscle tone caused by intoxication with botulinum toxin, floppy baby syndrome suggests that the baby has *Clostridium botulinum* bacteria resident in the intestines.

fluid percussion model The fluid percussion model induces a reproducible form of brain injury that mimics closed head injury. The brain is pressurized through a cannula that is sealed to the skull and contains fluid that can be pressurized. The pressure of this fluid upon the brain causes damage.

fluor-deoxy-glucose (FDG) FDG is an analog of glucose that bears a radioactive fluor isotope, 18-Fluor. It can be used as an imaging agent to measure glucose consumption in humans using positron emission tomography (PET). Glucose consumption is a surrogate measure for brain activity.

focal ischemic strokes Strokes in which only a discrete brain region is affected by a blocked artery, without bleeding, are called focal ischemic strokes.

focal seizures A focal seizure originates in one brain hemisphere and typically affects only a small part of the brain. These were previously called partial seizures. They can present with and without dyscognia (loss of cognition) and, depending on the affected brain region, may present with convulsive muscle contractions affecting different parts of the body.

forced swim test (FST) The FST is a commonly used test of antidepressant efficacy that evaluates behavioral abnormalities in rodents. A mouse or rat is placed in a water tank from which it cannot escape. The animal is then observed for swimming behaviors and/or immobility (floating). Animals with a depressive-like phenotype will give up more quickly than normal animals, and thus rapidly adopt an immobile floating response. Commonly utilized antidepressants increase swimming behavior or decrease floating behavior, and thus this test is frequently used to screen new compounds for antidepressant efficacy.

fornix The fornix is a fiber bundle that connects the hippocampus, thalamus, and nucleus accumbens.

fragile X Fragile X is the most common genetic form of intellectual disability. It is caused by an abnormal trinucleotide expansion on the X chromosome adjacent to the FMRP protein. The silencing of the FMRP gene, which encodes an RNA-binding protein that regulates protein biosynthesis, causes an aberrant production of proteins that lead to the symptoms of fragile X. In addition to the neurocognitive impairments, fragile X patients show an unusually elongated face and ears and large testes. They are hypersensitive, are impulsive, and suffer attention deficits. Autistic features are also common.

fragile X mental retardation protein (FMRP) Fragile X mental retardation protein (FRMP) is an RNA-binding protein that regulates the translation of mRNA into protein. It is under the control of metabotropic glutamate receptor 5 (mGluR5). Normally, FMRP acts as a repressor of protein synthesis. Its absence causes protein biosynthesis to run amok. This is seen in fragile X, where the synthesis of the FMRP protein is reduced due to an instable region on the X chromosome adjacent to the FMRP gene, resulting in altered protein synthesis.

Freud's adjuvant Freud's adjuvant is a mixture of mineral oil and bacterial exotoxin containing membranes that stimulate a cell-mediated immune response. When used in combination with antigens such as myelin proteins to inoculate laboratory animals, it enhances the immune response.

frontotemporal dementia (FTD) FTD is a progressive neurodegenerative disorder that primarily affects the frontal lobe. It is the leading cause of dementia in patients under 65. FTD causes profound personality changes, including a loss of executive function and often socially unacceptable behavior. Language articulation is often impaired, while language comprehension remains intact.

fused in sarcoma (FUS) Fused in sarcoma is a multifunctional DNA/RNA regulator that can bind and modify both DNA and RNA. Loss of function of FUS causes abnormal synaptic spine morphology, presumably by affecting specific synaptic proteins. Mutations in FUS account for 6% of familial cases of ALS.

G protein-coupled receptors (GPCRs) GPCR is an umbrella term for membrane-associated receptors that respond to extracellular signals. GPCRs translate this extracellular signal to changes in signaling events inside the cells via intermediate G proteins or guanosine nucleotide binding proteins. The resulting action can be inhibitory or excitatory.

GABA hypothesis The GABA hypothesis tries to explain the symptoms of schizophrenia through an impairment of GABAergic signaling, particularly a reduction in GABA content and density of GABAergic neurons. This in turn upsets the excitation–inhibition balance in the dorsolateral prefrontal cortex circuitry presumed to be critical in controlling proper association of different memory traces that form a narrative.

gamma oscillations Gamma oscillations are high-frequency, 30–200 Hz synchronized voltage changes that can be recorded by EEG or extracellular electrodes in neuronal networks. Gamma oscillations are due to many neurons firing in synchrony, which is thought to bind the information processed in different parts of the brain together into a cognitive narrative or memory trace. For example, remembering a friend telling you a story may entail the recall of a memory with visual, auditory, and speech language information. This memory would therefore require synchrony between neurons in the visual cortex, hippocampus, auditory cortex, and speech language areas. In schizophrenia, incorrect elements of such narrative may be bound together due to alterations in gamma oscillations, resulting in hearing voices that are attributed to the wrong source.

ganglioside(s) These membrane-associated molecules have a long glycosphingolipid chain conjugated with sialic acid. These molecules were first isolated from the membranes of ganglion cells in the brain, hence their name.

gap junction A gap junction is a channel-like connective pathway between adjacent cells. This pore allows for the exchange of small molecules <1 kD between cells. Gap junctions are formed by two connexons, one on each cell, which are heteromeric proteins containing six connexins.

gene dosage effect hypothesis The gene dosage effect hypothesis suggests that individuals with trisomy 21, or Down syndrome, present with disease because all genes on the duplicated chromosome are found in 1.5-fold quantity. Therefore, most of the resulting proteins are 50% in excess, leading to the observed symptoms.

gene multiplication A common occurrence in cancer is the amplification of certain genes, whereby multiple copies of the gene are found in the mutated DNA.

gene therapy Gene therapy is the introduction of a missing or corrected version of a gene known to cause disease into somatic cells. Gene therapy is performed in an effort to correct the deficit provided by the mutated gene and thereby correct the disease or syndrome itself.

generalized seizures Seizures that originate from abnormal synchronous neuronal activity of both hemispheres are called generalized seizures. Consequently, the behavioral effects of these seizures, such as convulsions, affect both sides of the body. Note that absence seizures without convulsion are also frequently generalized seizures, as they affect both hemispheres.

genetic anticipation A peculiar phenomenon observed in Huntington disease is genetic anticipation, where offspring who inherit the mutated Hungtinton gene from their affected father develop disease at an earlier age than their father did.

genetic/idiopathic epilepsy For the majority of epilepsy cases we do not know the cause; therefore, these are called idiopathic. The term idiopathic has been replaced with the term genetic, since we assume that an inborn genetic cause must exist, although none as yet has been recognized in a given patient. Genetic epilepsy contrasts with symptomatic/acquired epilepsy, for which the cause is known and may include a tumor, trauma, or brain malformation.

genome-wide association study (GWAS) GWAS is an unbiased search for genetic changes that are associated with a disease or disease syndrome. GWAS studies search the genome of many patients in an effort to find nonrandom single nucleotide polymorphisms (SNP) that predominantly characterize persons with disease but are rarely observed in the control population.

germ theory Initially proposed in the sixteenth century, this theory suggested that microorganisms can cause and spread disease in humans. Robert Koch established the criterion based on which one can unequivocally attribute disease to a microorganism. Later, more formal demonstrations by Louis Pasteur showed that microorganisms can be grown in bottles and can cause disease.

Glasgow Coma Scale (GCS) The GCS is a standardized test to determine the level of consciousness on a 15-point scale. Values in three categories (eye opening, verbal response, and motor response) are added, yielding a final score ranging from 3 for a comatose individual to 15 for a normal person. This scale is frequently used in assessing the severity of traumatic brain injury.

glial fibrillary acid protein (GFAP) An intermediate filament protein associated with the cytoskeleton of astrocytes, GFAP is thought to add stability to the astrocytic cytoskeleton. GFAP expression is upregulated upon injury and is one defining feature of reactive gliosis, the astrocytic response to injury. GFAP is the most commonly used marker for identification of astrocytes in tissue.

glial scarring/glial scar Glial scarring is a common glial response to injury whereby astrocytes form a tenacious scar composed of interwoven processes and dense extracellular matrix. The scar is believed to seal off a site of injured brain tissue.

glioblastoma (GBM) Glioblastoma is the highest grade (WHO grade IV) of primary brain tumor. This term is often used synonymously with the term glioma, which encompasses all primary brain tumors of presumed glial origin. GBMs derive from the malignant transformation of stem cells, progenitor cells, or glial cells, and frequently express antigens found in glial cells. GBMs are divided into primary and secondary. Primary GBMs arise de novo without any known intermediates, while secondary GBMs develop gradually from low-grade astrocytomas, acquiring additional mutations that make them progressively more malignant, turning a grade II tumor into a grade IV.

glioma(s) Gliomas are primary brain tumors suspected to derive from glial cells or their progenitor cells. They stain positive for glial antigens. Gliomas encompass all malignancy grades, ranging from WHO grade I astrocytomas to grade IV glioblastoma multiforme. Gliomas also include tumors of oligodendroglial origin called oligodendrogliomas and mixed gliomas, containing both astrocytic and oligodendrocytic tumors.

global ischemic strokes Stokes in which the entire brain is affected, most commonly due to cardiac insufficiency, are called global ischemic strokes.

glutamate excitotoxicity Glutamate excitotoxicity is a frequently observed pathway of neuronal death in injury and disease, attributed to a rise of extracellular glutamate. This rise of glutamate causes an overstimulation of neuronal glutamate receptors, which in turn causes the uncontrolled entry of Ca^{2+}. The excess influx of Ca^{2+} then activates a series of destructive enzymes and death pathways. The most important receptor in this process appears to be the NMDA receptor, as it has very high Ca^{2+} permeability.

glutamate hypothesis The glutamate hypothesis ascribes the major deficits in schizophrenia to reduced glutamatergic signaling in the frontal cortex. It originated from the finding that silencing NMDA receptors with antagonists such as ketamine and phencyclidine (PCP) induces some of the typical schizophrenia symptoms, including hallucinations and psychosis, in normal individuals.

glutamic acid decarboxylase (GAD) GAD is the enzyme responsible for the biosynthesis of GABA from glutamate by decarboxylation. Changes in GAD, particularly the 67 kD form (GAD67), are used as surrogate markers for changes in GABA availability in tissue.

glutamine and glutamine synthetase Glutamine synthetase is an enzyme preferentially expressed in astrocytes of the brain. It is an important component of the glutamate–glutamine cycle. Glutamate released by neurons is taken up and neutralized by astrocytes, where it is converted via glutamine synthetase to glutamine through energy-dependent condensation with ammonia. Glutamine is then transported to presynaptic terminals as a precursor for the synthesis of glutamate.

gram-negative bacteria Bacteria can be distinguished by their outermost membrane. Those with a thick peptidoglycan layer are gram positive; those without are gram negative. Gram-negative bacteria contain lipopolysaccharides (LPS), which can cause endotoxic shock.

granule neurons Granule neurons are very small neurons found throughout the brain, with a very high concentration in the cerebellum.

gray matter Brain tissue occupied primarily by neuronal cell bodies, unmyelinated processes, astrocytes, and microglial cells is referred to as gray matter. The lack of lipid-rich myelin, which characterizes the white matter, makes the tissue appear grayish and darker in color.

Guillain–Barre syndrome (GBS) GBS is an autoimmune disease in which the body's immune system attacks proteins in the myelin sheath of the peripheral motor axon that innervates a muscle. The cause for the autoimmune response is not known, although the disease often develops a few weeks after infection with a gastrointestinal virus. It presents with temporary weakness and sensory abnormalities, yet in most instances, the symptoms resolve spontaneously with no lasting disability.

gummas Gummas are an abnormal but benign growth of tissue in the skin or inner organs caused by the body's reaction to the spirochete bacterium that causes syphilis.

hallucinations Hallucinations occur when a person perceives images, faces, sounds, smells, tastes, or any combinations of any sensory modality in the absence of a real stimulus. Hallucinations occur in an awake state, unlike a dream, and are often vivid.

haloperidol Haloperidol is a common first-generation antipsychotic (neuroleptic) drug that acts as a dopamine D2 receptor antagonist. In spite of the availability of newer drugs with improved side-effect profiles, haloperidol remains a relatively cheap and widely used drug to treat psychosis associated with schizophrenia, or acute psychosis after use of recreational drugs such as LSD, ketamine, or amphetamine.

hemorrhagic stroke A cardiovascular insult in which a cerebral vessel bursts, a hemorrhagic stroke allows blood to enter the brain from the burst vessel.

herniation The process of pressure buildup in the skull due to vasogenic edema, herniation forces the brain downward toward the spinal column, into a constricted region at the base of the skull called the foramen magnum. This process is a serious complication that can lead to death.

herpes simplex virus 1 (HSV-1) Herpes simplex virus type 1 (HSV-1) is a neurotrophic virus that specifically infects neurons. It is a large double-stranded DNA virus that infects epithelial or mucosal tissue, where the virus replicates, and assembles virions, which cause herpes sores when they lyse the cells. Virions are transported to the nervous system, and thus genes packaged into the viral genome can be specifically delivered to the nervous system, making it useful in animal models of disease and gene therapy studies.

highly active antiretroviral therapy (HAART) HAART is the most effective drug regimen for attenuating HIV injection to prevent AIDS. This treatment usually involves three or more drugs, including a protease inhibitor and multiple reverse transcriptase inhibitors.

histone acetyltransferases (HATs) HATs are a group of enzymes that modify histone tails by adding acetyl groups, resulting in histone acetylation. This causes a loosening of the chromatin attached to the histone, giving better access for gene transcription.

histone deacetylases (HDACs) HDACs are enzymes that remove acetyl groups from histone tails, thereby aiding in the compaction of chromatin. The acetyl groups cause a relaxation of chromatin, allowing genes to be transcribed; hence, the removal of acetyl groups by HDACs leads to a compaction of the chromatin, causing transcriptional repression.

histones Histones serve to compact chromatin into nucleosomes to conserve space inside a cell. They are octameric proteins composed of two copies each of histones H2A, H2B, H3, and H4. Each histone octamer allows for 146 base pairs of DNA to be tightly wrapped like a spool of thread. The N-terminal tail of each histone contains sites for epigenetic marking via histone acetylation, methylation, and phosphorylation, which result in changes of the tightness of the wrap. Loosening of the compacted chromatin permits RNA polymerase to access and transcribe genes. Tightening the chromatin suppresses access and impairs gene transcription.

HIV protein transactivator (tat) A virally produced regulatory protein, tat, enhances the efficiency of viral transcription.

homunculus The homunculus ("little man") is a topographic map that identifies the areas of the somatosensory and motor cortices responsible for sensory and motor innervation of different parts of the body.

hormonal stress model The hormonal stress animal model of depression is induced by elimination of the glucocorticoid receptors (GR) in forebrain neurons through conditional deletion of the GR gene allele. This lack of GR causes depressive symptoms associated with enhanced basal serum glucocorticoid levels.

hot plate test The hot plate test is a simple behavioral test used to examine hypersensitivity of a test animal to heat. It measures the delay until an animal withdraws its paw from a heated metal plate, with a shortened delay suggestive of a hypersensitivity to pain.

htt htt is the gene encoding for huntingtin proteins.

HTT HTT indicates the huntingtin protein (see below).

human equivalent dose (HED) It is typical to establish drug doses in animal models to achieve physiological effects of drugs and interventions. These doses include minimal effective dose, maximal tolerated or safe dose, and half-effective dose. These doses obtained from animal experimentation are then extrapolated to the human equivalent through a formula that compares the body surface of a human to that of the animal in which the dosing was established, rather than comparing the body weight.

huntingtin HTT is a large protein that contains 3144 amino acids and does not have any homology to other proteins. Its functions are not well understood, yet it is essential during development, as elimination of the gene is embryonic lethal. In the adult, HTT is involved in regulating the production and release of brain-derived neurotrophic factor (BDNF). The HTT gene is silenced in Huntington disease through polyglutamine repeats near the HTT gene.

hyperexcitability Abnormally active nerve cells or circuits, characterized by an elevated frequency of neuronal impulses, are referred to as hyperexcitable. This can be the consequence of changes in a cell's resting membrane potential, the ion gradient across the membrane, or changes in the excitation–inhibition balance, whereby more excitatory neurotransmitters and fewer inhibitory transmitters are active.

hypothalamic–pituitary–adrenal (HPA) axis The HPA axis explains the neurobiology underlying stress-induced changes in mood and anxiety via the release of cortisol. Stress enhances the release of corticotrophin-releasing hormone (CRH or CRF) from the hypothalamus. CRH then causes the release of adrenocorticotrophic hormone (ACTH) from the pituitary, which in turn stimulates the production and release of cortisol from the adrenal gland, preparing the body to respond to a stressor. Altered responses of the HPA axis are often observed in mood and anxiety disorders like depression and post-traumatic stress disorder (PTSD).

ictal state The electrical brain activity of an epilepsy patient has two defined states on the electroencephalogram (EEG). The ictal state occurs during a seizure and shows rapid fluctuations in voltage, often called a spike and wave discharge, with a frequency of 3–5 Hz. The interictal state is the phase between seizures, during which the EEG is either normal or may show short-lasting single spikes of activity.

immunoglobulin These antibodies are produced by immune cells and serve to protect the body from pathogens.

inclusion bodies Protein aggregates that form intracellularly are called inclusion bodies. These appear in multiple neurological illnesses and have different names in different diseases, for example, Lewy bodies in Parkinson disease, Negri bodies in rabies and infections, and stress granules in ALS.

incomplete penetrance If not all carriers of a mutation or genetic alteration develop disease, we refer to this as incomplete penetrance. In Huntington disease, for example, polyglutamine repeats of 36–39 can cause disease, but do not always, and this is therefore an incomplete penetrance. Yet, 40 or more repeats always cause disease and thus result in complete penetrance.

independence hypothesis This hypothesis suggests that the death of upper and lower motor neurons occurs simultaneously and in a seemingly unrelated fashion.

indirect pathway One of two pathways in the basal ganglia that is involved in the control of movement, the indirect pathway is an inhibitory pathway, and therefore suppresses movement.

induced pluripotent stem cells (iPSCs) The latest generation of stem cells, iPSCs are derived from somatic cells, such as skin fibroblasts, by exposure to a cocktail of four specific transcription factors.

inflammation Inflammation is a complex response of vascular tissue to injury, disease, or infection. It generally serves to initiate healing. Nervous system inflammation is frequently associated with activation of microglial cells and astrocytes, the resident immune cells of the central nervous system.

Institutional Review Board (IRB) The IRB is a review body composed of scientists, physicians, and lay persons that reviews any proposal to conduct research of any kind using human subjects. IRB review seeks to assure that people participating in research are not harmed or exploited; that the study is sound; that it has an important objective; and that the potential benefits of a study clearly outweigh and justify any potential risk. The IRB assures that participants give informed consent; that is, they understand the research and its objectives, and that participants elect to participate voluntarily and are free to terminate participation without any need to justify their decision.

interferon beta Interferon beta is a cytokine that reduces the number of inflammatory cells entering the brain. It is used to treat multiple sclerosis, where it has been shown to reduce the relapse rate by up to 40%.

interictal state On the electroencephalogram (EEG) the electrical brain activity of an epilepsy patient has two defined states; the ictal state is during a seizure, and the interictal state is between seizures. The EEG may be normal during the interictal state or show some short-lasting single spikes of activity.

interneurons Interneurons are a subtype of neuron that are involved in local processing of information received from other neurons or receptor cells. These are often called relay neurons, since they relay information between other neurons. They typically lack the long axons found in projection neurons.

interpersonal therapy Interpersonal therapy is based on the assumption that depressions arise from unresolved interpersonal conflicts that need to be identified and resolved. Although different in approach, this works equally as well as cognitive behavioral therapy.

ischemic cascade The series of events that take place when brain tissue becomes ischemic and therefore has a reduced supply of oxygen and nutrients is called the ischemic cascade. These events include depletion of cellular ATP, the depolarization of the neuronal membrane potential, opening of NMDA channels, influx of Ca^{2+}, activation of destructive enzymes, and induction of cell death.

ischemic penumbra The brain region adjacent to a brain infarct in which the tissue has insufficient blood flow, but where the residual flow still supplies sufficient nutrients and oxygen to prevent neurons from dying, is called the ischemic penumbra. The neurons in the penumbra are likely unable to function normally, but can be rescued if blood flow is restored quickly. The ischemic penumbra is the primary target of stroke therapy. The goal of chemical or mechanical stenosis (opening of the vessel) is to rescue neurons in the penumbra.

ischemic stroke Most strokes are ischemic in nature, meaning that their symptoms are attributable to a loss in blood flow, or ischemia. However, in rare cases, a loss of oxygen supply may cause hypoxia without ischemia. For example, a diabetic patient may suffer from hypoglycemia without ischemia.

joint attention The earliest form of social interaction, whereby a child and parent or caregiver share their attention to objects or situations in the environment, is called joint attention. Most simply, a parent expects a child to follow his gaze or finger pointing if s(he) is pointing to words, an animal, or an object: "See that kitty." At a later stage, the child will initiate the same behavior, expecting the parent to join in his or her attention to a person, animal, or object. A deficit in joint attention is an early warning sign that a child may have autism.

Jonas Salk and Hilary Koprovski Jonas Salk and Hilary Koprovski are typically credited with the development of a vaccine for poliomyelitis. Salk used an inactivated "dead" virus for inoculation, whereas Koprovski used an attenuated live vaccine. Both proved to be effective.

Kennedy disease An X-linked inherited disease that presents with muscle weakness in males in their 30s and 40s, Kennedy disease is caused by a trinucleotide (CAG) expansion in the androgen receptor gene on the X chromosome. Patients have facial muscle twitches and hand tremors, although upper motor neurons of the cortex are unaffected.

ketamine Ketamine is an NMDA glutamate receptor antagonist that acts as an allosteric modulator. Ketamine is used as a local or general anesthetic and has been shown to reduce depression when used in sub-anesthetic doses. Recreational ketamine drug use can cause psychosis similar to that seen in schizophrenia patients.

ketogenic diet The ketogenic diet is a diet rich in fat and low in carbohydrates that is used by some patients to treat pharmacoresistant epilepsy. The absence of glucose, which is the body's preferred energy substrate, forces the body to use fat as the main energy substrate. In the liver, fat is converted to fatty acids or ketone bodies, giving this diet its name. Ketone bodies are used in the brain as an energy substrate in place of glucose. The classic ketogenic diet consists of 80% fat, with the remaining balance consisting of glucose and protein. The efficacy of the ketogenic diet has been confirmed in children through controlled clinical studies; however, long-term compliance with this diet is difficult, particularly in children.

killer T cell Also known as a cytotoxic T cell, a killer T cell is a lymphocyte or white blood cell that is produced in the thymus and expressed in the cluster of differentiation 8 (CD8) glycoprotein on its surface. CD8 binds to major histocompatibility complex I. Upon activation, killer T cells degrade and lyse the cell they bind to.

kindling This term is used to describe the initiation of synchronous brain activity that gives rise to seizures. Akin to kindling a fire with small pieces of scrap wood and a small flame, the repeated stimulation of neurons in the brain can set up networks of synchronous activity. This approach is frequently used to produce epilepsy in laboratory animals.

kuru Kuru is a rare disease among the Force tribe of Papua New Guinea that is caused by members of the tribe eating the brains of affected tribe members in a cannibalistic, religiously founded ritual. The brain contains misfolded mutated prion protein, a highly infectious particle that has also been identified as the cause for mad cow disease and Creutzfeldt-Jakob disease.

L-DOPA L-3,4 dihydroxyphenylalanine, is the precursor molecule for dopamine and is converted by DOPA decarboxylase to dopamine. It is sold as levodopa to treat Parkinson disease.

Lambert Eaton myotonia Lambert Eaton myotonia is an autoimmune disease in which the body's immune system attacks presynaptic Ca^{2+} channels required for the release of acetylcholine at the muscle synapse. This results in muscle weakness. The disease is relatively rare, but is frequently associated with small lung cell carcinoma. It is therefore called a paraneoplastic syndrome, or a condition associated with cancer. It is believed that cancer cells present antigens that are identical to the presynaptic Ca^{2+} channels, and thus prime the immune system to make antibodies, which then attack the muscle synapse.

lamination The layering of tissue into discrete functional or anatomic layers is called lamination. The cortex normally has six layers, and this lamination is readily visible in histological brain sections.

late-onset or sporadic AD The vast majority (~95%) of AD cases are late onset or sporadic, and thus the majority of patients develop dementia after the age of 65 with no known genetic cause. The disease gradually progresses, causing death within 10 years of diagnosis. Unequivocal diagnoses can only be made after death, as it requires detection of amyloid plaques and/or fibrillary tangles on autopsy.

learned helplessness When an individual or animal gives up any attempt to escape an unpleasant situation, assuming that no escape exists, it is demonstrating learned helplessness. This is often featured in animal models of depression and mood disorders.

lentivirus A lentivirus is a retrovirus that can infect nondividing cells and incorporate its DNA into the host cell. These are well suited to transport a large quantity of genetic material, and can be used to transfect neurons in the adult nervous system. Lentiviral vectors are commonly used in the laboratory and are used in experimental gene therapy studies in humans.

leucine-rich repeat kinase 2 (LRRK2) gene The LRRK2 gene encodes for a very large protein of unknown function; however, mutations in LRRK2 are the most frequent cause for autosomal dominant familial Parkinson disease.

leukoencephalopathy This term is used broadly for diseases that affect the white matter with unknown cause.

levodopa Levodopa is the medical name for L-DOPA, used as a precursor for dopamine in the treatment of Parkinson disease. It is often used in a combination preparation with carbidopa, which prevents the premature systemic conversion of the levodopa to dopamine to avoid peripheral dopamine effects.

Lewy bodies Protein aggregates found in the midbrain of patients with Parkinson disease, Lewy bodies are aggregates of mutated and misfolded alpha-synuclein protein.

ligand-gated channels Also known as ionotropic receptors, ligand-gated channels are transmembrane proteins that couple the binding of an extracellular neurotransmitter or peptide to the flux of ion across the cell membrane. This directly changes the membrane voltage. The receptors for acetylcholine, GABA, glycine, and glutamate are examples of ligand-gated channels.

limbic system A collection of brain structures that generate basic functions such as emotion, motivation, and sex drive, the limbic system include the thalamus, amygdala, hippocampus, and olfactory bulb.

lipopolysaccharide (LPS) LPS is an irritant or endotoxin found in the outer membrane of gram-negative bacteria. It causes inflammation and can lead to septic shock.

lipoprotein receptor-related protein 1 (LRP) LRP1 is a receptor protein expressed by neurons that bind lipoproteins such as apolipoprotein E, which transports cholesterol from the blood for the neuronal synthesis of cell membranes.

lithium Lithium is the chemical element with atomic number 3. It is an alkali metal that is highly reactive. It is usually ionic and, in biological systems, it exists as chloride salt. It has been shown to act as a mood stabilizer and benefits patients with major depression. Its biological activity is not entirely understood, but includes a number of pathways that regulate neurochemical balance in the brain.

long-term depression (LTD) LTD is a cellular form of associative learning or, more accurately, forgetting. It occurs following long-term stimulation of one cell at a low frequency, about once per second over 10–15 min. This causes a reduction in postsynaptic activity by decreasing the number of functional AMPA type glutamate receptors.

long-term potentiation (LTP) LTP is an associated form of cellular learning. It is initiated by high-frequency stimulation of a neuron, up to 100 Hz for several seconds, which causes its NMDA receptors to flux Ca^{2+}. This Ca^{2+} flux increases the number of AMPA receptors at the synapse, making responses to subsequent stimuli larger. It is believed to be a cellular form of learning when two stimuli are connected.

loss of function If a protein loses its function due to a mutation, we call this a loss of function. Note that proteins that are normal may lose their function if they get sequestered in protein aggregates.

lower motor neurons The motor neurons that send axons into the periphery to innervate muscle cells are called lower motor neurons. These are located either in the anterior horn of the spinal cord or in the brain stem nuclei of the cranial nerves.

lumbar vertebrae The human spinal column has five lumbar vertebrae between the rib cage and the pelvis, designated L1–L5. Nerves exiting between these vertebrae innervate the lower part of the body from the navel to the toe.

lysis The breakdown of the cell membrane, resulting in spillage of the cytosolic content, is called lysis.

lysosome The lysosome is the cell organelle involved in clearance of cellular debris.

macrophages Macrophages are white blood cells that phagocytose or ingest other cells or cellular debris for degradation. They are found throughout the body but are not present in the brain.

mad cow disease Also called bovine spongiform encephalopathy, mad cow disease is a rare neurodegenerative disease caused by an infection of the bovine brain with mutated prion protein. This is a relatively rare disease, yet a large outbreak was reported in people in Great Britain in the late twentieth century. This outbreak killed over 100 people who had consumed meat from infected animals.

magnetic resonance imaging (MRI) This noninvasive medical imaging technique, often abbreviated MRI, is the most sensitive imaging modality to identify lesions and brain abnormalities. It can readily distinguish changes in the water content of tissue caused by tissue lesions, edema, or inflammation associated with disease.

major depressive disorder Major depression is the medical term used for a clinically defined depression. A cardinal feature is the general lack of interest in anything pleasurable, which in medical terms is often called anhedonia. Other symptoms include sadness, hopelessness, guilt, fatigue, irritability, change in appetite, and difficulty sleeping. Depression is a very common condition affecting approximately 10% of the population. It is believed to be due to a chemical imbalance, notably affecting the concentration of monoamine transmitters such as serotonin and norepinephrine. Treatments therefore aim to increase the concentration of these transmitters by inhibiting their reuptake. These drugs are called selective serotonin reuptake inhibitors (SSRIs) or serotonin-norepinephrine reuptake inhibitors (SNRIs).

major histocompatibility complex (MHC) The MHC is made up of a number of surface proteins that cells use to display antigens to T cells. Antigens presented in the context of MHC class I are recognized by CD8 on the surface of killer T cells, initiating cell destruction, whereas antigen presentation with MHC class II are recognized by CD4 and activate helper T cells to establish specific immunity.

mammalian regulator of rapamycin (mTOR) mTOR is a serine/threonine protein kinase that regulates many cell processes, including cell growth, survival, process development, and protein synthesis. mTOR senses changes in cellular nutrient, oxygen, and energy levels. It is dysregulated in many diseases ranging from obesity to cancer. Altered mTOR signaling is implicated in a number of neurological conditions including autism, aging, and Alzheimer disease.

mania Mania is an abnormally elevated mood that is commonly associated with bipolar disorder (bipolar depression) but can also result from illicit drug use. Mania is typically associated with inflated self-esteem and grandiosity. During the manic state, patients are hyperactive, have lots of enthusiasm and great expectations, yet typically fail to carry out their plans. In bipolar depression an individual goes from manic episodes to depressed episodes, each lasting from several weeks to 12 months.

medical cytogenetics Medical cytogenetics is the medical application of the study of chromosomal structure. It is a quick way to discover coarse chromosomal abnormalities, such as a duplication of chromosomes as the underlying cause of Down syndrome. Similarly, it can readily detect an extra X chromosome, as seen in Kleinfelder syndrome.

medulloblastoma The most common primary brain tumor in children, medulloblastoma forms from mutations in neural progenitor cells in the cerebellum. The most common mutations are genes of the hedgehog signaling pathway involved in normal brain development.

meninges The meninges are the three tenacious membrane layers of connective tissue that cover the brain and spinal cord. The innermost is called the pia mater, the middle one the arachnoid mater, and the outermost is called the dura mater.

meningioma Meningioma is a cancer of the meninges, the membranes formed by connective tissue layers that cover the surface of the brain and spinal cord. Meningiomas are the most common primary brain tumor in adults, and are also the one with the best prognosis. Their superficial growth typically allows complete resection without recurrence.

meningococcus Meningococcus is the common name used for the gram-negative bacterium *Neisseria meningitides* that causes meningitis or inflammation of the meningeal coverings of the brain.

mesencephalon The mesencephalon, or midbrain, is comprised of the inferior and superior colliculi, the cerebral peduncle, midbrain tegmentum, periaqueductal gray, and the substantia nigra.

mesial-temporal lobe sclerosis The most common form of epilepsy in adults is temporal lobe epilepsy (TLE). This form of epilepsy is most commonly associated with a pathological loss of neurons and structural malformation in the mesial temporal lobe of the cortex, an area that also encompasses the hippocampus. Removal of all or part of the mesial temporal lobe, sometimes with partial resection of the hippocampus and amygdala on the affected hemisphere, provides excellent outcomes in many patients.

mesolimbic pathway The mesolimbic pathway encompasses neurons in the ventral tegmental area of the midbrain that project to the nucleus accumbens, amygdala, hippocampus, and frontal cortex. These projections are dopaminergic, and abnormal activity in this pathway is associated with addiction, schizophrenia, and depression.

metalloproteinases Metalloproteinases are a large family of zinc-dependent proteolytic enzymes that degrade extracellular matrix molecules. They are involved in tissue repair after injury, matrix degradation in cancer, and angiogenesis and host defense.

methyl-CpG-binding domain (MBD) MBD is a site on the DNA that recognizes cytosine residues. These cytosine residues mark DNA slated to be methylated and silenced by methyltransferases.

methyl-CpG-binding protein 2 (MECP2) MeCP2 is a protein that binds to methylated cytosine residues on DNA. Once bound to DNA, it recruits histone deacetylases that induce chromatin compaction, making the gene inaccessible for transcription. One of the affected genes is the brain-derived neurotrophic factor (BDNF) gene that is constitutively repressed by MeCP2. Mutations in MeCP2 cause Rett syndrome, a developmental intellectual disability.

methylation The addition of a methyl group (CH3) to a substrate is called methylation. It is catalyzed by enzymes such as methyltransferases and antagonized by demethylases. In biological systems, methylation is an important regulator of gene expression and, typically, methylated DNA results in gene silencing.

mhtt mhtt is the commonly used abbreviation for the mutated huntingtin gene.

mHTT mHTT is the commonly used abbreviation for the mutated huntingtin protein.

microdialysis Microdialysis is a technique in which a small, microscopic dialysis tube is used to sample fluid from tissue for chemical analysis. It can be used to sample fluid from brain tissue, which can then be chemically analyzed to measure levels of various neurotransmitters. The system can also be used to infuse molecules into the brain.

microglia Microglia are resident immune cells of the nervous system. They are related to macrophages in the periphery and share the same ability to ingest foreign particles and cells. They are constantly trolling the brain in search of evidence of injury or disease. Upon binding of foreign particles to Toll-like receptors, they become activated and engage in wound clearance. As part of their response, they release cytokines and chemokines that attract other microglia, macrophages, and blood-borne immune cells.

microRNAs (miRNA) microRNAs are recently discovered small noncoding RNAs. These 21–25 nucleotide miRNAs bind target sites in messenger RNAs and can regulate protein expression by either repressing mRNA translation or inducing a breakdown of the mRNA. Each microRNA can control hundreds of RNAs ribonucleotides that are involved in the regulation of the translation of mRNA into protein.

microtubules Microtubules are a key component of the cytoskeleton. Although they are found throughout the cell cytoplasm, they are particularly enriched in axonal processes, where they serve as guide rails for axonal transport. Microtubules are composed of dimers of alpha and beta tubulin stabilized by interacting proteins such as tau.

mild cognitive impairment (MCI) The smallest measurable deviation in memory function derived from memory tests, MCI is diagnosed by a score range of 24–29 on the Mini-Mental Status Examination, which scores mental function on a scale of 0–30 points. Mild cognitive decline may progress to Alzheimer disease, particularly in patients that have enhanced genetic risk.

mild traumatic brain injury (mTBI) Previously called concussion, mTBI is a closed-head injury, typically sports related. While it does not show visible signs of contusion through a brain scan, mTBI still causes transient neurological symptoms including headaches, memory loss, and difficulty concentrating.

mimetics Synthetic drugs that mimic a natural ligand and bind to the same receptor or enzyme are called mimetics. They are typically characterized by some enhanced properties such as increased specificity, affinity, stability, or bioavailability, or are simply easier to manufacture in sufficient quantity to use as a drug.

Mini-Mental Status Examination The Mini-Mental Status Examination is a standardized, quick, and sensitive bedside test of both working and episodic memory. The 30-point test includes knowledge of the patient's current whereabouts and names of objects and ability to recall a series of objects and words. A lower score is suggestive of cognitive impairment.

mirror system A neural circuit that becomes equally activated when we perform a task and when we watch another person perform the same task, the mirror system is believed to be a primate-specific circuit that enables us to learn by imitation. Mirror neurons have been identified in the visual cortex and other cortical regions.

moderate Parkinson disease (PD) The moderate stage of Parkinson disease is characterized by an intermittent occurrence of dyskinesia or hyperactive movements in patients on levodopa.

moderate traumatic brain injury (TBI) Moderate TBI presents with prolonged loss of consciousness, and brain injury is visible through brain scans. This classification of TBI requires immediate medical attention and may leave lasting neurological signs.

monoamine hypothesis The monoamine hypothesis is the proposal that a neurochemical imbalance of the brain monoamines, particularly norepinephrine and serotonin, is responsible for the behavioral changes characteristic of depression.

monoamine oxidase (MAO) MAO is the enzyme that catalyzes the oxidative conversion of monoamines such as serotonin and norepinephrine to ineffective aldehydes. Inhibitors of this enzyme (MAOIs) are used for treating depression.

monoclonal antibodies Monoclonal antibodies are derived from identical immune cells that recognize and bind with high specificity to the substance they were raised against. In contrast, polyclonal antibodies are made by multiple immune cells that recognize multiple different epitopes, possibly on different substances, and therefore lack specificity.

mosaic If genes are expressed in a mosaic way, wild-type genes or certain alleles will be expressed in some cells while mutated genes or different alleles will be expressed in others. The resulting tissue will have cells that differ in their gene expression.

motivational salience Salience is another word for relative importance. Motivational salience refers to having the drive to want something and being willing to work for it, such as for a reward. Dopamine is typically considered to be the reward transmitter that provides this desire.

MPTP MPTP is the abbreviation for 1-methyl-4-phenyl-1,2,3,6-tetrahydropyridine, a chemical that induces Parkinson-like symptoms in patients and animals. A byproduct of synthetically manufactured heroin, MPTP is converted in the brain to MPP+, which exerts toxic effects on mitochondria by impairing their ATP production.

multiple sclerosis (MS) MS is a chronic demyelinating disease of the central nervous system, and is believed to be due to an autoimmune response to components in the myelin produced by oligodendrocytes.

muscarinic receptors Acetylcholine receptors are divided into muscarinic and nicotinic. The muscarinic receptors are G protein-coupled receptors that signal via a number of intracellular signaling molecules. The nicotinic receptors are ligand-gated ion channels that flux Na+ and K+ ions and directly alter the cell's membrane potential.

muscular dystrophy Muscular dystrophy refers to a group of diseases characterized by muscle weakness due to cellular and molecular changes in the musculoskeletal system. Typically the nerve and neuromuscular synapse remain unaffected.

myasthenia gravis Myasthenia gravis is an autoimmune disease in which muscle weakness is the result of an autoimmune attack on the postsynaptic acetylcholine receptors in the muscle.

myelin A lipid-rich membrane that wraps around axons in a sheath that functions as an electrical insulator, myelin is produced by oligodendrocytes and Schwann cells. Bare regions, or nodes of Ranvier, along the axon lack myelin. These nodes become the sites at which action potentials are regenerated and jump from node to node across myelin segments. This saltatory conductance allows small-caliber axons to conduct with high velocity and without a loss in signal. Myelin also serves important trophic support roles for the axon, and demyelination impairs axonal health.

myelin basic protein (MBP) MBP is an abundant ~18.5 kD protein found in myelin produced by oligodendrocytes and Schwann cells. Immunization of mice, rats, or rabbits with fragments of MBP triggers an immune response called experimental autoimmune encephalomyelitis (EAE) that causes a similar myelin loss to that seen in multiple sclerosis (MS). This has led to the suggestion that an autoimmune response to myelin may be the cause for MS.

myelin oligodendrocyte glycoprotein (MOG) MOG is a myelin-associated adhesion protein expressed in the myelin sheath that glues adjacent myelin membranes together, enhancing the integrity of this important axonal insulator.

myelin-associated glycoprotein (MAG) MAG is a transmembrane glycoprotein expressed by myelinating oligodendrocytes and Schwann cells. It binds to receptors on the axon, including Nogo-66, where it has a dual role. It has a trophic role supporting normal axon–myelin interactions in the healthy brain, but also suppresses neurite outgrowth in the injured brain.

myelination The compacted sheaths of lipid-rich membrane that surround axons are called myelin, and are excellent insulators that help rapid signal conduction in the brain. The process by which the myelin sheath is applied to axons by oligodendrocytes is called myelination.

myotomes The muscles that are innervated by a single spinal nerve root leaving the spinal cord at a discrete spinal segment are called myotomes. For example, the nerves exiting at the first thoracic vertebrae, T1, are responsible for finger abduction (spreading). Those exiting at the fourth cervical vertebrae, C4, are responsible for shoulder elevation.

Na+/K+ ATPase Among the most abundant ion transport proteins found in essentially all cells of the body, the Na+/K+ ATPase transports three Na+ ions out of the cell and two K+ ions into the cell, both against their electrochemical gradient. This establishes the negative resting membrane potential found in all cells. It uses ATP in the process, and is among the most avid consumers of energy in the form of ATP in the brain to maintain ionic gradients across nerve cells.

Naegleria fowleri N. fowleri, the "brain-eating amoeba," is the amoeba that, when colonizing the brain, causes acute meningoencephalitis or PAM, a deadly disease.

negative gain of function If a mutation causes a negative or harmful function that is not present in the wild-type protein, we speak of a negative gain of function.

negative symptoms Behavioral symptoms that are typically classified in neuropsychiatric illnesses as negative include depression, anhedonia, loss of motivation, apathy, asocial behavior, loss of affect, and poor use and understanding of speech and language. They contrast to positive symptoms such as hallucinations, delusions, or paranoia.

Negri bodies As intracellular inclusion bodies found in cerebellar Purkinje cells and pyramidal neurons of patients infected with rabies, Negri bodies are believed to harbor the virus.

Neisseria meningitides *N. meningitides* is the gram-negative bacterium that causes meningitis.

neovascularization The formation of new blood vessels from existing ones, also called angiogenesis, neovascularization occurs most commonly in cancer. It frequently occurs in response to vascular endothelial growth factor (VEGF), which induces the vascular endothelial cells that form the wall of blood vessels to produce sprouts. The new vessels eventually supply a growing tumor with nutrients to support its growth. This process is the target of biological drugs such as avastin that suppress VEGF signaling.

nerve growth factor (NGF) NGF is the first identified member of a large family of growth factors that support brain function and development. NGF is a small protein that binds to the Tropomyosin receptor kinase A (TRKA). Its activity supports the growth and survival of its target neurons, often sensory sympathetic neurons. Since its discovery by Rita Levi-Montalcini and Stanley Cohen in the 1950, over 50 related neurotrophic molecules have been identified in the nervous system. I like to dub them the endogenous "multivitamins" of the brain.

neural progenitor cells Neural progenitor cells are immature cells of the subventricular zone that are pluripotent and can give rise to essentially all cells of the nervous system. Unlike stem cells, they are not omnipotent and cannot, for example, give rise to liver or lung cells.

neuregulin Neuregulins are a family of proteins that are related to the epidermal growth factor (EGF). They are secreted on membrane-bound receptors that bind to one of three receptors, tyrosine kinase receptors HER2, HER3, or HER4. Neuregulins serve important roles in nervous system development, particularly by influencing the differentiation of myelinating glial cells.

neurexins Neurexins are synaptic proteins that have both adhesive and signaling functions. They are mostly found on the presynaptic terminal, where they interact with extracellular ligands from postsynaptic sites, such as neuroligin. This interaction is believed to facilitate the development of synapses. Neurexins bind the black widow spider toxin, alpha-latrotoxin, which causes massive, catastrophic neurotransmitter release. Alterations in the neurexin genes are suspected of contributing to neurocognitive disorders including autism, Tourette syndrome, and schizophrenia.

neuro-AIDS Cognitive impairments and dementia that can present in a small number of AIDS patients that harbor HIV in the nervous system are called neuro-AIDS. Cases of neuro-AIDS have become rare with the more effective treatment of patients using highly active antiretroviral therapy (HAART).

neuroblastomas Primary tumors derived from neuroblasts or immature nerve cells in the peripheral nervous system, neuroblastoma is the most common tumor of infants, but often carries an excellent prognosis. In adults, neuroblastomas are exceedingly rare and difficult to treat.

neurocysticercosis Neurocysticercosis is a brain disorder caused by the infection and colonization of the brain with cysticerci. These tapeworm larvae cause calcified cysts in the brain that can cause seizures.

neurofibrillary tangles Tangles are intracellular aggregates of hyperphosphorylated tau protein and are a pathological hallmark of Alzheimer disease. While they are always found in the brain of patients with Alzheimer disease, it is unclear whether the tangles contribute to disease or are a coping strategy whereby cells aggregate nonfunctional proteins.

neurofibromatosis Neurofibromatosis is an autosomal dominant genetic disorder caused by mutations in the neurofibromatosis 1 or 2 genes. This rare disorder presents with tumors of the peripheral nerve sheaths and disfiguring benign skin lesions called café au lait spots. The disability varies according to the affected nerves and severity of disease.

neurofilaments Neurofilaments are a major component of the neuronal cytoskeleton. These 10 nm intermediate filaments are composed of polypeptides that are related to keratin found in skin cells. Antibodies to neurofilaments serve as tools to identify neurons in tissue sections.

neurogenesis The genesis or birth and development of nerve cells either during development or in the adult brain is called neurogenesis.

neuroleptic Neuroleptics are antipsychotic drugs used to suppress psychotic behavior. These include the first-generation "typical neuroleptics" that inhibit D2 dopamine receptors, such as chlorpromazine and haloperidol. They also include the newer "atypical neuroleptics" such as clozapine, risperidone, and olanzapine, which also affect serotonergic, noradrenergic, and histaminergic neurons.

neuromas This is an old medical term used to describe a tumor of the peripheral nerve sheath that is more accurately called a schwannoma. These are tumors of the myelin-forming Schwann cells, a type of peripheral glial cell. A common schwannoma affects the acoustic nerve and is called acoustic neuroma. These are typically benign and can be surgically cured.

neuromuscular junction The synapse between the peripheral motor axon and the muscle is called the neuromuscular junction. It is a complex synapse with a highly folded synaptic membrane. Each motor axon innervates multiple muscle fibers that form a motor unit, yet each muscle fiber only receives input from one motor axon.

neurophil Neurophil is brain tissue that primarily contains neuronal processes and glial cells but is devoid of neuronal cell bodies or myelinated axons.

neurosyphilis A late stage of untreated syphilis infection, neurosyphilis typically presents over 10 years after the initial infection. Patients show lesions in the spinal cord due to progressive degeneration of the spinal sensory nerve roots along the dorsal column, which essentially causes a functional differentiation.

neurotrophin A group of growth factors with trophic effect on neurons, neurotrophins help to increase neuronal survival. The classical neurotrophin is a nerve growth factor, which induces sprouting of sympathetic ganglion cells. An important neurotrophin in the nervous system is the brain-derived neurotrophic factor (BDNF). However, there are over 50 neurotrophins reported to exist in the brain.

nicotinic receptors Acetylcholine receptors are divided into muscarinic and nicotinic. The nicotinic receptors are ligand-gated ion channels that flux Na^+ and K^+ ions and directly alter the cell's membrane potential. By contrast, muscarinic receptors are G protein-coupled receptors.

NMDA receptor (NMDA-R) N-methyl-D-aspartate (NMDA) receptors are a family of ionotropic receptors normally activated by glutamate. They derive their name from the fact that they can also be activated by the amino acid NMDA. All NMDA-Rs are heterotetramers (composed of four different subunits) that share two obligatory GluN1 subunits, typically with a combination of two additional GluN2 subunits. These come in two isoforms, GluN2A and GluN2B. NMDA-R activation causes influx of Na^+ and Ca^{2+} ions. The NMDA-R is about 5-fold more permeable to Ca^{2+} than Na^+. Under physiological conditions, the NMDA-R is blocked by an intracellular Mg^{2+} ion lodged inside the pore. Depolarization transiently removes this block, enabling the channel to conduct. NMDA-Rs have a number of allosteric modulatory sites at which endogenous regulators such as glycine or D-serine bind, but also contain important sites for binding of drugs such as ketamine, PCP, or amphetamine.

nodes of Ranvier These are the sites at which action potentials are (re)generated as they travel along the axonal nerve process. They are the site of highest density of sodium channels, and are flanked by myelin on each side. Each node is about 1 µm wide, with nodes evenly spaced along the axon between 100 and 1500 µm apart depending on axon diameter. The action potential spreads in a saltatory way by jumping from node to node, which greatly improves signal conduction velocity even over long distances.

nogo-A Nogo is a repulsive protein expressed by oligodendrocytes on the surface of myelin. It binds to the nogo receptor on nerve processes or neurites, activating a signaling cascade that leads to the disassembly of the microtubules in the neurites and a collapse of the growth cone through which the neurites advance. As a result, neurite outgrowth is paralyzed. Nogo signaling is believed to be a major reason why axons in the central nervous system cannot regenerate.

non-cell autonomous This is emerging concept whereby neurological illnesses cannot be explained through a deficit in the affected neurons alone, but require consideration of changes in the surrounding non-neuronal cells as well. The most important cells to consider as non-cell autonomous contributors are astrocytes, oligodendrocytes, vascular cells, and microglial cells.

nondeclarative memory In contrast to declarative memories, which include facts, people, events, and names, nondeclarative memories are skills we perform without awareness, like riding a bike and using tools. They reside largely in the basal ganglia, amygdala, cerebellum, and sensory and motor cortices.

Notch Notch is an important and, evolutionarily, a highly conserved signaling molecule that mediates cell–cell signaling required for cell differentiation and overall establishment of polarity in a multicellular organism. Adjacent cells each express both the Notch receptor and a membrane-anchored ligand, which in mammals is either Jagged or Delta-like. Upon binding to Notch, the intracellular domain of Notch is cleaved by γ-secretase and then trafficked to the cell nucleus to regulate gene transcription.

nucleus accumbens The nucleus accumbens consists of two subcortical nuclei, one in each hemisphere, and is part of the basal ganglia within the striatum. The nucleus accumbens contains medium spiny GABAergic neurons that regulate the dopaminergic limbic centers in the brain. The accumbens nucleus is often called the brain's pleasure center, and its activity provides a drive for further pleasure seeking. Not surprisingly, this nucleus is implicated in drug seeking and addiction.

olfactory nerve The first cranial nerve is the olfactory nerve; it innervates the olfactory epithelium of the nose and conducts the signals from odorant receptors back to the olfactory bulb.

Oligoclonal bands Oligoclonal bands are seen when cerebrospinal fluid from multiple sclerosis (MS) patients is analyzed by Western blot. They resemble immunoglobulin antibodies produced by immune cells in the brain and are of diagnostic value to confirm a suspected MS diagnosis, as up to 90% of MS patients show these bands.

oligodendrocyte Oligodendrocytes are macroglial cells exclusively found in the central nervous system and are responsible for the formation of the insulating myelin sheath. The equivalent cell in the peripheral nervous system is the Schwann cell. One oligodendrocyte can provide myelin segments for up to 50 axons.

oligodendrocyte precursor cells (OPCs) Oligodendrocytes are embryologically derived from progenitor or precursor cells. Evidence supports a common precursor cell shared with astrocytes, the so-called oligodendrocyte type 2 astrocyte (O2A) progenitor cell. Another important precursor cell is the NG2 cell, which expresses the NG2 proteoglycan and is present throughout life.

oligodendroglioma An oligodendroglioma is a tumor arising from cells of the oligodendrocyte lineage, i.e., progenitor cells destined to become myelin-producing oligodendrocytes in the central nervous system.

paraneoplastic syndrome A disease or symptom that is associated with and caused by a distant cancer is a paraneoplastic syndrome.

paranode The myelinated axon can be divided into the following segments: the node of Ranvier, which is myelin-free and contains a high density of Na^+ channels; the internode, which spans the axonal membrane between two adjacent nodes of Ranvier and is covered by myelin; and the paranode, which is the region immediately flanking the nodes of Ranvier. This paranode region contains cell adhesion and cell recognition molecules involved in the organization of the nodes, such as contactin-associated protein (Caspr; also known as paranodin) and contactin. The juxtaparanode is the final region separating the paranode from the internode, and this region contains a high density of K^+ channels.

paraplegia Permanent loss of motor and sensory function in the lower extremities that spares the upper extremities is called paraplegia.

parasympathetic nervous system Together with the sympathetic and enteric nervous system the parasympathetic nervous system forms the autonomic nervous system, which controls most of the body's internal organs. The parasympathetic nervous system is tasked with the "rest-and-digest" functions that contrast with the "fight-or-flight" response mediated by the sympathetic nervous system. The parasympathetic nervous system uses acetylcholine as its principal neurotransmitter and acts through muscarinic and nicotinic receptors. The sympathetic nervous system primarily engages adrenaline (epinephrine) and noradrenaline (norepinephrine) acting via alpha and beta adrenergic receptors.

parenchyma The bulk tissue of an organ is the parenchyma. In the brain, this consists of neurons and glia.

Parkin Parkin is an E3 ligase that marks defective mitochondria for degradation by the lysosomes. The Parkin gene is mutated in some autosomal recessive familial forms of juvenile-onset Parkinson disease.

Parkinson disease (PD) A neurodegenerative disease in which dopaminergic neurons in the midbrain die, causing a progressive loss of motor control, Parkinson disease typically presents after age 60. Symptoms include rigidity, tremors, problems with balance, and overall slowness of movement. Dopamine replacement therapy is a mainstay in the clinical management of the disease.

Partial seizure (complex and simple) This is an old term for a seizure originating in one brain hemisphere and affecting a discrete brain region. This term has since been replaced with the term focal seizure. If the patient loses consciousness, it was previously called a complex partial seizure, and is now called a focal seizure with dyscognia; if the patient remains conscious throughout, it was called a simple partial seizure, and is now called a focal seizure without dyscognia (loss of cognition).

parvalbumin-expressing GABAergic interneurons Parvalbumin is a Ca^{2+}-binding protein that is typically expressed in GABAergic inhibitory interneurons. These fast-spiking cells probably require enhanced Ca^{2+} buffer capacity to assure that their high rate of firing does not elevate Ca^{2+} above physiologically tolerable levels.

Patient H.M H.M. (Henry Molaison) was a patient who suffered from intractable epilepsy and had both of his hippocampi surgically removed to treat his seizures. Although he became seizure free after the surgery, it left him unable to remember new information for longer than a few minutes. Yet, he was able to recall old memories acquired prior to his surgery, and maintained intact procedural memories. He was widely studied in attempts to localize the memory trace. His case supports the view that the hippocampus is essential to the formation of new declarative memories.

penicillin One of the earliest antibiotics, penicillin was first identified by Alexander Fleming in 1928 and is still widely used to treat bacterial infections, primarily from gram-positive bacteria such as *Streptococcus*.

pericyte A pericyte is a contractile brain cell that regulates blood flow through small capillaries. They are tightly associated with the entire vasculature tree, but are particularly prominent on small-diameter capillaries. During development, pericytes are required to induce the expression of tight-junction proteins that form the blood–brain barrier.

personalized medicine Personalized medicine seeks to identify the right drug (or intervention) at the right concentration for the right patient at the right time.

phagoptosis Phagoptosis is a recently recognized cell death mechanism whereby phagocytic cells such as macrophages or microglial cells engulf and digest a living cell.

Phenobarbital Phenobarbital is the most commonly used drug worldwide to stop convulsions associated with seizures and epilepsy. It belongs to the drug family of barbiturates, which have a depressant action on the nervous system. Phenobarbital prolongs the activity of GABA-A receptors, thereby enhancing GABAergic inhibition.

phospholipase C (PLC) Phospholipase C enzymes cleave phospholipids, converting them into a series of signaling molecules. An excellent example is the cleavage of phosphatidylinositol into 4,5,-bisphosphate (PIP2), diacyl glycerol (DAG), and inositol 1,4,5-triphosphate (IP3) by PLC. These cleavage products each act as a distinct second messenger inside the cell.

PiB or "Pittsburgh compound" The Pittsburgh compound, named for its discovery at the University of Pittsburgh, is a contrast agent with which amyloid plaques can be detected in patients by positron emission tomography (PET). It is comprised of a thioflavin derivative (thioflavin T) conjugated to a radioactive isotropic (typically 18-Fluor).

Pierre Paul Broca Broca was a nineteenth century neurologist practicing in France. He first identified the brain region primarily involved in the formation or articulation of speech. Located in the left inferior frontal gyrus, this region is now called Broca's region. Lesions in this region cause expressive aphasia, or a loss of the ability to produce language.

placebo effect A placebo effect is one that cannot be reasonably assigned to the drug or intervention studied, but instead is solely attributable to an inert substance or imaginary intervention. The placebo effect can be very strong, as humans have a desire to get better when ill. To determine how much the placebo effect weighs into the outcome of an intervention, clinical trials typically compare an active intervention to a placebo.

plasmapheresis A medical procedure that cleanses the blood of antibodies, plasmapheresis is used in diseases where autoantibodies are the suspected disease-causing mechanisms. The antibody-containing blood plasma is separated from blood cells and replaced with fresh plasma that is free of reactive antibodies.

pleiotropic Pleiotropic refers to something that affects multiple properties. Drugs such as beta blockers, for example, can have pleiotropic effects. They may reduce heart rate, affect breathing, and reduce sweating while also reducing anxiety.

poliomyelitis A rare viral disease that causes lesions of spinal motor neurons that innervate the legs, poliomyelitis typically develops in a few young children infected with the poliovirus. The disease was nearly eradicated in the Western world through mass vaccinations carried out in the 1960s and 1970s, but is making a comeback in the developing world.

polyglutamine repeats The huntingtin gene is flanked by polyglutamine repeats, a tract of repeated CAG codons that code for glutamine. These repeats only cause disease if the number of these repeats exceeds 39, presumably due to transcriptional repression of the huntingtin gene.

positional cloning Positional cloning is a method to identify a gene by narrowing down its chromosomal locus. This is done through studies of large families with multiple generations of affected individuals where common abnormalities on the chromosome can be identified.

Once the chromosomal locus is narrowed sufficiently it can be sequenced.

positive symptoms Behavioral symptoms that are typically classified in neuropsychiatric illnesses as positive include hallucinations, delusions, or paranoia. These contrast with negative symptoms, such as loss of motivation, apathy, asocial behavior, loss of affect, and poor use and understanding of speech and language.

positron emission tomography (PET) PET is a widely available imaging technique used to detect radiolabeled tracers or metabolites. It is frequently used in conjunction with fluorodeoxyglucose, a radioactive analog of glucose, to measure energy consumption. It receives its name from the active particle that is formed during the decay of the radioisotope used as a tracer. Short-lived radioisotopes are injected into the blood of a patient and allowed to incorporate into active molecules in the brain. The isotope gradually decays by beta decay, thereby emitting positrons (the antiparticle of an electron). As the positron travels over a very short distance in tissue, it decelerates and annihilates an electron. This process generates gamma-ray photons that can be detected by a scintillation device. By acquiring images in different focal planes, the entire imaged area can be reconstructed from these gamma rays and visualized by tomography in three dimensions.

post-translational modifications Changes made to proteins during protein biosynthesis but not encoded by the amino acid sequence are called post-translational modifications. For example, the cleavage of a protein segment turns an inactive proform into the mature, active form of a protein. Another example is the addition of sugar groups by glycosylation to make the protein more water soluble, or the addition of lipid groups to make the protein lipophilic.

postsynaptic density The postsynaptic membrane contains a high density of neurotransmitter receptors and regulatory proteins. These often make it readily identified on electron-microscopic tissue sections by an increased absorption of electrons. This increased absorption makes the membrane appear darker and denser on these images, and this region is therefore referred to as the postsynaptic density.

premotor stages of disease The premotor stages of disease are particularly relevant to Parkinson disease. It has been recognized that constipation, loss of olfaction, and abnormal sleep, particularly the acting out of dreams, may predict the future onset of disease. Importantly, 90% of PD patients have a measurably decreased sense of smell at time of diagnosis, suggesting that this may be an important predictor of disease prior to the onset of motor symptoms.

presenilin Presenilins are 8-transmembrane domain proteins that derive from two genes, PSEN1 and 2. They are essential parts of the protein complex that form secretases required for the cleavage of membrane-associated proteins such as the APP. Familial forms of Alzheimer disease are often linked to mutations in either of the two presenilin genes, PSEN1 or PSEN 2. The mutated presenilins cleave the APP protein in a way that results in an increasing number of plaque-forming Ab42.

primary injury In traumatic brain or spinal cord injury, primary injury refers to the changes in tissue structure and function resulting immediately from the applied force. Rupture of blood vessels and shearing or dissection of nerves would be examples of primary injury.

primary progressive multiple sclerosis (MS) A subtype of multiple sclerosis, primary progressive MS is characterized by a continued worsening of symptoms without intermittent phases of remission. This form is found in 15% of all cases.

prions A prion is a protein of unknown function, which, when mutated, can infect other wild-type proteins to assume its misfolded state. Prions impose their three-dimensional structure on other proteins, essentially serving as a mold to coax other proteins to assume its shape. Prions are linked to mad cow disease and Creutzfeldt-Jakob disease.

problem-solving therapy Problem-solving therapy operates on the premise that by learning problem-solving skills, a person with depression is better equipped to cope with his/her life situation and can resolve some of his/her specific stressors.

procedural memory Procedural memories are recollections on how to accomplish a task, such as riding a bicycle or playing an instrument, as well as using tools and devices. These are often called the "how" memories and do not involve the hippocampus.

prodrome The prodrome is a period of time that precedes the onset of disease by months or years. During this time, subtle signs or behavioral abnormalities can be detected that signal that an individual may be at risk of becoming symptomatic.

progressive muscular atrophy (PAM) A motor neuron disease that is less common than ALS and less well understood, PAM only affects the lower motor neurons and has a much better prognosis.

progressive/relapsing multiple sclerosis (MS) The rarest presentation of MS, the progressive/relapsing presentation occurs in 5% of patients and presents with progressive and significant worsening of symptoms that will not ease with time.

proprioception The perception of the relative position of our extremities in space is referred to as proprioception.

PrPSC PrPSC is the abbreviation for a mutated prion protein found in scrapies, a neurodegenerative disease in sheep.

proteases Enzymes that degrade proteins are called proteases.

proteasome The proteasome is a large intracellular protein complex that binds and enzymatically cleaves proteins. Its primary role is the clearance of damaged and nonfunctional proteins, which are identified through polyubiquitin marks applied by ubiquitin ligases that troll the cell searching for dysfunctional protein.

proteolipid protein (PLP) A major component of myelin, PLP is a 280 amino acid transmembrane protein that plays an important role in maintaining the compact myelin sheath.

proteostasis Proteostasis is a term that describes the maintenance of a constant protein content, or equilibrium between production and removal of proteins in a cell.

proteostasis network The many proteins and signaling molecules that participate in the maintenance of proteostasis or an equilibrium between protein synthesis and degradation is called the proteostasis network.

pruning In the context of neuroscience, pruning refers to a reduction in synapses during development. Much like a tree's branches are pruned to give the tree a better shape, activity of nerve cells during development leads to pruning of the synaptic tree by removing synapses that are inactive.

PSD-95/postsynaptic Density Protein Postsynaptic density protein of 95 kD molecular weight is a regulatory protein of postsynaptic glutamate receptors. It contains PDZ domains, characteristic protein–protein interaction domains often found in molecules that function as molecular scaffolds.

psychosis Psychosis is a relatively unspecific term for deranged, abnormal behavior. It implies that an individual has a problem in mentation that makes him/her dissociated from reality. A psychotic individual may experience delusions, hallucinations, or paranoia, and may misrepresent himself/herself as someone else.

psychotherapy An umbrella term for nonmedicinal therapies, psychotherapy is typically administered by a mental health professional to treat mood and anxiety disorders like depression.

PTEN-induced putative kinase 1 (PINK1) PINK1 is a gene that encodes for a mitochondrial associated protein that constantly monitors mitochondrial health. Mutations in PINK1 give rise to familial early-onset juvenile Parkinson disease.

ptosis Ptosis is the medical term for a drooping upper or lower eyelid.

pyramidal cells Also called "principal cells," pyramidal cells are found in the cortex, amygdala, and hippocampus. These are large cells with a triangular cell body shaped like a pyramid. They are the most numerous glutamatergic cells in the brain and are typically the major output cells, sending long axons to distant targets. For example, the layer 5 upper motor neurons that form the corticospinal tract are pyramidal cells. They also extend extensive dendritic arbors into the overlaying cortex. The dendrites are typically classified as basal and apical, as they originate from the base and apex of the cell, respectively.

pyramidal side effect Pyramidal side effects are attributable to indirect drug effects on nerves that do not travel through the "pyramids" of the brain stem. The pyramids harbor the corticobulbar and corticospinal tract, and are responsible for voluntary movements of extremities and cranial nerve innervation of the face, head, and neck. Improper use of antipsychotic drugs or narcoleptics can affect dopaminergic neurons in the basal ganglia, causing parkinsonian-like motor symptoms, such as bradykinesia or tremor. They can also affect dopaminergic neurons involved in emotions, leading to anxiety, distress, or paranoia.

quadriplegia (tetraplegia) Quadriplegia refers to a permanent impairment of motor and sensory function affecting all four extremities, and is most commonly caused by a spinal cord injury at or above cervical spine 5 (C5).

radial glial cells A specific glial cell type that is found during development, radial glia span from the ventricular zone, the innermost layer of the brain and wall of the neural tube, to the outermost pial surface or covering of the brain. The radial glial cells serve as guides for neurons to find their position within the cortex during development. Radial glia disappear in the cortex prior to birth, yet persist in the cerebellum, where neuronal migration occurs postnatally.

radicals Any molecule with unpaired electrons is called a radical. These are highly reactive and can alter many biological processes, typically by stealing electrons. An example of such an alteration is lipid peroxidation by a hydroxyl radical. A radical that reacts with a nonradical always produces another radical, creating a chain reaction that can only be stopped when two radicals react and produce a nonradical species. Important radicals in neurobiology include reactive oxygen (hydroxyl and superoxide radicals) and nitrogen (nitric oxide radicals).

Rasmussen's encephalitis Rasmussen's encephalitis is a rare and catastrophic form of acquired epilepsy in children. The root cause is an autoimmune response whereby the body makes autoantibodies to specific nervous system proteins, especially GluR3. Antibody removal by plasmapheresis, administration of intravenous immunoglobulin, or reducing inflammation with corticosteroids or corticosteroid-releasing hormones attenuates symptoms. However, surgical removal of the brain hemisphere most affected by the seizures is the most promising form of treatment.

reactive astrocytes Reactive astrocytes form the glial scar that seals off wounds in the central nervous system. The occurrence of reactive astrocytes is often called astrogliosis or reactive gliosis, and also characterizes brain infection and neurodegeneration. Reactive astrocytes are characterized by thickened processes and an upregulation of the intermediate filament protein GFAP. GFAP staining is often used to identify reactive astrocytes in the brain. Reactive astrocytes contribute to the resolution of disease by participating in the immune response through the release of cytokines and chemokines. They may, however, also provide an impermeable barrier to prevent neuronal regeneration.

reactive gliosis See also reactive astrocytes. The collection of morphological and biochemical changes in astrocytes as a response to injury or disease is called reactive gliosis.

reactive oxygen species (ROS) These highly reactive forms of oxygen have an unpaired electron, with the most important examples being the superoxide radical (O_2-), hydrogen peroxide (H_2O_2), and hydroxyl radical (-OH). ROS are produced in normal mitochondria by premature reduction of electrons passing through the electron chain. Up to 2% of electrons are prematurely reduced, allowing them to leak and bind to molecular oxygen, thereby generating the superoxide radical (O_2.-). Cells defend themselves against the potentially toxic effect of ROS by neutralizing them through binding to cellular antioxidants such as vitamin E or glutathione, and by enzymes such as SOD that degrade the superoxide radical to H_2O_2 and water. Superoxide radicals are also deliberately released by microglial cells to kill microbes entering the brain. These radicals are generated by NADPH oxidase, a membrane-bound enzyme that transfers electrons to molecular oxygen. ROS can also be produced through exposure to ultraviolet rays or ionizing radiation. ROS can interact with proteins and lipids, causing oxidative damage, with lipid peroxidation being one example. As such, ROS have been implicated in many neurological diseases, most notably stroke, amyotrophic lateral sclerosis, and Parkinson disease.

recombinant insulin-like growth factor 1 (IGF1) IGF1 is a recombinantly manufactured version of an endogenous human hormone that is similar in structure to insulin. It is a small 70 amino acid peptide produced in the liver that serves important roles during growth and development. It is a natural activator of the AKT signaling pathway, which is also the target of the TrkB receptor activated by BDNF. Hence, IGF1 can be used to bypass the effects of BDNF.

relapsing-remitting MS The most common form of multiple sclerosis, relapsing-remitting MS presents in 85% of patients and is characterized by periods of worsening symptoms (relapse), separated by prolonged periods of recovery (remission).

repolarization Repolarization is the recovery of the membrane potential from a state more positive than the resting potential back to the original resting potential. For example, repolarization returns the membrane to rest following activation of a neurotransmitter receptor.

retrograde Retrograde can be described as moving backward. The direction of information flow or transport of molecules is termed retrograde when it goes back to the cell body and anteriograde when it is moved away from the cell body.

retrovirus A retrovirus is a single-stranded RNA virus that stores its genetic codes as RNA rather than the more typical DNA code. The virus then reverse-transcribes RNA into DNA using reverse transcriptase. The most infamous retroviruses include HIV, herpes, and lentivirus.

Rett syndrome A relatively rare, X-linked intellectual disability that primarily affects girls, Rett syndrome is caused by mutations in the methyl-CpG 2-binding protein (MeCP2), a transcriptional regulator with multiple protein targets. Early normal development is halted by developmental stagnation or regression, with affected girls losing purposeful hand movements and demonstrating a regression in language. Since the disease has transient autistic features, it is categorized within the autism spectrum.

rotarod test The rotarod test is an animal behavioral test used to examine a loss of motor coordination in rodents. The animals are placed on a rotating rod and must balance their body to prevent falling from the rod as it moves. The longer they succeed in doing so without falling, the better their motor coordination is deemed to be.

saccade Saccades are quick, simultaneous movements of both eyes. When people look at objects or their environment, they focus on one object at a time, then quickly jump to a new focal point. This movement is called a saccade. Saccadic eye movements also occur when the environment is moving relative to the head, such as when a person is riding in a car. Here, the eye moves in order to stabilize the moving environment, yet will periodically jump back to a new point of focus through a quick saccade.

sacral column The five sacral vertebrae of the spinal column are fused into one single bone, the sacrum. Nerves travel through the center of the sacral column.

saltatory conductance The process by which action potentials travel along myelinated axons, salutatory conduction allows the signal to jump quickly from one node of Ranvier to the next, creating increased velocity even in small-caliber axons.

schizophrenia Schizophrenia is a neuropsychiatric disorder without known cause that affects approximately 1% of the world's population, making it the most common disorder with unknown cause. It is characterized by psychosis, apathy, and poor mentation, and is often diagnosed between the ages of 18 and 24.

schwannoma Schwannomas are tumors that form due to the malignant transformation of Schwann cells, the myelin-forming peripheral glial cells that make up the peripheral nerve sheath. These tumors are often called neuromas, as they affect the nerve.

sclerotic lesions Sclerotic lesions refer to an area of brain injury or disease characterized by scar tissue. These lesions typically occur as a result of astrocytes becoming reactive, multiplying, and constructing a tenacious mesh of processes and extracellular matrix molecules.

scotoma scotoma is a visual field defect, similar to a blind spot, caused by damage to the retina, optic nerve, or visual cortex. It can be caused by multiple sclerosis and other demyelinating diseases.

scrapie Scrapie is a neurodegenerative disease found in sheep caused by misfolded prion protein PrPsrc.

seasonal affective disorder A monopolar depressive illness that manifests only during a defined time of year, seasonal affective disorder typically occurs with the changes of season in the fall or winter months.

secondary injury While primary injury is mostly a result of physical forces, secondary injury is largely a biochemical response that presents with gradual changes in tissue composition and function. While primary injury, for example, includes rupture of blood vessels, secondary injury includes loss of ATP, accumulation of toxic molecules such as glutamate, reactive gliosis, and scarring, as well as invasion by immune cells.

secondary progressive multiple sclerosis (MS) Secondary progressive MS is a form or stage of MS characterized by progressive worsening of symptoms interspaced with periods of acute worsening from which the patient recovers only partially. Many patients with relapsing-remitting MS eventually convert to secondary progressive disease.

secretases Enzymes that cleave a transmembrane protein within the phospholipid membrane, creating a secreted protein fragment, are called secretases.

seizure The synchronous discharge of neuronal networks in the brain associated with a range of behavioral abnormalities is called a seizure. The abnormal electrical activity can be recorded with surface electrodes on an electroencephalogram (EEG). The behavioral abnormalities seen with seizure range from a blank stare into space to muscle convulsions. Seizures may be generalized, affecting both sides of the brain and body, or focal, affecting just a discrete region in only one of the brain hemispheres. They are further divided into seizures with or without dyscognia based on whether cognition is lost during the event.

selective serotonin reuptake inhibitors (SSRIs) SSRIs are a group of relatively new drugs that specifically inhibit the reuptake of serotonin in the brain, thereby prolonging its action on serotonergic synapses. SSRIs have become the most widely prescribed and advertised antidepressant drugs, and include Prozac, Celexa, and Zoloft.

semantic memories Memories that encompass general knowledge, such as the use of language, the number system, the purpose and use of tools, and classes of animals and plants, are called semantic memories. Semantic memory is typically retained for a very long time, and works in tandem with episodic memory to generate narrative, relating "what" happened, to "whom," "when," and "where."

serotonin–norepinephrine reuptake inhibitors (SNRIs) SNRI refers to a class of antidepressants that inhibits the reuptake of both serotonin and norepinephrine, prolonging their activity at synapses.

severe traumatic brain injury (TBI) Severe TBI is characterized by visible injury to the brain, often with fracture of the skull. These injuries cause long-lasting changes in brain function and may even include extended periods of coma.

shaken baby syndrome Shaken baby syndrome refers to a severe brain injury with subdural bleeding, retinal bleeding, and cerebral edema resulting from violent shaking of a baby's head. Early death or life-long disability is common.

shear force Unaligned forces that move different parts of the brain or spinal cord in different directions during injury are called shear forces. For example, a rapid movement of the head in a car accident can cause the spinal cord and white matter tracts in the brain to experience different force vectors and move in different directions, thereby shearing nerve tracts.

single nucleotide polymorphism (SNP) A difference in one or two of the three nucleotides encoding for the same amino acid is called an SNP. For example GTT, GTC, GTA, and GTG each encode for the amino acid valine, and thus are considered SNPs.

single-photon emission computed tomography (SPECT) Single-photon emission computed tomography is a medical imaging technology that uses gamma or X-rays. It typically requires administration of a radioactive compound, and allows for the generation of three-dimensional images by computer from a number of planar images.

small interfering RNA (siRNA) Small interfering RNAs are double-stranded RNAs of 25–30 base pairs. They interact with RNA, suppressing its translation into protein, and can be designed to specifically suppress or knock down the protein of interest.

SNARE SNARE is an abbreviation for "soluble NSF attachment protein receptor." SNARE proteins are a superfamily of proteins involved in orchestrating the docking and fusion of vesicles, including neurotransmitter-containing vesicles. The SNARE proteins of cholinergic muscle synapses are the target of botulinum toxin.

sodium–potassium–chloride transporter The sodium–potassium–chloride transporter refers to a family of membrane proteins that includes NKCC1 and NKCC2 and aids the active transport of Na$^+$, K$^+$, and Cl$^-$ across cell membranes. The transport harnesses the inward directed gradient for Na$^+$ to move 1 Na$^+$ with 1 K$^+$ and 2 Cl$^-$into the cell in an electroneutral fashion. NKCC1 is found throughout the body and brain, while NKCC2 is primarily found in the kidney.

sodium valproate Also called valproate and sold under the name Depakote, sodium valproate is a commonly used drug to treat epileptic seizures. Valproate has been recognized to also act as a histone deacetylase (HDAC) inhibitor. It should not be used during pregnancy, as it increases the risk for birth defects.

spinal cord The spinal cord is the nerve column that extends from the base of the brain through the vertebral column. It carries descending information to instruct voluntary muscle movements, as well as ascending sensory information to convey environmental signals to the brain. These ascending signals may include the position of limbs, temperature, tactile information, and pain. The spinal cord is also the center for many reflexes.

spinal cord contusion As a severe injury of the spinal cord where bleeding and tissue swelling is common, contusion injuries are typically the result of severe motor vehicle accidents.

spinal cord injury (SCI) Any damage to the spinal cord or its nerve roots is classified as a spinal cord injury. SCI causes transient or permanent loss of strength, sensation, or control over bodily functions.

spinal muscular atrophy (SMA) SMA is an autosomal recessive disease that presents with muscle weakness as a result of a mutation in the SMN gene family (survival of motor neuron). SMA affects the lower motor neurons in the anterior horn of the spinal cord but spares the cranial nerves. Therefore, the disease does not affect the bulbar muscles of the face and throat.

spines Spines are mushroom-shaped membrane protrusions on the dendrites of glutamatergic neurons. They are the sites of communication between nerve cells and contain postsynaptic neurotransmitter receptors. Spines are dynamic structures that contain actin molecular motors in their shaft and can therefore be retracted or extended. Spine dynamics may be involved in signal processing between cells. Abnormal spine morphology characterizes fragile X, Rett, and Down syndromes.

spinobulbar atrophy Spinobulbar atrophy is characterized by the progressive development of muscle weakness and wasting (atrophy) that affects the bulbar muscles of the face and throat as well as the limbs. The disease has adult onset and worsens slowly over time as the affected motor neurons in the spinal cord or brain stem die.

spinocerebellar ataxia Spinocerebellar ataxia refers to a group of inherited neurodegenerative disorders that share a progressive development of incoordination of gait, hands, and eye movements.

spirochete A spirochete is a type of bacterium that looks like a corkscrew; an example is *Treponema pallidum* a spirochete bacteria responsible for causing syphilis.

spongiform encephalopathy Also called mad cow disease, spongiform encephalopathy is caused by brain infection with mutated prion protein.

strabismus An eye disorder in which the eyes are not properly aligned, strabismus hampers binocular vision and proper depth perception.

status epilepticus Status epilepticus is a serious, life-threatening condition whereby a person suffers a convulsive seizure that lasts longer than 5 min. The seizure must be stopped as quickly as possible, usually through administration of diazepam (Valium). Status epilepticus causes 42,000 deaths each year in the United States.

Streptococcus pneumoniae Streptococcus pneumoniae refers to a gram-negative bacterium responsible for a number of infectious diseases, including pneumonia, bronchitis, meningitis, and many others.

stress granules Stress granules are cytoplasmic inclusion bodies that contain aggregated proteins in nerve tissue from ALS patients.

striatum The striatum is a component of the basal ganglia and is critically involved in the fine control of movements.

stroke belt The stroke belt is the geographic region in the United States with significantly increased incidence of stroke. It roughly spans from Louisiana to northern Virginia, including Mississippi, Tennessee, Georgia, and North and South Carolina.

subarachnoid space The subarachnoid space lies between the arachnoid mater and the pia mater, the middle-most and innermost layers of the three meningeal coverings of the brain, respectively.

subcortical dementia Subcortical dementia is forgetfulness and slowness in mental processing resulting from neuronal loss in subcortical structures such as the thalamus, basal ganglia, and brain stem nuclei. Subcortical dementia is commonly found in progressive supranuclear palsy and Huntington disease.

subgranular zone The subgranular zone is the proliferative cell layer in the hippocampus where neurons are born throughout life. This adult neurogenesis only occurs in a few places of the nervous system. Newly born neurons in the hippocampus are believed to participate in learning and memory.

subiculum Part of the cortex, this structure is situated between the hippocampus and the entorhinal cortex, immediately adjacent to the CA1 region of the hippocampus. It is believed to play a role in learning and memory, and abnormal activity in the subiculum has been linked to certain forms of epilepsy.

substantia nigra The susbtantia nigra is a brain structure of the midbrain that plays an important role in the control of movement. Its dopaminergic neurons are rich in neuromelanin, giving the structure a black appearance. These neurons project to the striatum, and their progressive loss causes Parkinson disease.

substantia nigra, pars compacta The pars compacta is one of two subregions of the substantia nigra, the other being the pars reticulata. The pars compacta contains the dopamine-producing neurons that project to the striatum.

subventricular zone (SVZ) A proliferative zone deep within the brain associated with the cells lining the third and fourth ventricle, the SVZ contains four cell layers. The innermost layer is compoised of the ependymal cells lining the ventricles. The second layer contains astrocytes, while the third layer contains astrocyte cell bodies intertwined with some oligodendrocytes and ependymal cells. These astrocytes are highly proliferative and form multipotent progenitor cells, making the SVZ one of the few regions in the adult brain where neurogenesis can occur.

sudden infant death syndrome (SIDS) SIDS is the unexpected and unexplained death of infants in their cribs. The cause of death is hypothesized to result from suffocation from the pillow when infants sleep on their bellies.

superior temporal gyrus The superior temporal gyrus is one of the major ridges found on the lateral cortex just above the ear. It contains the primary auditory cortex, required for processing sounds, and Wernicke's area, required for speech understanding.

superoxide dismutase-1 gene (SOD1) The SOD1 gene encodes for a protein involved in the neutralization of radicals produced in the mitochondria. SOD1 is a ubiquitous enzyme that is tasked with the conversion and neutralization of highly reactive superoxide (oxygen with an extra electron) to either oxygen (O_2) or hydrogen peroxide (H_2O_2).

sympathetic nervous system Together with the parasympathetic and enteric nervous system, the sympathetic nervous system forms the autonomic nervous system, which controls most of the body's internal organs. The sympathetic nervous system is tasked with placing the body in a state of alertness, often called the "fight-or-flight" response. The sympathetic nervous system primarily engages adrenaline (epinephrine) and noradrenaline (norepinephrine) acting via alpha and beta adrenergic receptors. By contrast, the parasympathetic nervous system uses acetylcholine as its principal neurotransmitter and acts through muscarinic and nicotinic receptors. It places the body in a "rest-and-digest" state.

symptomatic/acquired epilepsy Symptomatic or acquired epilepsy presents with a known or suspected cause. The single largest cause of acquired epilepsy is trauma to the head, such as closed or penetrating head wounds or concussions. Other causes are stroke, neurodegenerative diseases, or drug use. These account for the majority of epilepsies diagnosed in adults age 18–45. While the term acquired epilepsy implies that the patient was somewhat actively involved in the acquisition of disease, this is often not the case. Hence, the term symptomatic better describes epilepsies in which there is a well-described underlying cause.

synaptobrevin Synaptobrevin, also known as vesicle-associated membrane protein 2 (VAMP2), belongs to the SNARE family of synaptic proteins involved in the assembly of synaptic vesicles in the presynaptic terminal.

synaptogenesis Synaptogenesis is the genesis or formation of new synaptic contacts between neurons. This process includes synthesis of all the structural and functional components necessary to make these new synapses function properly, such as synaptic vesicles and proteins involved in regulated exocytosis. While synaptogenesis occurs throughout life, it is particularly pronounced in early life during critical periods of learning and development.

T helper cell One of the white blood cells involved in the body's immune response, T helper cells recognize foreign antigens and stimulate the production of antibodies by B cells.

T lymphocytes T lymphocytes are white blood cells that participate in the body's immune response. They encompass many different cell types produced by the thymus, including T helper cells, T regulatory cells, and killer T cells, among others.

T regulatory T regulatory cells refers to a population of T cells involved in modulating the immune response through the release of cytokines.

tail suspension test (TST) The TST is used to evaluate behavioral abnormalities related to depression in rodents. A mouse is suspended from its tail and its behavior is observed. Animals with signs of depression will give up any attempt to free themselves from this position more quickly than normal animals; common antidepressants thus increase the amount of time that animals spend struggling to free themselves.

tau Tau is a protein that is associated with the microtubules in axons. These serve a mechanical role to give the axon structure and rigidity, but also serve as guide rails for the axonal transport of molecules and organelles. The tau protein serves to stabilize the microtubules and has many phosphorylation sites, yet only a few are normally occupied. Hyperphosphorylation causes a dissociation of tau from the microtubules, leading to their disintegration. Tau can then aggregate into large intracellular protein complexes called fibrillary tangles, which are a pathological hallmark of Alzheimer disease.

tauopathies A number of diseases present with hyperphosphorylated tau that aggregates to form fibrillary tangles. These are, for the most part, neurodegenerative diseases, and are often called "tauopathies." They include, among others, Alzheimer, frontotemporal dementia (FTD), progressive supranuclear palsy (PSP), and chronic traumatic encephalopathy (CTE).

TDP-43 Transactive-region DNA-binding protein gene, TARDBP, also known as TDP-43, is an RNA-binding protein that acts to suppress transcription alongside other related regulators, such as MeCP2. TDP-43 affects the transcription of over 6000 RNAs, and mutations in this gene cause rare familial forms of ALS.

telencephalon The telencephalon is also called the forebrain and includes the cerebral cortex and the hippocampus.

temozolamide A DNA alkylating agent marketed under the name Temodar, temozolamide is used to treat malignant brain tumors, glioma, and melanoma.

tentorium "tent" describes a fold of the dura mater running along he back of the brain where it acts as a demarcating line that separates the cerebellum from the cerebrum. It is often used as a landmark to descibe the location of tumors of the cerebrum, which are supertentorial compared to tumors of the brain stem or cerebellum which are supratentorial.

tetanus Tetanus refers to spastic body contractions, commonly known as "lockjaw," that are caused by wound infection with tetanus toxin.

tetrabenazine Tetrabenazine inhibits the vesicular monoamine transporter 2 (VMAT2), which is responsible for loading dopamine and related monoamines into synaptic vesicles. It is the only drug that is specifically approved to suppress the chorea movements in Huntington disease.

Theiler murine encephalomyelitis virus A mouse virus that causes demyelination, Theiler murine encephalomyelitis virus presents similar to multiple sclerosis and is therefore frequently used as a mouse model to study this disease.

theory of mind The ability to read other people's emotions, desires, and beliefs and to show empathy is often called theory of mind. This ability is lacking or compromised in individuals suffering from autism or schizophrenia.

thoracic vertebrae The thoracic vertebrae consist of 12 vertebrae, T1–T12, flanked by the cervical and lumbar vertebrae. Lesions to the thoracic spinal cord typically leave patients paraplegic and unable to move their legs.

thrombus A thrombus is a blood clot resulting from buildup of plaque tissue on the inside of a blood vessel as a result of atherosclerosis. This deposit narrows the blood vessel and eventually occludes it completely, depriving the tissue downstream of nutrients and oxygen. Strokes that are caused by a thrombus are called thrombolytic strokes. A fragment breaking off from a thrombus is called an embolus.

thymus The thymus is a specialized organ in the immune system that produces immune cells called T lymphocytes. T cells undergo selection in the thymus to ensure that they do not react with and become activated by the body's own cells or proteins.

tight junction proteins These are cell–cell adhesion molecules expressed by the vascular endothelial cells of the brain. These molecules are hemophilic and interact with each other to fuse adjacent membranes together, thereby creating a barrier that is an impairment to water-soluble molecules. The major tight junction proteins are occludins and claudins, and they interact intracellularly through zona occludens proteins to link them to the cell's cytoskeleton. Tight junction proteins are essential to formation of the blood–brain barrier.

tissue plasminogen activator (tPA) tPA is a serine protease produced by blood vessel endothelial cells. It catalyzes the local cleavage of plasminogen, produced in the liver, to plasmin. Plasmin is a serine protease that degrades many proteins found in blood clots, including fibrin. tPA is an important regulator of clotting, and, consequently, recombinant tPA has emerged as an effective thrombolytic agent, often called a "clot buster."

Toll-like receptors **(TLRs)** TLRs are a family of surface receptors that participate in the innate immune response. They detect structurally conserved molecules typically produced by microbes, and subtypes of TLRs have different substrate specificity. For example, bacterial lipopolysaccharides are recognized by TLR4, viral DNA is recognized by TLR3, and TLR1 and 2 recognize bacterial peptides. Protein deposits, including amyloid plaques, are primarily detected via TLR4. TLR activation leads to the release of inflammatory cytokines that recruit immune cells.

Tonic–clonic seizure The most common seizure type diagnosed in adults is the tonic-clonic seizure. These are characterized by tonic muscle contraction, causing a clasping of upper and lower extremities that may also involve the jaw muscles and may cause a patient to bite his tongue. Within tens of seconds, the tonic phase gives way to the clonic phase, characterized by repeated jerking and muscle contractions lasting up to a minute. Ultimately, the patient is unresponsive, fatigued, and sweating, showing excessive salivation and possibly signs of incontinence.

transcranial magnetic stimulation (TMS) TMS is a noninvasive method to alter neuronal activity through local application of a magnetic field generated in a coil by weak electrical currents placed in close proximity to the brain.

transcriptional regulators Proteins that affect the transcription of DNA into mRNA, thereby enhancing or reducing the expression of new protein, are called transcriptional regulators.

transient ischemic attack (TIA) A "mini stroke" with short, transient symptoms such as tingling, loss of vision, and brief paralysis, a TIA does not result in a loss of consciousness. All symptoms spontaneously resolve in less than 24 h.

trepanation A practice that dates back to prehistoric times and popularized during the Middle Ages, trepanation involves drilling a burr hole into the skull for therapeutic or spiritual purposes.

tricarboxylic acid cycle (TCA cycle) Also known as the citric acid or Krebs cycle, the TCA cycle is the metabolic pathway used by aerobic cells to generate energy through the oxidation of carbohydrates, fat, and protein. The TCA cycle takes place in the mitochondria, where the produced NADH is directly fed into the synthesis of ATP through oxidative phosphorylation in the electron transport chain. The TCA cycle also provides precursors for the synthesis of some amino acids.

tricyclic antidepressant(s) The first drugs used to treat depression, tricyclic antidepressants are named after the three rings of interconnected atoms they contain. Examples include imipramine and amitriptyline.

trigeminal ganglion (plural, ganglia) The trigeminal ganglion refers to a collection of nerve cell bodies that form the fifth cranial nerve, the trigeminal nerve. This mixed sensory and motor nerve has three branches that innervate different parts of the head, face, and mouth. These branches are called ophthalmic V1, maxillary V2, and mandibular V3, respectively.

trinucleotide CGG CGG encodes for the amino acid arginine. Repeats of CGG in the untranscribed region of the X chromosome cause chromosomal instability. This leads to a silencing of the nearby gene encoding the fragile X mental retardation protein, and is the cause for fragile X syndrome.

trisynaptic circuit The trisynaptic circuit is a signaling pathway that involves three synapses in the hippocampus. Neurons in the dentate gyrus project excitatory terminals onto CA3 pyramidal cells, which then innervate the CA1 pyramidal neurons, which provide the major output of the hippocampus to the medial prefrontal cortex.

tropomyosin-related kinase type B (TrKB) TrKB is the receptor for the brain-derived neurotrophic factor (BDNF). It is a tyrosine kinase that signals via phosphorylation of tyrosine residues on downstream intermediate signaling molecules, eventually activating the external receptor kinases ERK1/2 or AKT (protein kinase B), which regulate gene transcription.

Ts65Dn mouse The Ts65Dn mouse is a transgenic mouse model of Down syndrome in which 150 genes of chromosome 16, the mouse equivalent to the human chromosome 21, are duplicated. This mouse replicates many of the salient phenotypes of Down syndrome in humans, including the dysmorphic facial features.

tuberous sclerosis complex Tuberous sclerosis complex refers to a group of genetic disorders characterized by noncancerous tumors in many parts of the body. The disease is caused by mutations in either the TSC1 or TSC2 gene. Affected individuals may have intellectual disabilities and autistic behavioral features.

tubulin A molecule of approximately 55,000 Da that is a building block for microtubules, tubulin is an important cytoskeletal element present in most cells. In the nervous system, microtubules formed from dimers of alpha and beta tubulin form microtubules spanning the length of the axon and serve as guide rails for axonal transport of cargo molecules.

tumor necrosis factor alpha (TNFα) Tumor necrosis factor alpha is the most well-known cytokine that induces cell death. It is released by monocytes in the body, where it causes septic shock. In the nervous system, TNF alpha can be released by microglia, macrophages, or astrocytes under certain conditions.

tumor suppressor genes Tumor suppressor genes encode for proteins that monitor defined steps of the cell cycle and interfere if errors to the DNA occur. Examples are the retinoblastoma gene RB and the p53 tumor suppressor gene. According to the popular two-hit hypothesis, cancer only forms if both allelic copies of a cell's tumor suppressor gene are mutated.

Tuskegee Syphilis Study The Tuskegee Syphilis Study was an infamous clinical study sponsored by the U.S. Department of Health and conducted in Tuskegee, Alabama, between 1932 and 1972. It examined the effects of untreated syphilis in 400 African-American men. Study participants were not informed that they were participating in a study, and researchers withheld treatment even when curative penicillin-G became widely available. This study led to the Belmont Report, which laid the foundation for the current oversight for clinical studies provided by institutional review boards (IRBs).

ubiquitin proteasome system (UPS) UPS refers to a protein complex in the cell that enzymatically degrades proteins into their amino acid constituents. Nonfunctional proteins are marked for degradation by E3 ligases (an enzyme that attaches ubiquitin groups to the protein). These marked proteins are then transported to the proteasome for degradation.

upper motor neurons Motor neurons in the motor cortex are called upper motor neurons. These never innervate muscles directly, but signal to lower motor neurons in the spinal cord or brain stem.

Varicella zoster Varicella zoster is the virus that causes chicken pox and shingles.

vagus nerve stimulation (VNS) Vagus nerve stimulation is a procedure that uses an implanted pacemaker to electrically stimulate the vagus nerve (10th cranial nerve), thereby altering nerve activity in the brain stem. Vagus nerve stimulation is used to treat epilepsy and treatment-resistant depression.

valproate sodium (Depakote) The most commonly used drug worldwide to treat seizure disorders and epilepsy is Depakote. Its mechanism of action is not fully understood, although it has been shown to reduce Na$^+$ channel activity and enhance GABAergic inhibition. It also has

been recognized as a histone-deacetylase inhibitor, and this activity may enhance gene transcription.

vascular co-option The early interaction of a cancer cell with the existing vasculature is called vascular co-option. This is in contrast to angiogenesis, which is the production of new blood vessels by tumor-released factors.

vascular edema Vascular edema refers to tissue swelling due to excessive entry of water into the extracellular brain spaces from leaky blood vessels.

vascular endothelial growth factor (VEGF) This growth factor coordinates the proliferation of the vascular endothelial cells that form the wall of blood vessels. By binding to tyrosine kinase receptors, VEGF induces the expression of genes that cause proliferation of endothelial cells that assemble to form new blood vessels. VEGF is normally produced by cells that are starved of oxygen in an effort to balance oxygen supply with demand. VEGF is also produced by tumors that stimulate angiogenesis to supply growing tumor masses with nutrients.

vascular thrombosis The buildup of plaque in the body's vasculature, as seen in veins of the leg or the pulmonary artery, is called vascular thrombosis. These plaques can shed fragments or emboli that can make their way to the brain, resulting in a stroke.

velocardiofacial or DiGeorge syndrome The deletion of a small part of chromosome 22q11.2 produces velocardiofacial syndrome, or DiGeorge syndrome. This condition affects the heart, face, brain, and other organs, including the immune system. Among the deleted genes are two microRNAs that control the protein synthesis of hundreds of proteins. Twenty-five percent of DiGeorge syndrome patients develop schizophrenia.

ventral horn The ventral horn is the anterior, front-facing part of the spinal cord that harbors the motor neurons that innervate the upper and lower extremities.

ventral tegmental area (VTA) The VTA is a group of neurons found at the base of the midbrain. These neurons release dopamine onto the targets of the VTA, primarily in the frontal lobe. The VTA is the natural reward center that is involved in motivation and is highly activated during orgasm.

ventricles The human brain contains four interconnected, fluid-filled spaces called the ventricles. The two largest ones, 3 and 4, are most centrally located underneath the left and right cortex.

vesicle A vesicle is a small organelle composed of a lipid membrane that encloses molecules that are trafficked or released from cells. These molecules may include peptides and amino acid transmitters.

Vesicle-associated membrane protein 2 (VAMP2) VAMP2, also known as synaptobrevin, is a presynaptic protein necessary for assembling synaptic vesicles for future release by the presynaptic terminal.

viral and bacterial meningitis Meningitis refers to inflammation of the membranes that cover the outside of the brain (the meninges) as a result of a bacterial or viral infection. Headaches, stiff neck, nausea, and impaired consciousness are typical clinical signs.

virions A viron is a complete virus in its infectious state. It contains DNA or RNA, as well as the viral coat proteins through which the virus enters cells.

vitamin D Vitamin D refers to a group of fat-soluble organic compounds required for the absorption of Ca^{2+} by the intestine for the calcification of bone. Vitamin D deficiency causes a softening of bones and is the cause of a childhood disease called rickets, which presents with fragile and deformed bones due to defective mineralization. Humans produce vitamin D in the skin when exposed to UVB rays from the sun, yet many people are still deficient in vitamin D.

Von Frey filaments Von Frey filaments are nylon bristles of varying diameter that are used to poke the skin with a defined force as the bristles bend. Small-diameter bristles exert a small force, while larger-diameter bristles exert a stronger force. These are used to test patients for hypersensitivity to tactile stimuli, and can also be used in animal models of spinal cord injury or sensory processing.

weight drop model The weight drop model is a clinically relevant animal model designed to mimic closed-head injury in anesthetized animals by dropping a calibrated weight from a defined height onto the exposed skull, often with a metal disk attached to prevent fracture. Similar paradigms can be used to model spinal cord injury.

white matter Brain regions that are rich in myelinated nerve processes appear lighter or whitish in tissue sections due to the high lipid content. This allows one to distinguish myelinated brain, the white matter, from the gray matter that has little myelin and instead harbors the cell bodies and unmyelinated dendritic processes of neurons.

Wilder Penfield Penfield was a twentieth century neurosurgeon practicing at the Montreal Neurological Institute. He pioneered epilepsy surgery and, in the process, mapped the function of the human cortex by stimulating defined brain regions during surgery in awake patients in order to spare important "eloquent" brain regions. From these studies he established the "homunculus," a map that identifies which areas of the cortex supply sensory and motor innervation of the body.

working memory A short-term memory that is modifiable and requires attention is called a working memory. Recounting a series of numbers forward and backward is an example of working memory. The prefrontal cortex plays an important role in working memory. Working memory declines with age and is one of the early deficits in Alzheimer disease.

Index

Note: Page numbers followed by "b", "f" and "t" indicate boxes, figures and tables respectively.

Printed in the United States
By Bookmasters